Results and Problems in Cell Differentiation

Volume 75

Results and Problems in Cell Differentiation is an up-to-date book series that presents and explores selected questions of cell and developmental biology. Each volume focuses on a single, well-defined topic. Reviews address basic questions and phenomena, but also provide concise information on the most recent advances. Together, the volumes provide a valuable overview of this exciting and dynamically expanding field. This series is indexed in PubMed.

Marta Halasa • Anna Wawruszak
Editors

Histone and Non-Histone Reversible Acetylation in Development, Aging and Disease

Springer

Editors
Marta Halasa
Transplant Immunology
Houston Methodist Research Institute
Houston, TX, USA

Department of Surgery
Houston Methodist Hospital
University of Texas
Houston, TX, USA

Anna Wawruszak
Department of Biochemistry
and Molecular Biology
Medical University of Lublin
Lublin, Poland

ISSN 0080-1844 ISSN 1861-0412 (electronic)
Results and Problems in Cell Differentiation
ISBN 978-3-031-91458-4 ISBN 978-3-031-91459-1 (eBook)
https://doi.org/10.1007/978-3-031-91459-1

This Springer imprint is published by the registered company Springer Nature Switzerland AG
The registered company address is: Gewerbestrasse 11, 6330 Cham, Switzerland

Preface

This volume explores various aspects of reversible acetylation of histone and non-histone proteins, focusing on their roles in development, aging, and disease progression. It examines the biological consequences of modulating acetylation levels through histone acetyltransferases (HATs) and histone deacetylases (HDACs). The authors present these manipulations at multiple levels, from in vitro and in vivo studies to clinical trials and FDA-approved therapies.

The first part provides a comprehensive overview of post-translational modifications in the context of development, aging, and neurodegenerative and cancer-related diseases. It highlights the role of histone acetylation in higher-order genome organization and explores the functions of chromatin readers, writers, and erasers involved in histone acetylation. This section also presents current strategies for categorizing the regulatory consequences of histone acetylation using bioinformatics.

The second part focuses on the modulation of cytoskeleton-associated proteins through reversible acetylation. It examines how acetylation influences cytoskeletal compartments, thereby regulating cellular structure and function, and explores microtubule acetylation, showing how it induces significant changes in the 3D organization of the cytoplasm, which in turn affects intracellular transport, cell division, polarity, and migration, and investigates the role of tubulin acetylation in response to mechanical and oxidative stress.

The third part explores acetylation in the context of development. It describes acetylation as a key regulator of early embryogenesis, influencing chromatin structure and gene expression. These dynamic changes are crucial for developmental processes but can be influenced by environmental factors, potentially affecting epigenetic inheritance. This part also further emphasizes the role of acetylation in shaping cell lineage specification and tissue development, particularly in stem cells and progenitor cells, underscoring its importance in maintaining developmental plasticity and its potential susceptibility to environmental perturbations.

The final part of the volume explores the physiological and pathological aspects of histone and non-histone protein acetylation and their modulation by activators and inhibitors. It describes the crucial role of reversible acetylation in regulating stem cell differentiation during remyelination, highlighting its importance in preserving cognitive function and preventing neurodegeneration in aging by controlling neural precursor cells and oligodendrocyte precursor cells activity in the central nervous system. It also discusses the role of HDACs in cardiac function and related diseases, providing insights into the molecular mechanisms underlying cardiac aging and the therapeutic potential of histone deacetylase inhibitors (HDACi) to reverse this process. The following chapters focus on immune-related events in both cancer and noncancer diseases. They analyze the epigenetic and non-epigenetic roles of Sirtuin 1 in macrophage polarization and function, linking these processes to longevity, aging, and immunosenescence, explore the repurposing of HDACi for solid organ rejection, focusing on their effects on T and B cells, and examine the acetylation-mediated mechanisms behind acute and persistent viral infections, explaining how HDACi can interrupt persistent viral infections by increasing acetylation levels.

The context- and cancer-type-dependent role of HATs and HDACs in tumor development and progression is well established. The subsequent chapters focus on cancer. They discuss the role of HATs in various cancers through the acetylation of histone and non-histone proteins, emphasizing the importance of HAT inhibitors. Finally, an in-depth explanation of lysine acetyltransferase 6 (KAT6A/KAT6B) in biological processes is provided, highlighting the consequences of its dysregulation in cancer. This topic is further explored by examining the recruitment of KAT6 by the reader bromodomain BRPF1 and the role of this complex in both cancer and neurodevelopmental disorders. Moreover, BRPF1 mutations and their implications in cancer progression are discussed, emphasizing their potential as biomarkers or therapeutic targets.

Houston, TX, USA — Marta Halasa
Lublin, Poland — Anna Wawruszak

Contents

About the Editors

Marta Halasa earned her PhD in medical and health science from the Medical University of Lublin, Poland, in 2022. Her research during her doctoral studies focused on epigenetic modifiers, particularly histone deacetylases, and their impact on cancer progression and metastasis. During her PhD, she completed a long-term internship at the University of Turku, Finland, where she gained expertise in 3D cell techniques. She also undertook an internal internship at the Military Institute of Medicine in Warsaw, Poland, where she investigated cell-gene therapy in Alzheimer's disease. After completing her PhD, she obtained a postdoctoral fellow position at the Houston Methodist Research Institute, USA. Currently, her research focuses on investigating the role of macrophages in chronic transplant rejection.

Anna Wawruszak got her PhD at the Department of Biochemistry and Molecular Biology at the Medical University in Lublin, Poland, in 2020. Since then, she has been working at the Medical University of Lublin as a postdoc, researcher, and, finally, an Associate Professor. Her primary expertise lies in translational medicine, epigenetics, and breast cancer research. Her work is mainly focused on the anticancer activity of histone deacetylase inhibitors in breast cancer models. She has also an interest in green chemistry and biotechnology approaches for the development of nature-based cosmetics.

Part I
Introduction to Reversible Acetylation and Its Biological Consequences

Chapter 1
Histone and Non-histone Reversible Acetylation in Development, Aging, and Disease

Sezgin Gunes, Neslihan Hekim, Sercan Ergun, Elzem Nisa Alkan, and Cansu Can

Abstract Post-translational modifications (PTM) involve chemical modifications of amino acid residues within histone and non-histone proteins and are chemically diverse. PTM plays a vital role in regulating the chromatin structure in the nucleus, thus gene regulation. Among the various PTM, reversible acetylation of histone non-histone proteins has fundamental functions in various cellular processes. In all organisms, histone acetylation of lysine residues is connected with transcription activation. Acetyltransferases and deacetylases are well-known enzymes in the acetylation of the histone and non-histone proteins. This chapter will review the latest progress in histone and non-histone reversible acetylation epigenetic alterations and mechanisms and summarize how they affect development, aging, and diseases.

1.1 Introduction

Epigenetic modifications have emerged as a critical factor in gene regulation, human health, and disease (Crafa et al. 2024). The genomic DNA of somatic cells is coiled into chromatin via various histone (H) proteins in arrays of nucleosomes. Nucleosomes harbor histone cores of DNA coiled around an octamer (two copies of H2A, H2B, H3, and H4) and linker H1 protein (Gunes et al. 2024).

S. Gunes (✉)
Department of Medical Biology, Medical Faculty, Ondokuz Mayis University, Samsun, Turkiye

Stem Cell Application and Research Center, Ondokuz Mayis University, Samsun, Turkiye
e-mail: sgunes@omu.edu.tr

N. Hekim · S. Ergun · E. N. Alkan · C. Can
Department of Medical Biology, Medical Faculty, Ondokuz Mayis University, Samsun, Turkiye

M. Halasa, A. Wawruszak (eds.), *Histone and Non-Histone Reversible Acetylation in Development, Aging and Disease*, Results and Problems in Cell Differentiation 75,
https://doi.org/10.1007/978-3-031-91459-1_1

Modifying histone proteins comprises various epigenetic alterations on histones that modulate gene expression but do not involve changes in DNA sequences. Major histone modifications are histone methylation, acetylation, phosphorylation, ubiquitination, SUMOylation, and ADP-ribosylation (Ntorla and Burgoyne 2021). Histone acetylation alters chromatin configuration; therefore, gene expression can be regulated. Protein acetylation and deacetylation form a reversible acetylation network. Two groups of enzymes regulate this modification process. These enzymes are lysine acetyltransferases (KATs) and lysine deacetylases (KDACs or HDACs). KATs and KDACs work together to regulate the acetylation of lysine in various physiological processes in the organism (Xia et al. 2020). Acetylation occurs mainly on lysine adding an acetyl group to the serine or a threonine hydroxyl side chain corresponding to O-acetylation (Kim et al. 2006; Ashton et al. 2024). O-acetylation was detected in only a few eukaryotic organisms compared to N-ter acetylation or K-acetylation (Ashton et al. 2024).

This chapter focuses on the general introduction and latest progress of histone and non-histone reversible acetylation epigenetic alterations, mechanisms, and regulation of chromatin structure and gene expression. Additionally, the chapter summarizes how non-histone reversible acetylation affects the development, aging, and multiple diseases.

1.2 Histone Reversible Acetylation and Deacetylation

Many processes related to gene expression and the maintenance of genomic integrity are influenced by the interactions between DNA and histones. The organization of chromatin structure, including the formation of heterochromatic and euchromatic configurations, is fundamental in regulating gene activation (Kouzarides 2007). One of the key processes involved in these regulations is the post-translational modifications (PTMs) of histone proteins, which are highly conserved across species (Piontkivska et al. 2002). Numerous histone modifications have been identified throughout the years, with histone acetylation being the first modification discovered (Allfrey et al. 1964).

The term "histone acetylation" denotes the attachment of acetyl groups derived from acetyl-CoA to the lysine residues on N-terminal regions of histone proteins (Allfrey et al. 1964) (Tables 1.1 and 1.2). In contrast, "histone deacetylation" describes the removal of these acetyl groups from acetylated histones. These reversible and highly dynamic processes are regulated by histone acetyltransferases (HATs) and histone deacetylases (HDACs).

Table 1.1 Classification of acetyltransferases (HATs)

Family	Protein	Function	Cellular localization
GNAT	HAT1	Catalyzes the acetylation of newly synthesized H4	Cytoplasm
	GCN5/PCAF	Coactivator of transcription	Nucleus
MYST	TIP60	Coactivator of transcription, repair	Cytoplasm
	MOZ	Regulation of transcription	Nucleus
	MORF	Activator/repressor	Nucleus
	HBO1	Acetylation of histone H3 and H4	Nucleus
	MOF	Dosage compensation	Nucleus
ORPHAN	P300/CBP	Coactivator of transcription	Nucleus/cytoplasm

Abbreviations in alphabetical order: *GCN5/PCAF* General Control Non-derepressible 5/P300/CBP-Associated Factor, *HAT1* Histone Acetyltransferase 1, *HBO1* Histone Acetyltransferase Binding to ORC1, *MOF* Males Absent On the First, *MORF* Monocytic Leukemia Zinc Finger Protein-Related Factor, *MOZ* Monocytic Leukemia Zinc Finger Protein, *P300/CBP* E1A-Associated Protein P300/CREB-Binding Protein, *TIP60* Tat-Interacting Protein 60

1.2.1 HATs and HDACs

Since HATs acetylate the ε-amino groups of lysine residues in the histone tails, they are also referred to as lysine acetyltransferases (KATs). There are different classifications for HATs, which have been evolutionarily conserved from yeast to humans (Lee and Workman 2007). Around 30 HATs have been discovered in humans to date, and they are categorized into two groups based on where they are found in the cell: nuclear and cytoplasmic (Gujral et al. 2020). Nuclear HATs, referred to as type A, are typically responsible for acetylation related to transcription, while type B cytoplasmic HATs catalyze the acetylation of newly synthesized histones (Roth et al. 2001). Another classification is based on sequence variations in their catalytic domains, which likely correspond to differences in substrate recognition and regulatory mechanisms (Lee and Grant 2019). There are five well-studied and distinct HAT subfamilies categorized in this way: Hat1 (KAT1), Gcn5/PCAF (KAT2A/KAT2B), p300/CBP (KAT3B/KAT3A), MYST (KAT5), and Rtt109 (KAT11) (Marmorstein and Zhou 2014). HATs are generally considered to have substrate specificity, as they tend to acetylate specific lysine residues on particular histone proteins. However, some HATs are capable of catalyzing the acetylation of both histones and non-histone proteins (Choudhary et al. 2009; Narita et al. 2019). Additionally, HATs often function as part of multi-subunit protein complexes, and participation in this complex formation regulates the substrate specificity of HATs (Wapenaar and Dekker 2016). Depending on their catalytic subunits, HAT complexes can target distinct acetylation sites on the same histone (Lee and Workman 2007) (Fig. 1.1). The Spt–Ada–Gcn5 Acetyltransferase Complex (SAGA), Ada Two-A Containing Complex (ATAC), Nucleosome Acetyltransferase of H4 Complex (NuA4), and Tat-interactive protein 60 kDa (Tip60) complexes are

Table 1.2 Classification of histone deacetylases (HDACs)

Class	Sub-class	Protein	Function	Primary cellular localization
Zn^{2+} dependent				
Class I		HDAC1	Regulates gene transcription in cell growth and cell death, including cell cycle progression, DNA repair, and apoptosis	Nucleus
		HDAC2	Regulates steroid function to inhibit inflammatory response in lungs	
		HDAC3	Modulating inflammatory responses	
		HDAC8	Removal from lysine residues in histones and non-histone proteins	
Class II	IIa	HDAC4	Regulates gene transcription, cell growth, survival, and proliferation	Nucleus/ Cytoplasm
		HDAC5	Regulates cellular proliferation	
		HDAC7	Modulates endothelial cell migration and angiogenesis	
		HDAC9	Cell viability and induces apoptosis	
	IIb	HDAC6	Regulates cell migration, apoptosis, and protein homeostasis	Cytoplasm
		HDAC10	Regulates cyclin A2 expression by deacetylating histones	
Class IV		HDAC11	Regulator of immune functions	Nucleus/ cytoplasm
NAD^{+} dependent				
Class III	I	SIRT1	Halts the cell cycle and induce DNA repair	Nucleus/ cytoplasm
		SIRT2	Regulates microtubule acetylation	Cytoplasm
		SIRT3	Assists mitochondria to maintain metabolic stability	Mitochondria
	II	SIRT4	Represses fatty acid oxidation, inhibits tumor growth, and promotes autophagy	Mitochondria
	III	SIRT5	Regulates various metabolic processes	Mitochondria
	IV	SIRT6	Regulates cellular homeostasis	Nucleus
		SIRT7	Facilitates the transcription of DNA	Nucleolus

well-studied examples of HAT complexes found in humans (Chen et al. 2024; Doyon et al. 2004; Krebs et al. 2011).

HDACs, which are the reversers of histone acetylation, are members of the eraser proteins group. In humans, 18 different histone deacetylases have been identified and categorized into four classes based on sequence similarities. Class I HDACs, similar to the Rdp3 protein in yeast, include HDAC1, HDAC2, HDAC3, and HDAC8. Class II HDACs with sequence similarity to the Hda1 protein in yeast include HDAC4, HDAC5, HDAC6, HDAC7, HDAC9, and HDAC10. The Nicotinamide Adenine Dinucleotide (NAD^{+})-dependent class III HDACs, which share sequence similarity with the Sir2 proteins in yeast, are also called the Sirtuin (SIRT) family and include SIRT1, SIRT2, SIRT3, SIRT4, SIRT5, SIRT6, and SIRT7. Finally, class IV, which is similar to class I and class II, consists only of

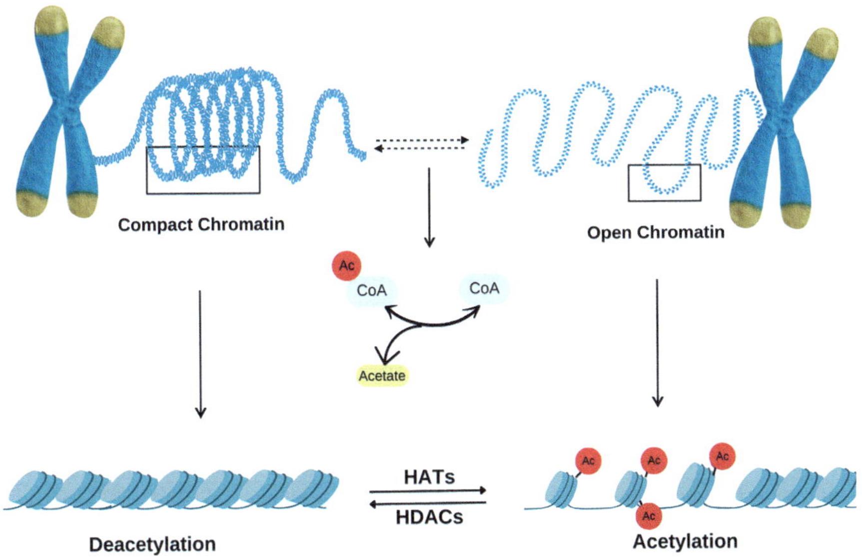

Fig. 1.1 Histone reversible acetylation and deacetylation

HDAC11 (Seto and Yoshida 2014). Like HATs, the substrates of HDACs are not limited to histones alone. As discussed below, non-histone proteins can also undergo acetylation, and certain HDACs can deacetylate these proteins as well. HDACs can contribute to cancer etiology through their roles in regulating the activity of oncogenes and tumor suppressor genes. Additionally, the deacetylation of various non-histone substrates by HDACs also promotes cancer development (Li and Seto 2016). Furthermore, evidence suggests that HDACs participate in tumorigenesis by being in a cross-talk with other epigenetic factors (Lee and R SH. 2013). Their effects in the regulation of the expression of genes involved in cell proliferation, cell cycle control, and apoptosis make HDACs a critical target for cancer. Therefore, HDAC inhibitors are frequently used as drugs in cancer treatment (Gujral et al. 2020).

The reversible biochemical processes carried out by HATs and HDACs are highly dynamic and occur continuously in association with various cellular activities. Acyl-CoA donors, which act as cofactors for HATs, are metabolites. Therefore, the cell's metabolic state significantly influences the targets and rate of acetylation (Shvedunova and Akhtar 2022).

1.2.2 Regulation of Gene Activity

The processes of histone acetylation and deacetylation are fundamental in regulating gene activity. The addition of acetyl groups to histone proteins through acetylation reduces the positive charge of the histone proteins. This results in a weakening of the electrostatic interaction between the negatively charged phosphate backbone of DNA and the histone protein. The loosened chromatin is now more accessible for binding transcription factors and RNA polymerase. This process facilitates gene expression in the specified region. The removal of acetyl groups during deacetylation, in turn, enhances the affinity between DNA and histones, resulting in chromatin condensation and the subsequent repression of gene expression.

Although acetylation can occur on all four canonical core histones—H2A, H2B, H3, and H4—histone H4 appears to be primarily responsible for significant changes in chromatin structure. The acetylation of lysine 16 on histone H4 (H4K16ac) has been proposed to disrupt nucleosome interactions (Shvedunova and Akhtar 2022). The effect of histone acetylation on gene expression is commonly explained as the loosening of chromatin, which facilitates the binding of transcription factors. However, there appears to be more to this process, and the underlying mechanisms remain a topic of ongoing debate. Promoters and enhancers, which are gene regulatory sequences, play critical roles in the initiation and elongation of transcription. Enhancers can interact with the target gene's promoter based on their epigenetic state, thereby increasing gene expression levels (Sungalee et al. 2021). The activity of enhancers is regulated by post-translational modifications of histones, and the acetylation of lysine 27 on histone H3 (H3K27ac) is essential for enhancer activation (Raisner et al. 2018).

Acetylated lysines also serve as signals for certain proteins that contain specific domains capable of recognizing these modifications. The histone code hypothesis suggests that these modifications on histones create a message that provides binding sites for reader proteins (Jenuwein and Allis 2001). The first recognition domains identified in reader proteins are bromodomains. Bromodomains are evolutionarily conserved amino acid sequences, and 61 bromodomains have been identified across 46 different proteins to date (Pan et al. 2023). In addition, YEATS domains and tandem plant homeodomains (DPF) are also known modules that recognize acetylated lysines (Khan et al. 2017). These proteins, which contain acetyl recognition domains, participate in the regulation of gene activity through functions such as chromatin remodeling and transcriptional activation (Gong et al. 2016) (Fig. 1.2).

1.2.3 DNA Repair Response

The acetylation of histones is linked not only to the activation of transcription but also to several cellular processes, including DNA replication and repair (Blakey and Litt 2015). The proper repair of double-strand breaks (DSBs) in DNA requires the reorganization of chromatin. Post-translational modifications of histones are among the key mechanisms involved in modulating chromatin during the repair response triggered by DNA damage (Van and Santos 2018). In particular, the acetylation of H3K56 and H4K16 facilitates the formation of a relaxed chromatin structure, which is crucial for the efficient repair of DSBs (Dhar et al. 2017; Vempati et al. 2010). As chromatin loosens, the repair factors involved in the process can access the DNA (Zhang et al. 2021). It has been shown that H3K56 acetylation increases chromatin

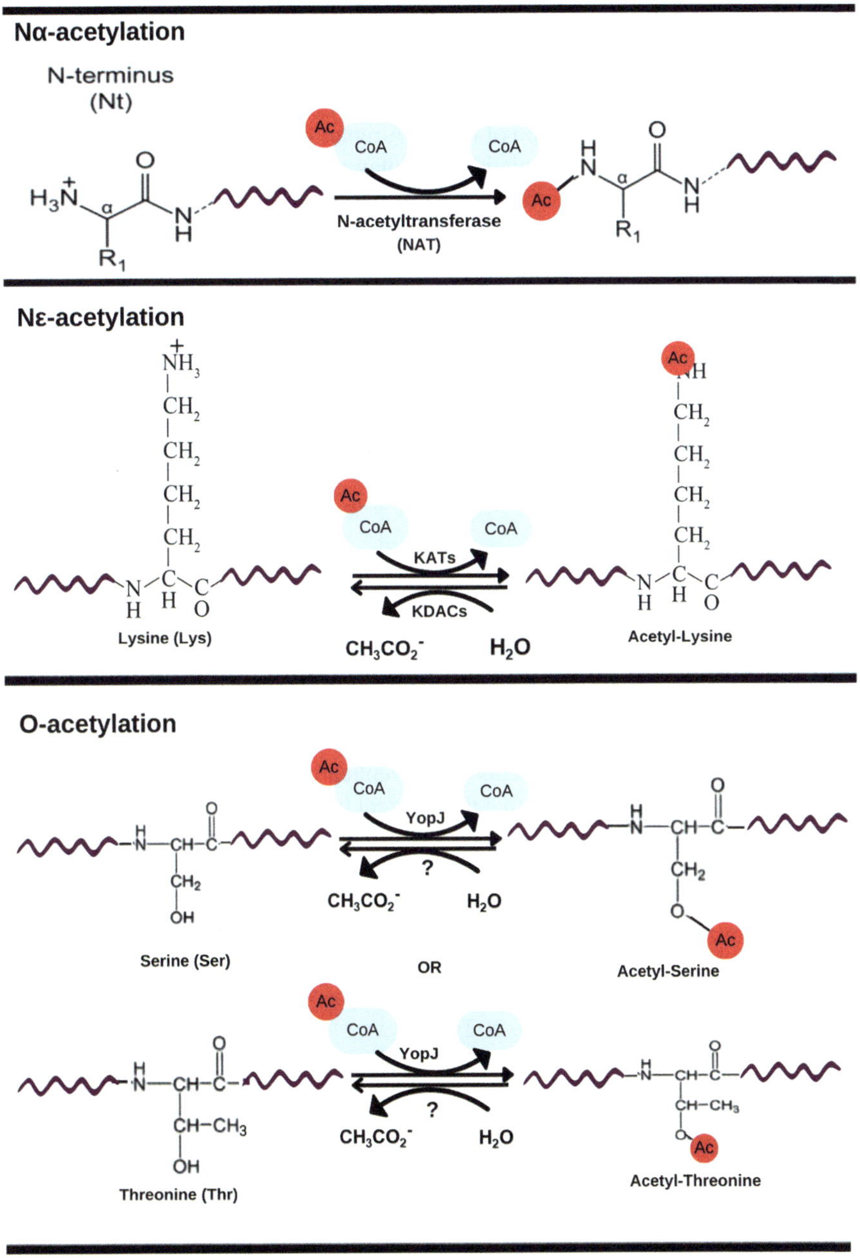

Fig. 1.2 Regulation of reversible lysine acetylation and deacetylation

accessibility during nucleotide excision repair (NER), in addition to its role in homologous recombination (HR) and non-homologous end joining (NHEJ), facilitating the recruitment of repair factors to the damaged site (Khan and Chaudhuri 2022). Just as histone acetylation facilitates DNA repair, deacetylation serves as a signal that directs the completion of the DNA repair process (Van and Santos 2018). Histone acetylation and deacetylation also play a role in determining the pathway chosen to repair DSBs (Aricthota et al. 2022). The acetylation of H4K16 and H2AK15 by TIP60 promotes the use of Breast Cancer Type 1 Susceptibility Protein (BRCA1) while inhibiting the role of Tumor Protein p53-Binding Protein 1 (53BP1), which promotes NHEJ, thereby steering the repair process towards HR (Jacquet et al. 2016; Tang et al. 2013). Conversely, deacetylation exerts the opposite effect and is necessary for repair through NHEJ. It has been shown that the deacetylation of H4K16Ac and H3K56Ac by HDAC1 and HDAC2 promotes NHEJ (Miller et al. 2010). Moreover, it has been suggested that HDACs are involved in all DNA repair pathways in some capacity (Roos and Krumm 2016).

1.2.4 Replication

Just as in DNA repair, numerous proteins are involved in the replication process, and their access to DNA relies on the decompaction of chromatin structure. The acetylation of histone H4K8 and H4K12 is essential for establishing the relaxed chromatin structure required during DNA replication (Ruan et al. 2015). The activation of origins, which are the points where replication begins, is closely related to acetylation. It has been reported that the efficient firing of origins during the S-phase requires acetylation of histone H3 and H4 tails (Unnikrishnan et al. 2010). On the other hand, the persistence of the acetylated state poses a risk to replication. The deacetylation of H3K56ac is particularly important for maintaining the activity of replication origins. The failure to remove acetyl groups from H3K56ac by specific HDACs, Hst3 and Hst4, after the S phase jeopardizes origin activity, leading to delays in the initiation of replication and increasing the vulnerability of cells to replicative stress (Tremblay et al. 2023).

1.3 Non-histone Reversible Acetylation and Deacetylation

The organization and functions of biological systems are highly complicated and diverse. The regulation of protein functions is essential for biological systems and involves various regulatory mechanisms, such as transcriptional and translational regulation and reversible and irreversible post-translational modifications. Reversible PTMs are known to be efficient mechanisms in regulating protein functions. The main advantage of PTMs is that they can be dynamically regulated quickly, with lower energy necessary than protein turnover, and can change protein function (Narita et al. 2019).

Non-histone acetylation and deacetylation are vital in maintaining cellular homeostasis (Batta et al. 2007). These reversible modifications affect the functions of proteins through varied mechanisms, such as by regulating enzymatic activity, protein stability, subcellular localization, and interplay with other post-translational modifications and by managing protein-protein and DNA-protein interactions (Narita et al. 2019). The history of protein modifications begins with Vincent Allfrey and colleagues first discovering acetylation on histones in 1964 (Allfrey et al. 1964). In the later studies, reversible acetylation was detected in high mobility group (HMG) proteins (Sterner et al. 1979) and alpha-tubulin proteins, which are non-histone proteins (L'Hernault and Rosenbaum 1983, 1985). In the late 1990s, the p53 transcription factor acetylation by its coactivator CBP/p300 was discovered (Gu and Roeder 1997). Three primary types of protein acetylation have been characterized based on the acetylation site of the protein (Fig. 1.3): N-terminal acetylation, commonly named Nα- acetylation; lysine acetylation or commonly known as K-acetylation (KAc) or Nε-acetylation and O-acetylation. Adding an acetyl group to the α-amino group of the N-terminal amino acid is called Nα-acetylation, an irreversible process responsible for modifying about 85% of human protein (Hollebeke et al. 2012). Nε-acetylation is the reversible addition of the acetyl group to the ε-amino group of a lysine residue (Farrugia and Puglielli 2018).

Over the past decade, proteomic studies have shown that non-histone proteins are frequently acetylated, constituting a significant part of the acetylome in mammalian cells. Non-histone acetylated/deacetylated proteins are involved in critical cellular pathways regarding gene transcription and protein folding, cell division, signal transduction, autophagy and metabolism, DNA damage repair, and cell structure (Narita et al. 2019; Peng and Seto 2011). Thus, non-histone reversible acetylation/deacetylation is a fundamental and conserved PTM mechanism in a wide range of organisms, from bacteria to humans (Xia et al. 2020).

Compared to histone protein, the acetylation/deacetylation of non-histone proteins has increasingly become a research focus due to its vital regulatory mechanisms and wide range of applications (Xia et al. 2020).

1.3.1 Discovery and Scope of Non-histone Protein Acetylation and Deacetylation

Acetylation was mostly identified on targets using conventional approaches until the beginning of the twenty-first century due to technological limitations (such as in vitro acetyltransferase assays using acetyl-lysine antibodies or with radioisotope labeled acetyl-CoA) (Narita et al. 2019). Proteomics studies used high-resolution mass spectrometry to identify numerous acetylation sites (Kim et al. 2006; Choudhary et al. 2009). The identification of thousands of acetylation sites has

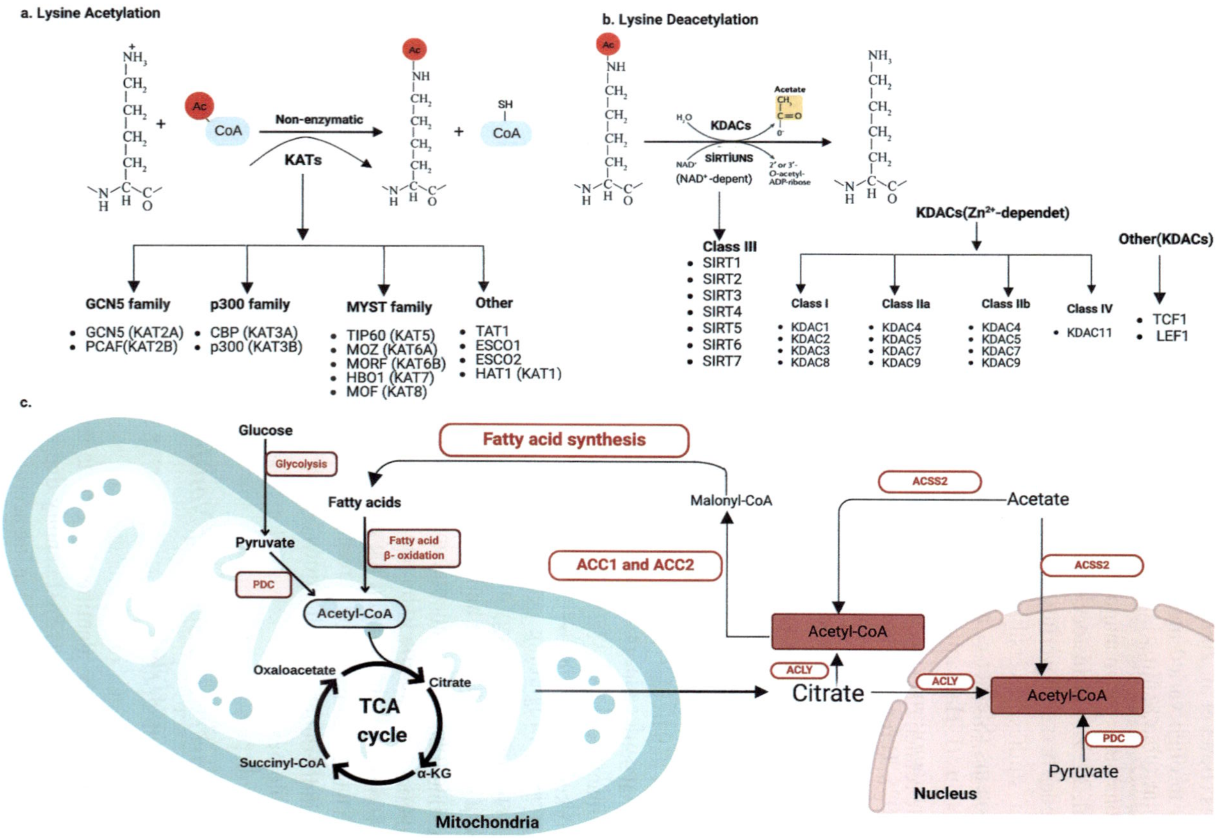

Fig. 1.3 Mechanisms of acetylation types: (**a**) Nα-acetylation: N-terminal acetyltransferases (NATs) catalyze the transfer of acetyl groups from acetyl-CoA to termini of α-amino groups, (**b**) Nε-acetylation or K-acetylation (KAc): Reversible post-translational protein modification of lysine catalyzed by lysine acetyltransferases (KATs) and lysine deacetylases (KDACs). (**c**) O-acetylation. Abbreviations (in alphabetical order): *Ac* acetylated, *ACC1* Acetyl-CoA Carboxylase 1, *ACC2* Acetyl-CoA Carboxylase 2, *ACSS2* Acyl-CoA Synthetase Short-Chain Family Member 2, *ACLY* ATP-Citrate Lyase, *Co-A* Coenzyme A, *PDC* Pyruvate Dehydrogenase Complex, *TCA cycle* Tricarboxylic Acid Cycle

increased interest in a wide range of research communities. Non-histone protein reversible acetylation and deacetylation have been found to play a role in all major biological processes. Biochemical and cellular functions of acetylation and deacetylation of non-histone substrates include DNA binding, transcriptional regulation, protein-protein interactions, protein stability and localization, protein folding, DNA repair (Batta et al. 2007), cell division, signal transduction, autophagy, and metabolism (Ashton et al. 2024; Narita et al. 2019).

Well-characterized targets of non-histone KAT/KDACs substrates include signaling mediators (β-catenin, SMAD Family Member 7 (SMAD7), Signal Transducer and Activator of Transcription 3 (STAT3), tumor suppressor proteins (p53 and Runt-related transcription factor 3 (RUNX3)), steroid receptors (androgens, estrogen), coregulators (Cellular Myelocytomatosis Viral Oncogene (c-Myc), E2F Transcription Factor 1 (E2F1), High Mobility Group Proteins (HMG), Yin Yang 1 (YY1), Erythroid Krüppel-like Factor (EKLF), GATA-binding transcription factors (GATA factors), Myogenic Differentiation 1 (MyoD), Nuclear Factor Kappa-Light-Chain-Enhancer of Activated B Cells (NF-κB), Hypoxia-Inducible Factor 1 Alpha (HIF-1α), Forkhead Box P3 (FoxP3)) and transcription factors, as well as structural (cell motility proteins), chaperone, and nuclear import proteins (α-tubulin, importin-α, Ku70, Heat Shock Protein 90 (HSP90)) (Spange et al. 2009; Mrakovcic et al. 2019; Glozak et al. 2005; Yang and Seto 2008).

1.4 Histone and Non-histone Reversible Acetylation in Development

The initial period of mammalian embryo development is marked by a sequence of specific synchronized landmarks. Protamines on the sperm-derived DNA are swapped into histone proteins in the cytoplasm as soon as fertilization occurs (Gunes et al. 2016). The steps of protamine-histone replacement are the earliest significant chromatin modification event encountered in the embryo. In this process, covalent modifications occurring in chromatin affect DNA packaging by histones, increasing or decreasing gene transcription. In mammalian cells, one of these covalent modifications is histone acetylation.

Histones are not acetylated in the chromatin of male and female reproductive cells before fertilization. Just following the entrance of the sperm to the oocytes, vital protamines on the paternal chromosomes gradually converted to the histone proteins throughout sperm decondensation, before histones become hyperacetylated (Cabot and Cabot 2018; Rybouchkin et al. 2006). Histone 4 on the paternal chromosome is substantially acetylated, which makes protamine inclusion easier in the paternal genome. When the S-phase starts in murine-fertilized eggs, hyperacetylation of histone 4 within male and female pronuclei reaches in equal amounts (Cabot and Cabot 2018).

HAT complexes are responsible for the acetylation of histones, including acetyltransferases and other proteins associated with chromatin. Recently, a new complex containing Ada2a-containing complex component 2 (ATAC2) was characterized beside the ATAC, which was previously known as a complex in mammals (Guelman et al. 2009). Another acetyltransferase, GCN5, is also involved in the structure of the new complex and plays an essential role in development. It was reported that histone H3 and histone H4 lysine acetylation were significantly reduced in *Atac2*-null mouse embryos (Guelman et al. 2009). In addition, it was observed that during early embryonic development, the mice showed growth retardation, higher apoptosis rate, and changes in the cell cycle (Guelman et al. 2009).

Since mutations causing the deletion of HATs may result in death during the embryonic period, studies were conducted with conditional mouse models to determine the effects of acetyltransferases on development (Nagarajan et al. 2013; Luense et al. 2019). Mice lacking the *Hat1* gene, a type B enzyme that acetylates the H4K5/12 ac, died without completing their embryonic development due to lung and skeletal defects (Nagarajan et al. 2013). Embryonic cell proliferation was not significantly affected in heterozygous mouse models, where protein levels were reduced approximately to half. However, the number of γ-H2AX foci increased in null mice, thus making embryonic cells more sensitive to DNA damage (Nagarajan et al. 2013). In another study, *Gcn5* knockout mice were created specifically for germ cells (Luense et al. 2019). This study observed an increase in the number of histones that were not replaced with protamines, which consequently led to a higher number of abnormal nucleosomes in the sperm of these mice (Luense et al. 2019). It was also observed that these nucleosome positions were maintained during the pre-implantation of mouse embryos. While the replacement of histones with protamines after hyperacetylation during spermiogenesis in the testes of control mice proceeded, hyperacetylation did not occur in knockout mice with deleted acetyltransferase genes. Therefore, the transition of histones with protamines could not be fully achieved (Luense et al. 2019). Thus, the researchers also showed that Gcn5, essential for embryonic development, plays a crucial role in chromatin reorganization and supports sperm function and fertility.

Acetylation of histone tails is a vital and effective tool in regulating transcription. However, other proteins and factors associated with histone-modifying enzymes can also affect gene activity via chromatin structure. Point mutations in the T-box transcription factor, which plays an important role in mouse embryonic development, have been demonstrated to alter acetylation in target regions (Beisaw et al. 2018). In vivo and in vitro studies with mouse embryos have revealed that H3K27ac via p300 is diminished resulting from a point mutation (Beisaw et al. 2018). This mutation was reported to convert a conserved tyrosine amino acid to alanine in T-box factor. Mutant embryos also showed alterations in hematopoietic and endothelial cell gene expression, as well as problems in neonatal hematopoietic progenitor differentiation (Beisaw et al. 2018).

Another study on reversible acetylation affecting post-translational activities of non-histone proteins was conducted with neonatal mouse and rat hearts (Collesi et al. 2018). Vectors containing a specific domain that carries out the Notch receptor's intracellular signaling have been administered to the cells. Notch signaling pathway is known to be important in the development of embryonic myocytes. Also, the consequence of Notch signaling on heart morphogenesis affects specific transcription factors and activators related to acetylation, including GCN5 and p300, after the signal is translocated to the nucleus (Collesi et al. 2018). Thus, researchers indicated that the cytoplasmic and subsequently released domain of the Notch receptor was acetylated after vector transduction, and this situation ensured the continuity of Notch signaling in the cells (Collesi et al. 2018). Deacetylation of the cytoplasmic domain of the receptor was carried out by Sirt1, a class III histone deacetylase, thus degrading the protein and leading to the termination of the signal (Collesi et al. 2018). Another study investigating the importance of post-translational acetylation of p53 during the embryonic period was carried out with human fetuses obtained from pregnancies terminated due to neural tube defects (Massa et al. 2020). The study has also used mutant mouse embryos with the unexpressed CBP/P300 interacting transactivator protein. The number of acetylated histone H4-positive cells was observed to be higher in normal early-stage fetuses. However, interestingly, histone H4 acetylation density was similar to normal fetuses in spina bifida fetuses (Massa et al. 2020). Although no difference in p53 expression was observed between wild-type and mutant mice, p53 acetylation contributed to the neural tube defect phenotype (Massa et al. 2020). Thus, the dynamic nature of the acetylation profile during neural tube formation in mammalian embryonal development was demonstrated. However, it has been shown that acetylation of histone variants, as histones, may be essential marks in embryonic development. Acetylation of H2A.Z in pre-implantation of mouse and bovine embryos was shown to be differentiated depending on the species and embryonic developmental stage H2A.Z was acetylated during pre-implantation in mouse blastomeres, which was predicted to lead to transcriptional activation. Interestingly, acetylation marking of this histone variant was not parallel with genome activation, and acetylation of H2A.Z could not be detected in mouse embryos at the 2-cell stage, which is the time of central embryonic genome activation (Boskovic et al. 2012). Modifications of histone variants at these different stages may be thought to contribute to embryonic development in a species-specific manner.

Studies conducted with model organisms showing changes in life stages during development may reveal the importance of histone acetylation, especially in morphological changes (Liu et al. 2024; Werner et al. 2023). In a study with *Pristionchus*, a short-life cycle nematode very similar to *Caenorhabditis elegans*, researchers underlined the importance of acetylation in the epigenetic control of developmental plasticity (Werner et al. 2023). It was revealed that the memory of the formation of one or two teeth in the mouth area, depending on the environmental conditions in the early larval period of the worm, depends on the acetylation status of the H4K5/12 (Werner et al. 2023). In a study published this year with an insect from the Lepidoptera (*Helicoverpa armigera*) how histone acetyltransferase acetylates the target histone tail was investigated (Liu et al. 2024). KAT8, one of the evolutionarily tightly conserved A-type histone acetyltransferases, was selected. Depending on the cell and tissue type, KAT8 is required for the acetylation of lysine 16 of histone H4, as well as for the acetylation of important non-histone targets, including p53. In addition, this acetyltransferase has an essential role in development, as it is influential in embryonic stem cell proliferation in both insects and mammals. In the study, the researchers reported that KAT8 increased midgut autophagy in the larval period by increasing the transcription of the autophagy-related gene 8 (Liu et al. 2024). However, the KAT8 level was also promoted by the transcription factor Forkhead box O (FOXO) (Liu et al. 2024).

1.5 Histone and Non-histone Reversible Acetylation in Aging

This section focuses on the role of histone and non-histone reversible acetylation in aging, with insights drawn from recent scientific research.

1.5.1 Histone Acetylation and Aging

Histone acetylation is a post-translational modification primarily regulated by HATs and HDACs, which add and remove acetyl groups, respectively. Acetylation neutralizes the positive charge of histones, loosening the interaction between histones and DNA, thereby facilitating transcriptional activation. Conversely, deacetylation promotes chromatin compaction and transcriptional repression (Cai et al. 2024).

Recent studies have shown that aging is accompanied by global changes in histone acetylation levels, leading to altered gene expression profiles. Sen et al. revealed that aging in human tissues is associated with a significant decline in histone H3K27 acetylation, which correlates with decreased expression of genes involved in tissue regeneration and maintenance. These changes were observed in liver and muscle tissues from aged individuals, highlighting the connection between histone acetylation and tissue-specific aging processes (Sen et al. 2016).

Studies in model organisms, such as Drosophila and mice, have shown that overexpression of specific HATs, including CREB-binding protein (CBP), can extend lifespan. In Drosophila, CBP enhances the acetylation of histone H3 at lysine 56 (H3K56), a mark associated with improved DNA repair and stress resistance, which counteracts age-related damage and promotes genome integrity. These findings highlight the potential of histone acetylation in regulating longevity by modulating chromatin structure and gene expression (Bradshaw 2021). Additionally, other acetylation marks like H3K9 and H3K27 are associated with controlling genes linked to aging and neuroinflammation, further reinforcing the role of histone acetylation in aging processes (Huang et al. 2023).

In mammals, histone deacetylase inhibition has been proposed as a therapeutic strategy to counteract age-related transcriptional changes. In research conducted on the role of nicotinamide mononucleotide (NMN) in restoring histone acetylation levels in aged mice, NMN supplementation increases NAD^+ levels in aged mouse liver, which restores sirtuin activity and modulates histone acetylation patterns, leading to improvements in the protein acetylome. By reversing aging-associated epigenetic alterations, NMN contributes to enhanced gene regulation and cellular homeostasis, addressing key mechanisms of age-related decline. These results underscore the importance of maintaining histone acetylation homeostasis to mitigate aging-related disorders (Luo et al. 2022).

Histone acetylation also contributes to cellular senescence, a hallmark of aging. Cellular senescence is characterized by an irreversible growth arrest and altered chromatin landscape, including the formation of senescence-associated heterochromatin foci (SAHF). These structures repress the expression of proliferation-promoting genes. A recent review emphasized that reduced histone acetylation at specific loci drives the formation of SAHF and silencing of E2F target genes, thereby enforcing the senescent phenotype (Qian et al. 2023).

1.5.2 Non-histone Acetylation and Aging

Non-histone protein acetylation also plays a crucial role in the regulation of aging. A growing body of evidence highlights the importance of acetylation in modulating the function of key regulatory proteins, including transcription factors, metabolic enzymes, and DNA repair proteins. The acetylation of non-histone proteins influences their stability, activity, subcellular localization, and interactions with other molecules (Imai and Guarente 2014).

One well-studied non-histone protein that undergoes acetylation is p53, a tumor suppressor involved in cell cycle regulation and DNA damage response. A study reported that the sirtuin1-p53 regulatory axis plays a crucial role in aging by modulating cellular stress responses, metabolism, and genomic stability, where Sirtuin1 deacetylates and inhibits p53 activity, thereby promoting cell survival and reducing senescence. Dysregulation of this axis contributes to aging-related pathologies by impairing DNA repair and increasing oxidative stress, highlighting its significance in maintaining cellular homeostasis and longevity (Ong and Ramasamy 2018).

FOXO transcription factors, particularly FOXO3, are also targets of reversible acetylation and play essential roles in regulating longevity and stress resistance. Acetylation of FOXO3 modulates its ability to activate genes involved in autophagy, antioxidant defense, and metabolic regulation. A study demonstrated that SIRT1, an NAD + -dependent deacetylase Class III, deacetylates FOXO3, enhancing its activity in promoting cellular stress resistance and longevity in aged mice. FOXO3 deacetylation was associated with improved mitochondrial function and reduced reactive oxygen species (ROS) accumulation, which are key factors in aging (Zhao and Liu 2021).

Sirtuins, particularly SIRT1 and SIRT6, are deacetylases known for their roles in regulating non-histone protein acetylation and age-related cellular processes. SIRT1 deacetylates multiple proteins involved in metabolic regulation, stress responses, and DNA repair. A study demonstrated that activating SIRT1 through pharmacological agents could promote longevity by improving metabolic health and reducing inflammation in aged organisms. SIRT1 activation also enhanced mitochondrial biogenesis through deacetylation of peroxisome proliferator-activated receptor gamma coactivator 1-alpha (PGC-1α), a key regulator of energy homeostasis (Chen et al. 2020).

SIRT6 has also garnered attention for its role in promoting healthy aging. It was shown that SIRT6 enhances DNA repair and reduces inflammation by deacetylating proteins involved in these pathways. In aged mice, overexpression of SIRT6 extended lifespan and improved physical activity levels, providing further evidence for the therapeutic potential of targeting sirtuin pathways to combat aging (Tasselli et al. 2017).

Non-histone acetylation also regulates metabolic processes that become dysregulated with age. For example, SIRT3, a mitochondrial deacetylase, modulates the activity of enzymes involved in the tricarboxylic acid (TCA) cycle and oxidative phosphorylation. Research demonstrated that SIRT3 deacetylates and activates isocitrate dehydrogenase 2 (IDH2), an enzyme critical for maintaining the mitochondrial redox balance. Reduced SIRT3 activity in aged mice led to increased mitochondrial dysfunction and oxidative stress, which contributed to the aging phenotype (Yu et al. 2012).

The reversible acetylation of histone and non-histone proteins is a key regulatory mechanism in aging. Histone acetylation affects chromatin structure and gene expression patterns, influencing cellular processes such as stress response, tissue regeneration, and metabolism. Non-histone acetylation, on the other hand, regulates the activity of proteins involved in metabolic regulation, DNA repair, and cellular stress responses. Collectively, these processes highlight the importance of acetylation in the aging process and suggest that targeting acetylation pathways may offer therapeutic strategies to delay aging and improve health span. Recent studies have provided valuable insights into how acetylation modifications affect aging, opening new avenues for interventions aimed at mitigating age-related diseases.

1.5.2.1 Huntington's Disease

Huntington's disease (HD) is an autosomal dominant neurodegenerative disorder resulting from a CAG triplet repeat expansion dynamic mutation in the huntingtin gene (MacDonald 1993). HDAC protein family was proposed to lead to transcriptional dysregulation in HD (Zhang et al. 2024). The higher levels of HDAC1 and HDAC5 proteins cause a decreased acetylation of related gene loci, leading to transcriptional silencing and dysregulation.

HDAC6 is a specific α-MT deacetylase participating in the deacetylation and ubiquitination of microtubules and other substrates. Studies regarding HDAC6 inhibitor treatment have been reported to be contradictory. HDAC6 inhibitor treatment has increased the acetylation of tubulin and brain-derived neurotrophic factor (BDNF) release by inhibiting HDAC6 compensation for aberrations in intracellular transport in HD (Simoes-Pires et al. 2013). On the other hand, Bobrowska et al.

found no improvement regarding the HDAC6 knockdown HD animal models (MacDonald 1993). In conclusion, the HDAC protein upregulation may provide great potential for treating HD.

1.5.2.2 Alzheimer's Disease

Alzheimer's disease (AD) is a leading progressive neurodegenerative illness and the primary cause of dementia. AD is characterized by pathologic amyloid-β accumulation that destroys synaptic junctions, inhibiting memory-related gene expression in the junctions and neurons involved in the calcium signaling pathway in organelles and tau hyperphosphorylation neuronal fibers (Scheltens et al. 2021). Histone modification's impact on the late-onset development of AD has received considerable attention (Kumar et al. 2022). HDACs cause transcriptional silencing through the deacetylation of histone tails, leading to neuroinflammation, neuronal apoptosis, and death. Studies showed a significant role of histone deacetylation in the progression of the disease; therefore, HDAC inhibitors are requested as potential drugs in the effective medication of AD. HDAC inhibition prevents Aβ protein deposition and induces memory improvement. HDAC knockdown mouse models have demonstrated that inhibition of these genes restores structural, functional, and synaptic plasticity and extinguishes neurodegeneration-associated memory impairments. The most widely studied HDAC subtype for AD is HDAC2, which increases during the disease, inhibits gene transcription required for learning and memory, and leads to impaired learning and memory. This shows that cognitive capacities are not lost completely, only impaired due to the epigenetic block (Graff et al. 2012).

On the other hand, overexpression of HDAC3 has been found to be associated with accumulation of Aβ amyloid protein, microglia aggravation, dendritic spine density, and spatial memory reduction. Similar findings were reported regarding the negative regulation of memory in the hippocampal nucleus of mice with overexpression of HDAC3. Therefore, inhibition of HDAC3 can improve these conditions (Zhu et al. 2017). Indeed, HDAC4 selective inhibition also reforms better expression of synaptic plasticity and memory genes (Chen et al. 2021). Although the deprivation of HDAC5 can worsen memory dysfunction in AD mouse models, it does not affect the pathogenesis of the disease (Agis-Balboa et al. 2013).

1.5.2.3 Mental Diseases

Elevated levels of HDACs lead to deacetylation of histones and alterations in chromatin structure in patients with psychiatric disorders. HDACs also cause the repression of gene expression, which is involved in the pathogenesis of psychiatric diseases (Tang et al. 2011). Indeed, the expression of HDAC2 was reported to be higher in the dorsolateral prefrontal cortex tissue of patients with clinical depression than in healthy individuals. 5-Hydroxytryptamine receptor 1A (5-HT1A) selective receptor agonists can treat depression-related behaviors by reducing the HDAC1,

HDAC2, HDAC4, and HDAC5 expression and increasing histone acetylation levels. An increase in histone acetylation and a decrease in HDAC proteins and HDAC4 isoforms was observed in patients with schizophrenia. Indeed, Martinostat PET analysis shows that HDAC expression in multiple parts of brain tissue in schizophrenic patients was reduced compared with controls, and the level of cognition was positively correlated with the HDAC level. Additionally, a case-controlled study performed in Korea also showed that expression of HDAC3 and HDAC4 is associated with the development of schizophrenia. Nevertheless, peripheral blood samples from mouse models and patients with schizophrenia have been showing overexpression of HDAC1 (Scheltens et al. 2021).

1.6 Conclusion

The human genome is a dynamic complex nucleoprotein containing DNA, histones, and non-histone proteins. Epigenetic modifications have emerged as critical factors in regulating all the DNA-templated pathways, human health, and disease. Epigenetics modulation is critical in interpreting epigenomic profiling regarding an organism's development, diseases, and aging. Notably, the pathogenetic mechanisms and interplay between histone acetylation and deacetylation may provide the actual cause of developmental abnormalities and aging at the molecular level. Reversible acetylation would become a significant biomarker in epigenomic analysis. This process is highly dynamic and continuously performed by HATs and HDACs with various cellular activities (Shvedunova and Akhtar 2022). Their effects on the regulation of the expression of genes are involved in cell cycle control via p53, cell proliferation, and apoptosis. In addition, the deacetylation of various non-histone substrates by HDACs also promotes multiple pathologies, including developmental aberrations and tumorigenesis, by being in a cross-talk with other epigenetic factors, making HDACs a critical target for cancer treatment. Therefore, HDAC inhibitors are frequently used as therapeutic agents in cancer treatment (Gujral et al. 2020).

Various model organisms have also shown that histone acetylation and deacetylation modifications play a role in regulating apoptosis, cell cycle regulation, gene expression regulations, and the development and aging of organisms.

In conclusion, while HDAC1 and HDAC2 are the most extensively studied proteins from this family, particularly HDAC1 and HDAC2 play a vital role, other members are also important for proliferation and differentiation of cells, pathogenesis of disease, aging, and medication of diseases.

References

Agis-Balboa RC, Pavelka Z, Kerimoglu C, Fischer A (2013) Loss of HDAC5 impairs memory function: implications for Alzheimer's disease. J Alzheimers Dis 33(1):35–44
Allfrey VG, Faulkner R, Mirsky AE (1964) Acetylation and methylation of histones and their possible role in the regulation of RNA synthesis. Proc Natl Acad Sci USA 51(5):786–794
Aricthota S, Rana PP, Haldar D (2022) Histone acetylation dynamics in repair of DNA double-strand breaks. Front Genet 13:926577
Ashton AW, Dhanjal HK, Rossner B, Mahmood H, Patel VI, Nadim M et al (2024) Acetylation of nuclear receptors in health and disease: an update. FEBS J 291(2):217–236
Batta K, Das C, Gadad S, Shandilya J, Kundu TK (2007) Reversible acetylation of non histone proteins: role in cellular function and disease. Chromatin Dis:193–214
Beisaw A, Tsaytler P, Koch F, Schmitz SU, Melissari MT, Senft AD et al (2018) BRACHYURY directs histone acetylation to target loci during mesoderm development. EMBO Rep 19(1):118–134
Blakey CA, Litt MD (2015) Chapter 2 – histone modifications—models and mechanisms. In: Huang S, Litt MD, Blakey CA (eds) Epigenetic gene expression and regulation. Academic Press, Oxford, pp 21–42
Boskovic A, Bender A, Gall L, Ziegler-Birling C, Beaujean N, Torres-Padilla ME (2012) Analysis of active chromatin modifications in early mammalian embryos reveals uncoupling of H2A.Z acetylation and H3K36 trimethylation from embryonic genome activation. Epigenetics 7(7):747–757
Bradshaw PC (2021) Acetyl-CoA metabolism and histone acetylation in the regulation of aging and lifespan. Antioxidants 10(4):572
Cabot B, Cabot RA (2018) Chromatin remodeling in mammalian embryos. Reproduction 155(3):R147–RR58
Cai J, Deng Y, Min Z, Li C, Zhao Z, Jing D (2024) Deciphering the dynamics: Exploring the impact of mechanical forces on histone acetylation. FASEB J 38(15):e23849
Chen C, Zhou M, Ge Y, Wang X (2020) SIRT1 and aging related signalling pathways. Mech Ageing Dev 187:111215
Chen YA, Lu CH, Ke CC, Chiu SJ, Chang CW, Yang BH et al (2021) Evaluation of class IIA histone deacetylases expression and in vivo epigenetic imaging in a transgenic mouse model of Alzheimer's disease. Int J Mol Sci 22(16)
Chen K, Wang L, Yu Z, Yu J, Ren Y, Wang Q et al (2024) Structure of the human TIP60 complex. Nat Commun 15(1):7092
Choudhary C, Kumar C, Gnad F, Nielsen ML, Rehman M, Walther TC et al (2009) Lysine acetylation targets protein complexes and co-regulates major cellular functions. Science 325(5942):834–840
Collesi C, Felician G, Secco I, Gutierrez MI, Martelletti E, Ali H et al (2018) Reversible Notch1 acetylation tunes proliferative signalling in cardiomyocytes. Cardiovasc Res 114(1):103–122
Crafa A, Cannarella R, Calogero AE, Gunes S, Agarwal A (2024) Behind the genetics: the role of epigenetics in infertility-related testicular dysfunction. Life (Basel) 14(7)
Dhar S, Gursoy-Yuzugullu O, Parasuram R, Price BD (2017) The tale of a tail: histone H4 acetylation and the repair of DNA breaks. Philos Trans R Soc Lond Ser B Biol Sci 372(1731):20160284
Doyon Y, Selleck W, Lane WS, Tan S, Cote J (2004) Structural and functional conservation of the NuA4 histone acetyltransferase complex from yeast to humans. Mol Cell Biol 24(5):1884–1896
Farrugia MA, Puglielli L (2018) Nepsilon-lysine acetylation in the endoplasmic reticulum – a novel cellular mechanism that regulates proteostasis and autophagy. J Cell Sci 131(22)
Glozak MA, Sengupta N, Zhang X, Seto E (2005) Acetylation and deacetylation of non-histone proteins. Gene 363:15–23
Gong F, Chiu LY, Miller KM (2016) Acetylation reader proteins: linking acetylation signalling to genome maintenance and cancer. PLoS Genet 12(9):e1006272

Graff J, Rei D, Guan JS, Wang WY, Seo J, Hennig KM et al (2012) An epigenetic blockade of cognitive functions in the neurodegenerating brain. Nature 483(7388):222–226

Gu W, Roeder RG (1997) Activation of p53 sequence-specific DNA binding by acetylation of the p53 C-terminal domain. Cell 90(4):595–606

Guelman S, Kozuka K, Mao Y, Pham V, Solloway MJ, Wang J et al (2009) The double-histone-acetyltransferase complex ATAC is essential for mammalian development. Mol Cell Biol 29(5):1176–1188

Gujral P, Mahajan V, Lissaman AC, Ponnampalam AP (2020) Histone acetylation and the role of histone deacetylases in normal cyclic endometrium. Reprod Biol Endocrinol 18(1):84

Gunes S, Arslan MA, Hekim GNT, Asci R (2016) The role of epigenetics in idiopathic male infertility. J Assist Reprod Genet 33(5):553–569

Gunes S, Metin Mahmutoglu A, Hekim N (2024) Epigenetics of nonobstructive azoospermia. Asian J Androl

Hollebeke J, Van Damme P, Gevaert K (2012) N-terminal acetylation and other functions of Nα-acetyltransferases, vol 393, p 291

Huang M, Malovic E, Ealy A, Jin H, Anantharam V, Kanthasamy A et al (2023) Microglial immune regulation by epigenetic reprogramming through histone H3K27 acetylation in neuroinflammation. Front Immunol 14:1052925

Imai S-i, Guarente L (2014) NAD+ and sirtuins in aging and disease. Trends Cell Biol 24(8):464–471

Jacquet K, Fradet-Turcotte A, Avvakumov N, Lambert JP, Roques C, Pandita RK et al (2016) The TIP60 complex regulates bivalent chromatin recognition by 53BP1 through direct H4K20me binding and H2AK15 acetylation. Mol Cell 62(3):409–421

Jenuwein T, Allis CD (2001) Translating the histone code. Science 293(5532):1074–1080

Khan P, Chaudhuri RN (2022) Acetylation of H3K56 orchestrates UV-responsive chromatin events that generate DNA accessibility during nucleotide excision repair. DNA Repair (Amst) 113:103317

Khan A, Bridgers JB, Strahl BD (2017) Expanding the reader landscape of histone acylation. Structure 25(4):571–573

Kim SC, Sprung R, Chen Y, Xu Y, Ball H, Pei J et al (2006) Substrate and functional diversity of lysine acetylation revealed by a proteomics survey. Mol Cell 23(4):607–618

Kouzarides T (2007) Chromatin modifications and their function. Cell 128(4):693–705

Krebs AR, Karmodiya K, Lindahl-Allen M, Struhl K, Tora L (2011) SAGA and ATAC histone acetyl transferase complexes regulate distinct sets of genes and ATAC defines a class of p300-independent enhancers. Mol Cell 44(3):410–423

Kumar V, Kundu S, Singh A, Singh S (2022) Understanding the role of histone deacetylase and their inhibitors in neurodegenerative disorders: current targets and future perspective. Curr Neuropharmacol 20(1):158–178

L'Hernault SW, Rosenbaum JL (1983) Chlamydomonas alpha-tubulin is posttranslationally modified in the flagella during flagellar assembly. J Cell Biol 97(1):258–263

L'Hernault SW, Rosenbaum JL (1985) Reversal of the posttranslational modification on Chlamydomonas flagellar alpha-tubulin occurs during flagellar resorption. J Cell Biol 100(2):457–462

Lee CY, Grant PA (2019) Chapter 1-1 – role of histone acetylation and acetyltransferases in gene regulation. In: McCullough SD, Dolinoy DC (eds) . Academic, Toxicoepigenetics, pp 3–30

Lee J, R SH. (2013) Cancer epigenetics: mechanisms and crosstalk of a HDAC inhibitor, vorinostat. Chemotherapy (Los Angel) 2(111)

Lee KK, Workman JL (2007) Histone acetyltransferase complexes: one size doesn't fit all. Nat Rev Mol Cell Biol 8(4):284–295

Li Y, Seto E (2016) HDACs and HDAC inhibitors in cancer development and therapy. Cold Spring Harb Perspect Med 6(10)

Liu TW, Zhao YM, Jin KY, Wang JX, Zhao XF (2024) KAT8 is upregulated and recruited to the promoter of Atg8 by FOXO to induce H4 acetylation for autophagy under 20-hydroxyecdysone regulation. J Biol Chem 300(3):105704

Luense LJ, Donahue G, Lin-Shiao E, Rangel R, Weller AH, Bartolomei MS et al (2019) Gcn5-mediated histone acetylation governs nucleosome dynamics in spermiogenesis. Dev Cell 51(6):745–58 e6
Luo C, Ding W, Zhu S, Chen Y, Liu X, Deng H (2022) Nicotinamide mononucleotide administration amends protein acetylome of aged mouse liver. Cells 11(10):1654
MacDonald ME et al (1993) A novel gene containing a trinucleotide repeat that is expanded and unstable on Huntington's disease chromosomes. The Huntington's Disease Collaborative Research Group. Cell 72(6):971–983
Marmorstein R, Zhou MM (2014) Writers and readers of histone acetylation: structure, mechanism, and inhibition. Cold Spring Harb Perspect Biol 6(7):a018762
Massa V, Avagliano L, Grazioli P, De Castro SCP, Parodi C, Savery D et al (2020) Dynamic acetylation profile during mammalian neurulation. Birth Defects Res 112(2):205–211
Miller KM, Tjeertes JV, Coates J, Legube G, Polo SE, Britton S et al (2010) Human HDAC1 and HDAC2 function in the DNA-damage response to promote DNA nonhomologous end-joining. Nat Struct Mol Biol 17(9):1144–1151
Mrakovcic M, Kleinheinz J, Fröhlich LF (2019) p53 at the crossroads between different types of HDAC inhibitor-mediated cancer cell death. Int J Mol Sci 20(10):2415
Nagarajan P, Ge Z, Sirbu B, Doughty C, Agudelo Garcia PA, Schlederer M et al (2013) Histone acetyl transferase 1 is essential for mammalian development, genome stability, and the processing of newly synthesized histones H3 and H4. PLoS Genet 9(6):e1003518
Narita T, Weinert BT, Choudhary C (2019) Functions and mechanisms of non-histone protein acetylation. Nat Rev Mol Cell Biol 20(3):156–174
Ntorla A, Burgoyne JR (2021) The Regulation and Function of Histone Crotonylation. Front Cell Dev Biol 9:624914
Ong AL, Ramasamy TS (2018) Role of Sirtuin1-p53 regulatory axis in aging, cancer and cellular reprogramming. Ageing Res Rev 43:64–80
Pan Z, Zhao Y, Wang X, Xie X, Liu M, Zhang K et al (2023) Targeting bromodomain-containing proteins: research advances of drug discovery. Mol Biomed 4(1):13
Peng L, Seto E (2011) Deacetylation of nonhistone proteins by HDACs and the implications in cancer. Histone Deacetyl Biol Clin Implicat:39–56
Piontkivska H, Rooney AP, Nei M (2002) Purifying selection and birth-and-death evolution in the histone H4 gene family. Mol Biol Evol 19(5):689–697
Qian J, Zhou X, Tanaka K, Takahashi A (2023) Alteration in the chromatin landscape during the DNA damage response: continuous rotation of the gear driving cellular senescence and aging. DNA Repair 131:103572
Raisner R, Kharbanda S, Jin L, Jeng E, Chan E, Merchant M et al (2018) Enhancer activity requires CBP/P300 bromodomain-dependent histone H3K27 acetylation. Cell Rep 24(7):1722–1729
Roos WP, Krumm A (2016) The multifaceted influence of histone deacetylases on DNA damage signalling and DNA repair. Nucleic Acids Res 44(21):10017–10030
Roth SY, Denu JM, Allis CD (2001) Histone acetyltransferases. Annu Rev Biochem 70:81–120
Ruan K, Yamamoto TG, Asakawa H, Chikashige Y, Kimura H, Masukata H et al (2015) Histone H4 acetylation required for chromatin decompaction during DNA replication. Sci Rep 5:12720
Rybouchkin A, Kato Y, Tsunoda Y (2006) Role of histone acetylation in reprogramming of somatic nuclei following nuclear transfer. Biol Reprod 74(6):1083–1089
Scheltens P, De Strooper B, Kivipelto M, Holstege H, Chetelat G, Teunissen CE et al (2021) Alzheimer's disease. Lancet 397(10284):1577–1590
Sen P, Shah PP, Nativio R, Berger SL (2016) Epigenetic mechanisms of longevity and aging. Cell 166(4):822–839
Seto E, Yoshida M (2014) Erasers of histone acetylation: the histone deacetylase enzymes. Cold Spring Harb Perspect Biol 6(4):a018713
Shvedunova M, Akhtar A (2022) Modulation of cellular processes by histone and non-histone protein acetylation. Nat Rev Mol Cell Biol 23(5):329–349

Simoes-Pires C, Zwick V, Nurisso A, Schenker E, Carrupt PA, Cuendet M (2013) HDAC6 as a target for neurodegenerative diseases: what makes it different from the other HDACs? Mol Neurodegener 8:7

Spange S, Wagner T, Heinzel T, Krämer OH (2009) Acetylation of non-histone proteins modulates cellular signalling at multiple levels. Int J Biochem Cell Biol 41(1):185–198

Sterner R, Vidali G, Allfrey VG (1979) Studies of acetylation and deacetylation in high mobility group proteins. Identification of the sites of acetylation in HMG-1. J Biol Chem 254(22):11577–11583

Sungalee S, Liu Y, Lambuta RA, Katanayeva N, Donaldson Collier M, Tavernari D et al (2021) Histone acetylation dynamics modulates chromatin conformation and allele-specific interactions at oncogenic loci. Nat Genet 53(5):650–662

Tang B, Dean B, Thomas EA (2011) Disease- and age-related changes in histone acetylation at gene promoters in psychiatric disorders. Transl Psychiatry 1(12):e64

Tang J, Cho NW, Cui G, Manion EM, Shanbhag NM, Botuyan MV et al (2013) Acetylation limits 53BP1 association with damaged chromatin to promote homologous recombination. Nat Struct Mol Biol 20(3):317–325

Tasselli L, Zheng W, Chua KF (2017) SIRT6: novel mechanisms and links to aging and disease. Trends Endocrinol Metabol 28(3):168–185

Tremblay R, Mehrjoo Y, Ahmed O, Simoneau A, McQuaid ME, Affar EB et al (2023) Persistent acetylation of histone H3 lysine 56 compromises the activity of DNA replication origins. Mol Cell Biol 43(11):566–595

Unnikrishnan A, Gafken PR, Tsukiyama T (2010) Dynamic changes in histone acetylation regulate origins of DNA replication. Nat Struct Mol Biol 17(4):430–437

Van HT, Santos MA (2018) Histone modifications and the DNA double-strand break response. Cell Cycle 17(21–22):2399–2410

Vempati RK, Jayani RS, Notani D, Sengupta A, Galande S, Haldar D (2010) p300-mediated acetylation of histone H3 lysine 56 functions in DNA damage response in mammals. J Biol Chem 285(37):28553–28564

Wapenaar H, Dekker FJ (2016) Histone acetyltransferases: challenges in targeting bi-substrate enzymes. Clin Epigenetics 8:59

Werner MS, Loschko T, King T, Reich S, Theska T, Franz-Wachtel M et al (2023) Histone 4 lysine 5/12 acetylation enables developmental plasticity of Pristionchus mouth form. Nat Commun 14(1):2095

Xia C, Tao Y, Li M, Che T, Qu J (2020) Protein acetylation and deacetylation: an important regulatory modification in gene transcription. Exp Ther Med 20(4):2923–2940

Yang X-J, Seto E (2008) Lysine acetylation: codified crosstalk with other posttranslational modifications. Mol Cell 31(4):449–461

Yu W, Dittenhafer-Reed KE, Denu JM (2012) SIRT3 protein deacetylates isocitrate dehydrogenase 2 (IDH2) and regulates mitochondrial redox status. J Biol Chem 287(17):14078–14086

Zhang J, Lu X, MoghaddamKohi S, Shi L, Xu X, Zhu WG (2021) Histone lysine modifying enzymes and their critical roles in DNA double-strand break repair. DNA Repair (Amst) 107:103206

Zhang LY, Zhang SY, Wen R, Zhang TN, Yang N (2024) Role of histone deacetylases and their inhibitors in neurological diseases. Pharmacol Res 208:107410

Zhao Y, Liu Y-S (2021) Longevity factor FOXO3: a key regulator in aging-related vascular diseases. Front Cardiovasc Med 8:778674

Zhu X, Wang S, Yu L, Jin J, Ye X, Liu Y et al (2017) HDAC3 negatively regulates spatial memory in a mouse model of Alzheimer's disease. Aging Cell 16(5):1073–1082

Chapter 2
Acetylation-Mediated Epigenetic Consequences for Biological Control and Cancer

Andrew J. Fritz, Kyle T. McKay, Haley W. Greenyer, Emory Pacht, Rabail H. Toor, Rahim Ullah, Jackson R. Del Porto, Abigail G. Person, Sadie J. Korzec, Kathleen E. Bright, Genevieve Brzoza, Jessica L. Heath, Prachi N. Ghule, Jonathan A. R. Gordon, Andre J. Van Wijnen, Seth E. Frietze, Karen C. Glass, Jane B. Lian, Janet L. Stein, and Gary S. Stein

Abstract Acetylation of histones epigenetically mediates transcriptional dynamics of gene activation and suppression in response to physiological regulatory signals. The acetylated states of histone proteins define the activities of gene promoter and enhancer elements by contributing to competency for regulatory protein interactions and control of chromatin organization including higher-order inter and intra-chromosomal interactions. Cell transformation and tumor progression are associated with and functionally related to histone acetylation. Targeting the regulatory machinery for histone acetylation provides treatment options for cancer-compromised gene expression with specificity and reduced off-target consequences.

A. J. Fritz · H. W. Greenyer · E. Pacht · R. H. Toor · R. Ullah
J. R. Del Porto · A. G. Person · S. J. Korzec · P. N. Ghule · J. A. R. Gordon · K. C. Glass
J. B. Lian · J. L. Stein · G. S. Stein (✉)
Department of Biochemistry, University of Vermont, Burlington, VT, USA

University of Vermont Cancer Center, University of Vermont, Burlington, VT, USA

University of Vermont Larner College of Medicine, University of Vermont, Burlington, VT, USA
e-mail: gary.stein@med.uvm.edu

M. Halasa, A. Wawruszak (eds.), *Histone and Non-Histone Reversible Acetylation in Development, Aging and Disease*, Results and Problems in Cell Differentiation 75,
https://doi.org/10.1007/978-3-031-91459-1_2

2.1 Introduction

The origins for mechanistically understanding the regulatory consequences of epigenetic control are discoveries in the 1960s that Post-Translational Modifications (PTMs) of histone tail residues correlate with gene expression. Initial findings suggested functional relationships of histone acetylation and phosphorylation with gene activation (Allfrey et al. 1964; Kleinsmith et al. 1966; Pogo et al. 1966; Pogo et al. 1968). The plasticity and reversibility of histone acetylation and phosphorylation by deacetylases and phosphatases, in response to physiological regulatory signals, reinforced the principle that PTMs are engaged in gene regulation. Despite an early report for reversibility of histone methylation (Byvoet 1972), initially histone methylation was considered irreversible, repressive, and long-term. More recent findings provide insight into reversibility of histone methylation and establish pivotal contributions of methylated histones to both gene activation and repression (Jambhekar et al. 2019).

K. T. McKay
University of Vermont Cancer Center, University of Vermont, Burlington, VT, USA

University of Vermont Larner College of Medicine, University of Vermont, Burlington, VT, USA

Department of Pharmacology, University of Vermont, Burlington, VT, USA

K. E. Bright · S. E. Frietze
Department of Biochemistry, University of Vermont, Burlington, VT, USA

University of Vermont Cancer Center, University of Vermont, Burlington, VT, USA

University of Vermont Larner College of Medicine, University of Vermont, Burlington, VT, USA

College of Nursing and Health Sciences, University of Vermont, Burlington, VT, USA

G. Brzoza
Department of Biochemistry, University of Vermont, Burlington, VT, USA

College of Nursing and Health Sciences, University of Vermont, Burlington, VT, USA

Department of Pediatrics, University of Vermont, Burlington, VT, USA

J. L. Heath
Department of Biochemistry, University of Vermont, Burlington, VT, USA

University of Vermont Cancer Center, University of Vermont, Burlington, VT, USA

University of Vermont Larner College of Medicine, University of Vermont, Burlington, VT, USA

Department of Pediatrics, University of Vermont, Burlington, VT, USA

A. J. Van Wijnen
Department of Biochemistry, University of Vermont, Burlington, VT, USA

2.2 Histone Acetylation as a Team Player: Functional Crosstalk between PTMs

Epigenetic parameters, including DNA methylation, histone PTMs including acetylation, chromatin remodeling, and accessibility are interrelated and functionally integrated (Kondo 2009; Lee et al. 2010; Du et al. 2015; Beltrao et al. 2013). Their interactions are context-dependent, allowing them to synergistically activate or repress chromatin or have antagonistic effects. For example, DNA methylation complexes may interact with histone-modifying enzymes, influencing their activity and altering gene expression outcomes. Additionally, histone acetylation generally promotes an open chromatin state, potentially allowing DNA demethylation (Choi et al. 2015; Lee et al. 2007). Conversely, the regulatory machinery driving DNA methylation complexes with enzymes responsible for histone deacetylation (Choi et al. 2015). This interplay between acetylation and methylation can also influence the activity of chromatin remodeling complexes, affecting gene accessibility and expression. The readout of these epigenetic chromatin states facilitates the binding of transcription factors and interactions with regulatory machinery and noncoding RNAs (lncRNAs) to control essential processes including silencing, pausing, initiation, elongation, and alternative splicing. This regulatory machinery in turn establishes the extent to which genes are transcriptionally active, repressed, or poised for expression. In addition to post-translation modifications of histones, PTMs of transcription factors and coregulatory proteins further mediate structural and functional properties of regulatory complexes that control transcriptional activity to accommodate transient, short-term, and sustained requirements for gene expression.

The combination of multiple PTMs on histone proteins is known as the histone code (Jenuwein and Allis 2001; Turner 2002; Rothbart and Strahl 2014; The Allis code 2023). This histone code is mediated through the activity of chromatin regulators that read, write, and erase specific PTMs to direct cellular outcomes (Taverna et al. 2007; Marmorstein and Trievel 2009). Individual modifications found on histone proteins can have a direct effect on chromatin compaction, or they may serve as signals to recruit additional proteins to carry out downstream regulatory processes (Struhl 1998). For example, addition of a negative acetyl group on histone proteins neutralizes their positive charge and weakens their interaction with DNA in the nucleosome (Grunstein 1997). Furthermore, multiple acetylated moieties within a chromatin region can induce charge-charge repulsion supporting a more open chromatin state that is potentially competent for transcription (Grunstein 1997; Tse et al. 1998). Documenting PTMs in response to physiological regulatory cues provides a cornerstone for investigating enzymatically-mediated epigenetic control of histone-DNA interactions and chromatin organization. While methylation of histones is involved in both inactivation and activation of chromatin depending on which residue is methylated, histone acetylation is linked with gene activation.

2.3 Histone Acetylation as a Key Regulator of Gene Activation

Physiological control of gene enhancers and promoters supports fidelity of phenotype as well as resistance to compromised gene expression that can result in the onset and progression of diseases including cancer. While physiological processes are regulated by RNA polymerase II (RNAPII; e.g., PLOR2A)-directed transcription, control of cell growth by ribosomal genes is mediated by RNA polymerase I (RNAPI; e.g., POLR1A)-directed transcription. A unique dimension to transcriptional regulation is the dual control of both RNAPI and RNAPII promoters by tissue-specific transcription factors (Young et al. 2007). This mechanism supports coordinated regulation of cell growth, cell cycle, and tissue-specific gene expression. The integration of these functional relationships is obligatory for biological control and is often compromised with tumor initiation and progression (Zaidi et al. 2010a, b, 2011).

Enhancers are regulatory DNA elements that significantly elevate the expression of target genes and can reside considerable distances from target promoters (Heinz et al. 2015; Panigrahi and O'Malley 2021). Specific regulatory elements are enriched for specific histone PTMs (Table 2.1). While histone 3 (H3) lysine (K) 27 acetylation (ac) and H3K36 trimethylation (H3K27ac and H3K4me1) create a permissive environment for the recruitment of transcription factors and coactivators (Barral and Dejardin 2023; Creyghton et al. 2010; Pradeepa 2017), poised enhancers are marked by monomethylation (H3K4me1) alone (Pradeepa 2017). Further sub-classification of enhancers demonstrates that the abundance of H3K4me1, H3K27ac, and H3K36 trimethylation (H3K36me3) and the phosphorylated serines 2 and 5 (pSer2/5) forms of RNA polymerase II correlates with expression of cell type-specific genes and inclusion of histone 3 (H3) lysine (K) 9 trimethylation (H3K9me3) and/or histone 3 (H3) lysine (K) 27 trimethylation (me) (27me3) indicate poised enhancers (Zentner et al. 2011). Acetylated active enhancers express enhancer RNAs (eRNAs) or enhancer-associated long noncoding RNAs (elncRNAs) (Statello et al. 2021). These eRNAs and elncRNAs aid in the recruitment of transcriptional machinery and chromatin remodelers (Kim et al. 2018; Jiao et al. 2018; Li et al. 2013; Amaral et al. 2018). They are ubiquitously expressed in different cell lines, outnumbering mRNA (Kristjansdottir et al. 2020), further highlighting the relevance of enhancers as an integral part of genetic regulation. Clusters of highly active enhancers, termed super-enhancers, integrate the expression of functionally related regulatory genes supporting coordinate transcriptional control. Super-enhancers are critical for maintaining cell identity (Whyte et al. 2013) and ensuring an appropriate response to environmental signals (Jia et al. 2020). The regulatory mechanisms that read, write, and erase histone PTMs at enhancers are key controllers of physiological processes including development (Long et al. 2016), cell differentiation (Heintzman et al. 2009), response to environmental stimuli (Heinz et al. 2015), cell growth (Tang et al. 2020), and cell cycle regulation (Liu et al. 2017).

Table 2.1 The readers, writers and erasers of histone PTMs engaged in gene activation

PTM	Regulation	Regulatory element	Writers	Erasers	Readers	Citation
H3K4me3	Active	Promoters	MLL1, MLL2, SET1A, SET1B	KDM5	TAF3, CHD1, ING1/2, BPTF, RAG2	Miziak et al. (2024), Kranz and Anastassiadis (2020), Ansari et al. (2008), Sadakierska-Chudy and Filip (2015), and Dantas et al. (2019)
H3K27ac	Active	Promoters and enhancers	CBP/p300	HDAC3, HDAC1, HDAC2	BRD2, BRD3, BRD4	Miziak et al. (2024), Gou and Zhang (2024), Park and Kim (2020), and Deeney et al. (2016)
H3K9ac	Active	Promoters and enhancers	GCN5, PCAF, CBP, p300	HDAC1, HDAC2, HDAC3, HDAC10	BRD2, BRD3, BRD4, CBP, p300	Miziak et al. (2024), Gou and Zhang (2024), Park and Kim (2020), Deeney et al. (2016), and Fullgrabe et al. (2011)
H3K4me2	Active	Promoters and enhancers	MLL1, MLL2	KDM1A (LSD1), KDM5	TAF3, CHD1, ING2, BPTF	Ansari et al. (2008), Xiao et al. (2024), Fang et al. (2019), Romani et al. (2019), van Ingen et al. (2008), and Richart et al. (2016)

(continued)

Table 2.1 (continued)

PTM	Regulation	Regulatory element	Writers	Erasers	Readers	Citation
H3K4me1	Poised active	Promoters and enhancers	MLL3, MLL4	KDM1A (LSD1), KDM6	TAF3, BPTF, DNMT3L	Fang et al. (2019), Romani et al. (2019), van Ingen et al. (2008), Bae and Lesch (2020), and Xie et al. (2023)
H3K36me3	Active	Gene bodies (potentially enhancers)	SETD2	KDM2A, KDM2B	MRG15, LEDGF, ZMYND11, DNMT3B	The Allis code (2023), Xiao et al. (2024), Sharda and Humphrey (2022), Vacik et al. (2018), Liedtke et al. (2021), Lian et al. (2024), and Gao et al. (2020)
H3K18ac	Active	Promoters and gene bodies	p300, CBP, GCN5	HDAC1, SIRT2, HDAC2	BRD4, BRD3, BRD2	Li and Seto (2016), Halasa et al. (2019), Nava et al. (2019), Li et al. (2022), Fox et al. (2024), and Contrepois et al. (2012)
H4K16ac	Active	Promoters and enhancers	MOF (MYST1/ KAT8)	SIRT1, SIRT2, HDAC1, HDAC2	BRD2, BRD4, 14-3-3 proteins, MOF	Contrepois et al. (2012) and Zippo et al. (2009)
H3K79me2	Active	Gene bodies	DOT1L		LEDGF, 53BP1, AF9, SIRT1	Li et al. (2014), Vlaming and van Leeuwen (2016), and Fradet-Turcotte et al. (2013)
H3K23ac	Active	Promoters and enhancers	CBP, p300, GCN5	HDAC1, HDAC2, HDAC3	TRIM24	Fox et al. (2024), Wu et al. (2020), and Tsai et al. (2010)

(continued)

Table 2.1 (continued)

PTM	Regulation	Regulatory element	Writers	Erasers	Readers	Citation
H3K56ac	Active	Gene bodies	CBP, p300	SIRT1, SIRT2, HDAC1	ASF1	Aricthota et al. (2022), Yuan et al. (2009), Gonzales-Cope et al. (2016), and Khan et al. (2021)
H4K5ac	Active	Promoters and enhancers	CBP, p300, GCN5	HDAC1, HDAC2	BRD2, BRD4	Henry et al. (2013), Ciurciu et al. (2006), Lazarchuk et al. (2020), Quaas et al. (2022), Kaji and Chihara (1990), Marmorstein and Zhou (2014), Umehara et al. (2010), and LeRoy et al. (2008)
H4K8ac	Active	Promoters and enhancers	CBP, p300, PCAF	HDAC1, HDAC2, HDAC3	BRD2	Lee et al. (2024a), Henry et al. (2013), Quaas et al. (2022), Umehara et al. (2010), Johnson et al. (2002), Kim et al. (2020a, b), and Lee et al. (2024b)
H4K12ac	Active	Promoters and enhancers	CBP, p300, GCN5	HDAC1, HDAC2	BRD2, BRD4	Henry et al. (2013), Umehara et al. (2010), Bhaskara et al. (2013), Guan et al. (2009), Kanno et al. (2004), and Nagarajan et al. (2015)
H2AK5ac	Active	Promoters and enhancers	CBP, p300	HDAC1, HDAC2	BRD2, BRD4	Josling et al. (2012) and Grozinger et al. (1999)

(continued)

Table 2.1 (continued)

PTM	Regulation	Regulatory element	Writers	Erasers	Readers	Citation
H2AZac	Active/ poised	Promoters and enhancers	CBP, p300	HDAC1, HDAC2	BRD2, BRD4	Colino-Sanguino et al. (2019) and Draker et al. (2012)
H3K14ac	Active	Promoters and enhancers	GCN5, PCAF, CBP, p300	HDAC1, HDAC2, HDAC3, SIRT1	BRD2, BRD4, 14-3-3 proteins	LeRoy et al. (2008), Kim et al. (2020a), Marmorstein (2001), Love et al. (2012), Bhaskara et al. (2010), Price et al. (2020), Jing and Lin (2015), Dey et al. (2003), Devaiah et al. (2016), and Winter et al. (2008)
H3K27me1	Active/ poised	Promoters and enhancers	EZH1, EZH2	KDM6A (UTX), KDM6B (JMJD3)	Polycomb complexes, ZFP57	Shen et al. (2008), Ferrari et al. (2014), and Pan et al. (2018)
H4K20me1	Active/ poised	Promoters and enhancers	PR-SET7/ SETD8	PHF8, KDM4B	L3MBTL1, 53BP1, BRCA1	Shoaib et al. (2021), Hyun et al. (2017), and Nakamura et al. (2019)

A list of some of the histone PTMs that are engaged in gene activation is provided. This demonstrates that acetylation and methylation are both involved in gene activation. Histone PTMs are read written and/or erased by key proteins including the Mixed Lineage Leukemia proteins (MLLs), SET domain containing family (SETs), Lysine Demethylases (KDMs), Chromodomain Helicase DNA Binding protein (CHDs), Inhibitor of Growth proteins (INGs), Bromodomain PHD finger transcription factor (BPTF), Recombination Activating Gene 2 (RAG2), General control Non-repressed 5 (GCN5), P300/CBP-Associated Factor (PCAF), Male Absent on the First (MOF)/ MYST histone acetyltransferase 1 (MYST1)/Lysine AcetylTransferase 8 (KAT8), DOT1-like, histone lysine methyltransferase (DOT1L), Lens Epithelium-Derived Growth Factor (LEDGF), Tumor protein P53 Binding Protein 1 (53BP1), Tripartite motif-containing protein 24 (TRIM24), Anti-Silencing Function 1 (ASF1), 14-3-3 protein family, Enhancer of Zeste Homolog 1 and 2 (EZH1, EZH2), Lysine demethylase 6A also known as Ubiquitously Transcribed tetratricopeptide repeat containing, X chromosome (UTX), Jumonji Domain containing 3 (JMJD3), and Lethal(3) Malignant Brain Tumor-Like protein 1 (L3MBTL1)

The integration of PTMs at these regulatory elements with higher-order genome structures has become increasingly apparent. The organization of ~2.2 meters of DNA within the confines of the ~6–10-micron cell nucleus necessitates extensive genomic configuration that accommodates plasticity, and at times rapid responsiveness, for remodeling to support biological control of gene expression. Equally relevant are contributions of histone modifications including acetylation to establishing gene regulatory networks that integrate regulatory machinery and genetic information encoded in noncontiguous sequences, including inter- and intra-chromosomal interactions and functional crosstalk. We provide indications for the strategic engagement of histone acetylation for multiple determinants of chromatin structure and function.

2.3.1 Chromatin Readers, Writers, and Erasers of Histone Acetylation

Dysregulated protein acetylation is increasingly recognized as a key driver in cancer. Abnormal patterns of histone acetylation can disrupt cellular processes such as proliferation, apoptosis, and DNA repair. Acetylation of different residues on the nucleosomal subunits has been determined to promote or inhibit tumorigenesis, while others are context-dependent (extensively reviewed in 44). Aberrant levels of histone acetylation significantly influence gene expression of key genes involved in cancer, as demonstrated by decreased levels of the 21 kilodalton protein (p21) gene, which regulates the cell cycle (Deepak et al. 2020). Modulating the activity of writers, readers, and erasers of histone acetylation provides the ability to restore normal acetylation patterns, re-establish control over gene expression, and offer new avenues for targeted cancer therapy described below.

2.3.2 Chromatin Readers Molecularly Interpret Acetylation

Histone acetylation modifications are often recognized directly by chromatin reader domains (Fig. 2.1) (Dyson et al. 2001). Conserved reader domains have evolved to recruit chromatin effector proteins to a local genomic region where they can carry out their molecular activities (Dhalluin et al. 1999). Although acetylation of chromatin is a broadly utilized epigenetic mechanism that is well known to play a critical role in the control of gene expression, surprisingly only a few conserved families of reader domains have been identified to recognize acetylated histones (Thorne et al. 1990). Included among the acetyl-lysine readers are the bromodomain (BRDs) (Dhalluin et al. 1999), the YEATS domain (an acronym formed by combining letters from Yaf9 (yeast homolog of AF9), ENL [Eleven-Nineteen Leukemia protein], AF9 [ALL1-fused gene from chromosome 9], Taf14 [TBP-associated factor 14 in S.cerevisiae], and Sas5 [a subunit of the yeast SAS complex]) that contain them) (Li et al. 2014), and the tandem plant homeodomain (PHD) finger (Wu et al. 2010) families. Bromodomains (BRDs), discovered in 1992 in the Brahma gene of Drosophila melanogaster, are evolutionarily conserved protein modules that consist of ~110-amino acids and were the first proteins demonstrated to possess the ability

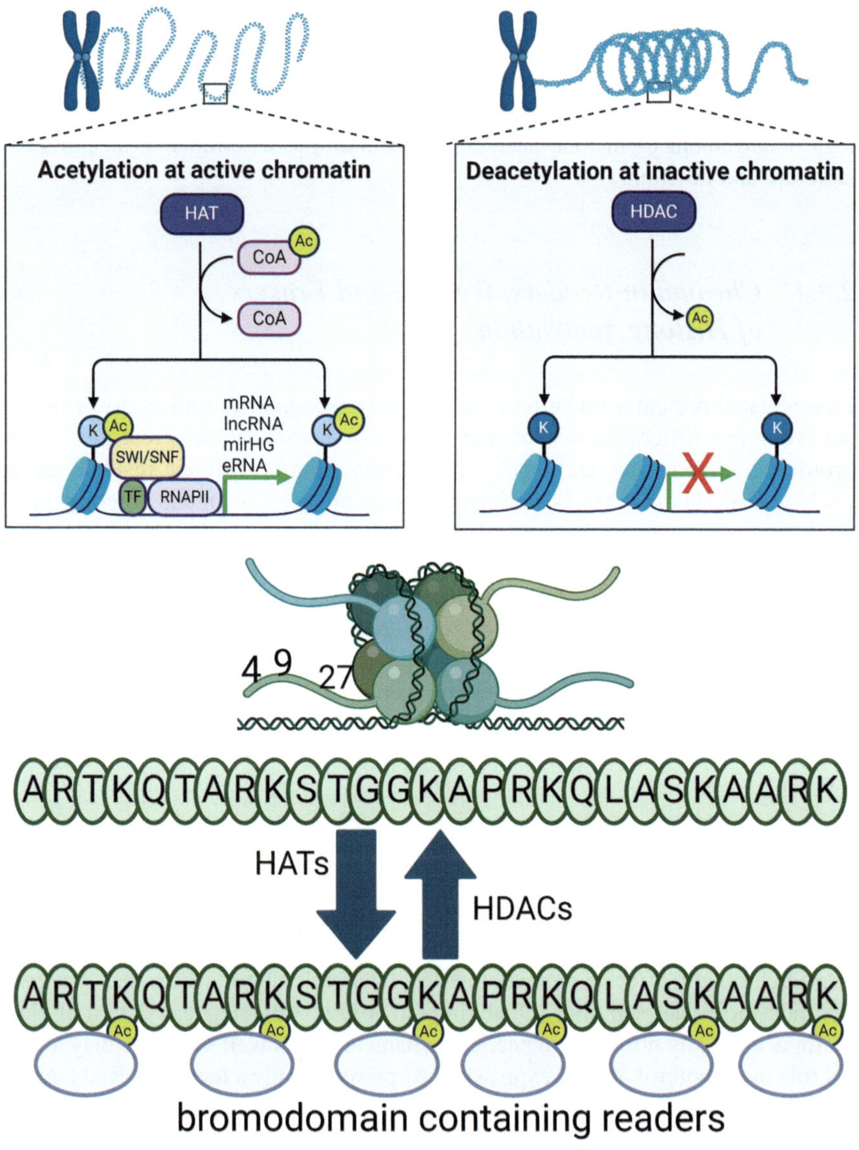

Fig. 2.1 Readers, writers, and erasers of histone acetylation. Schematically illustrated are the regulatory activities for readers, writers, and erasers engaged in epigenetic control of gene expression. (Created with BioRender.com)

to recognize acetylated histone lysine residues (Dhalluin et al. 1999; Kennison and Tamkun 1992; Haynes et al. 1992). To date, eight subfamilies comprised of 61 BRDs have been identified across 46 distinct bromodomain-containing proteins (BDPs) in the human genome (Filippakopoulos et al. 2012). Bromodomains are often found in epigenetic regulatory proteins such as histone acetyltransferases and chromatin remodeling complexes in order to support their associated protein interaction with chromatin (Dyson et al. 2001; Hudson et al. 2000; Jacobson et al. 2000). Bromodomains prefer to recognize acetylated lysine residues but can also bind to larger acyl groups, including propionyl, butyryl, and crotonyl groups (Flynn et al. 2015).

More recently, the YEATS domain-containing proteins (YDPs) have emerged as readers of acetylated lysine. The four most well-known human YDPs have long been implicated in gene regulation: ENL; AF9, YEATS domain in AF9/MLLT3 protein; YEATS2, YEATS domain-containing protein 2; and GAS41, glioma amplified sequence 41 (Li et al. 2023). The YEATS domain preferentially recognizes crotonylated histones but also binds larger acetylation moieties (Li et al. 2014, 2016; Li and Seto 2016). However, it wasn't until 2014 when the crystal structure of the ALL1-Fused gene from Chromosome 9 (AF9 YEATS domain) bound to mono-acetylated H3 (H3K9ac) was solved providing unambiguous evidence that YEATS domains function as acetyl-lysine readers (Li et al. 2014). Further, it was shown the AF9 YEATS domain binds H3K9ac using a mechanism that is unique from BRDs, such as bromodomain-containing protein 4 (BRD4) (Li et al. 2014; Andrews et al. 2016; Shanle et al. 2015). Rather than coordinating the acetyl moiety through a conserved Asn residue, which is found in nearly all BRDs, the AF9 YEATS domain uses a hydrogen bond network in combination with shape complementarity to form a serine-lined aromatic cage that optimizes hydrophobic contacts with acetylated histone H3K9ac (Li et al. 2014). Similar to BDPs, dysregulation of YDPs has been correlated with multiple disease states, particularly cancers, and their discovery provides new insight into epigenetic regulation through histone acetylation (Zhao et al. 2017; Wan et al. 2017; Erb et al. 2017).

The plant homeodomain (PHD) fingers typically recognize methylated lysine on chromatin, but in certain instances, two adjacent PHD fingers form a tandem domain that recognizes acetylated lysine (Zeng et al. 2010). PHD fingers are a large family of protein units related to E3 ubiquitin ligase RING (Really Interesting New Gene) fingers and phosphoinositide-binding FYVE domains (named after the Fab 1, yeast Fab 1; YOTB, yeast orthologue of PIKfyve; Vac 1, vesicle transport protein; and EEA1, early endosomal antigen-1 proteins that contain them). PHD finger domains are further divided into several subfamilies distinguished by their ability to recognize various post-translationally modified histone ligands. One major subset binds strongly to tri-methylated histone H3 (H3K4me3), while another preferentially recognizes unmodified histone H3. Another group forms a tandem PHD finger domain created by integrating two adjacent PHD fingers to enable a new functionality in acetyl-lysine recognition (Wu et al. 2010). The tandem PHD fingers of DPF3, MOZ (*mo*nocytic leukemia *z*inc-finger protein), and MORF (*MOZ*-*r*elated *f*actor) have been shown to bind to mono-acetylated histone 3 (H3K14ac) (Wu et al. 2010; Qiu et al. 2012).

Chromatin readers that recognize acetyl-lysine modifications play a significant role in the epigenetic regulation of DNA-templated processes such as transcription, replication, repair, and chromatin remodeling. Despite the relatively small number of protein families that recognize acetylated histones, the bromodomain, YEATS, and tandem PHD finger domains are associated with the pathophysiology of human disease and infectious diseases. As such, these domains are promising therapeutic targets to prevent and treat disease.

2.4 Acetylation Readers in Diseases Represent Promising Therapeutic Targets

Bromodomain-containing proteins play critical roles in epigenetic regulation of gene transcription and dysregulation of their activity contributes to the pathophysiology of disease (Filippakopoulos et al. 2012), (Boyson et al. 2021), (Wang et al. 2023). For example, the four conserved members of the human BET (Bromodomain and Extra Terminal domain) family—BRD2, bromodomain-containing protein 2; BRD3, bromodomain-containing protein 3; BRD4, bromodomain-containing protein 4; and BRDT, bromodomain testis-specific protein—regulate critical oncogenes through the modulation of pathways including Janus kinase-signal transducer and activator of transcription (JAK-STAT) and nuclear factor-kappa beta (NF-kβ) pathways (Pecharroman et al. 2023; Chan et al. 2015). In gastric cancer, accumulation of histone acetylation at the promoter of the oncogenic aldehyde dehydrogenase 1A3 gene (ALDH1A3), following neoadjuvant chemotherapy, leads to H3K27ac- and BRD4-dependent drug-tolerant persister (DTP) cells early in treatment (Lee et al. 2024a). In cells treated with 5-fluorouracil (5-FU), BRD4 was recruited to the acetylated promoter of ALDH1A3. Knockdown of BRD4 reduced the drug-induced upregulation of ALDH1A3. Treatment with the BET inhibitor biraresib (OTX015) blocked BRD4 recruitment to the gene promoter, though minimal changes in acetylation were observed (Lee et al. 2024a). Combined treatment with 5-fluorouracil (5-FU) and OTX015 decreased tumor growth and DTP cell survival in vivo, synergistically enhancing the chemotherapeutic efficacy (Lee et al. 2024a). Another example of acetylation reader proteins that are overexpressed in multiple cancers and linked to poor patient outcomes are AAA domain containing 2 (ATAD2) and AAA domain containing 2B (ATAD2B), which are members of the Adenosine 5'-TriPhoposphate (ATPase) family containing ATPases associated with diverse cellular activities (AAA+ domains). These proteins have a unique configuration with an AAA+ ATPase enzyme fused to a C-terminal bromodomain (Wang et al. 2023; Kalashnikova et al. 2010; Lloyd and Glass 2018; Shang et al. 2015; Wan et al. 2014; Baggiolini et al. 2021). Thus, despite efficacy and toxicity challenges, potent inhibitors of human BDPs have emerged as promising cancer treatments (Wang et al. 2023; Filippakopoulos et al. 2010; Liang et al. 2023).

Beyond cancer, BDPs are implicated in other pathological contexts including diseases where they have been identified in parasitic organisms (Jeffers et al. 2017). The genomes of *Plasmodium falciparum*, the organism responsible for the most lethal form of malaria (Zekar and Sharman 2024), and *Toxoplasma gondii*, which is the leading cause of death secondary to foodborne illness in the United States (Almeria and Dubey 2021; Scallan et al. 2011) have genes that encode seven and twelve suspected bromodomain-containing proteins, respectively (Fleck et al. 2023; Hanquier et al. 2020; Josling et al. 2015; Singh et al. 2022). Although epigenetic regulatory processes such as histone acetylation are generally conserved in Apicomplexan parasites, their bromodomain-containing proteins are parasitic-specific, and typically do not have orthologs in higher-level eukaryotes. However, they do play essential roles in regulating gene expression throughout the parasite life cycle. One example is the *P. falciparum* Bromodomain Protein 1 (PfBDP1), which has been shown to contribute to parasite proliferation by regulating the expression of invasion-related genes (Josling et al. 2015; Singh et al. 2022). Unfortunately, the underlying mechanisms controlling these processes are not well understood. Nevertheless, parasite-specific BDPs represent new targets for the treatment of prevalent diseases such as malaria and toxoplasmosis (Tallant et al. 2021).

2.4.1 HATs as Writers of Histone Acetylation

As previously described, histone acetylation is critical in promoting chromatin accessibility and represents a key epigenetic signal for readers acting in the complex systems of control over transcription that mediates processes including proliferation, differentiation, and DNA damage repair. Histone acetyltransferase (HAT) enzymes facilitate the transfer of an acetyl group to lysine residues of histone proteins (Fig. 2.1), neutralizing and weakening the histone-DNA interactions, thus creating a relaxed chromatin structure. HATs are classified into two main families based on their location within the cell: Type A and Type B HATs (Shvedunova and Akhtar 2022).

Type A HATs include the nuclear proteins that acetylate nucleosome histone proteins promoting chromatin relaxation and are categorized into five main groups: GNAT (GCN5-related acetyltransferases), PCAF (p300/CREB binding protein associated factors), MYST (MOZ, YBF2/SAS3, SAS2, and TIP60 protein), general transcription factor family, and the SRC/NCoA (steroid/ nuclear receptor co-activators) family (Biswas and Rao 2018). Type B HATs include the cytoplasmic HAT proteins, which acetylate newly synthesized histones to be chaperoned into the nucleus for incorporation into nascent chromatin. Histone acetyltransferase 1 (HAT1) was the first identified cytoplasmic HAT (Parthun et al. 1996) and acts in a complex to facilitate the incorporation of newly synthesized histone H4 into nascent chromatin (Parthun 2012). However, since its discovery, there has been evidence supporting localization in the nucleus as well (Imhof and Wolffe 1999; Ai and Parthun 2004).

2.5 HATs Off: Dysregulation of Acetylation Writers as Targets in Disease

Aberrant activity of HATs, such as p300 overexpression, has been linked to cancer malignancies, including drug resistance. Both adenovirus early gene 1A (E1A) Binding Protein 300 kilodalton (EP300) and cyclic adenosine monophosphate (cAMP) element (CRE) binding protein (CREBBP) are HATs that function through acetylation of histone 3 lysine 18 (H3K18) and histone 3 lysine 27 (H3K27) at gene promoters and enhancers and are considered transcriptional co-activators central to tumor development. For example, overactivity of HATs such as adenovirus early gene 1A (E1A) Binding Protein 300 kilodalton (p300/CBP) enhances the expression of oncogenes, driving cancer progression (Chen et al. 2022; Waddell et al. 2021). Knockdown of EP300 has recently been shown to reduce malignant growth behaviors of lobaplatin-resistant triple-negative breast cancer cells. Recent advancements have led to the creation of small-molecule HAT inhibitors targeting both enzymatic and BRD functions, which have been recently reviewed (Laurberg et al. 1985). Early and novel inhibitors targeting p300, such as C646 and MJP6412, have shown promising therapeutic effects in several cancers (Bowers et al. 2010; Gu et al. 2016a, b Ma et al. 2024). Additional HATs of recent interest include the histone acetyltransferases 6A and 6B (HAT6A/B). CTx-648 (PF-9363), a highly potent, selective inhibitor of HAT6A/B histone acetyltransferases effectively suppresses the growth of HAT6A-high expressing ER+ breast cancer models, including those resistant to endocrine therapy. This is primarily through inhibition of histone (Pogo et al. 1966) lysine (Zaidi et al. 2010a) H3K23 acetylation by HAT6A (Sharma et al. 2023). Additionally, a recent phase 1 trial evaluated the selective HAT6A/B inhibitor, PF-07248144, in ER+/HER2− metastatic breast cancer patients. The inhibitor demonstrated a tolerable safety profile and notable clinical efficacy, particularly when combined with fulvestrant (Mukohara et al. 2024). Studies such as these establish HAT6A/B as promising targets for therapeutic intervention in endocrine therapy-resistant breast cancer.

While early emphasis was placed on the role of Type A HATs in cancer development and progression, more recent evidence suggests the role of HAT1 in replication and promoting cell proliferation has made it a target for cancer therapies. Knockdown of HAT1 leads to cell cycle arrest in embryonic stem cells (Xue et al. 2014). Additionally, HAT1 has been identified as a contributing factor in hepatocellular carcinoma progression and cisplatin resistance (Zhu et al. 2022; Jin et al. 2017) through acetylation of the histone (Pogo et al. 1968) lysine (Lee et al. 2007) (H4K5) and histone (Pogo et al. 1968) lysine 12 (H4K12) residues, which recruit the bromodomain BRD4 for transcriptional regulation of immune response genes such as PD-L1 (Jin et al. 2017). While targeting histone acetyltransferases with BRD domains (HAT BRDs) and catalytic domains as therapeutic strategies is promising (Laurberg et al. 1985), given the critical functions of HATs in development, great care has been taken to pursue selective targeting. Approaches to balance efficacy while minimizing off-target effects are necessary (Laurberg et al. 1985).

2.5.1 Erasing Acetylation: The Role of Histone Deacetylases in Gene Regulation

Histone PTMs are dynamically written and removed to shape the epigenetic landscape that mediates cellular outcomes. Histone deacetylases (HDACs) are enzymatic erasers of histone acetylation modifications and function in opposition to the HAT enzymes that write them (Fig. 2.1) (Turner 2000). HDACs are broken into four classes based on their enzymatic properties and homology (Biswas et al. 2018; Xi et al. 2011). Three of the four classes are zinc-dependent HDACs including class I, which contains histone deacetylases (HDAC) 1, 2, 3, 3; and 8, Histone deacetylase 8; class II, containing histone deacetylases (HDAC) 4, 5, 6, 7, 9, and 10; and class IV, containing histone deacetylases 11 (HDAC11). Alternatively, class III HDACs rely on a nicotinamide adenine dinucleotide (NAD+) cofactor to carry out their enzymatic mechanism and are structural homologs with the yeast Sirtuin 2 (Sir2) deacetylase (Blander and Guarente 2004). As described above, once the acetyl group is removed, the lysine sidechain regains its positive charge. This results in a more compact chromatin state that consequently reduces transcriptional activity (Chen et al. 2015). However, separately, HDACs also deacetylate many non-histone proteins. This includes transcription factors, chaperones, and signaling molecules and results in changes to protein stability, protein-protein interactions, and protein-DNA interactions (Glozak et al. 2005; Jin et al. 2004).

2.6 Erasing Dysregulation of Histone Acetylation as a Key Therapeutic in Disease

Clinical evidence indicates that abnormal histone acetylation correlates with poorer outcomes in cancer. High levels of HDAC1 and HDAC2 are associated with more advanced tumor stages, increased metastasis, and reduced survival rates (Li and Seto 2016; Chen et al. 2015; Yang et al. 2022a; Alseksek et al. 2022; Zhuang et al. 2020; Jo et al. 2023; Li et al. 2021a; Tsilimigras et al. 2018; Yang et al. 2022b). Furthermore, elevated HDAC levels, particularly HDAC2, HDAC3, HDAC8, and Sirtuin 1 (SIRT1), have been found in breast cancer tissues, contributing to chemoresistance and aggressive tumor behavior (Yang et al. 2022a; Jo et al. 2023; An et al. 2020; Xu et al. 2024). HDAC overexpression (particularly HDAC1 and HDAC2) silences tumor suppressor genes like tumor protein (p53), allowing uncontrolled cancer cell growth (Gong and Miller 2013; Barrett et al. 2024; Hai et al. 2021; Mrakovcic et al. 2019). As a result, HDACs have become significant targets for therapeutic intervention (Chen et al. 2015; Li et al. 2021a; Yang et al. 2022b). In 2006, Vorinostat (Zolinza) became the first histone deacetylase inhibitor (HDACi) approved for cutaneous T-cell leukemia by the FDA (Siegel et al. 2009). HDAC inhibitors, including Vorinostat and Romidepsin, have shown promise in clinical trials, especially for Triple Negative Breast Cancer (TNBC, lacking expression of

Estrogen Receptor/Progesterone Receptor/Human Epidermal Growth Factor Receptor 2) patients (Maccallini et al. 2022; Li et al. 2021b). These findings suggest that HDAC inhibitors could be effective treatments for aggressive cancer subtypes that have limited treatment options such as TNBC (Garmpis et al. 2017) or colorectal cancer with resistance to Tumor necrosis factor-Related Apoptosis-Inducing Ligand (TRAIL, 130). Treatment of TRAIL-resistant cells with HDAC inhibitors leads to hyperacetylation of histones and subsequently upregulates Death Receptor 5. Ultimately, upregulation of the intrinsic apoptosis pathway promotes TRAIL sensitivity and increases TRAIL-induced apoptosis (Kim et al. 2024).

Histone acetylation also plays a role in DNA repair by facilitating the recruitment of repair proteins to damaged DNA sites (Hunt et al. 2013; Van and Santos 2018; Aricthota et al. 2022). However, in breast cancer, HDAC overactivity can impair this repair process, leading to the accumulation of mutations that promote tumor progression (Lian et al. 2023). For instance, in human breast cancer cell line (MCF-7), HDAC inhibition by Trichostatin A (TSA) increases p21 expression, halting the cell cycle through a p53-dependent pathway (Meng et al. 2012; Bouyahya et al. 2022; Hrgovic et al. 2016). In late-stage breast cancer cell line (MDA-MB-231), which lack functional p53, TSA and panobinostat (LBH589) induce p21 expression by acetylating the histone variant 1 (H2A.Z) at the transcription start site. This acetylation promotes the recruitment of RNA polymerase II and other transcriptional machinery to the tumor protein (p21), highlighting the role of histone acetylation in p21 activation through p53-independent mechanisms (Giaimo et al. 2019; Seto and Yoshida 2014; Tate et al. 2012).

Histone acetylation significantly influences alternative splicing, which contributes to the diversity of proteins expressed from a single gene. In cancer, aberrant splicing is a key factor in tumorigenesis, metastasis, and drug resistance (reviewed in (Zhang et al. 2021)). Histone acetylation affects splicing outcomes by modulating the recruitment of splicing factors to chromatin (Markouli et al. 2021; Rahhal and Seto 2019; Zhou et al. 2014; Imbriano and Belluti 2022; Yang et al. 2019). For example, the cluster of differentiation 44 (CD44) gene can produce protein isoforms that either promote cancer spread or suppress tumors. Histone acetylation controls which isoform is produced, and HDAC inhibitors can shift the balance toward tumor-suppressive forms (Chen et al. 2015; Zhang et al. 2019). A similar mechanism is observed with the fibroblast growth factor receptor 2 gene (FGFR2), where different splicing variants can promote or suppress cancer (Zhu et al. 2009). Targeting histone acetylation with HDAC inhibitors could restore normal splicing patterns, offering a novel therapeutic approach (Zhu et al. 2009).

Molecular docking analysis and molecular dynamic simulations between HDAC structures from the Protein Data Bank and ligands from the ZINC database have recently identified two potential stable HDAC inhibitors and have been proposed for future studies to treat cervical cancer (Ahmad et al. 2024). Both DNA ligase 1 (LIG1) and DNA ligase 2 (LIG2) interact stably with HDAC catalytic active sites and alter the structural dynamics of the proteins, particularly in loop regions. Importantly, these ligands show low toxicity to physiological systems and good passive absorption. Further research in vitro and in vivo is required to deduce the effectiveness of these ligands, but this work emphasizes the power of computational tools and the promising power of artificial intelligence for future strategies in identifying precision cancer treatments targeting histone acetylation (Ahmad et al. 2024).

Targeting histone acetylation and interacting proteins has been proven effective in vitro and in vivo to work synergistically with other treatments to overcome drug resistance in cancer (Lee et al. 2024a; Kim et al. 2024). Combining HDAC inhibitors with chemotherapy agents like doxorubicin has been shown to enhance cancer cell death, offering a promising therapeutic strategy for aggressive breast cancers (Smoots et al. 2024; Mandapati and Lukong 2023). These drugs can reactivate silenced tumor suppressor genes and trigger cancer cell death. Moreover, research is increasingly exploring the combination of HDAC inhibitors with other therapies, such as poly (ADP-ribose) polymerase, or PARP inhibitors targeting DNA repair pathways. This is important given the role of histone acetylation in mediating DNA damage and competency and fidelity of repair outlined above. This combination therapy is especially promising for patients with breast cancer protein 1 (BRCA) mutations, who have compromised DNA repair mechanisms (Ramos et al. 2023; Yin et al. 2018; Valdez et al. 2024).

2.7 Histone Acetylation-Mediated Remodeling of the Tumor Microenvironment

In addition to its role within cancer cells, histone acetylation also affects the tumor microenvironment (TME) in TNBC, impacting immune responses and tumor progression (Zhuang et al. 2020; Zhou and Yu 2024; Lu et al. 2021). For example, HDAC2 has been shown to upregulate PD-L1 (Programmed Death Ligand 1) expression in TNBC, contributing to immune evasion. Inhibiting HDAC2 reduces PD-L1 expression, enhances the recruitment of tumor-infiltrating lymphocytes (TILs), and increases the efficacy of immune checkpoint blockade (ICB) therapies (Xu et al. 2021; Terranova-Barberio et al. 2017; Burke et al. 2020; Yamaguchi et al. 2022; Cha et al. 2019). HDAC6 is another key player in the TME and is associated with inflammatory responses and cell motility. HDAC6 inhibition remodels the TME, promoting immune cell recruitment and limiting tumor growth (Zhang et al. 2023; Passaro et al. 2021). This highlights the potential of HDAC inhibitors to reshape the TME and improve cancer treatment outcomes.

Similarly in bladder cancer, the HDAC inhibitor CM-1758 was shown to combine effectively with immunotherapies to treat metastasis (Nunes et al. 2024). Bladder cancer cell lines show an increase in total H3 acetylation, H3K9, and H3K27 acetylation and a decrease in DNA methylation following treatment with CM-1758. Not only is an increase in cancer cell death observed following treatment, but there are also remarkable cell cycle effects across cell lines (Nunes et al. 2024). Notably, CM-1758 treatment leads to increases in promoter methylation of PD-L1 and subsequently increased mRNA and membrane protein PD-L1 levels.

Mouse models were used to evaluate the effectiveness of CM-1758 alone and in combination with an anti-PD-L1 antibody (Nunes et al. 2024). Treatment with CM-1758 alone and with immune checkpoint inhibitors (ICIs) effectively inhibit tumor growth and modulate the immune system. Treatment with CM-1758 harnesses histone acetylation to offer a promising alternative for bladder cancer patients who do not respond to ICIs due to low PD-L1 expression or reduced immune cell infiltration (Nunes et al. 2024). Future research should focus on identifying specific histone acetylation patterns that could serve as biomarkers for treatment response. A precision medicine strategy targeting multiple epigenetic mechanisms in a context-specific manner will further improve outcomes for cancer patients.

2.8 Histone Acetylation as a Therapeutic in Diseases beyond Cancer

Beyond cancer, HDAC research has extended to disorders affecting the nervous system such as Huntington's and Alzheimer's diseases, as well as psychiatric disorders including depression, drug addiction, schizophrenia, and various anxiety disorders (Ziemka-Nalecz et al. 2018; Butler and Bates 2006). In both neurodegenerative and psychiatric diseases, the homeostasis of histone acetylation is greatly disturbed, skewing toward the hypoacetylated state. Hyperacetylation of histone proteins has been found to provide neuroprotective effects and has generated an increased interest in the development of HDAC inhibitors (HDACi) (Lee et al. 2007). Generating HDAC inhibitors that have isoform selectivity will provide the ability to target and treat specific disease states. As with bromodomain inhibitors, the development of HDACi will provide important therapeutic compounds that can be used for further investigations into the fundamental roles of HDACs in downstream gene regulation.

2.9 Biologically Responsive Chromatin Remodeling Supports the Requirement for Genome Organization

These chromatin readers, writers, and erasers of histone acetylation work together to modulate chromatin structure and accessibility. Chromatin remodeling complexes interact with these enzymes, actively facilitating the transition between closed to open chromatin states in an ATP-dependent mechanism, thereby ensuring that chromatin structure can adapt to the cellular requirements allowing the regulatory machinery and other coregulatory complexes to access the DNA for gene expression, DNA replication, repair, and genome stability (Gangaraju and Bartholomew 2007). While pioneering transcription factors have the unique ability to bind to inaccessible chromatin and recruit remodeling complexes, other transcription factors require chromatin to be accessible prior to binding (Mayran and Drouin 2018).

2.10 Role of SWI/SNF Complexes in Chromatin Remodeling

The Switch/Sucrose Non-Fermentable (SWI/SNF) family of chromatin remodeling complexes plays an essential role in maintaining the dynamic nature of chromatin. These complexes use ATP hydrolysis to remodel nucleosomes, which are the basic units of chromatin. By shifting the positions of nucleosomes, SWI/SNF complexes can expose or hide specific DNA sequences, thereby regulating the accessibility of transcription factors and other regulatory proteins to the underlying DNA (Chen et al. 2023; Morse et al. 2024; Thorvaldsdottir et al. 2013). During cell differentiation, SWI/SNF complexes are particularly important. They facilitate the activation of genes specific to particular cell types by making the regulatory regions of these genes more accessible. This is achieved through the repositioning or eviction of nucleosomes, allowing transcription factors to bind to their target sites. Similarly, in response to signaling pathways, SWI/SNF complexes can dynamically adjust chromatin accessibility, either promoting or repressing the expression of genes necessary for the cellular response (Saha et al. 2024; Sahu et al. 2020).

2.11 Methylation-Mediated Chromatin Remodeling

In contrast to euchromatic open chromatin, heterochromatin represents a more compact and less accessible chromatin state, which contributes to gene silencing and genome stability. This compact structure is associated with several repressive modifications and features (Morrison and Thakur 2021). Histone methylation and chromatin condensation is a key modification associated with heterochromatin. For example, the methylation of histone H3 lysine 9 (H3K9me3) and lysine 27 (H3K27me3) marks heterochromatin regions. These modifications serve as binding sites for chromatin-associated proteins, such as Heterochromatin Protein 1 (HP1) and Polycomb group proteins, which further condense the chromatin structure, making the DNA less accessible for transcription (Nicetto and Zaret 2019; Rose and Klose 2014; Schlesinger et al. 2007).

DNA methylation is another critical feature of inactive chromatin. This process involves the addition of methyl groups to the 5′ position of cytosine residues, creating 5mC, and is enriched in CpG dinucleotides. DNA methylation located near gene promoters inhibits gene expression by preventing the binding of transcription factors and recruiting repressive complexes, such as methyl-CpG binding domain proteins (MBDs), which further contribute to chromatin compaction reinforcing gene silencing (Moore et al. 2013). Heterochromatin plays several important roles in cellular processes. For example, X-chromosome inactivation in female mammals is a process by which one of the X chromosomes is transcriptionally silenced to achieve dosage compensation between males (XY) and females (XX). This silencing is mediated by the formation of heterochromatin, which compacts the X chromosome into a transcriptionally inactive state (Prothero et al. 2009; Marahrens et al. 1998). Another example is the role of pericentromeric heterochromatin in maintaining chromosome structure and stability during cell division. This region of heterochromatin is essential for proper chromosome segregation, as it prevents chromosomal abnormalities by ensuring the correct attachment of spindle fibers to the centromere during mitosis (Allshire and Madhani 2018).

Chromatin remodeling complexes are not only involved in establishing active chromatin states but also in the transition between active and inactive states. The interplay between these complexes is crucial for maintaining the dynamic nature of the chromatin landscape. Polycomb Repressive Complex 2 (PRC2) is a key player in maintaining inactive chromatin states. This complex catalyzes the trimethylation of histone H3 at lysine 27 (H3K27me3), a mark associated with gene silencing. PRC2-mediated H3K27me3 recruits additional repressive proteins, such as Polycomb Repressive Complex 1 (PRC1), which further condenses chromatin and maintains the silenced state of target genes. PRC2 is particularly important during development, where it helps establish and maintain the repression of genes that should remain inactive in specific cell lineages (Boros et al. 2014; Sparmann and van Lohuizen 2006).

SWI/SNF complexes can counteract repressive marks established by complexes like PRC2. By remodeling nucleosomes and displacing repressive histone modifications, SWI/SNF complexes facilitate transitions to an active chromatin state when necessary. This activity is essential for the activation of genes that need to be expressed in response to environmental cues or during specific stages of development (Bracken et al. 2019; Mittal and Roberts 2020; Kadoch et al. 2016).

2.12 Histone Acetylation and Higher-Order Genome Organization

Establishing functionality for packaging the ~2.2 meters of DNA within a ~ 6–10 micrometer human cell nucleus necessitates a high degree of structural organization (Fig. 2.2). At the lowest level, DNA is wrapped around the nucleosome. The structure formed by repeated nucleosomes along DNA is termed the chromatin fiber or "beads-on-a-string" (Maeshima et al. 2014; Hansen et al. 2018) Nucleosome positioning and the accessibility of chromatin within regulatory elements are interdependent and responsive to external stimuli and developmental cues (Klemm et al. 2019). Maintenance or alteration of accessibility is dynamically regulated by chromatin-binding complexes, PTMs including acetylation, and nucleosomal positioning (as described above). During transcription, the stability of the nucleosome immediately upstream of the transcription start site (TSS) is affected by histone

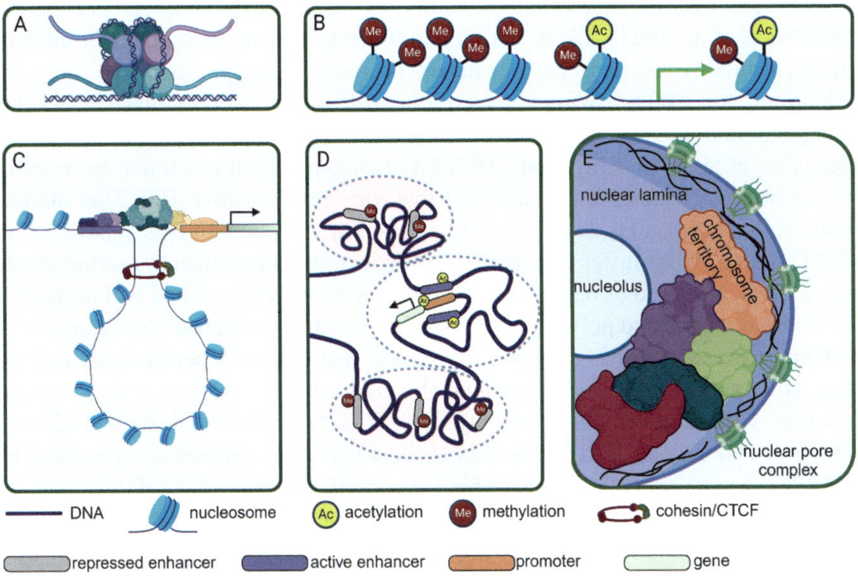

Fig. 2.2 The interrelated nature of histone acetylation and higher-order chromatin organization. Multiple levels of chromatin organization support the architectural organization of gene regulatory machinery and accessibility of genomic sequences to mediators of transcriptional control. DNA is wrapped around 2 sets of 4 core histone components (octamer) of the nucleosome (**a**) which are methylated and acetylated on their N-terminal tails to render genomic sequences accessible or inaccessible, either transcriptionally activated or suppressed respectively (yellow circles—acetylation, red circles—methylation, (**b**) Genomic loops representing higher-order chromatin organization (red rings—cohesin/CTCF complexes, (**c**) illustrated as a composite of contiguous predominantly inactive or transcriptionally accessible active domains which are phase separated (yellow circles—acetylation, red circles—methylation, lavender rectangles—active enhancers, mustard rectangle—promoter, grey rectangles—repressed enhancers, phase-separated condensates—light green and red circles, (**d**) Phase-separated domains representing nuclear microenvironments where transcriptionally inactive and active regions reside. At the highest level, chromosome territories are depicted (light blue circle—nucleolus, red, green, blue, salmon, and purple domains—chromosome territories, (**e**). (Created with BioRender.com)

replacement, acetylation and methylation, translational repositioning, and finally by eviction upon the formation of the pre-initiation complex (Marino-Ramirez et al. 2005). The chromatin fiber is also influenced by PTMs such as H3K27 acetylation at active enhancers to promote accessibility for transcription factors and coregulatory complexes (Konuma and Zhou 2024). As a consequence of the range of potential activity across the genome, the beads-on-a-string structures in nuclei are not only 10 nm but vary in diameter (Risca et al. 2017) with the chromatin fiber folding into topologically associating domains (TADs, (Lieberman-Aiden et al. 2009; Dekker and Heard 2015; Beagan and Phillips-Cremins 2020; Szabo et al. 2019)). The boundaries of these TADs are essential for normal genomic functioning. Deletion of TADs results in disrupted organismal phenotypes including altered chromatin interactions, gene expression, and reduced viability, and in most cases results in increased embryonic lethality or other developmental phenotypes (Rajderkar et al. 2023). TADs coalesce into compartments that are either euchromatic, termed A compartments, or heterochromatic B compartments (Lieberman-Aiden et al. 2009). Within these two categories, studies have identified computationally distinct subcompartments that are enriched for specialized activities (Falk et al. 2019; Fritz et al. 2019a). For example, high-resolution matrices of interaction frequencies have identified that virtually all active H3K27ac marked transcription start sites (TSSs) are found within A compartments (Harris et al. 2023). These TSSs cluster within different subcompartments than transcription end sites. Bivalent/poised or bistable chromatin at promoters is enriched within specific subcompartments, and polycomb repressed chromatin is enriched within specific B subcompartments (Harris et al. 2023). Nichols and Corces experimentally defined regulatory organization by compartmentalization, which is reflected by a strong correlation with transcriptional activity and H3K9me3 followed by H3K27me3 (Nichols and Corces 2021). This model identified four compartments marked by H3K27 acetylation, H3K9me3 (which was cell type-specific), H3K27me3, or devoid of these PTMs but enriched in H3K9 dimethylation (H3K9me2) (Nichols and Corces 2021). At the highest level, chromosomes are present within distinct regions within the nucleus, termed chromosome territories, which are non-randomly distributed within the nucleus in a cell type-dependent manner (Fritz et al. 2019b).

Proper genome regulation is essential to maintain cellular identity or to respond to environmental cues or stressors. Despite the need for stringent control whereby enhancers must interact with the correct gene upon acetylation-mediated activation, enhancers can contact genes irrespective of orientation or location relative to the gene of interest (Sahu et al. 2022). For example, while enhancers are often found nearby their cognate promoter or within the introns of that gene, they can also be located over one million base pairs away from the gene of interest (Panigrahi and O'Malley 2021). Multiple models of E-P communication (i.e., "looping", "scanning", "linking", or "conformation") have been proposed and are reviewed in (Popay and Dixon 2022). These models may not be mutually exclusive. The loop-extrusion model (Sanborn et al. 2015; Fudenberg et al. 2016; Nasmyth 2001; Alipour and Marko 2012; Ganji et al. 2018) hypothesizes that cohesin or condensin extrudes DNA until it encounters CCTC-binding factor (CTCF) proteins that have a convergent orientation relative to each other (Tang et al. 2015). Scenarios wherein RNAPII or cohesin/condensin/CTCF "scanning"-like mechanisms could provide promiscuity of interactions within a local domain to allow "linking" of interactions between the coregulatory machinery (i.e., specific transcription factors and the mediator-POLII preinitiation complex) to establish the correct E-P pair. Additionally, some enhancers may communicate through alterations in the "conformation" of DNA, while others establish interaction by "scanning" or "linking". The higher-order conformation of chromatin is complicated by the fact that the looping of these E-P pairs has been experimentally defined to skip over the nearest gene in approximately ~40–60% of cases (Yang and Hansen 2024; Chen et al. 2024). Furthermore, while multiple enhancers can cooperatively coregulate one gene, singular enhancers can regulate multiple genes simultaneously (Panigrahi and O'Malley 2021). This complex functional organization of genes and regulatory sequences constitutes gene regulatory networks.

It is unclear whether E-P interactions drive acetylation and transcription, vice versa, or if there is a dynamic interplay between these parameters. Studies have identified significant physical distances between enhancers and their cognate promoters using microscopic and molecular approaches (Heist et al. 2019; Alexander et al. 2019), and even an increased distance upon activation during development from Mouse Embryonic Stem Cells (mESCs) into Neural Progenitor Cells (NPCs, (Benabdallah et al. 2019)). In contrast, the 3D interactomes of 935 bona fide developmental enhancers across 10 mouse embryonic tissues were identified, and that the majority of these enhancers displayed tissue-specific conformations with increased physical interactions and decreased physical distances in their activated, presumably acetylated, state (Chen et al. 2024). Moreover, forcing or disrupting interactions has been shown to be sufficient to activate or repress transcription, respectively (reviewed in 216). Some E-P contacts have been shown to exist prior to gene activation during development (Freire-Pritchett et al. 2017; Ghavi-Helm et al. 2014). Despite this, Chen et al. determined that out of the 935 enhancers studied only 14% were invariable between tissues (Chen et al. 2024). Remarkably similar E-P conformations were identified between myogenic and neuronal lineages regardless of the activity state during cell-fate specification. However, during terminal differentiation E–P proximity was associated with a gain in activity and vice versa. The predominantly permissive E–P configurations could facilitate plasticity during cell-fate decisions while switching to a more instructive mode at later stages may ensure terminal tissue diversification (Pollex et al. 2024). More recently, it was discovered that during development from mESC into NPCs, prior to transcriptional activation many interactions between distal regulatory elements (DREs) and promoters are present and then lost, while de novo contacts also follow transcriptional activation (Mahara et al. 2024). Sequential interactions between DREs and their cognate promoters occur during gene induction. The magnitude of transcriptional activation is associated with different topological patterns of these DREs and promoters, including a novel class of DRE that only transiently interacts with promoters prior to activation (Mahara et al. 2024).

While the mechanisms driving E-P contacts are critical for gene regulation, the correlation between E-P interaction and transcriptional activation is not absolute. Different mechanisms of higher-order chromatin organization mediated gene activation may be required in a context-dependent manner. For example, increased distances between regulatory regions may be required to activate some genes, while physical E-P contacts are required for others. Furthermore, as described, the temporal nature of interactions leading to gene activation is another possible explanation for the incongruency. Some E-P contacts have been shown to exist prior to gene activation during development (Zabidi and Stark 2016) (Ghavi-Helm et al. 2014). Surprisingly similar E-P topologies were identified between myogenic and neuronal lineages during cell-fate specification regardless of the activity state. However, at later stages, during terminal tissue differentiation E–P proximity is associated with a gain in activity and vice versa. The alternative usage of predominantly permissive

E–P topologies to more instructive regulation at later stages may enable plasticity during cell-fate decisions while ensuring diversification during terminal tissue differentiation (Pollex et al. 2024). Some loops might not drive active transcription but establish a permissive environment to provide the potential for rapid transcription upon environmental stimuli. Furthermore, loops could be established by previous processes and are without need to be disrupted therefore remain. For example, the 14% of E-P interactions that were invariable between mESC lineages were proposed to be due to CTCF-mediated looping independent of enhancer activity (Chen et al. 2024). Interactions could also be dynamic and therefore only present in a small fraction of nuclei at any time. E-P interactions for upregulated genes could be rapidly and persistently dynamic, and more frequently disassociated, and consequently may not be captured due to the transient nature of the interaction. Similarly, upregulated gene expression could be present within a small fraction of cells. Additionally, measuring steady-state levels of RNA may not capture active transcription as many processes result in pools of RNA including transcriptional elongation, RNA processing, and RNA stability. Given that most experiments have been performed as a population average on steady-state RNAs, accurate measurements of these processes have been limited. Despite this, recent reports provide some interesting clues regarding these questions.

Oh et al. identified enhancer-promoter (E-P) pairs that are induced upon estradiol stimulation. Deletion of targeted promoters resulted in increased eRNA expression at their enhancer suggesting that these enhancers are activated prior to contacting their promoter pairs. These eRNAs were similarly upregulated upon the knockdown radiation-sensitive 21 cohesin complex component (RAD21), a key component of cohesin that is essential for looping, while simultaneously interfering with E-P contacts. Knockout of the promoter of trefoil factor 1 (TFF1) resulted in an increased interaction with the trefoil factor 3 (TFF3) promoter and an 18-fold increase in the expression of TFF3. A double knockout of TFF1's promoter and its enhancer restored the level of TFF3 to normal levels indicating the importance of the chromatin interactions in driving the aberrant level of TFF3 (Oh et al. 2021). Given these findings, it is expected that the acetylation at the enhancer and the activity level of the enhancer precede the E-P contact, but this does not preclude that the E-P contact may also drive transcriptional activity. Indeed, single-cell capturing of nascent RNAs using single-cell Global run-on sequencing (scGROseq) demonstrated the increased eRNA expression prior to bursting of expression at promoters (Mahat et al. 2024). Taken together with the given supposition that the elevated level of eRNA expression at the TFF1 gene is due to an uncoupling of the expression at the enhancer and the associated bursting of expression at the promoter, E-P contact prior to gene activation might be a more global phenomenon (Mahat et al. 2024). Interestingly, a screen using clustered regularly interspaced short palindromic repeats (CRISPR)-mediated gene interference (CRISPRi) screen on promoters that are frequently mutated in cancer determined that nearby oncogenes were activated in eight out of 25 cases, while only 1 was downregulated. This suggests that the enhancers that normally interact with these non-oncogenic promoters are aberrantly retargeted to these oncogenic promoters when the correct promoter has been mutated (Oh et al. 2021).

Another emerging concept is that the physical properties of the macromolecules comprising microenvironments drive liquid-like phase separation (Sabari et al. 2018, 2020). This hypothesis states that droplets or condensates form via weak multivalent interactions between macromolecules (Contrepois et al. 2012; Zippo et al. 2009; Vlaming and van Leeuwen 2016). For example, multivalent interactions occur in enhancers at transcription factor binding sites and binding of coregulatory proteins during transcription (Panigrahi and O'Malley 2021; Tang et al. 2022). The formation of these liquid-liquid phase separation (LLPS) is highly correlated with different factors such as protein concentration within the region of modification and enhancer action. The occupancy of transcription factors at active genes within condensates depends upon their macromolecular interactions with transcriptional machinery (Wu et al. 2020). Transcription factors often have Intrinsically Disordered Regions (IDRs) that induce the formation of liquid-like droplets (Sabari et al. 2018; Sabari et al. 2020). It has been shown that highly specific concentrations of RNA act locally to strongly promote the formation of phase condensates and subsequent recruitment of other phase elements as mentioned (Henninger et al. 2021). Interestingly, titration of low levels of RNA results in the formation or reentrant phase behavior whereas high RNA titration results in phase suppression (Henninger et al. 2021). These reactions are mediated by multivalent interactions of enhancer regulatory elements which all contribute to the cascade effect of recruitment to the phase condensates initiated by the growing protein conglomerates. In addition to the role that RNA concentrations play in the genesis of phase condensation, chemical modifications of RNA molecules also affect phase condensation. For example, N6-methyladenosine (m6A) is one common RNA and eRNA modification that promotes the formation of transcriptional condensates and subsequent gene activation (Tang et al. 2022).

The acetylation of chromatin at these active genes and the binding of readers of acetylation are critical for LLPS. While microinjection of reconstituted chromatin into nuclei results in its condensation, nucleosomal arrays with acetylated histones do not readily form condensates. Droplet formation for these acetylated nucleosomal arrays was rescued upon the addition of the Bromodomain-Containing Protein 4 (BRD4) prior to microinjection. This acetylated and BRD4-containing chromatin associates but does not coalesce with droplets formed using unacetylated chromatin (Boija et al. 2018). Within these condensates, enhancers could drive transcription at a distance without physical E-P interactions due to the localized concentration of the necessary coregulatory complexes. Such condensates have been shown to occur at super-enhancers that synergistically drive transcription of critical cell type-specific genes, tumor suppressors, and oncogenes. This suggests that the status of chromatin including acetylation drives LLPS.

Subnuclear trafficking (the movement of regulatory molecules between and within nuclear microenvironments) (Zaidi et al. 2010b) and LLPS are not mutually exclusive. Both hypotheses propose that the physical properties of macromolecular complexes drive higher-order nuclear organization. LLPS of the regulatory complexes into compartments includes the speckles, histone locus bodies, Cajal bodies, nucleoli, and nuclear lamina, taken together contribute to the overall nuclear organization. Recognition signals that have been shown to target transcription factors for subnuclear trafficking may drive their inclusion and/or retention within microenvironments coordinately with their other physical properties such as IDRs that drive LLPS.

It is clear, however, that many questions remain about enhancers and their mechanisms of action, and new methodologies and techniques are required to overcome current barriers and significant limitations to research to address these questions. One area requiring further investigation is the elucidation of the dynamic activity of phase condensates and LLPS as it was shown they are dynamic in multifaceted ways such as size, activity, and temporality (Henninger et al. 2021). Another question is why and how enhancers start transcribing eRNAs? There is also a requirement for investigation into the enhancer's counterpart, the silencer. Further research is needed to understand how silencers operate within the context of the genome to accommodate LLPS. Given the overwhelming amount of high-throughput data being generated, analytical strategies for studying acetylation-mediated genome regulation are outlined below.

2.13 Analytical Strategies to Functionally Categorize the Regulatory Consequences of Histone Acetylation

As the volume of epigenomic data has expanded, bioinformatics tools have become indispensable for extracting meaningful insights from complex datasets. These tools enable researchers to analyze patterns of acetylation and methylation across the genome, integrate this information with other omics data, and understand how these modifications influence cellular function and phenotype.

Epigenetic data analysis, particularly for acetylation and methylation marks, requires several computational steps to extract meaningful biological insights. For acetylation data, analysis typically begins with aligning sequenced reads to a reference genome, followed by peak calling to identify regions enriched with specific histone modifications (Nakato and Sakata 2021; Bailey et al. 2013). Alignment is commonly performed using tools like Bowtie2 (Langmead and Salzberg 2012) and BWA (Li and Durbin 2009), with MACS2 (Zhang et al. 2008) as the preferred tool for peak calling.

Advanced downstream analysis of acetylation data is supported by a wide variety of tools designed for differential analyses, peak annotation, and motif discovery. Differential analyses can identify changes in acetylation patterns across conditions, offering insights into how acetylation contributes to dynamic gene regulation in response to biological stimuli (Lesluyes et al. 2014). Popular tools for differential binding analysis include DiffReps (Shen et al. 2013), Diffbind, DESeq2 (Love et al. 2014), and edgR (Nikolayeva and Robinson 2014). Following the identification of peaks, motif discovery and enrichment analysis can reveal DNA sequence motifs enriched in those peaks, using tools like HOMER (Heinz et al. 2010) and DREME (Bailey 2011). HOMER, along with popular R packages ChIPSeeker (Yu et al. 2015), and ChIPpeakAnno (Zhu et al. 2010) can also be used to perform peak annotation, which associates peaks with functionally relevant genomic elements. To integrate multiple histone modification datasets, ChromHMM (Ernst and Kellis 2012) and Segway (Hoffman et al. 2012) are utilized for chromatin state annotation, systematically identifying combinations of histone modification patterns to detect large-scale variations across the genome.

For methylation data, particularly bisulfite sequencing, the analysis process begins with alignment and methylation calling. This involves aligning bisulfite-treated reads to a reference genome and determining the methylation status of individual cytosines, resulting in a high-resolution map of DNA methylation across the genome (Bock 2012; Raulusevicíute et al. 2019). Tools like Bismark (Krueger and Andrews 2011), BS Seeker (Chen et al. 2010), and BSMAP (Xi and Li 2009) can be utilized for efficient processing of large bisulfite sequencing datasets. Following the generation of a methylation map, differential methylation analysis is conducted to compare methylation levels across different conditions, identifying genome regions where methylation changes are linked to changes in gene expression or cellular phenotype (Piao et al. 2021). Differentially methylated sites or regions can be identified with the widely used R packages DSS (Feng and Wu 2019) and MethylKit (Akalin et al. 2012).

Integrative multi-omics approaches are essential for understanding the regulatory networks influenced by acetylation and methylation. These approaches often involve integrating transcriptomic data to link epigenetic modifications directly to gene expression. By mapping acetylation and methylation marks across the genome and correlating them with RNA-Seq data, researchers can identify which genes are affected by specific modifications, providing a detailed view of how epigenetic changes drive gene expression under different conditions (Crespo et al. 2020; Itai et al. 2023; Abnizova et al. 2024).

An integral aspect of multi-omics approaches is the integration of chromatin accessibility data with histone modification and methylation profiles. By combining ATAC-Seq or DNase-Seq data with ChIP-Seq and bisulfite sequencing, researchers can identify regulatory elements that are not only marked by specific histone modifications but also exist in an accessible state, indicating their active role in gene regulation (Bai et al. 2021). This layered analysis is particularly useful for pinpointing enhancer regions critical for gene activation in response to environmental or developmental cues.

Integrating epigenomic data with three-dimensional genome organization provides deeper insights into gene regulation. Understanding the spatial relationships between different genomic regions is crucial for deciphering how distant regulatory elements, such as enhancers, interact with promoters to control gene expression (Song et al. 2022). Sequence-based profiling technologies like Hi-C and ChIA-PET can be utilized to generate interaction maps that, when combined with acetylation and methylation data, offer insights into how these modifications influence chromatin architecture (Tang et al. 2017; Kumar et al. 2021). This analysis often focuses on identifying long-range chromatin interactions stabilized by specific epigenetic marks, revealing the role of the 3D genome structure in gene regulation. Such integrative approaches are vital for unraveling the complex regulatory networks that govern cellular identity and function, especially in development and disease.

Visualization is essential for interpreting histone modification and methylation data, providing intuitive ways to explore complex datasets. Genome browsers like the UCSC Genome Browser (Nassar et al. 2023) and Integrative Genomics Viewer (IGV) (Thorvaldsdottir et al. 2013) are commonly used to visualize epigenetic sequencing data. These tools allow researchers to examine specific genomic regions in detail, overlaying various data types such as acetylation marks, methylation levels, and gene annotations. By visualizing these data layers together, researchers can better understand how epigenetic modifications are distributed across the genome and their potential functional implications.

In addition to genome browsers, heatmaps and pathway visualization tools such as Cytoscape and the R package ComplexHeatmap are invaluable for summarizing and interpreting large-scale epigenetic data. For example, ComplexHeatmap can be used to display patterns of acetylation and methylation across different gene sets, helping to identify clusters of genes that share similar epigenetic modifications (Gu et al. 2016a). Cytoscape, on the other hand, enables the construction of networks that represent the relationships between acetylation-modified genes and their associated pathways (Shannon et al. 2003). These visualizations are critical for making sense of the intricate networks of epigenetic regulation and for generating hypotheses about the functional roles of acetylation and methylation in various biological processes.

As the field of epigenetics advances, the integration of increasingly complex datasets presents both exciting opportunities and significant challenges. The advent of single-cell technologies and high-resolution multi-omics data has heightened the need for tools capable of managing vast datasets while ensuring accurate, reproducible results across various platforms and studies. Although integrative approaches have become more sophisticated, deciphering the complex interactions between different epigenetic marks and their combined impact on gene regulation remains a daunting task. Future research will likely focus on leveraging machine learning and artificial intelligence to enhance the utility of multi-omics studies (Yang and Hansen 2024; Wang et al. 2024).

2.14 Strategies for the Identification of Coregulatory Epigenetic Machinery

In addition to these multi-omic studies, another highly informative and rapidly evolving technology involves proteomics to determine the protein-protein interactions between transcription factors, coregulatory complexes, and chromatin-modifying enzymes. Bio-ID technology has the potential to significantly enhance our understanding of acetylation-responsive epigenetic mechanisms by providing a detailed view of protein interactions and cellular dynamics. This approach utilizes the addition of a biotin ligase enzyme, such as biotin retention protein A (BirA), to a candidate protein of interest. Upon the addition of biotin, proteins that are proximal to the protein of interest are biotinylated and subsequently enriched using a streptavidin isolation protocol. By generating comprehensive interaction maps, Bio-ID can offer critical insights into the functional relationships between proteins and the role of acetylation in modulating these relationships (Sears et al. 2019; Roux et al. 2018; Guo et al. 2023). One promising application of Bio-ID in epigenetic research is the mapping of acetylation-responsive protein interactions. By tagging acetylation-sensitive proteins and their interacting partners, researchers can potentially uncover how acetylation modulates the composition and function of chromatin-associated protein complexes. This mapping could provide valuable insights into how acetylation influences the assembly and disassembly of these complexes and their impact on gene expression (Sears et al. 2019; Roux et al. 2018; Guo et al. 2023).

Recent advancements in Bio-ID technology could further refine its application in studying acetylation. Innovations in biotin ligase enzymes and tagging methods are likely to enhance the sensitivity and resolution of Bio-ID experiments, allowing for more precise identification of acetylation-responsive interactions. The integration of Bio-ID with mass spectrometry can facilitate comprehensive proteomic analyses, revealing specific acetylation-sensitive proteins and their interaction networks. Additionally, newer versions of Bio-ID technology are expected to offer cell type-specific and temporal resolution, enabling the study of acetylation effects across different cell types and developmental stages (Guo et al. 2023; Liu et al. 2020). Advances in bioinformatics and data analysis will also play a crucial role in interpreting Bio-ID data. Enhanced algorithms and databases will facilitate the analysis of complex interaction networks, enabling researchers to identify key players in acetylation-responsive epigenetic control and understand their roles in regulating chromatin dynamics (Liu et al. 2020).

2.15 Prospects and Perspectives: Acetylation-Mediated Epigenetic Control

The power of increased understanding of epigenetic control in regulating biology and pathology is rapidly expanding. And the pivotal engagement of histone acetylation in these processes is becoming increasingly evident. There is an evolving appreciation for acetylation-mediated control within the context of histone and DNA modifications, higher-order chromatin organization, and dynamic structure-function remodeling, which supports physiologic responsiveness and is compromised with the onset and progression of disease. Deeper understanding of the mechanisms governing acetylation-mediated control and how these processes are disrupted in disease will provide novel therapeutic strategies.

Grant Support Research reported in this review was supported by the following grants: P01 CA240685 (GSS, JLS), U54GM115516 (GSS), R01GM129338 (SF), R01CA230618 (SF), DE029311 (JBL, GSS, JLS), F32CA220935 (AJF), Arthur Perelman Professorship (GSS), Charlotte Perelman Research Fund (GSS), R01AR084561 (GSS, JLS, JBL).

References

Abnizova I, Stapel C, Boekhorst RT, Lee JTH, Hemberg M (2024) Integrative analysis of transcriptomic and epigenomic data reveals distinct patterns for developmental and housekeeping gene regulation. BMC Biol 22(1):78

Ahmad B, Saeed A, Al-Amery A, Celik I, Ahmed I, Yaseen M et al (2024) Investigating potential cancer therapeutics: insight into histone deacetylases (HDACs) inhibitions. Pharmaceuticals (Basel) 17(4)

Ai X, Parthun MR (2004) The nuclear Hat1p/Hat2p complex: a molecular link between type B histone acetyltransferases and chromatin assembly. Mol Cell 14(2):195–205

Akalin A, Kormaksson M, Li S, Garrett-Bakelman FE, Figueroa ME, Melnick A et al (2012) methylKit: a comprehensive R package for the analysis of genome-wide DNA methylation profiles. Genome Biol 13(10):R87

Alexander JM, Guan J, Li B, Maliskova L, Song M, Shen Y et al (2019) Live-cell imaging reveals enhancer-dependent Sox2 transcription in the absence of enhancer proximity. elife:8

Alipour E, Marko JF (2012) Self-organization of domain structures by DNA-loop-extruding enzymes. Nucleic Acids Res 40(22):11202–11212

Allfrey VG, Faulkner R, Mirsky AE (1964) Acetylation and methylation of histones and their possible role in the regulation of RNA synthesis. Proc Natl Acad Sci USA 51(5):786–794

Allshire RC, Madhani HD (2018) Ten principles of heterochromatin formation and function. Nat Rev Mol Cell Biol 19(4):229–244

Almeria S, Dubey JP (2021) Foodborne transmission of Toxoplasma gondii infection in the last decade: an overview. Res Vet Sci 135:371–385

Alseksek RK, Ramadan WS, Saleh E, El-Awady R (2022) The role of HDACs in the response of cancer cells to cellular stress and the potential for therapeutic intervention. Int J Mol Sci 23(15)

Amaral PP, Leonardi T, Han N, Vire E, Gascoigne DK, Arias-Carrasco R et al (2018) Genomic positional conservation identifies topological anchor point RNAs linked to developmental loci. Genome Biol 19(1):32

An P, Chen F, Li Z, Ling Y, Peng Y, Zhang H et al (2020) HDAC8 promotes the dissemination of breast cancer cells via AKT/GSK-3beta/snail signals. Oncogene 39(26):4956–4969
Andrews FH, Shanle EK, Strahl BD, Kutateladze TG (2016) The essential role of acetyllysine binding by the YEATS domain in transcriptional regulation. Transcription 7(1):14–20
Ansari KI, Mishra BP, Mandal SS (2008) Human CpG binding protein interacts with MLL1, MLL2 and hSet1 and regulates Hox gene expression. Biochim Biophys Acta 1779(1):66–73
Aricthota S, Rana PP, Haldar D (2022) Histone acetylation dynamics in repair of DNA double-strand breaks. Front Genet 13:926577
Bae S, Lesch BJ (2020) H3K4me1 distribution predicts transcription state and poising at promoters. Front Cell Dev Biol 8:289
Baggiolini A, Callahan SJ, Montal E, Weiss JM, Trieu T, Tagore MM et al (2021) Developmental chromatin programs determine oncogenic competence in melanoma. Science 373(6559):eabc1048
Bai D, Peng J, Yi C (2021) Advances in single-cell multi-omics profiling. RSC Chem Biol 2(2):441–449
Bailey TL (2011) DREME: motif discovery in transcription factor ChIP-seq data. Bioinformatics 27(12):1653–1659
Bailey T, Krajewski P, Ladunga I, Lefebvre C, Li Q, Liu T et al (2013) Practical guidelines for the comprehensive analysis of ChIP-seq data. PLoS Comput Biol 9(11):e1003326
Barral A, Dejardin J (2023) The chromatin signatures of enhancers and their dynamic regulation. Nucleus 14(1):2160551
Barrett AK, Shingare MR, Rechtsteiner A, Rodriguez KM, Le QN, Wijeratne TU et al (2024) HDAC activity is dispensable for repression of cell-cycle genes by DREAM and E2F:RB complexes. Nat Commun 15(1):4450
Beagan JA, Phillips-Cremins JE (2020) On the existence and functionality of topologically associating domains. Nat Genet 52(1):8–16
Beltrao P, Bork P, Krogan NJ, van Noort V (2013) Evolution and functional cross-talk of protein post-translational modifications. Mol Syst Biol 9:714
Benabdallah NS, Williamson I, Illingworth RS, Kane L, Boyle S, Sengupta D et al (2019) Decreased enhancer-promoter proximity accompanying enhancer activation. Mol Cell 76(3):473–84 e7
Bhaskara S, Knutson SK, Jiang G, Chandrasekharan MB, Wilson AJ, Zheng S et al (2010) Hdac3 is essential for the maintenance of chromatin structure and genome stability. Cancer Cell 18(5):436–447
Bhaskara S, Jacques V, Rusche JR, Olson EN, Cairns BR, Chandrasekharan MB (2013) Histone deacetylases 1 and 2 maintain S-phase chromatin and DNA replication fork progression. Epigenetics Chromatin 6(1):27
Biswas S, Rao CM (2018) Epigenetic tools (the writers, the readers and the erasers) and their implications in cancer therapy. Eur J Pharmacol 837:8–24
Biswas S, Reddy ND, Jayashree BS, Rao CM (2018) Evaluation of novel 3-Hydroxyflavone analogues as HDAC inhibitors against colorectal cancer. Adv Pharmacol Sci 2018:4751806
Blander G, Guarente L (2004) The Sir2 family of protein deacetylases. Annu Rev Biochem 73:417–435
Bock C (2012) Analysing and interpreting DNA methylation data. Nat Rev Genet 13(10):705–719
Boija A, Klein IA, Sabari BR, Dall'Agnese A, Coffey EL, Zamudio AV et al (2018) Transcription factors activate genes through the phase-separation capacity of their activation domains. Cell 175(7):1842–55 e16
Boros J, Arnoult N, Stroobant V, Collet JF, Decottignies A (2014) Polycomb repressive complex 2 and H3K27me3 cooperate with H3K9 methylation to maintain heterochromatin protein 1alpha at chromatin. Mol Cell Biol 34(19):3662–3674
Bouyahya A, El Omari N, Bakha M, Aanniz T, El Menyiy N, El Hachlafi N et al (2022) Pharmacological properties of Trichostatin A, focusing on the anticancer potential: a comprehensive review. Pharmaceuticals (Basel) 15(10)

Bowers EM, Yan G, Mukherjee C, Orry A, Wang L, Holbert MA et al (2010) Virtual ligand screening of the p300/CBP histone acetyltransferase: identification of a selective small molecule inhibitor. Chem Biol 17(5):471–482

Boyson SP, Gao C, Quinn K, Boyd J, Paculova H, Frietze S et al (2021) Functional roles of Bromodomain proteins in cancer. Cancers (Basel) 13(14)

Bracken AP, Brien GL, Verrijzer CP (2019) Dangerous liaisons: interplay between SWI/SNF, NuRD, and Polycomb in chromatin regulation and cancer. Genes Dev 33(15–16):936–959

Burke B, Eden C, Perez C, Belshoff A, Hart S, Plaza-Rojas L et al (2020) Inhibition of histone deacetylase (HDAC) enhances checkpoint blockade efficacy by rendering bladder cancer cells visible for T cell-mediated destruction. Front Oncol 10:699

Butler R, Bates GP (2006) Histone deacetylase inhibitors as therapeutics for polyglutamine disorders. Nat Rev Neurosci 7(10):784–796

Byvoet P (1972) In vivo turnover and distribution of radio-N-methyl in arginine-rich histones from rat tissues. Arch Biochem Biophys 152(2):887–888

Cha JH, Chan LC, Li CW, Hsu JL, Hung MC (2019) Mechanisms controlling PD-L1 expression in cancer. Mol Cell 76(3):359–370

Chan CH, Fang C, Yarilina A, Prinjha RK, Qiao Y, Ivashkiv LB (2015) BET bromodomain inhibition suppresses transcriptional responses to cytokine-Jak-STAT signaling in a gene-specific manner in human monocytes. Eur J Immunol 45(1):287–297

Chen PY, Cokus SJ, Pellegrini M (2010) BS seeker: precise mapping for bisulfite sequencing. BMC Bioinformatics 11:203

Chen HP, Zhao YT, Zhao TC (2015) Histone deacetylases and mechanisms of regulation of gene expression. Crit Rev Oncog 20(1–2):35–47

Chen Q, Yang B, Liu X, Zhang XD, Zhang L, Liu T (2022) Histone acetyltransferases CBP/p300 in tumorigenesis and CBP/p300 inhibitors as promising novel anticancer agents. Theranostics 12(11):4935–4948

Chen K, Yuan J, Sia Y, Chen Z (2023) Mechanism of action of the SWI/SNF family complexes. Nucleus 14(1):2165604

Chen Z, Snetkova V, Bower G, Jacinto S, Clock B, Dizehchi A et al (2024) Increased enhancer-promoter interactions during developmental enhancer activation in mammals. Nat Genet 56(4):675–685

Choi HJ, Park JH, Park M, Won HY, Joo HS, Lee CH et al (2015) UTX inhibits EMT-induced breast CSC properties by epigenetic repression of EMT genes in cooperation with LSD1 and HDAC1. EMBO Rep 16(10):1288–1298

Ciurciu A, Komonyi O, Pankotai T, Boros IM (2006) The drosophila histone acetyltransferase Gcn5 and transcriptional adaptor Ada2a are involved in nucleosomal histone H4 acetylation. Mol Cell Biol 26(24):9413–9423

Colino-Sanguino Y, Cornett EM, Moulder D, Smith GC, Hrit J, Cordeiro-Spinetti E et al (2019) A read/write mechanism connects p300 bromodomain function to H2AZ acetylation. iScience 21:773–788

Contrepois K, Thuret JY, Courbeyrette R, Fenaille F, Mann C (2012) Deacetylation of H4-K16Ac and heterochromatin assembly in senescence. Epigenetics Chromatin 5(1):15

Crespo M, Damont A, Blanco M, Lastrucci E, Kennani SE, Ialy-Radio C et al (2020) Multi-omic analysis of gametogenesis reveals a novel signature at the promoters and distal enhancers of active genes. Nucleic Acids Res 48(8):4115–4138

Creyghton MP, Cheng AW, Welstead GG, Kooistra T, Carey BW, Steine EJ et al (2010) Histone H3K27ac separates active from poised enhancers and predicts developmental state. Proc Natl Acad Sci USA 107(50):21931–21936

Dantas A, Al Shueili B, Yang Y, Nabbi A, Fink D, Riabowol K (2019) Biological functions of the ING proteins. Cancers (Basel) 11(11)

Deeney JT, Belkina AC, Shirihai OS, Corkey BE, Denis GV (2016) BET Bromodomain proteins Brd2, Brd3 and Brd4 selectively regulate metabolic pathways in the pancreatic beta-cell. PLoS One 11(3):e0151329

Deepak KGK, Vempati R, Nagaraju GP, Dasari VR, Rao DN et al (2020) Tumor microenvironment: challenges and opportunities in targeting metastasis of triple negative breast cancer. Pharmacol Res 153:104683
Dekker J, Heard E (2015) Structural and functional diversity of topologically associating domains. FEBS Lett 589(20 Pt A):2877-84
Devaiah BN, Case-Borden C, Gegonne A, Hsu CH, Chen Q, Meerzaman D et al (2016) BRD4 is a histone acetyltransferase that evicts nucleosomes from chromatin. Nat Struct Mol Biol 23(6):540–548
Dey A, Chitsaz F, Abbasi A, Misteli T, Ozato K (2003) The double bromodomain protein Brd4 binds to acetylated chromatin during interphase and mitosis. Proc Natl Acad Sci USA 100(15):8758–8763
Dhalluin C, Carlson JE, Zeng L, He C, Aggarwal AK, Zhou MM (1999) Structure and ligand of a histone acetyltransferase bromodomain. Nature 399(6735):491–496
Draker R, Ng MK, Sarcinella E, Ignatchenko V, Kislinger T, Cheung P (2012) A combination of H2A.Z and H4 acetylation recruits Brd2 to chromatin during transcriptional activation. PLoS Genet 8(11):e1003047
Du J, Johnson LM, Jacobsen SE, Patel DJ (2015) DNA methylation pathways and their crosstalk with histone methylation. Nat Rev Mol Cell Biol 16(9):519–532
Dyson MH, Rose S, Mahadevan LC (2001) Acetyllysine-binding and function of bromodomain-containing proteins in chromatin. Front Biosci 6:D853–D865
Erb MA, Scott TG, Li BE, Xie H, Paulk J, Seo HS et al (2017) Transcription control by the ENL YEATS domain in acute leukaemia. Nature 543(7644):270–274
Ernst J, Kellis M (2012) ChromHMM: automating chromatin-state discovery and characterization. Nat Methods 9(3):215–216
Falk M, Feodorova Y, Naumova N, Imakaev M, Lajoie BR, Leonhardt H et al (2019) Heterochromatin drives compartmentalization of inverted and conventional nuclei. Nature 570(7761):395–399
Fang Y, Liao G, Yu B (2019) LSD1/KDM1A inhibitors in clinical trials: advances and prospects. J Hematol Oncol 12(1):129
Feng H, Wu H (2019) Differential methylation analysis for bisulfite sequencing using DSS. Quant Biol 7(4):327–334
Ferrari KJ, Scelfo A, Jammula S, Cuomo A, Barozzi I, Stutzer A et al (2014) Polycomb-dependent H3K27me1 and H3K27me2 regulate active transcription and enhancer fidelity. Mol Cell 53(1):49–62
Filippakopoulos P, Qi J, Picaud S, Shen Y, Smith WB, Fedorov O et al (2010) Selective inhibition of BET bromodomains. Nature 468(7327):1067–1073
Filippakopoulos P, Picaud S, Mangos M, Keates T, Lambert JP, Barsyte-Lovejoy D et al (2012) Histone recognition and large-scale structural analysis of the human bromodomain family. Cell 149(1):214–231
Fleck K, McNutt S, Chu F, Jeffers V (2023) An apicomplexan bromodomain protein, TgBDP1, associates with diverse epigenetic factors to regulate essential transcriptional processes in Toxoplasma gondii. MBio 14(4):e0357322
Flynn EM, Huang OW, Poy F, Oppikofer M, Bellon SF, Tang Y et al (2015) A subset of human Bromodomains recognizes Butyryllysine and Crotonyllysine histone peptide modifications. Structure 23(10):1801–1814
Fox GC, Poncha KF, Smith BR, van der Maas LN, Robbins NN, Graham B, et al (2024) H3K18 & H3K23 acetylation directs establishment of MLL-mediated H3K4 methylation. bioRxiv
Fradet-Turcotte A, Canny MD, Escribano-Diaz C, Orthwein A, Leung CC, Huang H et al (2013) 53BP1 is a reader of the DNA-damage-induced H2A Lys 15 ubiquitin mark. Nature 499(7456):50–54
Freire-Pritchett P, Schoenfelder S, Varnai C, Wingett SW, Cairns J, Collier AJ et al (2017) Global reorganisation of cis-regulatory units upon lineage commitment of human embryonic stem cells. elife:6

Fritz AJ, Gillis NE, Gerrard DL, Rodriguez PD, Hong D, Rose JT et al (2019a) Higher order genomic organization and epigenetic control maintain cellular identity and prevent breast cancer. Genes Chromosomes Cancer 58(7):484–499
Fritz AJ, Sehgal N, Pliss A, Xu J, Berezney R (2019b) Chromosome territories and the global regulation of the genome. Genes Chromosomes Cancer 58(7):407–426
Fudenberg G, Imakaev M, Lu C, Goloborodko A, Abdennur N, Mirny LA (2016) Formation of chromosomal domains by loop extrusion. Cell Rep 15(9):2038–2049
Fullgrabe J, Kavanagh E, Joseph B (2011) Histone onco-modifications. Oncogene 30(31):3391–3403
Gangaraju VK, Bartholomew B (2007) Mechanisms of ATP dependent chromatin remodeling. Mutat Res 618(1–2):3–17
Ganji M, Shaltiel IA, Bisht S, Kim E, Kalichava A, Haering CH et al (2018) Real-time imaging of DNA loop extrusion by condensin. Science 360(6384):102–105
Gao L, Emperle M, Guo Y, Grimm SA, Ren W, Adam S et al (2020) Comprehensive structure-function characterization of DNMT3B and DNMT3A reveals distinctive de novo DNA methylation mechanisms. Nat Commun 11(1):3355
Garmpis N, Damaskos C, Garmpi A, Kalampokas E, Kalampokas T, Spartalis E et al (2017) Histone deacetylases as new therapeutic targets in triple-negative breast cancer: progress and promises. Cancer Genomics Proteomics 14(5):299–313
Ghavi-Helm Y, Klein FA, Pakozdi T, Ciglar L, Noordermeer D, Huber W et al (2014) Enhancer loops appear stable during development and are associated with paused polymerase. Nature 512(7512):96–100
Giaimo BD, Ferrante F, Herchenrother A, Hake SB, Borggrefe T (2019) The histone variant H2A.Z in gene regulation. Epigenetics Chromatin 12(1):37
Glozak MA, Sengupta N, Zhang X, Seto E (2005) Acetylation and deacetylation of non-histone proteins. Gene 363:15–23
Gong F, Miller KM (2013) Mammalian DNA repair: HATs and HDACs make their mark through histone acetylation. Mutat Res 750(1–2):23–30
Gonzales-Cope M, Sidoli S, Bhanu NV, Won KJ, Garcia BA (2016) Histone H4 acetylation and the epigenetic reader Brd4 are critical regulators of pluripotency in embryonic stem cells. BMC Genomics 17:95
Gou P, Zhang W (2024) Protein lysine acetyltransferase CBP/p300: a promising target for small molecules in cancer treatment. Biomed Pharmacother 171:116130
Grozinger CM, Hassig CA, Schreiber SL (1999) Three proteins define a class of human histone deacetylases related to yeast Hda1p. Proc Natl Acad Sci USA 96(9):4868–4873
Grunstein M (1997) Histone acetylation in chromatin structure and transcription. Nature 389(6649):349–352
Gu Z, Eils R, Schlesner M (2016a) Complex heatmaps reveal patterns and correlations in multidimensional genomic data. Bioinformatics 32(18):2847–2849
Gu ML, Wang YM, Zhou XX, Yao HP, Zheng S, Xiang Z et al (2016b) An inhibitor of the acetyltransferases CBP/p300 exerts antineoplastic effects on gastrointestinal stromal tumor cells. Oncol Rep 36(5):2763–2770
Guan JS, Haggarty SJ, Giacometti E, Dannenberg JH, Joseph N, Gao J et al (2009) HDAC2 negatively regulates memory formation and synaptic plasticity. Nature 459(7243):55–60
Guo J, Guo S, Lu S, Gong J, Wang L, Ding L et al (2023) The development of proximity labeling technology and its applications in mammals, plants, and microorganisms. Cell Commun Signal 21(1):269
Hai R, He L, Shu G, Yin G (2021) Characterization of histone deacetylase mechanisms in cancer development. Front Oncol 11:700947
Halasa M, Wawruszak A, Przybyszewska A, Jaruga A, Guz M, Kalafut J et al (2019) H3K18Ac as a marker of cancer progression and potential target of anti-cancer therapy. Cells 8(5)
Hanquier J, Gimeno T, Jeffers V, Sullivan WJ Jr (2020) Evaluating the GCN5b bromodomain as a novel therapeutic target against the parasite Toxoplasma gondii. Exp Parasitol 211:107868

Hansen JC, Connolly M, McDonald CJ, Pan A, Pryamkova A, Ray K et al (2018) The 10-nm chromatin fiber and its relationship to interphase chromosome organization. Biochem Soc Trans 46(1):67–76

Harris HL, Gu H, Olshansky M, Wang A, Farabella I, Eliaz Y et al (2023) Chromatin alternates between A and B compartments at kilobase scale for subgenic organization. Nat Commun 14(1):3303

Haynes SR, Dollard C, Winston F, Beck S, Trowsdale J, Dawid IB (1992) The bromodomain: a conserved sequence found in human, drosophila and yeast proteins. Nucleic Acids Res 20(10):2603

Heintzman ND, Hon GC, Hawkins RD, Kheradpour P, Stark A, Harp LF et al (2009) Histone modifications at human enhancers reflect global cell-type-specific gene expression. Nature 459(7243):108–112

Heinz S, Benner C, Spann N, Bertolino E, Lin YC, Laslo P et al (2010) Simple combinations of lineage-determining transcription factors prime cis-regulatory elements required for macrophage and B cell identities. Mol Cell 38(4):576–589

Heinz S, Romanoski CE, Benner C, Glass CK (2015) The selection and function of cell type-specific enhancers. Nat Rev Mol Cell Biol 16(3):144–154

Heist T, Fukaya T, Levine M (2019) Large distances separate coregulated genes in living Drosophila embryos. Proc Natl Acad Sci USA 116(30):15062–15067

Henninger JE, Oksuz O, Shrinivas K, Sagi I, LeRoy G, Zheng MM et al (2021) RNA-mediated feedback control of transcriptional condensates. Cell 184(1):207–25 e24

Henry RA, Kuo YM, Andrews AJ (2013) Differences in specificity and selectivity between CBP and p300 acetylation of histone H3 and H3/H4. Biochemistry 52(34):5746–5759

Hoffman MM, Buske OJ, Wang J, Weng Z, Bilmes JA, Noble WS (2012) Unsupervised pattern discovery in human chromatin structure through genomic segmentation. Nat Methods 9(5):473–476

Hrgovic I, Doll M, Kleemann J, Wang XF, Zoeller N, Pinter A et al (2016) The histone deacetylase inhibitor trichostatin a decreases lymphangiogenesis by inducing apoptosis and cell cycle arrest via p21-dependent pathways. BMC Cancer 16(1):763

Hudson BP, Martinez-Yamout MA, Dyson HJ, Wright PE (2000) Solution structure and acetyl-lysine binding activity of the GCN5 bromodomain. J Mol Biol 304(3):355–370

Hunt CR, Ramnarain D, Horikoshi N, Iyengar P, Pandita RK, Shay JW et al (2013) Histone modifications and DNA double-strand break repair after exposure to ionizing radiations. Radiat Res 179(4):383–392

Hyun K, Jeon J, Park K, Kim J (2017) Writing, erasing and reading histone lysine methylations. Exp Mol Med 49(4):e324

Imbriano C, Belluti S (2022) Histone Marks-dependent effect on alternative splicing: new perspectives for targeted splicing modulation in cancer? Int J Mol Sci 23(15)

Imhof A, Wolffe AP (1999) Purification and properties of the Xenopus Hat1 acetyltransferase: association with the 14-3-3 proteins in the oocyte nucleus. Biochemistry 38(40):13085–13093

Itai Y, Rappoport N, Shamir R (2023) Integration of gene expression and DNA methylation data across different experiments. Nucleic Acids Res 51(15):7762–7776

Jacobson RH, Ladurner AG, King DS, Tjian R (2000) Structure and function of a human TAFII250 double bromodomain module. Science 288(5470):1422–1425

Jambhekar A, Dhall A, Shi Y (2019) Roles and regulation of histone methylation in animal development. Nat Rev Mol Cell Biol 20(10):625–641

Jeffers V, Yang C, Huang S, Sullivan WJ Jr (2017) Bromodomains in protozoan parasites: evolution, function, and opportunities for drug development. Microbiol Mol Biol Rev 81(1)

Jenuwein T, Allis CD (2001) Translating the histone code. Science 293(5532):1074–1080

Jia Q, Chen S, Tan Y, Li Y, Tang F (2020) Oncogenic super-enhancer formation in tumorigenesis and its molecular mechanisms. Exp Mol Med 52(5):713–723

Jiao W, Chen Y, Song H, Li D, Mei H, Yang F et al (2018) HPSE enhancer RNA promotes cancer progression through driving chromatin looping and regulating hnRNPU/p300/EGR1/HPSE axis. Oncogene 37(20):2728–2745
Jin YH, Jeon EJ, Li QL, Lee YH, Choi JK, Kim WJ et al (2004) Transforming growth factor-beta stimulates p300-dependent RUNX3 acetylation, which inhibits ubiquitination-mediated degradation. J Biol Chem 279(28):29409–29417
Jin X, Tian S, Li P (2017) Histone acetyltransferase 1 promotes cell proliferation and induces cisplatin resistance in hepatocellular carcinoma. Oncol Res 25(6):939–946
Jing H, Lin H (2015) Sirtuins in epigenetic regulation. Chem Rev 115(6):2350–2375
Jo H, Shim K, Kim HU, Jung HS, Jeoung D (2023) HDAC2 as a target for developing anti-cancer drugs. Comput Struct Biotechnol J 21:2048–2057
Johnson CA, White DA, Lavender JS, O'Neill LP, Turner BM (2002) Human class I histone deacetylase complexes show enhanced catalytic activity in the presence of ATP and co-immunoprecipitate with the ATP-dependent chaperone protein Hsp70. J Biol Chem 277(11):9590–9597
Josling GA, Selvarajah SA, Petter M, Duffy MF (2012) The role of bromodomain proteins in regulating gene expression. Genes (Basel) 3(2):320–343
Josling GA, Petter M, Oehring SC, Gupta AP, Dietz O, Wilson DW et al (2015) A Plasmodium falciparum Bromodomain protein regulates invasion gene expression. Cell Host Microbe 17(6):741–751
Kadoch C, Copeland RA, Keilhack H (2016) PRC2 and SWI/SNF chromatin remodeling complexes in health and disease. Biochemistry 55(11):1600–1614
Kaji H, Chihara K (1990) Mechanism of generating neuropeptide family--selective regulation of gene transcription and processing enzyme. Nihon Rinsho 48(5):902–907
Kalashnikova EV, Revenko AS, Gemo AT, Andrews NP, Tepper CG, Zou JX et al (2010) ANCCA/ATAD2 overexpression identifies breast cancer patients with poor prognosis, acting to drive proliferation and survival of triple-negative cells through control of B-Myb and EZH2. Cancer Res 70(22):9402–9412
Kanno T, Kanno Y, Siegel RM, Jang MK, Lenardo MJ, Ozato K (2004) Selective recognition of acetylated histones by bromodomain proteins visualized in living cells. Mol Cell 13(1):33–43
Kennison JA, Tamkun JW (1992) Trans-regulation of homeotic genes in Drosophila. New Biol 4(2):91–96
Khan H, Tiwari P, Kaur A, Singh TG (2021) Sirtuin acetylation and deacetylation: a complex paradigm in neurodegenerative disease. Mol Neurobiol 58(8):3903–3917
Kim YJ, Xie P, Cao L, Zhang MQ, Kim TH (2018) Global transcriptional activity dynamics reveal functional enhancer RNAs. Genome Res 28(12):1799–1811
Kim S, Piquerez SJM, Ramirez-Prado JS, Mastorakis E, Veluchamy A, Latrasse D et al (2020a) GCN5 modulates salicylic acid homeostasis by regulating H3K14ac levels at the 5′ and 3′ ends of its target genes. Nucleic Acids Res 48(11):5953–5966
Kim JJ, Lee SY, Choi JH, Woo HG, Xhemalce B, Miller KM (2020b) PCAF-mediated histone acetylation promotes replication fork degradation by MRE11 and EXO1 in BRCA-deficient cells. Mol Cell 80(2):327–44 e8
Kim SL, Shin M, Jin BC, Seo S, Ha GW, Kim SW (2024) Acquired tumor necrosis factor-related apoptosis-inducing ligand (TRAIL) resistance of human colorectal cancer cells is linked to histone acetylation and is synergistically ameliorated by combination with HDAC inhibitors. Dig Dis Sci 69(9):3305–3317
Kleinsmith LJ, Allfrey VG, Mirsky AE (1966) Phosphorylation of nuclear protein early in the course of gene activation in lymphocytes. Science 154(3750):780–781
Klemm SL, Shipony Z, Greenleaf WJ (2019) Chromatin accessibility and the regulatory epigenome. Nat Rev Genet 20(4):207–220
Kondo Y (2009) Epigenetic cross-talk between DNA methylation and histone modifications in human cancers. Yonsei Med J 50(4):455–463

Konuma T, Zhou MM (2024) Distinct histone H3 lysine 27 modifications dictate different outcomes of gene transcription. J Mol Biol 436(7):168376
Kranz A, Anastassiadis K (2020) The role of SETD1A and SETD1B in development and disease. Biochim Biophys Acta Gene Regul Mech 1863(8):194578
Kristjansdottir K, Dziubek A, Kang HM, Kwak H (2020) Population-scale study of eRNA transcription reveals bipartite functional enhancer architecture. Nat Commun 11(1):5963
Krueger F, Andrews SR (2011) Bismark: a flexible aligner and methylation caller for Bisulfite-Seq applications. Bioinformatics 27(11):1571–1572
Kumar S, Kaur S, Seem K, Kumar S, Mohapatra T (2021) Understanding 3D genome organization and its effect on transcriptional gene regulation under environmental stress in plant: a chromatin perspective. Front Cell Dev Biol 9:774719
Langmead B, Salzberg SL (2012) Fast gapped-read alignment with Bowtie 2. Nat Methods 9(4):357–359
Laurberg P, Torring J, Weeke J (1985) A comparison of the effects of propylthiouracil and methimazol on circulating thyroid hormones and various measures of peripheral thyroid hormone effects in thyrotoxic patients. Acta Endocrinol 108(1):51–54
Lazarchuk P, Hernandez-Villanueva J, Pavlova MN, Federation A, MacCoss M, Sidorova JM (2020) Mutual balance of histone deacetylases 1 and 2 and the acetyl reader ATAD2 regulates the level of acetylation of histone H4 on nascent chromatin of human cells. Mol Cell Biol 40(9)
Lee JS, Shukla A, Schneider J, Swanson SK, Washburn MP, Florens L et al (2007) Histone crosstalk between H2B monoubiquitination and H3 methylation mediated by COMPASS. Cell 131(6):1084–1096
Lee JS, Smith E, Shilatifard A (2010) The language of histone crosstalk. Cell 142(5):682–685
Lee J, Mashima T, Kawata N, Yamamoto N, Morino S, Inaba S et al (2024a) Pharmacologic targeting of histone H3K27 acetylation/BRD4-dependent induction of ALDH1A3 for early-phase drug tolerance of gastric cancer. Cancer Res Commun 4(5):1307–1320
Lee SY, Lee SH, Choi NH, Kim JY, Kweon JH, Miller KM et al (2024b) PCAF promotes R-loop resolution via histone acetylation. Nucleic Acids Res 52(15):8643–8660
LeRoy G, Rickards B, Flint SJ (2008) The double bromodomain proteins Brd2 and Brd3 couple histone acetylation to transcription. Mol Cell 30(1):51–60
Lesluyes T, Johnson J, Machanick P, Bailey TL (2014) Differential motif enrichment analysis of paired ChIP-seq experiments. BMC Genomics 15(1):752
Li H, Durbin R (2009) Fast and accurate short read alignment with burrows-wheeler transform. Bioinformatics 25(14):1754–1760
Li Y, Seto E (2016) HDACs and HDAC inhibitors in cancer development and therapy. Cold Spring Harb Perspect Med 6(10)
Li W, Notani D, Ma Q, Tanasa B, Nunez E, Chen AY et al (2013) Functional roles of enhancer RNAs for oestrogen-dependent transcriptional activation. Nature 498(7455):516–520
Li Y, Wen H, Xi Y, Tanaka K, Wang H, Peng D et al (2014) AF9 YEATS domain links histone acetylation to DOT1L-mediated H3K79 methylation. Cell 159(3):558–571
Li Y, Sabari BR, Panchenko T, Wen H, Zhao D, Guan H et al (2016) Molecular coupling of histone crotonylation and active transcription by AF9 YEATS domain. Mol Cell 62(2):181–193
Li X, Su X, Liu R, Pan Y, Fang J, Cao L et al (2021a) HDAC inhibition potentiates anti-tumor activity of macrophages and enhances anti-PD-L1-mediated tumor suppression. Oncogene 40(10):1836–1850
Li M, Viswanadhapalli S, Santhamma B, Pratap UP, Luo Y, Liu J et al (2021b) LIFR inhibition enhances the therapeutic efficacy of HDAC inhibitors in triple negative breast cancer. Commun Biol 4(1):1235
Li Z, Zhang H, Cai C, Lin Z, Zhen Z, Chu J et al (2022) Histone acetyltransferase GCN5-mediated lysine acetylation modulates salt stress aadaption of Trichoderma. Appl Microbiol Biotechnol 106(8):3033–3049
Li X, Liu S, Li X, Li XD (2023) YEATS domains as novel epigenetic readers: structures, functions, and inhibitor development. ACS Chem Biol 18(4):994–1013

Lian B, Chen X, Shen K (2023) Inhibition of histone deacetylases attenuates tumor progression and improves immunotherapy in breast cancer. Front Immunol 14:1164514
Lian C, Zhang C, Tian P, Tan Q, Wei Y, Wang Z et al (2024) Epigenetic reader ZMYND11 noncanonical function restricts HNRNPA1-mediated stress granule formation and oncogenic activity. Signal Transduct Target Ther 9(1):258
Liang H, Yang C, Zeng R, Song Y, Wang J, Xiong W et al (2023) Targeting CBX3 with a dual BET/PLK1 inhibitor enhances the antitumor efficacy of CDK4/6 inhibitors in prostate cancer. Adv Sci (Weinh) 10(36):e2302368
Lieberman-Aiden E, van Berkum NL, Williams L, Imakaev M, Ragoczy T, Telling A et al (2009) Comprehensive mapping of long-range interactions reveals folding principles of the human genome. Science 326(5950):289–293
Liedtke V, Schroder C, Roggenbuck D, Weiss R, Stohwasser R, Schierack P et al (2021) LEDGF/p75 is required for an efficient DNA damage response. Int J Mol Sci 22(11)
Liu Y, Chen S, Wang S, Soares F, Fischer M, Meng F et al (2017) Transcriptional landscape of the human cell cycle. Proc Natl Acad Sci USA 114(13):3473–3478
Liu X, Salokas K, Weldatsadik RG, Gawriyski L, Varjosalo M (2020) Combined proximity labeling and affinity purification-mass spectrometry workflow for mapping and visualizing protein interaction networks. Nat Protoc 15(10):3182–3211
Lloyd JT, Glass KC (2018) Biological function and histone recognition of family IV bromodomain-containing proteins. J Cell Physiol 233(3):1877–1886
Long HK, Prescott SL, Wysocka J (2016) Ever-changing landscapes: transcriptional enhancers in development and evolution. Cell 167(5):1170–1187
Love IM, Sekaric P, Shi D, Grossman SR, Androphy EJ (2012) The histone acetyltransferase PCAF regulates p21 transcription through stress-induced acetylation of histone H3. Cell Cycle 11(13):2458–2466
Love MI, Huber W, Anders S (2014) Moderated estimation of fold change and dispersion for RNA-seq data with DESeq2. Genome Biol 15(12):550
Lu J, He X, Zhang L, Zhang R, Li W (2021) Acetylation in tumor immune evasion regulation. Front Pharmacol 12:771588
Ma M, Li M, Zhang C, Yang Z, Chen X, Lu P et al (2024) Discovery of a highly potent PROTAC degrader of p300/CBP proteins for the treatment of enzalutamide-resistant prostate cancer. J Med Chem
Maccallini C, Ammazzalorso A, De Filippis B, Fantacuzzi M, Giampietro L, Amoroso R (2022) HDAC inhibitors for the therapy of triple negative breast cancer. Pharmaceuticals (Basel) 15(6)
Maeshima K, Imai R, Tamura S, Nozaki T (2014) Chromatin as dynamic 10-nm fibers. Chromosoma 123(3):225–237
Mahara S, Prussing S, Smialkovska V, Krall S, Holliman S, Blum B et al (2024) Transient promoter interactions modulate developmental gene activation. Mol Cell
Mahat DB, Tippens ND, Martin-Rufino JD, Waterton SK, Fu J, Blatt SE et al (2024) Single-cell nascent RNA sequencing unveils coordinated global transcription. Nature 631(8019):216–223
Mandapati A, Lukong KE (2023) Triple negative breast cancer: approved treatment options and their mechanisms of action. J Cancer Res Clin Oncol 149(7):3701–3719
Marahrens Y, Loring J, Jaenisch R (1998) Role of the Xist gene in X chromosome choosing. Cell 92(5):657–664
Marino-Ramirez L, Kann MG, Shoemaker BA, Landsman D (2005) Histone structure and nucleosome stability. Expert Rev Proteomics 2(5):719–729
Markouli M, Strepkos D, Basdra EK, Papavassiliou AG, Piperi C (2021) Prominent role of histone modifications in the regulation of tumor metastasis. Int J Mol Sci 22(5)
Marmorstein R (2001) Structure of histone acetyltransferases. J Mol Biol 311(3):433–444
Marmorstein R, Trievel RC (2009) Histone modifying enzymes: structures, mechanisms, and specificities. Biochim Biophys Acta 1789(1):58–68
Marmorstein R, Zhou MM (2014) Writers and readers of histone acetylation: structure, mechanism, and inhibition. Cold Spring Harb Perspect Biol 6(7):a018762

Mayran A, Drouin J (2018) Pioneer transcription factors shape the epigenetic landscape. J Biol Chem 293(36):13795–13804

Meng J, Zhang HH, Zhou CX, Li C, Zhang F, Mei QB (2012) The histone deacetylase inhibitor trichostatin A induces cell cycle arrest and apoptosis in colorectal cancer cells via p53-dependent and -independent pathways. Oncol Rep 28(1):384–388

Mittal P, Roberts CWM (2020) The SWI/SNF complex in cancer – biology, biomarkers and therapy. Nat Rev Clin Oncol 17(7):435–448

Miziak P, Baran M, Borkiewicz L, Trombik T, Stepulak A (2024) Acetylation of histone H3 in cancer progression and prognosis. Int J Mol Sci 25(20)

Moore LD, Le T, Fan G (2013) DNA methylation and its basic function. Neuropsychopharmacology 38(1):23–38

Morrison O, Thakur J (2021) Molecular complexes at Euchromatin, heterochromatin and Centromeric chromatin. Int J Mol Sci 22(13)

Morse K, Bishop AL, Swerdlow S, Leslie JM, Unal E (2024) Swi/Snf chromatin remodeling regulates transcriptional interference and gene repression. Mol Cell 84(16):3080–97 e9

Mrakovcic M, Kleinheinz J, Frohlich LF (2019) p53 at the crossroads between different types of HDAC inhibitor-mediated cancer cell death. Int J Mol Sci 20(10)

Mukohara T, Park YH, Sommerhalder D, Yonemori K, Hamilton E, Kim SB et al (2024) Author correction: inhibition of lysine acetyltransferase KAT6 in ER(+)HER2(−) metastatic breast cancer: a phase 1 trial. Nat Med 30(8):2371

Nagarajan S, Benito E, Fischer A, Johnsen SA (2015) H4K12ac is regulated by estrogen receptor-alpha and is associated with BRD4 function and inducible transcription. Oncotarget 6(9):7305–7317

Nakamura K, Saredi G, Becker JR, Foster BM, Nguyen NV, Beyer TE et al (2019) H4K20me0 recognition by BRCA1-BARD1 directs homologous recombination to sister chromatids. Nat Cell Biol 21(3):311–318

Nakato R, Sakata T (2021) Methods for ChIP-seq analysis: a practical workflow and advanced applications. Methods 187:44–53

Nasmyth K (2001) Disseminating the genome: joining, resolving, and separating sister chromatids during mitosis and meiosis. Annu Rev Genet 35:673–745

Nassar LR, Barber GP, Benet-Pages A, Casper J, Clawson H, Diekhans M et al (2023) The UCSC genome browser database: 2023 update. Nucleic Acids Res 51(D1):D1188–D1D95

Nava M, Dutta P, Zemke NR, Farias-Eisner R, Vadgama JV, Wu Y (2019) Transcriptomic and ChIP-sequence interrogation of EGFR signaling in HER2+ breast cancer cells reveals a dynamic chromatin landscape and S100 genes as targets. BMC Med Genet 12(1):32

Nicetto D, Zaret KS (2019) Role of H3K9me3 heterochromatin in cell identity establishment and maintenance. Curr Opin Genet Dev 55:1–10

Nichols MH, Corces VG (2021) Principles of 3D compartmentalization of the human genome. Cell Rep 35(13):109330

Nikolayeva O, Robinson MD (2014) edgeR for differential RNA-seq and ChIP-seq analysis: an application to stem cell biology. Methods Mol Biol 1150:45–79

Nunes SP, Morales L, Rubio C, Munera-Maravilla E, Lodewijk I, Suarez-Cabrera C et al (2024) Modulation of tumor microenvironment by targeting histone acetylation in bladder cancer. Cell Death Discov 10(1):1

Oh S, Shao J, Mitra J, Xiong F, D'Antonio M, Wang R et al (2021) Enhancer release and retargeting activates disease-susceptibility genes. Nature 595(7869):735–740

Pan MR, Hsu MC, Chen LT, Hung WC (2018) Orchestration of H3K27 methylation: mechanisms and therapeutic implication. Cell Mol Life Sci 75(2):209–223

Panigrahi A, O'Malley BW (2021) Mechanisms of enhancer action: the known and the unknown. Genome Biol 22(1):108

Park SY, Kim JS (2020) A short guide to histone deacetylases including recent progress on class II enzymes. Exp Mol Med 52(2):204–212

Parthun MR (2012) Histone acetyltransferase 1: more than just an enzyme? Biochim Biophys Acta 1819(3–4):256–263
Parthun MR, Widom J, Gottschling DE (1996) The major cytoplasmic histone acetyltransferase in yeast: links to chromatin replication and histone metabolism. Cell 87(1):85–94
Passaro E, Papulino C, Chianese U, Toraldo A, Congi R, Del Gaudio N et al (2021) HDAC6 inhibition extinguishes autophagy in cancer: recent insights. Cancers (Basel) 13(24)
Pecharroman I, Sole L, Alvarez-Villanueva D, Lobo-Jarne T, Alonso-Maranon J, Bertran J et al (2023) IkappaB kinase-alpha coordinates BRD4 and JAK/STAT signaling to subvert DNA damage-based anticancer therapy. EMBO J 42(21):e114719
Piao Y, Xu W, Park KH, Ryu KH, Xiang R (2021) Comprehensive evaluation of differential methylation analysis methods for bisulfite sequencing data. Int J Environ Res Public Health 18(15)
Pogo BG, Allfrey VG, Mirsky AE (1966) RNA synthesis and histone acetylation during the course of gene activation in lymphocytes. Proc Natl Acad Sci USA 55(4):805–812
Pogo BG, Pogo AO, Allfrey VG, Mirsky AE (1968) Changing patterns of histone acetylation and RNA synthesis in regeneration of the liver. Proc Natl Acad Sci USA 59(4):1337–1344
Pollex T, Rabinowitz A, Gambetta MC, Marco-Ferreres R, Viales RR, Jankowski A et al (2024) Enhancer-promoter interactions become more instructive in the transition from cell-fate specification to tissue differentiation. Nat Genet 56(4):686–696
Popay TM, Dixon JR (2022) Coming full circle: on the origin and evolution of the looping model for enhancer-promoter communication. J Biol Chem 298(8):102117
Pradeepa MM (2017) Causal role of histone acetylations in enhancer function. Transcription 8(1):40–47
Price AJ, Manjegowda MC, Kain J, Anandh S, Bochkis IM (2020) Hdac3, Setdb1, and Kap1 mark H3K9me3/H3K14ac bivalent regions in young and aged liver. Aging Cell 19(2):e13092
Prothero KE, Stahl JM, Carrel L (2009) Dosage compensation and gene expression on the mammalian X chromosome: one plus one does not always equal two. Chromosom Res 17(5):637–648
Qiu Y, Liu L, Zhao C, Han C, Li F, Zhang J et al (2012) Combinatorial readout of unmodified H3R2 and acetylated H3K14 by the tandem PHD finger of MOZ reveals a regulatory mechanism for HOXA9 transcription. Genes Dev 26(12):1376–1391
Quaas CE, Lin B, Long DT (2022) Transcription suppression is mediated by the HDAC1-Sin3 complex in Xenopus nucleoplasmic extract. J Biol Chem 298(11):102578
Rahhal R, Seto E (2019) Emerging roles of histone modifications and HDACs in RNA splicing. Nucleic Acids Res 47(10):4911–4926
Rajderkar S, Barozzi I, Zhu Y, Hu R, Zhang Y, Li B et al (2023) Topologically associating domain boundaries are required for normal genome function. Commun Biol 6(1):435
Ramos L, Truong S, Zhai B, Joshi J, Ghaidi F, Lizardo MM et al (2023) A bifunctional PARP-HDAC inhibitor with activity in Ewing sarcoma. Clin Cancer Res 29(17):3541–3553
Rauluseviciute I, Drablos F, Rye MB (2019) DNA methylation data by sequencing: experimental approaches and recommendations for tools and pipelines for data analysis. Clin Epigenetics 11(1):193
Richart L, Real FX, Sanchez-Arevalo Lobo VJ (2016) c-MYC partners with BPTF in human cancer. Mol Cell Oncol 3(3):e1152346
Risca VI, Denny SK, Straight AF, Greenleaf WJ (2017) Variable chromatin structure revealed by in situ spatially correlated DNA cleavage mapping. Nature 541(7636):237–241
Romani M, Daga A, Forlani A, Pistillo MP, Banelli B (2019) Targeting of histone demethylases KDM5A and KDM6B inhibits the proliferation of Temozolomide-resistant glioblastoma cells. Cancers (Basel) 11(6)
Rose NR, Klose RJ (2014) Understanding the relationship between DNA methylation and histone lysine methylation. Biochim Biophys Acta 1839(12):1362–1372
Rothbart SB, Strahl BD (2014) Interpreting the language of histone and DNA modifications. Biochim Biophys Acta 1839(8):627–643
Roux KJ, Kim DI, Burke B, May DG (2018) BioID: a screen for protein-protein interactions. Curr Protoc Protein Sci 91:19 23 1-19 23 15

Sabari BR, Dall'Agnese A, Boija A, Klein IA, Coffey EL, Shrinivas K et al (2018) Coactivator condensation at super-enhancers links phase separation and gene control. Science 361(6400)
Sabari BR, Dall'Agnese A, Young RA (2020) Biomolecular condensates in the nucleus. Trends Biochem Sci 45(11):961–977
Sadakierska-Chudy A, Filip M (2015) A comprehensive view of the epigenetic landscape. Part II: histone post-translational modification, nucleosome level, and chromatin regulation by ncRNAs. Neurotox Res 27(2):172–197
Saha D, Animireddy S, Bartholomew B (2024) The SWI/SNF ATP-dependent chromatin remodeling complex in cell lineage priming and early development. Biochem Soc Trans 52(2):603–616
Sahu RK, Singh S, Tomar RS (2020) The mechanisms of action of chromatin remodelers and implications in development and disease. Biochem Pharmacol 180:114200
Sahu B, Hartonen T, Pihlajamaa P, Wei B, Dave K, Zhu F et al (2022) Sequence determinants of human gene regulatory elements. Nat Genet 54(3):283–294
Sanborn AL, Rao SS, Huang SC, Durand NC, Huntley MH, Jewett AI et al (2015) Chromatin extrusion explains key features of loop and domain formation in wild-type and engineered genomes. Proc Natl Acad Sci USA 112(47):E6456–E6465
Scallan E, Griffin PM, Angulo FJ, Tauxe RV, Hoekstra RM (2011) Foodborne illness acquired in the United States--unspecified agents. Emerg Infect Dis 17(1):16–22
Schlesinger Y, Straussman R, Keshet I, Farkash S, Hecht M, Zimmerman J et al (2007) Polycomb-mediated methylation on Lys27 of histone H3 pre-marks genes for de novo methylation in cancer. Nat Genet 39(2):232–236
Sears RM, May DG, Roux KJ (2019) BioID as a tool for protein-proximity labeling in living cells. Methods Mol Biol 2012:299–313
Seto E, Yoshida M (2014) Erasers of histone acetylation: the histone deacetylase enzymes. Cold Spring Harb Perspect Biol 6(4):a018713
Shang P, Meng F, Liu Y, Chen X (2015) Overexpression of ANCCA/ATAD2 in endometrial carcinoma and its correlation with tumor progression and poor prognosis. Tumour Biol 36(6):4479–4485
Shanle EK, Andrews FH, Meriesh H, McDaniel SL, Dronamraju R, DiFiore JV et al (2015) Association of Taf14 with acetylated histone H3 directs gene transcription and the DNA damage response. Genes Dev 29(17):1795–1800
Shannon P, Markiel A, Ozier O, Baliga NS, Wang JT, Ramage D et al (2003) Cytoscape: a software environment for integrated models of biomolecular interaction networks. Genome Res 13(11):2498–2504
Sharda A, Humphrey TC (2022) The role of histone H3K36me3 writers, readers and erasers in maintaining genome stability. DNA Repair (Amst) 119:103407
Sharma S, Chung CY, Uryu S, Petrovic J, Cao J, Rickard A et al (2023) Discovery of a highly potent, selective, orally bioavailable inhibitor of KAT6A/B histone acetyltransferases with efficacy against KAT6A-high ER+ breast cancer. Cell Chem Biol 30(10):1191–210 e20
Shen X, Liu Y, Hsu YJ, Fujiwara Y, Kim J, Mao X et al (2008) EZH1 mediates methylation on histone H3 lysine 27 and complements EZH2 in maintaining stem cell identity and executing pluripotency. Mol Cell 32(4):491–502
Shen L, Shao NY, Liu X, Maze I, Feng J, Nestler EJ (2013) diffReps: detecting differential chromatin modification sites from ChIP-seq data with biological replicates. PLoS One 8(6):e65598
Shoaib M, Chen Q, Shi X, Nair N, Prasanna C, Yang R et al (2021) Histone H4 lysine 20 mono-methylation directly facilitates chromatin openness and promotes transcription of housekeeping genes. Nat Commun 12(1):4800
Shvedunova M, Akhtar A (2022) Modulation of cellular processes by histone and non-histone protein acetylation. Nat Rev Mol Cell Biol 23(5):329–349
Siegel D, Hussein M, Belani C, Robert F, Galanis E, Richon VM et al (2009) Vorinostat in solid and hematologic malignancies. J Hematol Oncol 2:31

Singh AK, Phillips M, Alkrimi S, Tonelli M, Boyson SP, Malone KL et al (2022) Structural insights into acetylated histone ligand recognition by the BDP1 bromodomain of Plasmodium falciparum. Int J Biol Macromol 223(Pt A):316–326
Smoots SG, Schreiber AR, Jackson MM, Bagby SM, Dominguez ATA, Dus ED et al (2024) Overcoming doxorubicin resistance in triple-negative breast cancer using the class I-targeting HDAC inhibitor bocodepsin/OKI-179 to promote apoptosis. Breast Cancer Res 26(1):35
Song Q, Hou Y, Zhang Y, Liu J, Wang Y, Fu J et al (2022) Integrated multi-omics approach revealed cellular senescence landscape. Nucleic Acids Res 50(19):10947–10963
Sparmann A, van Lohuizen M (2006) Polycomb silencers control cell fate, development and cancer. Nat Rev Cancer 6(11):846–856
Statello L, Guo CJ, Chen LL, Huarte M (2021) Gene regulation by long non-coding RNAs and its biological functions. Nat Rev Mol Cell Biol 22(2):96–118
Struhl K (1998) Histone acetylation and transcriptional regulatory mechanisms. Genes Dev 12(5):599–606
Szabo Q, Bantignies F, Cavalli G (2019) Principles of genome folding into topologically associating domains. Sci Adv 5(4):eaaw1668
Tallant C, Bamborough P, Chung CW, Gamo FJ, Kirkpatrick R, Larminie C et al (2021) Expanding Bromodomain targeting into neglected parasitic diseases. ACS Infect Dis 7(11):2953–2958
Tang Z, Luo OJ, Li X, Zheng M, Zhu JJ, Szalaj P et al (2015) CTCF-mediated human 3D genome architecture reveals chromatin topology for transcription. Cell 163(7):1611–1627
Tang B, Cheng X, Xi Y, Chen Z, Zhou Y, Jin VX (2017) Advances in genomic profiling and analysis of 3D chromatin structure and interaction. Genes (Basel) 8(9)
Tang F, Yang Z, Tan Y, Li Y (2020) Super-enhancer function and its application in cancer targeted therapy. NPJ Precis Oncol 4:2
Tang SC, Vijayakumar U, Zhang Y, Fullwood MJ (2022) Super-enhancers, phase-separated condensates, and 3D genome organization in cancer. Cancers (Basel) 14(12)
Tate CR, Rhodes LV, Segar HC, Driver JL, Pounder FN, Burow ME et al (2012) Targeting triple-negative breast cancer cells with the histone deacetylase inhibitor panobinostat. Breast Cancer Res 14(3):R79
Taverna SD, Li H, Ruthenburg AJ, Allis CD, Patel DJ (2007) How chromatin-binding modules interpret histone modifications: lessons from professional pocket pickers. Nat Struct Mol Biol 14(11):1025–1040
Terranova-Barberio M, Thomas S, Ali N, Pawlowska N, Park J, Krings G et al (2017) HDAC inhibition potentiates immunotherapy in triple negative breast cancer. Oncotarget 8(69):114156–114172
The Allis code (2023) Nat Genet 55(4):521
Thorne AW, Kmiciek D, Mitchelson K, Sautiere P, Crane-Robinson C (1990) Patterns of histone acetylation. Eur J Biochem 193(3):701–713
Thorvaldsdottir H, Robinson JT, Mesirov JP (2013) Integrative genomics viewer (IGV): high-performance genomics data visualization and exploration. Brief Bioinform 14(2):178–192
Tsai WW, Wang Z, Yiu TT, Akdemir KC, Xia W, Winter S et al (2010) TRIM24 links a non-canonical histone signature to breast cancer. Nature 468(7326):927–932
Tse C, Sera T, Wolffe AP, Hansen JC (1998) Disruption of higher-order folding by core histone acetylation dramatically enhances transcription of nucleosomal arrays by RNA polymerase III. Mol Cell Biol 18(8):4629–4638
Tsilimigras DI, Ntanasis-Stathopoulos I, Moris D, Spartalis E, Pawlik TM (2018) Histone deacetylase inhibitors in hepatocellular carcinoma: a therapeutic perspective. Surg Oncol 27(4):611–618
Turner BM (2000) Histone acetylation and an epigenetic code. BioEssays 22(9):836–845
Turner BM (2002) Cellular memory and the histone code. Cell 111(3):285–291
Umehara T, Nakamura Y, Wakamori M, Ozato K, Yokoyama S, Padmanabhan B (2010) Structural implications for K5/K12-di-acetylated histone H4 recognition by the second bromodomain of BRD2. FEBS Lett 584(18):3901–3908

Vacik T, Ladinovic D, Raska I (2018) KDM2A/B lysine demethylases and their alternative isoforms in development and disease. Nucleus 9(1):431–441

Valdez BC, Tsimberidou AM, Yuan B, Nieto Y, Baysal MA, Chakraborty A et al (2024) Synergistic cytotoxicity of histone deacetylase and poly-ADP ribose polymerase inhibitors and decitabine in pancreatic cancer cells: implications for novel therapy. Oncotarget 15:361–373

van Ingen H, van Schaik FM, Wienk H, Ballering J, Rehmann H, Dechesne AC et al (2008) Structural insight into the recognition of the H3K4me3 mark by the TFIID subunit TAF3. Structure 16(8):1245–1256

Van HT, Santos MA (2018) Histone modifications and the DNA double-strand break response. Cell Cycle 17(21–22):2399–2410

Vlaming H, van Leeuwen F (2016) The upstreams and downstreams of H3K79 methylation by DOT1L. Chromosoma 125(4):593–605

Waddell A, Mahmud I, Ding H, Huo Z, Liao D (2021) Pharmacological inhibition of CBP/p300 blocks estrogen receptor alpha (ERalpha) function through suppressing enhancer H3K27 acetylation in luminal breast cancer. Cancers (Basel) 13(11)

Wan WN, Zhang YX, Wang XM, Liu YJ, Zhang YQ, Que YH et al (2014) ATAD2 is highly expressed in ovarian carcinomas and indicates poor prognosis. Asian Pac J Cancer Prev 15(6):2777–2783

Wan L, Wen H, Li Y, Lyu J, Xi Y, Hoshii T et al (2017) ENL links histone acetylation to oncogenic gene expression in acute myeloid leukaemia. Nature 543(7644):265–269

Wang F, Feng X, He Q, Li H, Li H (2023) The Saccharomyces cerevisiae Yta7 ATPase hexamer contains a unique bromodomain tier that functions in nucleosome disassembly. J Biol Chem 299(2):102852

Wang X, Wu X, Hong N, Jin W (2024) Progress in single-cell multimodal sequencing and multi-omics data integration. Biophys Rev 16(1):13–28

Whyte WA, Orlando DA, Hnisz D, Abraham BJ, Lin CY, Kagey MH et al (2013) Master transcription factors and mediator establish super-enhancers at key cell identity genes. Cell 153(2):307–319

Winter S, Simboeck E, Fischle W, Zupkovitz G, Dohnal I, Mechtler K et al (2008) 14-3-3 proteins recognize a histone code at histone H3 and are required for transcriptional activation. EMBO J 27(1):88–99

Wu ZL, Zheng SS, Li ZM, Qiao YY, Aau MY, Yu Q (2010) Polycomb protein EZH2 regulates E2F1-dependent apoptosis through epigenetically modulating Bim expression. Cell Death Differ 17(5):801–810

Wu D, Qiu Y, Jiao Y, Qiu Z, Liu D (2020) Small molecules targeting HATs, HDACs, and BRDs in cancer therapy. Front Oncol 10:560487

Xi Y, Li W (2009) BSMAP: whole genome bisulfite sequence MAPping program. BMC Bioinform 10:232

Xi Q, Wang Z, Zaromytidou AI, Zhang XH, Chow-Tsang LF, Liu JX et al (2011) A poised chromatin platform for TGF-beta access to master regulators. Cell 147(7):1511–1524

Xiao CK, Ren Y, Chen Q, Yang Y, Tang L, Xu L et al (2024) H4K20me3, H3K4me2 and H3K9me2 mediate the effect of ER on prognosis in breast cancer. Epigenetics 19(1):2343593

Xie G, Lee JE, Senft AD, Park YK, Jang Y, Chakraborty S et al (2023) MLL3/MLL4 methyltransferase activities control early embryonic development and embryonic stem cell differentiation in a lineage-selective manner. Nat Genet 55(4):693–705

Xu P, Xiong W, Lin Y, Fan L, Pan H, Li Y (2021) Histone deacetylase 2 knockout suppresses immune escape of triple-negative breast cancer cells via downregulating PD-L1 expression. Cell Death Dis 12(8):779

Xu M, Hou Y, Li N, Yu W, Chen L (2024) Targeting histone deacetylases in head and neck squamous cell carcinoma: molecular mechanisms and therapeutic targets. J Transl Med 22(1):418

Xue L, Hou J, Wang Q, Yao L, Xu S, Ge D (2014) RNAi screening identifies HAT1 as a potential drug target in esophageal squamous cell carcinoma. Int J Clin Exp Pathol 7(7):3898–3907

Yamaguchi H, Hsu JM, Yang WH, Hung MC (2022) Mechanisms regulating PD-L1 expression in cancers and associated opportunities for novel small-molecule therapeutics. Nat Rev Clin Oncol 19(5):287–305
Yang JH, Hansen AS (2024) Enhancer selectivity in space and time: from enhancer-promoter interactions to promoter activation. Nat Rev Mol Cell Biol 25(7):574–591
Yang Q, Zhao J, Zhang W, Chen D, Wang Y (2019) Aberrant alternative splicing in breast cancer. J Mol Cell Biol 11(10):920–929
Yang J, Song C, Zhan X (2022a) The role of protein acetylation in carcinogenesis and targeted drug discovery. Front Endocrinol (Lausanne) 13:972312
Yang Y, Zhang M, Wang Y (2022b) The roles of histone modifications in tumorigenesis and associated inhibitors in cancer therapy. J Natl Cancer Cent 2(4):277–290
Yin L, Liu Y, Peng Y, Peng Y, Yu X, Gao Y et al (2018) PARP inhibitor veliparib and HDAC inhibitor SAHA synergistically co-target the UHRF1/BRCA1 DNA damage repair complex in prostate cancer cells. J Exp Clin Cancer Res 37(1):153
Young DW, Hassan MQ, Pratap J, Galindo M, Zaidi SK, Lee SH et al (2007) Mitotic occupancy and lineage-specific transcriptional control of rRNA genes by Runx2. Nature 445(7126):442–446
Yu G, Wang LG, He QY (2015) ChIPseeker: an R/Bioconductor package for ChIP peak annotation, comparison and visualization. Bioinformatics 31(14):2382–2383
Yuan J, Pu M, Zhang Z, Lou Z (2009) Histone H3-K56 acetylation is important for genomic stability in mammals. Cell Cycle 8(11):1747–1753
Zabidi MA, Stark A (2016) Regulatory enhancer-Core-promoter communication via transcription factors and cofactors. Trends Genet 32(12):801–814
Zaidi SK, Young DW, Montecino M, Lian JB, Stein JL, van Wijnen AJ et al (2010a) Architectural epigenetics: mitotic retention of mammalian transcriptional regulatory information. Mol Cell Biol 30(20):4758–4766
Zaidi SK, Young DW, Montecino MA, Lian JB, van Wijnen AJ, Stein JL et al (2010b) Mitotic bookmarking of genes: a novel dimension to epigenetic control. Nat Rev Genet 11(8):583–589
Zaidi SK, Young DW, Montecino M, van Wijnen AJ, Stein JL, Lian JB et al (2011) Bookmarking the genome: maintenance of epigenetic information. J Biol Chem 286(21):18355–18361
Zekar L, Sharman T (2024) Plasmodium falciparum Malaria. StatPearls, Treasure Island (FL)
Zeng L, Zhang Q, Li S, Plotnikov AN, Walsh MJ, Zhou MM (2010) Mechanism and regulation of acetylated histone binding by the tandem PHD finger of DPF3b. Nature 466(7303):258–262
Zentner GE, Tesar PJ, Scacheri PC (2011) Epigenetic signatures distinguish multiple classes of enhancers with distinct cellular functions. Genome Res 21(8):1273–1283
Zhang Y, Liu T, Meyer CA, Eeckhoute J, Johnson DS, Bernstein BE et al (2008) Model-based analysis of ChIP-Seq (MACS). Genome Biol 9(9):R137
Zhang H, Brown RL, Wei Y, Zhao P, Liu S, Liu X et al (2019) CD44 splice isoform switching determines breast cancer stem cell state. Genes Dev 33(3–4):166–179
Zhang Y, Qian J, Gu C, Yang Y (2021) Alternative splicing and cancer: a systematic review. Signal Transduct Target Ther 6(1):78
Zhang QQ, Zhang WJ, Chang S (2023) HDAC6 inhibition: a significant potential regulator and therapeutic option to translate into clinical practice in renal transplantation. Front Immunol 14:1168848
Zhao D, Li Y, Xiong X, Chen Z, Li H (2017) YEATS domain-a histone acylation reader in health and disease. J Mol Biol 429(13):1994–2002
Zhou L, Yu CW (2024) Epigenetic modulations in triple-negative breast cancer: therapeutic implications for tumor microenvironment. Pharmacol Res 204:107205
Zhou HL, Luo G, Wise JA, Lou H (2014) Regulation of alternative splicing by local histone modifications: potential roles for RNA-guided mechanisms. Nucleic Acids Res 42(2):701–713
Zhu X, Asa SL, Ezzat S (2009) Histone-acetylated control of fibroblast growth factor receptor 2 intron 2 polymorphisms and isoform splicing in breast cancer. Mol Endocrinol 23(9):1397–1405
Zhu LJ, Gazin C, Lawson ND, Pages H, Lin SM, Lapointe DS et al (2010) ChIPpeakAnno: a Bioconductor package to annotate ChIP-seq and ChIP-chip data. BMC Bioinform 11:237

Zhu W, Zhang X, Yu M, Zhang Y, Li S, Yu C (2022) Profiles of acetylation regulation genes contribute to malignant progression and have a clinical prognostic impact on liver cancer. Dis Markers 2022:1724301

Zhuang J, Huo Q, Yang F, Xie N (2020) Perspectives on the role of histone modification in breast cancer progression and the advanced technological tools to study epigenetic determinants of metastasis. Front Genet 11:603552

Ziemka-Nalecz M, Jaworska J, Sypecka J, Zalewska T (2018) Histone deacetylase inhibitors: a therapeutic key in neurological disorders? J Neuropathol Exp Neurol 77(10):855–870

Zippo A, Serafini R, Rocchigiani M, Pennacchini S, Krepelova A, Oliviero S (2009) Histone crosstalk between H3S10ph and H4K16ac generates a histone code that mediates transcription elongation. Cell 138(6):1122–1136

Part II
Acetylation of Cytoskeleton-Associated Proteins

Chapter 3
Acetylation and Deacetylation of Cytoskeleton-Associated Proteins

Dale D. Tang

Abstract The cytoskeleton of eukaryotic cells undergoes a reorganization in response to intracellular and extracellular cues, which plays an essential role in orchestrating various cell functions including migration, development, differentiation, tissue homeostasis, contractility, proliferation, gene expression, cancer cell invasion, and airway/vascular remodeling. Acetylation occurs on the cytoskeletal components, such as microtubules, actin, and vimentin, which regulate cellular functions. Moreover, remodeling of the cytoskeleton is regulated by acetylation and deacetylation of regulatory proteins, including adapter proteins and protein kinases. Therefore, protein acetylation and deacetylation are critical mechanisms for cytoskeletal reorganization in response to changes of intracellular and extracellular environments.

The cytoskeleton of eukaryotic cells undergoes remodeling that plays a critical role in regulating the functions of the cells and contributes to the pathogenesis of diseases. More importantly, recent novel findings suggest that cytoskeletal reorganization is modulated by protein acetylation, which influences disease progression such as cancer cell migration and tissue remodeling. The major goal of this chapter is to summarize our current understanding of how protein acetylation affects the reorganization of the cytoskeleton, cell functions, and disease progression. Moreover, I will discuss the translational potential of this research area.

Nε-lysine acetylation on histone was found six decades ago (Allfrey et al. 1964). Subsequently, acetylation was discovered in high-mobility group (HMG) protein, tubulin, and p53 (Gu and Roeder 1997; L'Hernault and Rosenbaum 1983; Sterner et al. 1979). Since the beginning of the twenty-first century, thousands of acetylated proteins have been discovered (Narita et al. 2019). Lysine is a positively charged amino acid, which may facilitate the interaction of the charged proteins with other

D. D. Tang (✉)
Department of Molecular and Cellular Physiology, Albany Medical College, New York, USA
e-mail: tangd@amc.edu

M. Halasa, A. Wawruszak (eds.), *Histone and Non-Histone Reversible Acetylation in Development, Aging and Disease*, Results and Problems in Cell Differentiation 75,
https://doi.org/10.1007/978-3-031-91459-1_3

molecules, including DNA and proteins. Lysine acetylation results in charge-neutralization (i.e., no positive or negative charge) domains, which may alter the association of acetylated proteins with other molecules and, thus, cell functions (Narita et al. 2019; Li et al. 2014). Protein acetylation is catalyzed by histone acetyltransferases (HATs), while protein deacetylation is mediated by histone deacetylases (HDACs). Because many acetylated proteins are non-histone proteins, HATs and HDACs are also known as lysine acetyltransferases (KATs) and lysine deacetylases (KDACs), respectively.

Moreover, N-α-terminal (Nt) acetylation is a process by which an acetyl group (CH3CO) is covalently attached to the free α-amino group (NH3+) at the N-terminal end of a protein. Nt-acetylation has been implicated in the stability, folding characteristics, and binding properties of the acetylated protein. Nt-acetylation of protein is catalyzed by N-terminal acetyltransferases (NATs). Currently, no enzyme for Nt-deacetylation has been found (Ree et al. 2018; Drazic et al. 2018).

The cytoskeleton consists of actin filaments, intermediate filaments, and microtubules, which maintain cell integrity, shape, structure, and localization of organelles/nuclei. Moreover, the cytoskeleton undergoes the reorganization in response to changes in their environments. Acetylation may occur at cytoskeletal components (e.g., tubulin, actin, and vimentin), which regulates the reorganization of the cytoskeleton. Furthermore, acetylation may occur in certain proteins (e.g., adapter proteins and kinases), which subsequently modulates the remodeling of the cytoskeleton that plays an important role in regulating cell migration, development, differentiation, tissue homeostasis, contractility, proliferation, gene expression, cancer cell invasion, and airway/vascular remodeling (Hu et al. 2005; Cleary et al. 2014; Anfinogenova et al. 2007; Chen and Tang 2014; Wang et al. 2016; Wang 2014; Allen et al. 2011; Tang 2015, 2018; Tang and Gerlach 2017; Gunst and Zhang 2008; Saphirstein et al. 2015).

3.1 Role of Microtubule Acetylation in Trafficking and Migration

Microtubules are hollow and polarized tubes approximately 25 nm in diameter. They undergo dynamic assembly and disassembly in response to biochemical and mechanical cues within or around the cells to alter their spatial distribution and shape. These changes eventually influence cell morphology and a wide spectrum of cell functions including motility, trafficking, and mitosis (Janke and Montagnac 2017).

The properties of microtubules are regulated by post-translational modifications (PTMs) of their tubulin subunits. Acetylation of microtubules occurs at Lys-40 (K40) of α-tubulin (LeDizet and Piperno 1987), which is exposed to the internal surface of microtubules. In contrast, other PTMs including tyrosination, detyrosination, glutamylation, and glycosylation take place on the surface of microtubules (Song and Brady 2015). Although K40 acetylation was discovered more than three decades ago, its impact on the structure and functions of microtubules is still under investigation.

3.1.1 *Regulation of α-Tubulin by HATs and HDACs*

Although acetylation of α-tubulin was first discovered in more than 40 years ago (L'Hernault and Rosenbaum 1983), it took nearly 30 years for scientists to identify the enzyme that catalyzes α-tubulin acetylation. In 2010, MEC-17 (a protein related to the Gcn5 histone acetyltransferases) was identified as α-tubulin acetyltransferase (αTAT) (Akella et al. 2010). Moreover, biochemical studies have suggested that K40 acetylation only occurs on polymerized microtubules. Consistent with this finding, the catalytic activity of αTAT is much higher toward polymerized microtubules as compared to its unpolymerized state (Janke and Montagnac 2017). It is believed that αTAT1 enters the microtubule lumen from the open ends and cracks, which subsequently cuts the lattices, and acetylates K40 (Janke and Montagnac 2017; Coombes et al. 2016) (Fig. 3.1).

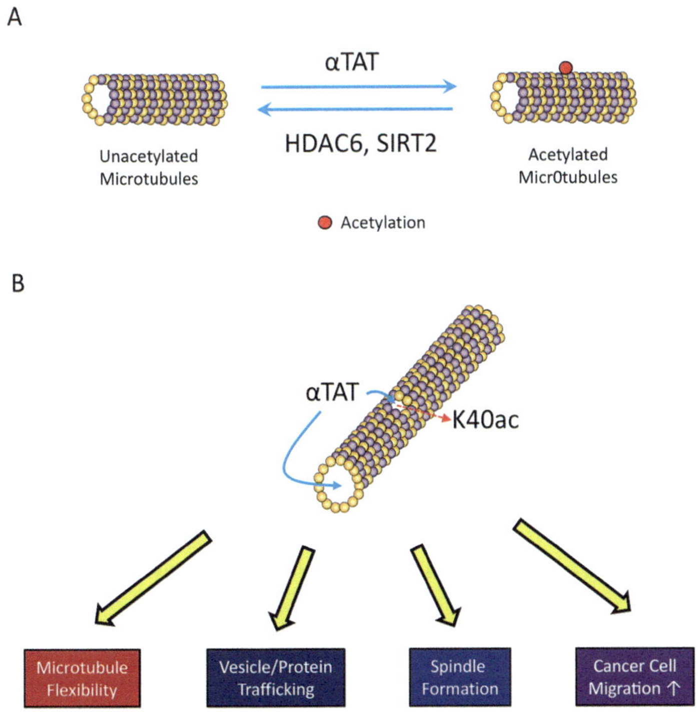

Fig. 3.1 Microtubules and acetylation. (**a**) α-Tubulin acetyltransferase 1 (αTAT1) catalyzes α-tubulin acetylation while histone deacetylase 6 (HDAC6) and sirtuin type 2 deacetylase (SIRT2) promote tubulin deacetylation. (**b**) αTAT1 enters the microtubule lumen from the open ends and cracks, which subsequently cuts the lattices, and acetylates K40 at the luminal surface. Potential roles of acetylated microtubules include: (1) increasing its flexibility and resisting mechanical stress; (2) promoting vesicle and protein trafficking; (3) facilitating spindle formation during mitosis; (4) enhancing cancer cell migration and invasion

What enzymes deacetylase tubulin? Histone deacetylase 6 (HDAC6), also known as KADC6, and sirtuin type 2 deacetylase (SIRT2) are the best two candidates that induce tubulin deacetylation (North et al. 2003; Hubbert et al. 2002). HDAC6 is localized in the cytoplasm and interacts with microtubules via a microtubule motor complex. The overexpression of HDAC6 in NIH3T3 cells leads to a global deacetylation of α-tubulin. In contrast, knockdown of HDAC6 in cells or treatment with HDAC6 inhibitors increases α-tubulin acetylation. In vitro biochemical experiments showed that purified HDAC6 deacetylates α-tubulin in assembled microtubules (Hubbert et al. 2002). Similarly, SIRT2 is a cytoplasmic protein that colocalizes with polymerized microtubules. In vitro and in vivo studies demonstrate that SIRT2 deacetylates K40 of α-tubulin. Knockdown of SIRT2 increases tubulin acetylation (North et al. 2003) (Fig. 3.1).

3.1.2 Impact of K40 Acetylation on Microtubule Structure

K40 acetylation was found to be associated with long-lived subsets of microtubules in cells (Maruta et al. 1986). It is still debatable whether acetylation promotes microtubule stability or whether stable microtubules facilitate acetylation. One proposed model is that K40 acetylation is a result of microtubule stabilization. This is because the chemical stabilization of microtubules resulted in K40 acetylation, whereas artificial tubulin acetylation did not stabilize microtubules (Palazzo et al. 2003).

Another model suggests that α-tubulin acetylation at K40 may affect the structural rearrangement of microtubules. It has been shown that K40 acetylation leads to weaker lateral interactions between protofilaments, which renders microtubules more flexible (Portran et al. 2017). Microtubules in living cells are often subjected to mechanical forces, which can cause nicks on microtubules and depolymerization. K40 acetylation may increase microtubule flexibility and allow them to resist mechanical stress (Xu et al. 2017) (Fig. 3.1).

3.1.3 Microtubule Acetylation at K40 Promotes Trafficking

Vesicle and protein trafficking plays an essential role in regulating cell functions including transport of neurotransmitters, endocytosis, protein redistribution, and gene expression. Current models believe that intracellular trafficking is driven by kinesin and dynein motor proteins that transport cargo along microtubule tracks. In general, kinesins travel toward the plus ends of microtubules in the cell periphery. In contrast, cytoplasmic dynein moves to the minus ends of microtubules in the cell center.

The transport of vesicles and proteins requires the coordination of microtubules and the molecular motors kinesin and dynein. In a study, K40 acetylation promoted the interaction of microtubules with kinesin-1. Loss of this site acetylation reduced the binding of kinesin-1 to microtubules and the motility of kinesin-1. Moreover, the increase of microtubule acetylation by pharmacological tools facilitated the transport of kinesin-1 cargo to neurite tips in vivo (Reed et al. 2006). Another study has shown that K40 acetylation increased dynein motility on microtubules whereas deacetylation at this site decreased the speed of axonemal dynein (Alper et al. 2014). In addition, microtubule acetylation enhanced the transport of lysosomal vesicles to the cell periphery for the formation of neutralized lysosomal vesicles (Jeong et al. 2023). Moreover, tubulin acetylation regulated glucocorticoid-induced nuclear import of glucocorticoid receptors in the corticotrophs of the anterior pituitary in vivo (Nakakura et al. 2017). Furthermore, tubulin acetylation modulates autophagosome formation and their fusion with lysosomes and trafficking (Janke and Montagnac 2017; Geeraert et al. 2010) (Fig. 3.1).

3.1.4 Impact of Tubulin Acetylation on Cancer Cell Migration

Acetylation of α-tubulin has been shown to be associated with cancer cell migration and invasion. In metastatic breast cancer cells, a high level of α-tubulin acetylation was detected and enriched in microtentacle protrusions. K40R mutation (arginine substitution at lysine-40) is acetylation-resistant but can still incorporate into the microtubule polymer. This mutation at K40 reduced the reattachment and migration of breast cancer cells without affecting proliferation. Moreover, α-tubulin acetylation was higher in many primary and metastatic tumors from patients. Proteomic analysis of breast cancer specimens unveiled the relationship between increased a-tubulin acetylation and the aggressive behaviors of breast cancers (Boggs et al. 2015) (Fig. 3.1).

Interestingly, the role of α-tubulin acetylation in cancer cells may be cell-type dependent. For instance, αTAT is known to catalyze K40 acetylation in various cell types. However, overexpression of αTAT diminished pseudopod formation, migration, invasion, and metastasis in an animal with lung cancer (Lee et al. 2018).

3.1.5 Tubulin Acetylation and Cell Division

Division of somatic cells involves mitotic spindle formation which ensures equal distribution of duplicated chromosomes into two daughter cells. Tubulin acetylation was found to be associated with spindle structure and mitosis of a human neuroblastoma cell line. Tubulin acetylation in this cell line may be regulated by the coordination of kindlin-2 (a focal adhesion protein) with microtubule-associated HDAC6 (Tan and Tan 2020). Moreover, alpha-tubulin N-acetyltransferase (ATAT1) was found to be colocalized with spindle poles and fibers in human fibroblast, suggesting a role for tubulin acetylation in spindle structure (Nekooki-Machida et al. 2018).

However, aTAT1-deficient mice did not show gross developmental and physiological abnormalities although no K40-acetylated microtubules were detected (Janke and Montagnac 2017; Kim et al. 2013). Thus, more studies are required to clarify the role of tubulin acetylation in spindle formation during mitosis.

3.2 Actin Acetylation Regulates Filament Polymerization

Thus far, there is no evidence that actin acetylation occurs at lysine residues. In contrast, actin undergoes Nt-acetylation, i.e., acetylation takes place at the free α-amino group of actin N-terminus (Ree et al. 2018; Drazic et al. 2018). Nt-acetylation of actin was described over 40 years ago. The acetylated process may assist in actin maturation (Redman and Rubenstein 1981; Varland et al. 2019). Recently, NAA80 has been identified as actin's N-terminal acetyltransferase (NAT) and regulates cytoskeleton remodeling and cell migration (Drazic et al. 2018). Lack of Nt-acetylation leads to reduced rates of actin filament polymerization in cells. Deficiency of Nt-acetylation also reduced actin filament elongation, including elongation driven by formins. In contrast, Nt-acetylation does not affect nucleation by the Arp2/3 complex. Furthermore, knockout of NAA80 reduced Nt-acetylation of actin, which affected the motility of a cell line derived from a patient with human chronic myelogenous leukemia (Drazic et al. 2018) (Fig. 3.2). However, it is unknown whether Nt-acetylation of actin is reversible in these cells.

How NAA80 catalyzes Nt-acetylation of actin? It appears that NAA80 does not directly associate with F-actin in cells. Instead, NAA80 interacts with the monomeric actin-profilin complex and acetylates the actin-profilin. It has been proposed that profilin functions as a chaperone for actin Nt-acetylation (Rebowski et al. 2020).

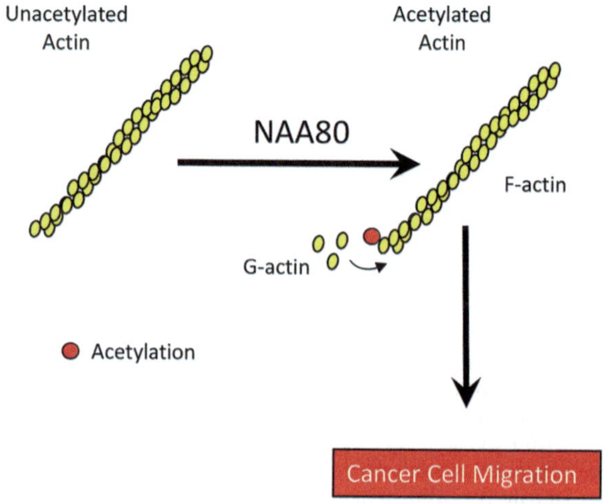

Fig. 3.2 N-terminal acetylation of actin. Actin acetylation occurs at the N-terminus of the molecule, which is catalyzed by NAA80 (one of N-terminal acetyltransferases). Acetylated actin facilitates actin filament polymerization and elongation, which subsequently affects leukemia cell migration

3.3 Acetylation of Vimentin Filaments Affects Migration and EMT

Vimentin is a type III intermediate filament protein that is ubiquitously expressed in a variety of cell types. At the molecular level, vimentin consists of an N-terminal "head" domain, a central α-helical "rod" domain, and a C-terminal "tail" domain. Vimentin spontaneously assembles into 10-nm diameter filaments, which are distributed from the plasma membrane to the nucleus, providing mechanical and structural integrity for the cell (Ostrowska-Podhorodecka et al. 2022; Tang 2008; Li et al. 2009; Li and Tang 2009). The vimentin framework also plays an important role in modulating various cell functions including migration, proliferation, contraction, and epithelial-mesenchymal transition (EMT) (Tang and Gerlach 2017; Ostrowska-Podhorodecka et al. 2022; Tang 2008; Leduc and Etienne-Manneville 2015). The studies on vimentin acetylation just began to emerge.

The acetylation of vimentin has been reported in cancer cells. Sirt5 is downregulated in hepatocellular carcinoma (HCC) cells and tissues. Sirt5 binds to and deacetylases vimentin at K120. Vimentin acetylation at this residue promotes HCC cell migration as evidenced by mutagenesis analysis. In addition, downregulation of Sirt5 increases EMT in this cancer cell type, which could be associated with increased vimentin acetylation (Guo et al. 2018) (Fig. 3.3).

Vimentin acetylation has also been linked to EMT, migration, and invasion of trophoblasts, which are important for normal pregnancy. Sirt1 deficiency was found in the placenta of patients and mice with advanced maternal age. Knockdown of

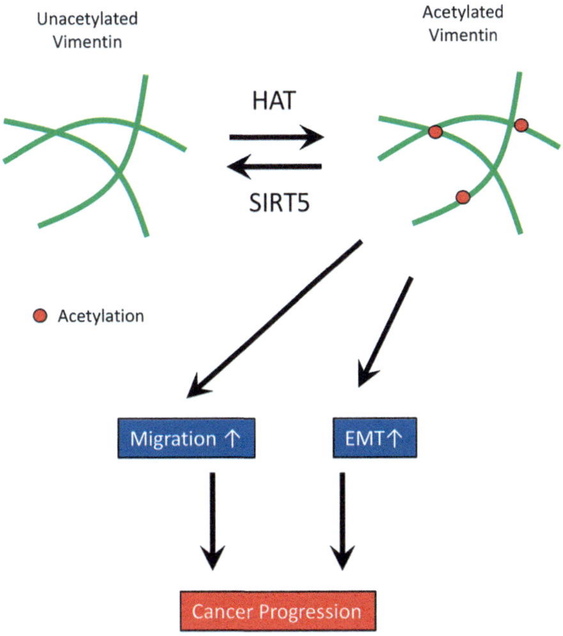

Fig. 3.3 K120 acetylation of vimentin intermediate filaments. Vimentin acetylation takes place at K120. Currently, we do not know which HATs catalyze vimentin acetylation. However, SIRT5 is identified as a HDAC to deacetylate vimentin. K120 acetylation of vimentin facilitates hepatocellular carcinoma cell migration and epithelial-mesenchymal transition (EMT), which contributes to cancer progression

Sirt1 enhances senescence and reduces migration and invasion of trophoblast. Srt1 can mediate vimentin acetylation and EMT in this cell type. Thus, Sirt1 is proposed to promote trophoblast EMT and invasiveness by modulating vimentin acetylation (Xiong et al. 2021).

3.4 Regulation of Cytoskeleton by Acetylated Proteins

As described earlier, the remodeling of the cytoskeleton is also regulated by acetylation and deacetylation of cytoskeleton-associated proteins such as adapter proteins. Adaptor proteins lack any intrinsic enzymatic activity but possess various protein-binding domains that link protein-binding partners together and form multi-protein signaling complexes. These signaling complexes subsequently regulate the activity, expression, and localization of other molecules, which affect cellular processes, including cytoskeletal reorganization, proliferation, apoptosis, contractility, migration, immune response, and transport (Tang 2018; Boning et al. 2019; Wang et al. 2013; Wang et al. 2022). The binding capability of adapter proteins can be regulated by acetylation and deacetylation.

3.4.1 Cortactin

Cortactin is an adapter protein that is able to regulate actin filament assembly, adhesion, migration, endocytosis, and contraction (Ammer and Weed 2008; Wang et al. 2014; Lapetina et al. 2009; Schnoor et al. 2018). Cortactin modulates the actin filament network by regulating the functional state of neuronal Wiskott-Aldrich Syndrome Protein (N-WASP), the Arp2/3 complex, profilin-1 (Pfn-1), and Nck (Ammer and Weed 2008; Wang et al. 2014; Lapetina et al. 2009). Cortactin also binds to filamentous actin (F-actin) to stabilize and/or regulate the organization of the actin network (Zhang et al. 2007).

Cortactin contains an N-terminal acidic domain (NTA), a repeat domain, and a C-terminal Src homology 3 (SH3) domain (Ammer and Weed 2008; Cosen-Binker and Kapus 2006). The repeat region of cortactin can bind F-actin (Ammer and Weed 2008; Cosen-Binker and Kapus 2006). In vitro biochemical and cellular studies identify nine lysine acetylation sites (Lys-87, Lys-124, Lys-161, Lys-189, Lys-198, Lys-235, Lys-272, Lys-309, and Lys-319) were present in the cortactin repeat region. Cortactin acetylation may be catalyzed by PCAF (p300/CBP-associated factor, an acetyltransferase). This is because co-expression of PCAF with cortactin increased the level of acetylated cortactin in Hela cells. However, co-expression of p300 with cortactin did not enhance cortactin acetylation in this cell type. This was also verified in an in vitro biochemical assay (Zhang et al. 2007; Zhang et al. 2009) (Fig. 3.4).

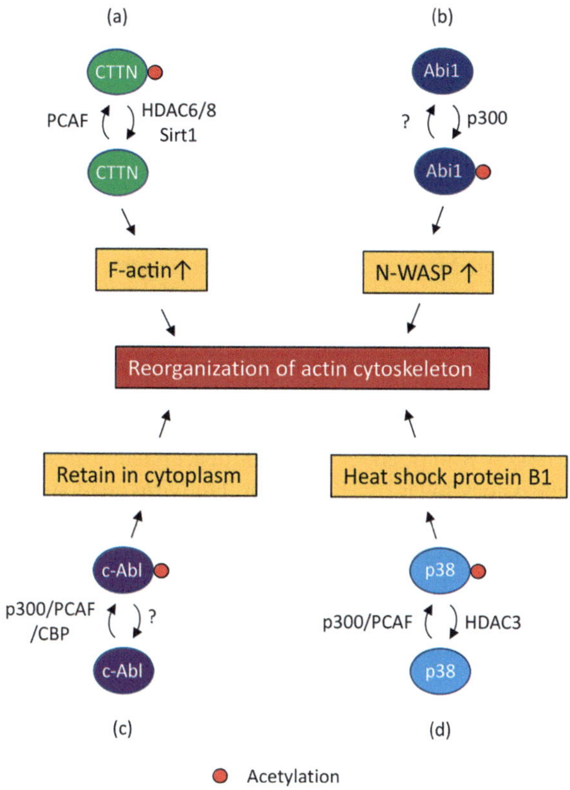

Fig. 3.4 Lysine acetylation-dependent regulation of actin cytoskeleton. Reorganization of the actin cytoskeleton is regulated by the acetylation of adapter proteins and kinases. (**a**) Cortactin (CTTN) acetylation at 9 lysines of the repeat region is catalyzed by PDAF, whereas CTTN is deacetylated by HDAC6, HDAC8, and Sirt1. Deacetylated CTTN stabilizes and/or promotes F-actin. (**b**) Abi1 acetylation at Lys-416 is mediated by p300, which associates with and activates N-WASP. The enzymes for Abi1 deacetylation have not been identified. (**c**) Acetylation of c-Abl tyrosine kinase at Lys-921 renders it in the cytoplasm, which may facilitate phosphorylation of its downstream targets in the cytoplasm and regulate the remodeling of the actin cytoskeleton. c-Abl is acetylated by p300, CBP, and PCAF. It is currently unknown which enzymes deacetylate c-Abl. (**d**) p38MAPK is acetylated by p300 and PCAF, and deacetylated by HDAC3. Acetylation of p38MAPK increases its activity, which phosphorylates heat shock protein B1 and regulates actin cytoskeletal remodeling

Acetylation on these nine lysines creates a charge-neutralization local environment and decreases the interaction of cortactin with F-actin, which increases the formation of globular actin (G-actin) (Zhang et al. 2007). In contrast, deacetylated cortactin restores the positively charged domain and increases its binding to F-actin, which stabilizes and/or promotes actin filament assembly (Li et al. 2014; Zhang et al. 2007; Zhang et al. 2009). Stimulation with growth factors induces cortactin deacetylation in cells, which may promote actin cytoskeletal reorganization and cell migration (Zhang et al. 2007). Cortactin deacetylation in NIH 3 T3 cells is mediated by HDAC6 (Zhang et al. 2007) whereas cortactin deacetylation in human ovarian epithelial cancer cell lines is regulated by SIRT1 (Sirtuin Type 1) (Zhang et al. 2009). Additionally, cortactin deacetylation mediated by SIRT1 may promote the reorganization of the actin cytoskeleton and membrane resealing of muscle cells (Iwahara et al. 2022).

In smooth muscle, contractile stimulation promotes the incorporation of globular actin to existing actin filaments (a process called actin polymerization), which is critical for smooth muscle contraction (Tang 2018; Gunst and Zhang 2008; Wang et al. 2021; Yamin and Morgan 2012). Moreover, contractile stimulation triggers cortactin deacetylation coordinately with increase of actin polymerization (an index of actin cytoskeletal reorganization) in smooth muscle tissues (Li et al. 2014; Tang 2018). Although there are many isoforms of HDACs in mammalian cells, HDAC8 is a major isoform in smooth muscle tissues. HDAC8 is localized both in the cytoplasm and the nucleus of mouse and human smooth muscle cells. This raises the possibility that HDAC8 may affect cortactin deacetylation in smooth muscle tissues. Indeed, the knockdown of HDAC8 by lentivirus-mediated shRNA inhibits cortactin deacetylation (Li et al. 2014; Tang 2018). Moreover, HDAC8 knockdown attenuates actin polymerization and force development in smooth muscle tissues (Li et al. 2014; Tang 2018). Thus, HDAC8-mediated cortactin deacetylation regulates the remodeling of the actin cytoskeleton and smooth muscle contractility (Fig. 3.4).

3.4.2 Abelson Interactor 1 (Abi1)

The adapter protein Abi1 plays a critical role in orchestrating a wide spectrum of cellular processes, including actin cytoskeletal remodeling (Innocenti et al. 2005), intercellular adhesion (Ryu et al. 2009), cardiovascular development (Ring et al. 2011), cell migration, and contraction (Wang et al. 2013, 2021; Yu et al. 2008; Stradal et al. 2001; Wang et al. 2020). Upon alterations of the environment (surrounding cells), Abi1 binds to N-WASP, which subsequently promotes actin filament branching and growth in cells (Tang 2018; Tang 2015; Gunst and Zhang 2008; Wang et al. 2013; Kim et al. 2010).

Structural analysis unveils that the SH3 domain of Abi1 associates with the proline-rich domain of N-WASP, which facilitates the formation of the Abi1/N--WASP protein complex (Tang 2018; Innocenti et al. 2005; Pollard 2007). The binding of Abi1 to N-WASP activates the latter and promotes actin polymerization in vitro (Innocenti et al. 2005).

Recent studies demonstrate that contractile stimulation increases Abi1 acetylation in smooth muscle. Mutagenesis analysis reveals Lys-416 in the SH3 domain as a major acetylation site (Wang et al. 2021). Substitution of Lys-416 with glutamine (K416Q) facilitates the interaction of Abi1 with N-WASP, actin polymerization, and smooth muscle contraction (Wang et al. 2021). Moreover, p300 is the lysine acetyltransferase that catalyzes Abi1 acetylation at Lys-416. p300 knockout inhibits actin polymerization and the contraction of smooth muscle tissues. These results unveil the important role of p300-mediated Abi1 acetylation in the remodeling of actin cytoskeleton and cellular functions (Fig. 3.4). Furthermore, actin cytoskeletal reorganization is known to regulate migration and growth. Thus, Abi1 acetylation likely contributes to airway remodeling, a key feature of asthma (Tang 2018; Wang et al. 2021, 2022).

3.5 Regulation of Cytoskeleton by Acetylated Kinases

Reorganization of the cytoskeleton is regulated by protein kinases including tyrosine kinases and serine/threonine kinases. In general, protein kinases catalyze the phosphorylation of other proteins including adapter proteins, which in turn modulates restructuring of the cytoskeleton. Interestingly, the activity of protein kinases themselves is regulated by acetylation.

3.5.1 C-Abl Tyrosine Kinase

c-Abl (Abelson tyrosine kinase) is a nonreceptor tyrosine kinase that plays an important role in regulating cell functions including adhesion, migration, cytokinesis, gene expression, and contraction (Hu et al. 2005; Cleary et al. 2014; Anfinogenova et al. 2007; Chen and Tang 2014; Wang et al. 2016; Wang 2014; Allen et al. 2011).

c-Abl function is dependent on its subcellular localization. In vitro experiments have identified c-Abl as a substrate for p300, CBP (CREB-binding protein), and PCAF (p300/CBP-associated factor) lysine acetyltransferases (di Bari et al. 2006). Site-directed mutagenesis unveils Lys-730 (located in the second nuclear localization signal) as a major residue for p300-mediated acetylation. Acetylation of c-Abl alters its subcellular localization (di Bari et al. 2006). c-Abl has been shown to accumulate in the cytoplasm during myogenic differentiation (Puri et al. 2002). c-Abl protein is acetylated at the early stages of myogenic differentiation. Acetylation on Lys-730 drives c-Abl accumulation in the cytoplasm and promotes differentiation (di Bari et al. 2006).

c-Abl mediates the tyrosine phosphorylation of several adapter proteins including p130 Crk-associated substrate (p130CAS), cortactin, and paxillin, which subsequently regulates the remodeling of the actin cytoskeleton, migration, and contraction (Tang 2015, Tang 2018; Wang et al. 2013, 2014, 2018; Lapetina et al. 2009; Lewis and Schwartz 1998). Since acetylation renders c-Abl in cytoplasmic distribution, acetylation at Lys-730 likely facilitates the cytoplasmic localization of c-Abl, which subsequently mediates the phosphorylation of these adapter proteins and actin cytoskeletal reorganization (Fig. 3.4).

Another research group has shown that DNA damage initiates c-Abl acetylation at Lys-921, which is mediated by Tip60 in cancer cells. Expression of the nonacetylatable K921R c-Abl mutant attenuates apoptosis in response to DNA damage

(Jiang et al. 2011). However, the role of acetylation at Lys-921 in cytoskeletal reorganization was not investigated in this study.

3.5.2 *p38 Mitogen-Activated Protein Kinase*

p38 mitogen-activated protein kinase (p38MAPK) serves as a node for signal transduction and plays an essential role in many biological processes. This includes cytoskeletal remodeling, cell proliferation, migration, differentiation, glucose and lipid metabolism, secretion, senescence, stress responses, apoptosis, and autophagy (Kudaravalli et al. 2022).

Exposure of cardiomyocytes to mechanical stress stimulates p38 MAPK acetylation, which is catalyzed by PCAF and p300 (Pillai et al. 2011). HDAC3 interacts with and deacetylases p38 MAPK in Hela cells. Acetylation of Lys-53 increases its binding to ATP and thus enhances the activity of p38 MAPK (Pillai et al. 2011). p38 MAPK can regulate the actin cytoskeleton via heat shock protein B1 (Hoffman et al. 2017). Therefore, acetylation-dependent activation of p38 may regulate the remodeling of the actin cytoskeleton (Fig. 3.4).

3.6 Targeting HATs and HDACs in Cancer Therapy

As described earlier, protein acetylation is increased by HATs and decreased by HDACs. In general, increased protein acetylation promotes cancer cell migration, invasion, EMT, and growth. In contrast, reducing protein acetylation may inhibit these responses. Thus, targeting HATs and HDACs is a promising avenue to prevent or treat cancer progression.

Several HDAC inhibitors have been approved for the treatment of hematological cancer. For example, panobinostat is a potent oral pan HDAC inhibitor. A multicenter, randomized, double-blind phase 3 trial showed that panobinostat was effective for the treatment of relapsed or relapsed and refractory multiple myeloma (San-Miguel et al. 2014). In addition, belinostat is another HDAC inhibitor, which was approved by FDA for the treatment of patients with relapsed or refractory peripheral T-cell lymphoma in 2014 (Lee et al. 2015).

In addition to hematological cancer, recent studies have shown that HAT inhibitors are promising to treat solid tumor. PF-07248144 is a selective pharmacological inhibitor of HAT6A and HAT6B. A phase I clinical trial showed that PF-07248144 was safe and effective to treat ER + HER2− metastatic breast cancer (Mukohara et al. 2024).

SPOP (speckle-type POZ protein) is a cancer-associated gene, which is frequently mutated in patients with primary prostate cancer. Its mutation rate ranges from 10 to 15% of human prostate cancer and is resistant to conventional chemotherapy. Preclinical studies have shown that NEO2734, a novel small molecule inhibitor of CBP/p300, inhibited SPOP-mutated prostate cancer cell growth in cell culture, organoid, and mouse xenograft models (Yan et al. 2019).

3.7 Conclusions and Perspectives

In addition to protein phosphorylation, reversible acetylation has emerged as a pivotal cellular mechanism that regulates the reorganization of the cytoskeleton in cells. There are two forms of acetylation: lysine acetylation and Nt-acetylation. Lysine acetylation and deacetylation of adapter proteins and protein kinases modulate the remodeling and polymerization of the actin network. Lysine acetylation of microtubules and vimentin are involved in migration, trafficking, and division. Nt-acetylation of actin may promote actin polymerization and maturation. Acetylation-dependent regulation of cytoskeleton-associated proteins contributes to the pathogenesis of diseases (Table 3.1). Moreover, HATs catalyze protein acetylation at lysine residues, while HDACs deacetylate protein at these residues. Targeting HATs and HDACs have been used to treat hematological cancer and potentially

Table 3.1 Acetylation regulation of cytoskeleton-associated proteins

Protein	Role	Disease/pathology	References
Microtubules	Trafficking	Neurological disorder	Reed et al. (2006), Alper et al. (2014), and Jeong et al. (2023)
	Migration	Breast cancer	Boggs et al. (2015)
	Spindle formation	Neuroblastoma	Tan and Tan (2020)
Actin	Actin assembly	Leukemia	Drazic et al. (2018), Redman and Rubenstein (1981), and Varland et al. (2019)
	Migration		
Vimentin	EMT	Liver cancer	Guo et al. (2018) and Xiong et al. (2021)
	Migration		
Cortactin	F-actin stabilization	Cancer	Zhang et al. (2007) and Zhang et al. (2009)
	Migration		
	Contraction	Hyperactivity	Li et al. (2014) and Tang (2018)
Abi1	Contraction		
	Migration	Airway remodeling	Tang (2018), Wang et al. (2022), and Wang et al. (2021)
	Proliferation		
c-Abl	Cytoplasm accumulation	Cancer	di Bari et al. (2006), Puri et al. (2002), and Jiang et al. (2011)
	DNA damage		
p38	Cardiomyocyte hypertrophy	Cardiomyopathy	Pillai et al. (2011) and Hoffman et al. (2017)
	Kinase activity		

some solid cancer. However, there are still unanswered questions in this research field. For example, are there other acetylated adapter proteins and kinases? Does reversible acetylation of adapter proteins and kinases regulate the restructuring of microtubules and vimentin filaments? Answers to these questions should advance our knowledge regarding the nature and role of reversible protein acetylation, which may disclose new biotargets for the development of novel clinical therapies.

Acknowledgments This work was supported by NHLBI Grants HL-110951 (to Dale D. Tang) and HL-130304 (to Dale D. Tang).

References

Akella JS, Wloga D, Kim J, Starostina NG, Lyons-Abbott S, Morrissette NS et al (2010) MEC-17 is an alpha-tubulin acetyltransferase. Nature 467(7312):218–222

Allen JC, Talab F, Zuzel M, Lin K, Slupsky JR (2011) c-Abl regulates Mcl-1 gene expression in chronic lymphocytic leukemia cells. Blood 117(8):2414–2422

Allfrey VG, Faulkner R, Mirsky AE (1964) Acetylation and methylation of histones and their possible role in the regulation of RNA synthesis. Proc Natl Acad Sci USA 51:786–794

Alper JD, Decker F, Agana B, Howard J (2014) The motility of axonemal dynein is regulated by the tubulin code. Biophys J 107(12):2872–2880

Ammer AG, Weed SA (2008) Cortactin branches out: roles in regulating protrusive actin dynamics. Cell Motil Cytoskelet 65(9):687–707

Anfinogenova Y, Wang R, Li QF, Spinelli AM, Tang DD (2007) Abl silencing inhibits CAS-mediated process and constriction in resistance arteries. Circ Res 101(4):420–428

Boggs AE, Vitolo MI, Whipple RA, Charpentier MS, Goloubeva OG, Ioffe OB et al (2015) Alpha-tubulin acetylation elevated in metastatic and basal-like breast cancer cells promotes microtentacle formation, adhesion, and invasive migration. Cancer Res 75(1):203–215

Boning MAL, Trittel S, Riese P, van Ham M, Heyner M, Voss M et al (2019) ADAP promotes degranulation and migration of NK cells primed during in vivo listeria monocytogenes infection in mice. Front Immunol 10:3144

Chen S, Tang DD (2014) C-Abl tyrosine kinase regulates cytokinesis of human airway smooth muscle cells. Am J Respir Cell Mol Biol 50(6):1076–1083

Cleary RA, Wang R, Waqar O, Singer HA, Tang DD (2014) Role of c-Abl tyrosine kinase in smooth muscle cell migration. Am J Physiol Cell Physiol 306(8):C753–C761

Coombes C, Yamamoto A, McClellan M, Reid TA, Plooster M, Luxton GW et al (2016) Mechanism of microtubule lumen entry for the alpha-tubulin acetyltransferase enzyme alphaTAT1. Proc Natl Acad Sci USA 113(46):E7176–E7E84

Cosen-Binker LI, Kapus A (2006) Cortactin: the gray eminence of the cytoskeleton. Physiology(Bethesda) 21:352–361

di Bari MG, Ciuffini L, Mingardi M, Testi R, Soddu S, Barila D (2006) C-Abl acetylation by histone acetyltransferases regulates its nuclear-cytoplasmic localization. EMBO Rep 7(7):727–733

Drazic A, Aksnes H, Marie M, Boczkowska M, Varland S, Timmerman E et al (2018) NAA80 is actin's N-terminal acetyltransferase and regulates cytoskeleton assembly and cell motility. Proc Natl Acad Sci USA 115(17):4399–4404

Geeraert C, Ratier A, Pfisterer SG, Perdiz D, Cantaloube I, Rouault A et al (2010) Starvation-induced hyperacetylation of tubulin is required for the stimulation of autophagy by nutrient deprivation. J Biol Chem 285(31):24184–24194

Gu W, Roeder RG (1997) Activation of p53 sequence-specific DNA binding by acetylation of the p53 C-terminal domain. Cell 90(4):595–606

Gunst SJ, Zhang W (2008) Actin cytoskeletal dynamics in smooth muscle: a new paradigm for the regulation of smooth muscle contraction. AJP Cell Physiol 295(3):C576–CC87
Guo D, Song X, Guo T, Gu S, Chang X, Su T et al (2018) Vimentin acetylation is involved in SIRT5-mediated hepatocellular carcinoma migration. Am J Cancer Res 8(12):2453–2466
Hoffman L, Jensen CC, Yoshigi M, Beckerle M (2017) Mechanical signals activate p38 MAPK pathway-dependent reinforcement of actin via mechanosensitive HspB1. Mol Biol Cell 28(20):2661–2675
Hu H, Bliss JM, Wang Y, Colicelli J (2005) RIN1 is an ABL tyrosine kinase activator and a regulator of epithelial-cell adhesion and migration. Curr Biol 15(9):815–823
Hubbert C, Guardiola A, Shao R, Kawaguchi Y, Ito A, Nixon A et al (2002) HDAC6 is a microtubule-associated deacetylase. Nature 417(6887):455–458
Innocenti M, Gerboth S, Rottner K, Lai FP, Hertzog M, Stradal TE et al (2005) Abi1 regulates the activity of N-WASP and WAVE in distinct actin-based processes. Nat Cell Biol 7(10):969–976
Iwahara N, Azekami K, Hosoda R, Nojima I, Hisahara S, Kuno A (2022) Activation of SIRT1 promotes membrane resealing via cortactin. Sci Rep 12(1):15328
Janke C, Montagnac G (2017) Causes and consequences of microtubule acetylation. Curr Biol CB 27(23):R1287–R1R92
Jeong J, Kim OH, Shim J, Keum S, Hwang YE, Song S et al (2023) Microtubule acetylation induced by oxidative stress regulates subcellular distribution of lysosomal vesicles for amyloid-beta secretion. J Cell Physiol 238(12):2812–2826
Jiang Z, Kamath R, Jin S, Balasubramani M, Pandita TK, Rajasekaran B (2011) Tip60-mediated acetylation activates transcription independent apoptotic activity of Abl. Mol Cancer 10:88
Kim HR, Graceffa P, Ferron F, Gallant C, Boczkowska M, Dominguez R et al (2010) Actin polymerization in differentiated vascular smooth muscle cells requires vasodilator-stimulated phosphoprotein. AJP Cell Physiol 298(3):C559–CC71
Kim GW, Li L, Ghorbani M, You L, Yang XJ (2013) Mice lacking alpha-tubulin acetyltransferase 1 are viable but display alpha-tubulin acetylation deficiency and dentate gyrus distortion. J Biol Chem 288(28):20334–20350
Kudaravalli S, den Hollander P, Mani SA (2022) Role of p38 MAP kinase in cancer stem cells and metastasis. Oncogene 41(23):3177–3185
Lapetina S, Mader CC, Machida K, Mayer BJ, Koleske AJ (2009) Arg interacts with cortactin to promote adhesion-dependent cell edge protrusion. J Cell Biol 185(3):503–519
LeDizet M, Piperno G (1987) Identification of an acetylation site of Chlamydomonas alpha-tubulin. Proc Natl Acad Sci USA 84(16):5720–5724
Leduc C, Etienne-Manneville S (2015) Intermediate filaments in cell migration and invasion: the unusual suspects. Curr Opin Cell Biol 32:102–112
Lee HZ, Kwitkowski VE, Del Valle PL, Ricci MS, Saber H, Habtemariam BA et al (2015) FDA approval: Belinostat for the treatment of patients with relapsed or refractory peripheral T-cell lymphoma. Clin Cancer Res 21(12):2666–2670
Lee CC, Cheng YC, Chang CY, Lin CM, Chang JY (2018) Alpha-tubulin acetyltransferase/MEC-17 regulates cancer cell migration and invasion through epithelial-mesenchymal transition suppression and cell polarity disruption. Sci Rep 8:8
Lewis JM, Schwartz MA (1998) Integrins regulate the association and phosphorylation of paxillin by c-Abl. J Biol Chem 273(23):14225–14230
L'Hernault SW, Rosenbaum JL (1983) Chlamydomonas alpha-tubulin is posttranslationally modified in the flagella during flagellar assembly. J Cell Biol 97(1):258–263
Li QF, Tang DD (2009) Role of p47(phox) in regulating Cdc42GAP, vimentin, and contraction in smooth muscle cells. Am J Physiol Cell Physiol 297(6):C1424–C1C33
Li QF, Spinelli AM, Tang DD (2009) Cdc42GAP, reactive oxygen species, and the vimentin network. AJP Cell Physiol 297(2):C299–C309
Li J, Chen S, Cleary RA, Wang R, Gannon OJ, Seto E et al (2014) Histone deacetylase 8 regulates cortactin deacetylation and contraction in smooth muscle tissues. Am J Physiol Cell Physiol 307(3):C288–C295

Maruta H, Greer K, Rosenbaum JL (1986) The acetylation of alpha-tubulin and its relationship to the assembly and disassembly of microtubules. J Cell Biol 103(2):571–579
Mukohara T, Park YH, Sommerhalder D, Yonemori K, Hamilton E, Kim SB et al (2024) Inhibition of lysine acetyltransferase KAT6 in ER(+)HER2(−) metastatic breast cancer: a phase 1 trial. Nat Med 30(8):2242–2250
Nakakura T, Suzuki T, Torii S, Asano-Hoshino A, Nekooki-Machida Y, Tanaka H et al (2017) ATAT1 is essential for regulation of homeostasis-retaining cellular responses in corticotrophs along hypothalamic-pituitary-adrenal axis. Cell Tissue Res 370(1):169–178
Narita T, Weinert BT, Choudhary C (2019) Functions and mechanisms of non-histone protein acetylation. Nat Rev Mol Cell Biol 20(3):156–174
Nekooki-Machida Y, Nakakura T, Nishijima Y, Tanaka H, Arisawa K, Kiuchi Y et al (2018) Dynamic localization of alpha-tubulin acetyltransferase ATAT1 through the cell cycle in human fibroblastic KD cells. Med Mol Morphol 51(4):217–226
North BJ, Marshall BL, Borra MT, Denu JM, Verdin E (2003) The human Sir2 ortholog, SIRT2, is an NAD+-dependent tubulin deacetylase. Mol Cell 11(2):437–444
Ostrowska-Podhorodecka Z, Ding I, Norouzi M, McCulloch CA (2022) Impact of Vimentin on regulation of cell signaling and matrix remodeling. Front Cell Dev Biol 10:869069
Palazzo A, Ackerman B, Gundersen GG (2003) Cell biology: tubulin acetylation and cell motility. Nature 421(6920):230
Pillai VB, Sundaresan NR, Samant SA, Wolfgeher D, Trivedi CM, Gupta MP (2011) Acetylation of a conserved lysine residue in the ATP binding pocket of p38 augments its kinase activity during hypertrophy of cardiomyocytes. Mol Cell Biol 31(11):2349–2363
Pollard TD (2007) Regulation of actin filament assembly by Arp2/3 complex and formins. Ann Rev Biophys Biomol Struct 36:451–477
Portran D, Schaedel L, Xu Z, Thery M, Nachury MV (2017) Tubulin acetylation protects long-lived microtubules against mechanical ageing. Nat Cell Biol 19(4):391–398
Puri PL, Bhakta K, Wood LD, Costanzo A, Zhu J, Wang JY (2002) A myogenic differentiation checkpoint activated by genotoxic stress. Nat Genet 32(4):585–593
Rebowski G, Boczkowska M, Drazic A, Ree R, Goris M, Arnesen T et al (2020) Mechanism of actin N-terminal acetylation. Sci Adv 6(15):eaay8793
Redman K, Rubenstein PA (1981) NH2-terminal processing of Dictyostelium discoideum actin in vitro. J Biol Chem 256(24):13226–13229
Ree R, Varland S, Arnesen T (2018) Spotlight on protein N-terminal acetylation. Exp Mol Med 50(7):1–13
Reed NA, Cai D, Blasius TL, Jih GT, Meyhofer E, Gaertig J et al (2006) Microtubule acetylation promotes kinesin-1 binding and transport. Curr Biol CB 16(21):2166–2172
Ring C, Ginsberg MH, Haling J, Pendergast AM (2011) Abl-interactor-1 (Abi1) has a role in cardiovascular and placental development and is a binding partner of the alpha4 integrin. Proc Natl Acad Sci USA 108(1):149–154
Ryu JR, Echarri A, Li R, Pendergast AM (2009) Regulation of cell-cell adhesion by Abi/diaphanous complexes. Mol Cell Biol 29(7):1735–1748
San-Miguel JF, Hungria VT, Yoon SS, Beksac M, Dimopoulos MA, Elghandour A et al (2014) Panobinostat plus bortezomib and dexamethasone versus placebo plus bortezomib and dexamethasone in patients with relapsed or relapsed and refractory multiple myeloma: a multicentre, randomised, double-blind phase 3 trial. Lancet Oncol 15(11):1195–1206
Saphirstein RJ, Gao YZ, Lin QQ, Morgan KG (2015) Cortical actin regulation modulates vascular contractility and compliance in veins. J Physiol 593(17):3929–3941
Schnoor M, Stradal TE, Rottner K (2018) Cortactin: cell functions of a multifaceted actin-binding protein. Trends Cell Biol 28(2):79–98
Song Y, Brady ST (2015) Post-translational modifications of tubulin: pathways to functional diversity of microtubules. Trends Cell Biol 25(3):125–136

Sterner R, Vidali G, Allfrey VG (1979) Studies of acetylation and deacetylation in high mobility group proteins. Identification of the sites of acetylation in HMG-1. J Biol Chem 254(22):11577–11583
Stradal T, Courtney KD, Rottner K, Hahne P, Small JV, Pendergast AM (2001) The Abl interactor proteins localize to sites of actin polymerization at the tips of lamellipodia and filopodia. Curr Biol 11(11):891–895
Tan HF, Tan SM (2020) The focal adhesion protein kindlin-2 controls mitotic spindle assembly by inhibiting histone deacetylase 6 and maintaining alpha-tubulin acetylation. J Biol Chem 295(18):5928–5943
Tang DD (2008) Invited review: intermediate filaments in smooth muscle. Am J Physiol Cell Physiol 294(4):C869–CC78
Tang DD (2015) Critical role of actin-associated proteins in smooth muscle contraction, cell proliferation, airway hyperresponsiveness and airway remodeling. Respir Res 16:134
Tang DD (2018) The dynamic actin cytoskeleton in smooth muscle. Adv Pharmacol 81:1–38
Tang DD, Gerlach BD (2017) The roles and regulation of the actin cytoskeleton, intermediate filaments and microtubules in smooth muscle cell migration. Respir Res 18(1):54
Varland S, Vandekerckhove J, Drazic A (2019) Actin post-translational modifications: the Cinderella of cytoskeletal control. Trends Biochem Sci 44(6):502–516
Wang JY (2014) The capable ABL: what is its biological function? Mol Cell Biol 34(7):1188–1197
Wang T, Cleary RA, Wang R, Tang DD (2013) Role of the adapter protein Abi1 in actin-associated signaling and smooth muscle contraction. J Biol Chem 288(28):20713–20722
Wang R, Cleary RA, Wang T, Li J, Tang DD (2014) The association of cortactin with profilin-1 is critical for smooth muscle contraction. J Biol Chem 289(20):14157–14169
Wang J, Rouse C, Jasper JS, Pendergast AM (2016) ABL kinases promote breast cancer osteolytic metastasis by modulating tumor-bone interactions through TAZ and STAT5 signaling. Sci Signal 9(413):ra12
Wang Y, Rezey AC, Wang R, Tang DD (2018) Role and regulation of Abelson tyrosine kinase in Crk-associated substrate/profilin-1 interaction and airway smooth muscle contraction. Respir Res 19(1):4
Wang R, Liao G, Wang Y, Tang DD (2020) Distinctive roles of Abi1 in regulating actin-associated proteins during human smooth muscle cell migration. Sci Rep 10(1):10667
Wang Y, Liao G, Wang R, Tang DD (2021) Acetylation of Abelson interactor 1 at K416 regulates actin cytoskeleton and smooth muscle contraction. FASEB J 35(9):e21811
Wang R, Wang Y, Liao G, Chen B, Panettieri RA Jr, Penn RB et al (2022) Abi1 mediates airway smooth muscle cell proliferation and airway remodeling via Jak2/STAT3 signaling. iScience 25(2):103833
Xiong L, Ye X, Chen Z, Fu H, Li S, Xu P et al (2021) Advanced maternal age-associated SIRT1 deficiency compromises trophoblast epithelial-mesenchymal transition through an increase in Vimentin acetylation. Aging Cell 20(10):e13491
Xu Z, Schaedel L, Portran D, Aguilar A, Gaillard J, Marinkovich MP et al (2017) Microtubules acquire resistance from mechanical breakage through intralumenal acetylation. Science 356(6335):328–332
Yamin R, Morgan KG (2012) Deciphering actin cytoskeletal function in the contractile vascular smooth muscle cell. J Physiol 590(Pt 17):4145–4154
Yan Y, Ma J, Wang D, Lin D, Pang X, Wang S et al (2019) The novel BET-CBP/p300 dual inhibitor NEO2734 is active in SPOP mutant and wild-type prostate cancer. EMBO Mol Med 11(11):e10659
Yu W, Sun X, Clough N, Cobos E, Tao Y, Dai Z (2008) Abi1 gene silencing by short hairpin RNA impairs Bcr-Abl-induced cell adhesion and migration in vitro and leukemogenesis in vivo. Carcinogenesis 29(9):1717–1724
Zhang X, Yuan Z, Zhang Y, Yong S, Salas-Burgos A, Koomen J et al (2007) HDAC6 modulates cell motility by altering the acetylation level of cortactin. Mol Cell 27(2):197–213
Zhang Y, Zhang M, Dong H, Yong S, Li X, Olashaw N et al (2009) Deacetylation of cortactin by SIRT1 promotes cell migration. Oncogene 28(3):445–460

Chapter 4
Tubulin Acetylation: A Critical Regulator of Microtubule Function

Inês L. S. Delgado, Bruno Carmona, Sofia Nolasco, Rita Marques, João Gonçalves, and Helena Soares

Abstract The cytoskeleton is conserved throughout the eukaryotic lineage and consists of a complex dynamic network mainly composed of three distinct polymers: microtubules (MTs), actin filaments, and intermediate filaments. MTs are polymers of α/β-tubulin heterodimers, playing a myriad of distinct cellular functions and are the main components of complex structures like the mitotic spindle, cilia, and centrioles. Post-translational modifications (PTMs) regulate the function and increase the complexity of the α/β-tubulin heterodimer pools. One of the PTMs

I. L. S. Delgado
CIISA—Centro de Investigação Interdisciplinar em Sanidade Animal, Faculdade de Medicina Veterinária, Universidade de Lisboa, Lisbon, Portugal

Laboratório Associado para Ciência Animal e Veterinária (AL4AnimalS), Lisbon, Portugal

Faculdade de Medicina Veterinária, Universidade Lusófona-Centro Universitário de Lisboa, Lisbon, Portugal

B. Carmona · H. Soares (✉)
Centro de Química Estrutural, Institute of Molecular Sciences, Faculdade de Ciências, Universidade de Lisboa, Lisbon, Portugal

Escola Superior de Tecnologia da Saúde de Lisboa, Instituto Politécnico de Lisboa, Lisbon, Portugal
e-mail: mhsoares@ciencias.ulisboa.pt; helena.soares@estesl.ipl.pt

S. Nolasco
CIISA—Centro de Investigação Interdisciplinar em Sanidade Animal, Faculdade de Medicina Veterinária, Universidade de Lisboa, Lisbon, Portugal

Laboratório Associado para Ciência Animal e Veterinária (AL4AnimalS), Lisbon, Portugal

Escola Superior de Tecnologia da Saúde de Lisboa, Instituto Politécnico de Lisboa, Lisbon, Portugal

R. Marques
Centro de Química Estrutural, Institute of Molecular Sciences, Faculdade de Ciências, Universidade de Lisboa, Lisbon, Portugal

J. Gonçalves
Evotec, Toulouse, France

M. Halasa, A. Wawruszak (eds.), *Histone and Non-Histone Reversible Acetylation in Development, Aging and Disease*, Results and Problems in Cell Differentiation 75,
https://doi.org/10.1007/978-3-031-91459-1_4

that has been extensively studied is the acetylation of lysine 40 (K40) on α-tubulin, which specifically occurs inside the MT lumen.

Acetylation plays a crucial role in controlling the stability and function of MTs, in response to signals from within and outside the cell. It impacts the cytoplasm's 3D arrangement and important cellular activities like intracellular transport, cell division, polarity, and migration. Recent research has also emphasized the significance of this PTM in regulating the mechanical properties of MTs and cellular sensing. The levels and activity of MT acetyltransferases and deacetylases are tightly regulated through various transcriptional, post-transcriptional, and post-translational mechanisms, including miRNAs, phosphorylation, protein-protein interactions, and regulated localization between the nucleus and cytoplasm. These regulatory processes involve components of diverse signaling pathways, and their deregulation has been implicated in numerous diseases, including neurological disorders, cancer, and cardiac conditions.

4.1 Overview of Post-Translational Modifications and Impact in Cell Function

The function of the eukaryotic cell depends on a series of tightly regulated events, from gene transcription to the execution of a role by a protein and its degradation. These events are controlled at various levels by complex integrated systems. Chromatin organization regulates access of the transcription machinery to the genome (Hübner et al. 2013). The translation of mRNA to protein is tightly regulated by cell signaling pathways such as a target of rapamycin (TOR) and mitogen-activated protein kinase (MAPK) (Roux et al. 2012). Post-translational modifications (PTMs)—chemical changes to protein amino acid sequences—represent a key mechanism of cellular regulation. Occurring after protein synthesis, PTMs influence a wide range of processes, including chromatin organization, mRNA translation, and cell signaling. It is now well established that PTMs are reversible and dynamic, contributing to an expanded functional proteome. For humans, it is estimated that more than 1,000,000 proteoforms arise from the expression of approximately 20,000 genes (Aebersold et al. 2018). They are a powerful mechanism to regulate protein functions in response to different extra and intracellular challenges and signals therefore contributing to cellular homeostasis (Beltrao et al. 2013; Kitamura and Galligan 2023). Moreover, the occurrence of distinct and/or combinatory PTM sites, or the existence of different PTMs for the same protein contributes to protein multi-functionality, regulatory versatility, and dynamics, as well as distinct cellular localization (Hunter 2007; Leutert et al. 2021). Well-known illustrating examples include the p53 (Gu and Zhu 2012; Liu et al. 2019b), histones (Cavalieri 2021; Millán-Zambrano et al. 2022), and tubulins (Wloga et al. 2017; McKenna et al. 2023). These modifications may be spontaneous or, more frequently, mediated by enzymes (Walsh et al. 2005). PTMs consist in the addition or removal of functional groups or specific amino acid residues, affecting various protein features such as folding, activity state, interactions, cellular localization, and degradation rate. Various functional groups have been identified as modifiers of proteins, including acetyl, glutamate, phosphate, SUMO (small ubiquitin-related modifier), and ubiquitin. Different types of enzymes are responsible for each type of modification, with reversible modifications usually also requiring specific enzymes to reverse the alteration (Walsh et al. 2005). Additionally, one protein may feature varying levels of the same PTM and different PTMs at once. This generates a multitude of modification patterns which enable a wide array of signals to be produced in the cell in response to its needs. The epidermal growth factor receptor (EGFR) and the necrosis factor-α receptor (TNFR) signaling pathways, and the DNA damage response pathways are well-studied examples of the coordination of various PTMs, such as phosphorylation, ubiquitination, acetylation, and SUMOylation, for the transmission and regulation of cellular signaling (Deribe et al. 2010).

4.1.1 The Specific Post-Translational Modifications of Tubulin

The cytoskeleton is a universal structure present in eukaryotic cells. This structure plays an essential role in maintaining and organizing the spatial arrangement of the cell and is involved in numerous cellular functions, such as maintaining/remodeling the cell's shape, cell migration, intracellular transport, cell division, and signal transduction. The eukaryotic cell cytoskeleton is highly conserved and includes three main classes of filamentous components: microtubules (MTs), actin filaments, and intermediate filaments. MTs are the largest diameter filaments of the cytoskeleton, with functions that overlap with those of the cytoskeleton itself. They form the essential scaffold for specialized organelles such as mitotic spindle, cilia, and centrioles.

MTs are hollow, tube-shaped polymers with the highly conserved α/β-tubulin heterodimer as their fundamental monomer. In vivo, MT dynamics rely on the availability of competent tubulin heterodimers, which can be synthesized de novo or recycled from existing MTs (Gonçalves et al. 2010). This tubulin dimer is assembled with the assistance of molecular chaperones, prefoldin, and cytoplasmic chaperone protein of TCP-1 (CCT) (Gao et al. 1992), and tubulin cofactors (TBCs), TBCA, TBCB, TBCC, TBCD, and TBCE (Gao et al. 1993; Bhamidipati et al. 2000; Lopez-Fanarraga et al. 2001). Beyond assisting in tubulin folding, these tubulin cofactors play a critical role in the assembly and disassembly of tubulin heterodimers as well as their degradation (see Fig. 4.1) (Lewis et al. 1997; Lopez-Fanarraga et al. 2001; Kortazar et al. 2007; Voloshin et al. 2010). Consequently, tubulin cofactors are vital for maintaining the quality of tubulin pools and regulating the recycling and degradation of tubulin heterodimers in vivo.

Briefly, the CCT captures tubulin folding intermediates, either directly from ribosomes or with the help of the hetero-hexameric chaperone prefoldin (Fig. 4.1) (for review Serna and Zabala 2016). Upon release from CCT, α- and β-tubulin

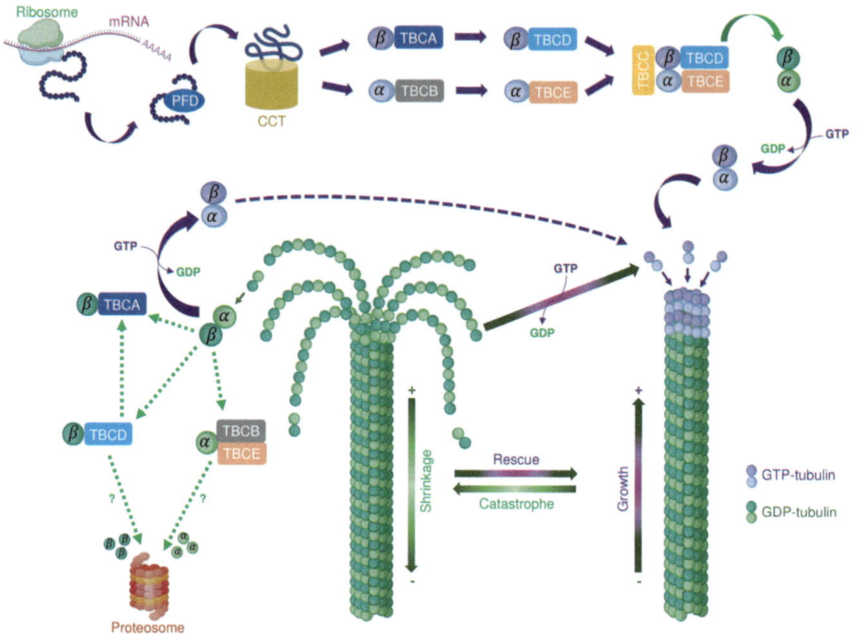

Fig. 4.1 Microtubule dynamic instability and the tubulin folding and recycling pathways. MTs are dynamic polymers assembled from α- and β-tubulin heterodimers. These heterodimers polymerize in a head-to-tail fashion, forming linear protofilaments. Typically, 13 protofilaments associate laterally to create the hollow, cylindrical MT wall. MTs exhibit structural polarity: one end, termed the minus end, is capped by α-tubulin, while the opposite plus end terminates with β-tubulin. The plus end of a MT has a lower critical concentration of free tubulin heterodimers required for polymerization compared to the minus end, making it more favorable for growth. Polymerization occurs only when tubulin heterodimers are bound to GTP. Upon incorporation into the MT lattice, the GTP bound to the β-tubulin subunit is hydrolyzed to GDP. This hydrolysis destabilizes the polymer by weakening the lateral interactions between adjacent protofilaments, contributing to microtubule dynamics. MTs grow when a GTP cap is present at the plus end. However, if this cap is lost, the MT undergoes rapid depolymerization, a process known as "catastrophe." MTs alternate stochastically between phases of growth and shrinkage, a phenomenon termed "dynamic instability." A schematic representation of tubulin folding, quality control, and heterodimer recycling pathways is also shown. Newly synthesized tubulin, emerging from ribosomes, interacts with prefoldin, which delivers this protein to the cytosolic chaperonin containing TCP1 (CCT). α-tubulin, in a quasi-native 3D structure, is then accepted by tubulin-binding cofactor B (TBCB), whereas β-tubulin is accepted by tubulin-binding cofactor A (TBCA). The two tubulins continue to follow separate routes and are delivered to tubulin-binding cofactor E (TBCE) and tubulin-binding cofactor D (TBCD), respectively. The two routes converge in a super-complex where α- and β-tubulin are assembled together, and tubulin-binding cofactor C (TBCC) stimulates the GTPase activity of β-tubulin, releasing the native tubulin heterodimer. Upon replacement of GDP with GTP, the tubulin heterodimer becomes competent to polymerize. Tubulin cofactors also disassemble heterodimers for recycling or degradation. TBCD and TBCE independently dissociate dimers, with TBCE's activity enhanced by TBCB. Recycled tubulin regulates free tubulin concentration and therefore MT dynamics (adapted from Camelo et al. 2016; Roll-Mecak 2020; Carmona et al. 2023). (Schematic representation created using BioRender)

monomers follow distinct pathways: α-tubulin is captured by TBCB, while β-tubulin is captured by TBCA. Subsequently, TBCE and TBCD bind to α- and β-tubulin, respectively. After TBCC binding, a super-complex is formed, which stimulates GTP hydrolysis by β-tubulin and leads to the release of α/β-tubulin-GDP heterodimers. The exchange of GDP for GTP generates a functional α/β-tubulin dimer that is competent for polymerization into a MT. In addition, tubulin heterodimers released from MTs can be dissociated by cofactors and either recycled or degraded. Both TBCD and TBCE can dissociate the tubulin heterodimer independently; however, TBCE's dissociation activity is significantly enhanced by the presence of TBCB. During this process, β-tubulin is retained by TBCD while α-tubulin is stabilized by the TBCB/TBCE complex. Notably, TBCA primarily receives β-tubulin from the dissociation of pre-existing heterodimers rather than from newly synthesized tubulin. Both pathways may ultimately lead to the degradation of tubulin monomers via the ubiquitin-proteasome system through unknown mechanisms. The TBCE/TBCB + TBCA system plays a crucial role in recycling tubulin heterodimers, thereby regulating the critical concentration of free tubulin heterodimers and controlling MT dynamics within cells (Nolasco et al. 2021).

Folded tubulin dimers are assembled head-to-tail into elongated protofilaments. These in turn assemble laterally in groups, most commonly of 13, forming the MTs. The basic scaffold of MTs is highly unchanging, but these filaments are also functionally diverse. Different types of tubulin isotypes are encoded in the genome by the distinct members of tubulin gene families and may be incorporated in MTs in varied proportions. For example, neuronal function is dependent on the appropriate expression of specific tubulin isotypes (Poirier et al. 2010; Tischfield et al. 2010). Additionally, MTs are affected by a wide range of mostly reversible but also irreversible PTMs. The interaction of these diverse factors with MTs, in various combinations, originates the "tubulin code" (Verhey and Gaertig 2007) which greatly expands the structural properties and functional diversity of MTs. Furthermore, distinct PTMs are preferentially found in differing MT subpopulations (Wloga and Gaertig 2010) and diverse tubulin isotypes and PTMs may be combined in the same MT (Roll-Mecak 2019), adding increased complexity to MTs and expanding their functional diversity. The "tubulin code" concept was first introduced by Verhey and Gaertig (2007), in analogy to the previously proposed "histone code" and has been reviewed recently in more detail by Janke and Magiera (2020) and McKenna et al. (2023) (Verhey and Gaertig 2007; Janke and Magiera 2020; McKenna et al. 2023).

The most well-characterized tubulin PTMs include the reversible modifications acetylation, detyrosination/re-tyrosination (Barra et al. 1973; Raybin and Flavin 1975; Hallak et al. 1977; Prota et al. 2013; Aillaud et al. 2017; Nieuwenhuis et al. 2017; Landskron et al. 2022), glutamylation (Garnham et al. 2015; Szczesna et al. 2022; Genova et al. 2023; Delgado et al. 2024), and glycylation (Redeker et al. 2005; Wloga et al. 2009; Rogowski et al. 2009; Szczesna et al. 2022), and the irreversible modifications Δ2- and Δ3-tubulin (Paturle-Lafanechère et al. 1991; Aillaud et al. 2016). Apart from these PTMs, tubulin is also the target of other PTMs, still less studied. Arginylation (MacTaggart et al. 2023), methylation (Kearns et al.

2021), palmitoylation (Caron 1997), phosphorylation (Fourest-Lieuvin et al. 2006), polyamination (Song et al. 2013), SUMOylation (Feng et al. 2021), and ubiquitylation (Huang et al. 2009) have been described modifying tubulin. However, little is known of how these modifications influence MT structure and function.

The acetylation PTM consists in the addition of a functional acetyl group (CH_3CO) to specific amino acid residues, most notably lysines (K). Tubulin acetylation can occur at various K residues (Choudhary et al. 2009), but the most well-characterized site for this modification is K40 of α-tubulin, which has been extensively studied (L'Hernault and Rosenbaum 1985; LeDizet and Piperno 1987). Tubulin K40 acetylation is a reversible modification catalyzed by tubulin N-acetyltransferase 1 (αTAT1) (Akella et al. 2010; Shida et al. 2010) and is deacetylated by histone deacetylase 6 (HDAC6) (Hubbert et al. 2002) and sirtuin 2 (SIRT2) (North et al. 2003).

In this chapter, we will focus on tubulin acetylation, examining its debated role in MT properties and cellular functions from an evolutionary perspective. We will explore the regulation of tubulin lysine acetyltransferases and deacetylases, and how this system enables cells to respond to environmental cues. Finally, we will review the current understanding of these processes and their implications for human disease.

4.2 Microtubule Acetylation: Mechanisms and Cellular Functions

MTs are highly dynamic structures characterized by a rapid exchange of tubulin dimers at their ends, which enables them to quickly transition between phases of stable growth and rapid shrinkage. This behavior, known as dynamic instability, allows for rapid reorganization in response to cellular challenges (Mitchison and Kirschner 1984; Horio and Hotani 1986; Burbank and Mitchison 2000). Cells also possess a set of stable MTs (Schulze and Kirschner 1987) which present a slow dimer exchange rate. Stable MTs are often enriched in PTMs.

In general, PTMs are found most frequently in the carboxyl-terminal tail (CTT) of α- and β-tubulin, located at the outer surface of MTs, and the incorporation of the PTM usually occurs in tubulin heterodimers already assembled into MTs (Verhey and Gaertig 2007). Initially, it was believed that these PTMs directly contributed to MT stabilization. However, more recent research suggests that their enrichment is more likely a result of stabilization, acquired over time, making them more common in stable MTs (Verhey and Gaertig 2007). In cells, acetylation of α-tubulin K40 often occurs in stable cytoplasmatic MTs, as well as complex stable MT structures like cilia axonemes and centrioles (Orbach and Howard 2019). For a long time, it was debated whether MT stabilization was correlative or causative. In fact, some studies show that acetylation takes place after MT stabilization (Janke and Bulinski 2011), and does not protect against depolymerization (Palazzo et al. 2003).

The question of whether K40 α-tubulin acetylation regulates MT dynamics has been explored since early studies. However, the present evidence is conflicting, and the available data has been inadequate to resolve this matter. Initial research indicated that K40 α-tubulin acetylation does not impact the dynamicity of MTs (Table 4.1). A study using calf brain purified tubulin did not find significant influence of acetylation on polymerization or depolymerization in vitro (Maruta et al. 1986). However, this study had some experimental limitations, as it used neuronal tubulin, which is known to possess additional post-translational modifications. This complicates the ability to rule out their potential interference with the observed results (Garnham and Roll-Mecak 2012). Additionally, other studies suggest that tubulin acetylation is not required for cilia assembly. For instance, in the ciliate protozoa *Tetrahymena* and the algae *Chlamydomonas reinhardtii*, mutants lacking αTAT1 (Alpha-Tubulin Acetyltransferase 1) or expressing a non-acetylatable K40R α-tubulin exhibit relatively normal cilia (Kozminski et al. 1993; Gaertig et al. 1995; Akella et al. 2010). Conversely, K40 α-tubulin acetylation is not detected across tissues and cellular structures of mutant mice lacking αTAT1 that exhibit subfertility, suggesting that acetylation is essential for sperm motility (Kalebic et al. 2013). Furthermore, various ciliopathies and other conditions, such as Bardet-Biedl syndrome and Huntington's disease, present ciliary dysfunction associated with altered tubulin acetylation (Loktev et al. 2008; Keryer et al. 2011).

Interestingly, MT acetylation is regulated by substrate rigidity through talin, a component of focal adhesion complexes, and actomyosin-dependent mechanosensing (Seetharaman et al. 2022). In previous studies, the authors demonstrated that αTAT1 and acetylated MTs localize at focal adhesions, promoting cell migration (Bance et al. 2019). In their most recent work, they show that αTAT1 at focal adhesions enhances their turnover. Mechanostimulation further promotes the recruitment of αTAT1 to focal adhesions, resulting in MT acetylation and the translocation of Yes-associated protein (YAP), a RhoGTPase-dependent transcriptional activator, to the nucleus (Seetharaman et al. 2022). Strikingly, YAP, a key component of the Hippo-YAP signaling pathway, plays a critical role in coordinating organ growth and homeostasis with nutrition and metabolism (Ibar and Irvine 2020). The MT acetylation also promotes the release of the guanine nucleotide exchange factor (GEF-H1) from MTs activating RhoA, actomyosin contractility, and traction forces (Seetharaman et al. 2022). In this view, the cytoskeleton interplay, the nature of the extracellular matrix (ECM), the presence of other tubulin PTMs, and cell adhesion, all have the potential to modulate the tubulin acetylation which in turn affects all these cellular functions.

To date, tubulin is the only known substrate of the lysine acetyltransferase αTAT1, while the lysine deacetylases HDAC6 and SIRT2 act on a variety of other substrates (discussed below). When comparing MTs to soluble tubulin dimers, αTAT1 shows a preference for MTs over soluble tubulin (Maruta et al. 1986; Akella et al. 2010; Kormendi et al. 2012). In contrast, HDAC6 favors soluble tubulin (Miyake et al. 2016; Skultetyova et al. 2017), while SIRT2 modifies both MTs and soluble tubulin without preference (North et al. 2003). Interestingly, methylation,

Table 4.1 Acetylation exerts a wide impact on MTS, influencing its properties, functions, and dynamics

K40 α-tubulin acetylation association with MT structures and roles	References
Specialized MT structures are enriched in acetylation	
Highly acetylated MTs are found: i) in cellular subpopulations of MTs, e.g., cytoplasmatic MTs around nucleus and Golgi apparatus and in axons (presenting a positive gradient toward the terminal end); ii) in complex structures such as ciliary axonemes, kinetochores, centrioles, the mitotic spindle, and the midbody.	Piperno and Fuller (1985), Cambray-Deakin and Burgoyne (1987), Wilson and Forer (1989), Ahmad et al. (1993), Brown et al. (1993), Barisic et al. (2015), and Orbach and Howard (2019)
ΔαTAT1 mice lack K40 α-tubulin acetylation.	Kalebic et al. (2013)
Acetylation supports the function of specialized MT structures	
αTAT1 and acetylated MTs localize at focal adhesions promoting their turnover.	Bance et al. (2019) and Seetharaman et al. (2022)
Various ciliopathies and other conditions present ciliary dysfunction associated with altered tubulin acetylation.	Loktev et al. (2008) and Keryer et al. (2011)
ΔαTAT1 mice lack K40 α-tubulin acetylation in tissues, exhibiting subfertility.	Kalebic et al. (2013)
Acetylation regulates the interaction of MTs, MAPs, and motors	
Tau presents preferential binding to acetylated MTs.	Saragoni et al. (2000)
Overexpression of tau, MAP1B, or MAP2c correlates with increased K40 α-tubulin acetylation levels.	Lee and Rook (1992) and Takemura et al. (1992)
Katanin severing activity is enhanced by acetylation.	Sudo and Baas (2010) and Mao et al. (2014)
Fidgetin presents a higher affinity for acetylated MTs in *Drosophila,* but for deacetylated MTs in mice.	Leo et al. (2015)
K40 α-tubulin acetylation enhances the motility rate of axonemal dynein.	Alper et al. (2014)
Acetylation directly influences the intrinsic structure of MTs	
α-Tubulin K40 acetylation increases MT flexibility.	Portran et al. (2017)
Acetylation protects MTs from mechanical breakage.	Xu et al. (2017)
Acetylation facilitates self-repair by promoting the incorporation of new GTP-tubulin heterodimers.	Schaedel et al. (2015), Portran et al. (2017); Janke and Montagnac (2017), and Eshun-Wilson et al. (2019)
αK40 acetylation affects two processes dependent on inter-protofilament interactions, reducing the nucleation frequency and accelerating the shrinkage rate of MTs.	Portran et al. (2017)
MT damage sites contribute to the establishment of acetylation by allowing access of αTAT1 to the MT lumen.	Andreu-Carbó et al. (2024)
Effects of acetylation were not evident in some biological systems	
No effect on polymerization or depolymerization rates of tubulin from calf brain (heavily acetylated) tubulin under various conditions.	Maruta et al. (1986)
Tetrahymena thermophila mutants lacking αTAT1 exhibit relatively normal motile cilia.	Gaertig et al. (1995) and Akella et al. (2010)
Chlamydomonas reinhardtii mutants expressing a non-acetylatable K40R α-tubulin lack gross phenotype.	Kozminski et al. (1993)

Abbreviations (in alphabetical order): *αK40* Acetylation of Alpha-Tubulin at Lysine 40, *αTAT1* Alpha-Tubulin Acetyltransferase 1, *Ax* Axon, *Dynein* Dynein, *GTP* Guanosine Triphosphate, *K40* Lysine 40, *MAP1B* Microtubule-Associated Protein 1B, *MAP2c* Microtubule-Associated Protein 2c, *MTs* Microtubules

involving the addition of a methyl group, can also occur at α-tubulin K40 (Kearns et al. 2021). Tubulin methylation is catalyzed by histone methyltransferase set domain-containing 2 (SETD2) (Park et al. 2016). In the case of tubulin, SETD2 preferentially modifies tubulin dimers (Kearns et al. 2021). Thus, the balance between tubulin acetylation and methylation is dependent on MT dynamics.

In contrast to most tubulin PTMs which occur on the C-terminal tail, K40 acetylation of α-tubulin uniquely occurs within the MT lumen (Soppina et al. 2012). Still, substantial evidence suggests that this PTM influences the interactions of MT-associated proteins (MAPs) and regulates MT molecular motors (Table 4.1). Tau protein, for instance, has been shown to preferentially bind to acetylated and tyrosinated MTs (Saragoni et al. 2000). Moreover, Tau overexpression increases K40 α-tubulin acetylation levels, protecting MTs against depolymerizing agents (Lee and Rook 1992; Takemura et al. 1992). Beyond Tau, the overexpression of other MAPs, such as MAP1B and MAP2c, has also been linked to increased levels of K40 α-tubulin acetylation levels (Takemura et al. 1992). In vitro studies suggest that this modification enhances the motility rate of dynein from the outer arm of the axoneme (Alper et al. 2014). In neurons, K40 α-tubulin acetylation is more abundant in MTs from axons, forming a gradient toward the axon terminal (Ahmad et al. 1993; Brown et al. 1993), compared to dendritic MTs (Piperno and Fuller 1985; Cambray-Deakin and Burgoyne 1987; Ahmad et al. 1993). Dendritic MTs are antiparallel exhibiting distinct polarities, with the heavily acetylated MTs (in bundles) associated with retrograde transport driven by kinesin-1, while lightly acetylated MTs are linked to anterograde transport driven by kinesin-3 (Tas et al. 2017). During cell division, K40 α-tubulin acetylation is abundant in the mitotic spindle, midbody, and kinetochore MTs (Wilson and Forer 1989; Barisic et al. 2015).

Acetylation also influences the MT-severing proteins katanin and fidgetin (Table 4.1). Katanin severing activity is enhanced by acetylation in the neuronal and non-neuronal systems of *Drosophila* (Mao et al. 2014) and in rat fibroblasts and neurons (Sudo and Baas 2010). In rat neurons, the katanin-stimulating effect of acetylation was observed in dendrites but not in the significantly more acetylated axons, a difference due to the increased binding of Tau protein to MTs of axons (Sudo and Baas 2010). In *Drosophila*, fidgetin also shows higher affinity for acetylated MTs; however, in mouse neurons, fidgetin shows preference for deacetylated MTs, presenting thus differing effects in different organisms (Leo et al. 2015). It is still unclear why MT severing proteins show preference for acetylated MTs, but this modification appears to regulate MT severing activity and thus influence MT stability.

Inside the MTs, numerous MT inner proteins (MIPs) form both lateral and longitudinal interactions with the tubulin protofilaments, creating an internal scaffold that provides mechanical support (Ma et al. 2019; Wang et al. 2021). MIPs can indirectly affect the external surface of the tubulin lattice by altering the angle between protofilaments (Ichikawa et al. 2017) and/or extending to the external surface through openings in the lattice between adjacent protofilaments (Ma et al.

2019). The MT lumen is non-uniformly covered by MIPs, which may confer variable flexibility within the tubulin lattice (Ma et al. 2019). Although many MIPs bind the K40 α-tubulin region and may recognize acetylation (Ma et al. 2019), the relationship between K40 α-tubulin acetylation and MIPs requires further investigation. To date, only one MIP, MAP6, has been identified in mammalian cytosolic MTs (Cuveillier et al. 2020). However, aside the cilia/basal body system, MIPs have also been observed in cortical MTs of *Toxoplasma gondii* (Wang et al. 2021) and in MTs within the ventral disk from *Giardia lamblia* (Schwartz et al. 2012). These findings suggest that MIPs may be more important for stable/structural MTs than for dynamic interphase MTs.

Numerous studies have demonstrated that αTAT1 is essential for maintaining the stability and organization of higher-order MT arrays and for regulating protofilament number in vivo (Cueva et al. 2012; Topalidou et al. 2012; Neumann and Hilliard 2014). These findings support the idea that K40 acetylation mechanically stabilizes MTs (Table 4.1). Later, microfluidic manipulations revealed that tubulin αK40 acetylation enhances MT resilience by increasing their flexibility, making them more resistant to repeated mechanical stresses (Portran et al. 2017). Acetylated MTs are thus more durable, while deacetylated ones accumulate damage over time. Acetylation alone has been shown to protect MTs from mechanical breakage (Xu et al. 2017). This increased flexibility is believed to result from reduced inter-protofilament interactions, allowing αTAT1 to fine-tune the load-bearing capacity of MTs and mechanical resistance to stress (Eshun-Wilson et al. 2019). In contrast, deacetylated MTs, with enhanced lateral contacts, are more rigid and brittle, making them more prone to breakage under stress (Portran et al. 2017; Xu et al. 2017). Additionally, it facilitates self-repair by promoting the incorporation of new GTP-tubulin heterodimers (Schaedel et al. 2015; Portran et al. 2017; Janke and Montagnac 2017; Eshun-Wilson et al. 2019). It was also found that acetylated tubulin assembles more slowly and disassembles three times faster than deacetylated tubulin, though polymerization rates are similar once MTs are formed suggesting that acetylation plays a role in the regulation of MT dynamics (Portran et al. 2017).

Recently, it was shown that damage sites, caused by various factors, contribute to the establishment of the acetylation pattern within cells (Andreu-Carbó et al. 2024) (Table 4.1). Andreu-Carbó et al. (2024) demonstrated that the number of damage sites on a MT is independent of kinesin-1 and is sufficient for the small αTAT1 to access the MT shaft and diffuse rapidly throughout the MT lumen (Andreu-Carbó et al. 2024) (Fig. 4.2). Additionally, they found that overexpression of kinesin-1 leads to MT deacetylation, mediated by the HDAC6 enzyme, which diffuses poorly and primarily deacetylates regions around the damage sites. This perspective overlooks the possibility that a specific subpopulation of acetylated MTs with distinct functions may arise from αTAT1 accessing the polymer ends, ultimately contributing to the observed acetylation pattern (Fig. 4.2).

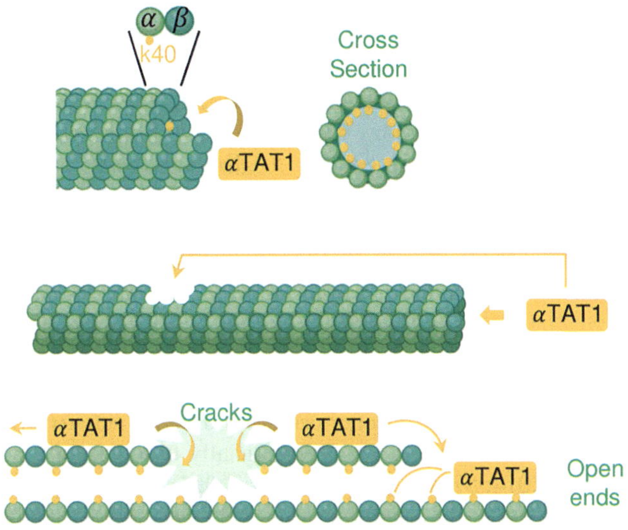

Fig. 4.2 Proposed model for tubulin K40 acetylation in a microtubule (MT) by αTAT1. The position of α-tubulin K40 is highlighted in orange within the MT lumen, as seen in a cross-section of the polymer. Potential entry points for αTAT1 into the MT lumen are indicated. αTAT1 can access the lumen through the open ends of the MT or via transient cracks in the MT wall. These cracks may be generated by mechanical stress or interactions with motor proteins. (Schematic representation created using BioRender)

4.3 Tubulin Acetylation: Key Enzymes and Their Regulatory Mechanisms

4.3.1 Tubulin Lysine Acetyltransferases and Deacetylases

The lysine residue K40 acetylation of α-tubulin is the most extensively studied acetylated modification and is likely the primary acetylation site in MTs. However, other potential acetylation sites were already identified by spectrometry analysis in both α- and β-tubulins, namely the lysine residue 394 (Lys394) and lysine residue 252 (Lys252), respectively (Choudhary et al. 2009; Weinert et al. 2011; Lundby et al. 2012; Liu et al. 2015a, b; Hansen et al. 2019). The K40 acetylation occurs when α-tubulin acetyltransferase αTAT1 catalyzes the transfer of an acetyl group from acetyl-coenzyme A (acetyl CoA) to the ε-amino group of K40 residue (Friedmann et al. 2012; Taschner et al. 2012). This enzyme activity (αTAT1/MEC17) can be found across diverse organisms within the eukaryotic lineage such as in the unicellular protozoa ciliate *Tetrahymena* (Gaertig et al. 1995; Akella et al. 2010) and the algae *Chlamydomonas* (Greer et al. 1985; L'Hernault and Rosenbaum 1985; LeDizet and Piperno 1987). It is also present in multicellular organisms like the nematode *Caenorhabditis elegans*, zebrafish embryos, and human cells (Akella et al. 2010; Shida et al. 2010). In human cells, a deficiency in αTAT1 abolishes the K40 acetylation (Kalebic et al. 2013).

αTAT1 is a member of the Gcn5-related N-acetyltransferase (GNAT) superfamily (Steczkiewicz et al. 2006; Friedmann et al. 2012). It features an N-terminal catalytic domain, which shares homology with other acetyltransferases, and a disordered C-terminal domain that contains cellular localization signals (Friedmann et al. 2012; Iuzzolino et al. 2024). As previously mentioned, the K40 residue is located within the lumen of the MT. αTAT1 accesses the MT lumen either through the open ends of the tubule or via lateral bends and breaks in the MT lattice. The enzyme's mobility within the MT lumen is regulated by its affinity for acetylation sites (Coombes et al. 2016). Before acetylation occurs, αTAT1 concentrations reach an equilibrium between the inside and outside of the MT lumen. Its slow catalytic rate functions as a timer for MT stability (Szyk et al. 2014). Dynamically unstable MTs are disassembled before αTAT1 can perform its acetylation activity, which explains why higher levels of acetylation are predominantly observed in long-lived MTs (Kull and Sloboda 2014; Szyk et al. 2014; Janke 2014). In animal cells, in addition to αTAT1 activity, other lysine acetyltransferases are responsible for minor α-tubulin Lys40 acetylation or the acetylation of different tubulin lysine residues. These lysine acetyltransferases, including αTAT1, and their key features related to MT and tubulin acetylation (while recognizing that many of these proteins also have other cellular functions) are summarized in Table 4.2.

Tubulin acetylation is a reversible process, with the acetyl group being removed from the lysine residue by specific enzymes known as deacetylases (Table 4.2). Histone deacetylases (HDACs) are a group of enzymes responsible for removing acetyl groups from histones, as well as other proteins, including tubulin. HDACs are conserved from yeast to human and can be classified into distinct classes based on their structural and functional similarities. However, the two primary enzymes responsible for tubulin deacetylation are HDAC6 and SIRT2. HDAC6, a member of the class II histone deacetylase family with homology to yeast Hda1 (Park and Kim 2020), is mainly localized in cytoplasm, perinuclear regions, and at the cell's leading edge, co-localizing with MTs. This deacetylase catalyzes deacetylation of α-tubulin and contains two HDAC domains and a C-terminal zinc finger domain, which binds ubiquitin (Hubbert et al. 2002; Zhang et al. 2003). The inhibition of HDAC6 activity, either through trichostatin A (TSA), tubastin A, or by gene silencing leads to an increase of acetylated α-tubulin. HDAC6 appears to preferentially catalyze the deacetylation of tubulin dimers rather than in MT polymers in contrast to acetylation, which predominantly occurs in MT polymers (Matsuyama et al. 2002).

The Sirtuin 2 (SIRT2) deacetylase is a member of the class III histone deacetylases and exhibits nicotinamide adenine dinucleotide (NAD^+)-dependent deacetylase activity, similar to the yeast Sir2 protein. Enzymes in this class are commonly referred to as sirtuins (Gray and Ekström 2001; Zhang et al. 2003). SIRT2 is predominantly cytoplasmic and co-localizes with α-tubulin and HDAC6. SIRT2 is expressed in two isoforms: isoform 1, which encodes a 389 amino acid residues protein, and isoform 2, which lacks the first 37 N-terminal amino acid residues

Table 4.2 Main tubulin lysine acetylases and deacetylases

Tubulin lysine acetyl-transferases	Family	Key features	Target lysine (K) residue and tubulin	Main functions associated with or resulting from MTS Acetylation	References
αTAT1	Member of the Gcn5-related N-acetyltransferase (GNAT) superfamily	Possesses an N-terminal catalytic domain and a disordered C-terminal domain with both nuclear export and import signals.	K40 in α-tubulin in MTs	Main acetylase for K40 α-tubulin residue, the primary acetylation site in MTs.	Steczkiewicz et al. (2006), Friedmann et al. (2012), Taschner et al. (2012), and Roy et al. (2022)
ARD1-NAT1 complex	N-acetyltransferase	ARD1 acts as the catalytic subunit and NAT1 acts as a cofactor;	Mts (?)[a]	Required for dendritic arborization during neuronal development.	Ohkawa et al. (2008)
GCN5	Member of the GNAT superfamily	GCN5 forms a complex with Myc-nick[b]; Myc-nick recruits GCN5 to MTs	α-Tubulin	GCN5-mediated MT acetylation. Promotes breast cancer cell migration. It is involved in muscle differentiation.	Conacci-Sorrell et al. (2010) and Li et al. (2015)
ELP3	Member of the GNAT superfamily	Is the catalytic subunit of the highly conserved acetyltransferase Elongator complex; contains a radical S-adenosyl-L-methionine (SAM) domain and a putative histone acetyltransferase domain. It localizes in both the nucleus and cytoplasm neurons.	α-Tubulin	Regulates cell migration, motor-based trafficking, and neuron branching by acetylating α-tubulin in mouse neurons. It also influences motor neuron axonal branching and length in zebrafish. Whether elongator functions as a true acetyltransferase for α-tubulin remains debated.	Gardiner et al. (2007), Creppe et al. (2009), Wynshaw-Boris (2009), Solinger et al. (2010), and Dalwadi and Yip (2018)
NAT10	Member of the GNAT superfamily	Is a nucleolar protein that localizes in the midbody during telophase.	α-Tubulin	Catalyzes α-tubulin acetylation enhancing the stability of MTs to regulate cytokinesis and promoting cell migration. The cytoplasmic and membranous localization of NAT10 mutants is related to human cancer	Shen et al. (2009) and Tan et al. (2018)

(continued)

Table 4.2 (continued)

Tubulin lysine acetyl-transferases	Family	Key features	Target lysine (K) residue and tubulin	Main functions associated with or resulting from MTS Acetylation	References
San	N-acetyltransferase	Contains an unrelated N-acetyltransferase domain. It modifies tubulin heterodimers but not polymerized MTs in vitro.	K252 in free β-tubulin	K252 affects the assembly of the α/β-tubulin heterodimer and MT polymerization and dynamics. In human HeLa cells, san depletion increases the regrowth rate of MTs after nocodazole treatment.	Chu et al. (2011)
HDAC6	Member of the histone deacetylase (HDAC) family	Contains two hdac domains and a C-terminal zinc finger domain that binds ubiquitin. Co-localizes with MTs, being present at the cytoplasm, perinuclearly and in the cell's leading edge.	α-Tubulin (preference for tubulin dimers)	HDAC6 α-tubulin deacetylation regulates MTs and cytoskeleton-mediated cell motility. In vivo, decreased HDAC6 activity leads to increased α-tubulin and MT acetylation whereas the overexpression of HDAC6 produces global α-tubulin deacetylation.	Hubbert et al. (2002), Matsuyama et al. (2002), and Zhang et al. (2003)
SIRT2	Member of the NAD-dependent sirtuin protein deacetylase family	Possess a conserved catalytic core and two distinct end domains, a Rossman fold domain and a zinc-binding domain. Expressed in two isoforms, both with deacetylase activity. Present in the cytoplasm, colocalizing with MTs and HDAC6, but also localizing to the nucleus mitotic structures during interphase and mitosis.	α-Tubulin K40 in MTs and soluble tubulin	Functions interdependently of HDAC6 in tubulin deacetylation. siRNA SIRT2 knockdown leads to tubulin hyperacetylation.	North et al. (2003) and North and Verdin (2007a, b)

[a]The specific K residue and which tubulin is acetylated is not known

[b]Myc-nick is a cytoplasmic, transcriptionally inactive member of the MYC oncoprotein family, generated by a proteolytic cleavage of full-length MYC

Abbreviations (in alphabetical order): *αTAT1* Alpha Tubulin Acetyltransferase 1, *ARD1-NAT1 complex* Arrest Defective 1-N-Acetyltransferase 1 Complex, *ELP3* Elongator Protein 3, *GCN5* General Control Non-derepressible 5, *HDAC6* Histone Deacetylase 6, *K252* Lysine 252 in β-tubulin, *K40* Lysine 40 in α-tubulin, *MTs* Microtubules, *NAT10* Nucleolar Protein 10, *SAM* S-Adenosyl-L-Methionine, *San* San Acetyltransferase, *SIRT2* Sirtuin 2

(North and Verdin 2007a). Despite this difference, both isoforms are capable to deacetylate tubulin. Like other sirtuins, SIRT2 features a conserved catalytic core of approximately 275 amino acid residues. It also contains two distinct domains, namely a highly conserved Rossman fold domain on one end, and a zinc-binding domain on the other end. These domains are connected by an acylated substrate-binding cleft and a NAD^+ co-substrate binding pocket (Bursch et al. 2024). Similar to HDAC6, SIRT2 knockdown leads to tubulin hyperacetylation. However, unlike HDAC6, SIRT2 is inhibited by nicotinamide but not by TSA (North et al. 2003).

4.3.2 *Regulatory Mechanisms*

Tubulin lysine acetyltransferases and deacetylases are tightly regulated through multiple mechanisms, including transcriptional control, post-transcriptional regulation by miRNAs, protein-protein interactions, PTMs such as phosphorylation, and their subcellular localization (Legube and Trouche 2003). Additionally, the activity of these enzymes is critically influenced by the intracellular availability of acetyl-CoA, which serves as the acetyl group donor for lysine acetylation catalyzed by αTAT1, and by the dependence of the SIRT2 deacetylase on NAD+ as a cofactor.

4.3.2.1 Transcriptional Regulation

The expression of αTAT1 gene in retinal pigment epithelial cells (RPE-1 cells) is regulated by the MRTF/SRF transcriptional pathway, which binds directly to the promoter region to control transcription (Fernández-Barrera et al. 2018) (See Table 4.3). This regulation relies on formin INF2, a protein that not only stimulates actin assembly, but also affects MT dynamics by binding to MTs (Chesarone et al. 2010). INF2 knockdown or knockout significantly reduces MT acetylation levels, due to accentuated decrease of αTAT1 mRNA expression. In this system formins modulate tubulin acetylation by both stabilizing MTs and activating the transcription of the α-TAT1 gene while remodeling actin cytoskeleton (Chesarone et al. 2010). This highlights the intricate interplay and mutual regulation between the actin and MT cytoskeletal networks.

The transcription of the αTAT1 gene is also regulated by DNA methyltransferase 3 alpha (DNMT3A), whose expression is influenced by a N6-methyladenosine (m6A) post-transcriptional modification of its mRNA, a process dependent on the activity of methyltransferase-like protein 3 (METTL3). DNMT3A typically binds to the promoter region of the αTAT1 gene, promoting its expression while inhibiting the trimethylation of histone H3 at lysine 27 (H3K27me3) mediated by EZH2 (the catalytic subunit of the polycomb repressive complex 2). Notably, in bone marrow-derived macrophages, reduced expression of METTL3 leads, as expected, to a loss of α-tubulin acetylation. This reduction has been linked to enhanced macrophage

Table 4.3 Factors involved in transcription regulation of tubulin lysine acetyltransferases and deacetylases

Gene name	Transcription regulators	Biological model	Major features	References
ATAT1 (codes αTat1)	MRTF/SRF	RPE-1 cells	Increases αTAT1 mRNA expression and depends on formin INF2.	Fernández-Barrera et al. (2018)
	DNMT3A	Mice bone marrow-derived macrophages	DNMT3A promotes αTAT1 transcription by inhibiting the H3K27me3 mediated by EZH2. DNMT3A is posttranscriptional regulated by METTL3.	Wu et al. (2010) and Yin et al. (2023)
	p300 (transcriptional coactivator of distinct DNA binding proteins)	HeLa cells	Negative regulator of αTAT1 expression; upon stress conditions is sequestered in the cytoplasm by hyperacetylated MTs.	Mackeh et al. (2014)
HDAC6	hnRNPK	HEK 293 cells	Regulates HDAC6 mRNA synthesis or metabolism; hnRNPK is a DNA/RNA binding protein.	Li et al. (2018)
SIRT2	β-Catenin		Directly binds to the Sirt2 gene promoter inhibiting SIRT2 expression.	Li et al. (2021)

Abbreviations (in alphabetical order): *ATAT1* Alpha-Tubulin Acetyltransferase 1, *β-catenin* Beta-Catenin, *DNMT3A* DNA Methyltransferase 3 Alpha, *EZH2* Enhancer of Zeste Homolog 2, *H3K27me3* Trimethylation of histone H3 at lysine 27, *HDAC6* Histone Deacetylase 6, *HEK 293 cells* Human Embryonic Kidney 293 cells, *hnRNPK* Heterogeneous Nuclear Ribonucleoprotein K, *INF2* Inverted Formin 2, *METTL3* Methyltransferase-Like Protein 3, *MRTF/SRF* Myocardin-Related Transcription Factor/Serum Response Factor, *MTs* Microtubules, *p300* Histone acetyltransferase p300 (transcriptional coactivator), *SIRT2* Sirtuin 2

migration, which is associated with symptom improvement in Alzheimer's disease through increased clearance of Aβ aggregates (Wu et al. 2010; Yin et al. 2023).

The complex interplay between distinct factors regulating MT acetylation is also highlighted by the regulation of αTAT1 expression by the histone acetyltransferase p300, a transcriptional coactivator that integrates signals from various pathways (Talukdar and Chatterji 2023). The mRNA levels of αTAT1 are negatively regulated by the p300, as evidenced by hyperacetylation of tubulin in p300 knockout cells (Mackeh et al. 2014). Under stress conditions, p300 translocates from the nucleus to the cytoplasm where it is sequestered by binding to hyperacetylated MTs (Mackeh et al. 2014; Iuzzolino et al. 2024). This sequestration relieves the negative regulation of αTAT1. Notably, p300 repression is small ubiquitin-like modifier (SUMO)-dependent and involves the recruitment and interaction with HDAC6 which mediates transcriptional repression (Girdwood et al. 2003; Legube and Trouche 2003). Furthermore, p300 acetylates HDAC6 at 16 lysine residues, suppressing its deacetylation activity and preventing the deacetylation of tubulin (Han et al. 2009).

Similarly, p300 interacts with SIRT2 triggering its acetylation and subsequent downregulation of its deacetylation activity (Han et al. 2008).

The balance between tubulin acetylation and deacetylation plays a crucial role in fine-tuning the levels of tubulin/MT acetylation within cells, particularly in response to external signals. This equilibrium is strictly dependent on the regulatory mechanisms controlling the activity of acetyltransferases and deacetylases. When these mechanisms are compromised, cellular and tissue homeostasis can be significantly disrupted. The transcriptional deregulation of HDAC6 and SIRT2 genes plays a crucial role in cancer cells, influencing key processes such as cell motility, metastasis, and cell cycle progression. For example, in the estrogen-receptor-positive breast cancer cell line MCF-7 disrupted estrogen signaling increases HDAC6 expression, leading to enhanced cell motility and a shift in cell morphology from round to axon-like, contributing to a more aggressive phenotype (Saji et al. 2005). Furthermore, HDAC6 is regulated by the heterogeneous nuclear ribonucleoprotein K (hnRNPK), a DNA/RNA binding protein that shuttles between the nucleus and the cytoplasm. This protein has been shown to negatively regulate HDAC6 mRNA and protein levels, and consequently affecting autophagy under normal nutrient conditions (Li et al. 2018).

Meanwhile, SIRT2 gene expression is negatively regulated by the Wnt/β-catenin signaling pathway in human colon cancer cells, where SIRT2 acts as a tumor suppressor. β-catenin directly binds to the SIRT2 promoter, inhibiting its mRNA and protein expression. In β-catenin knockdown cells SIRT2 expression increases (Li et al. 2021). While tubulin acetylation levels were not evaluated in these studies, the observed alterations likely involve changes in MT cytoskeleton.

4.3.2.2 miRNAs

Micro ribonucleic acids (miRNAs) are small, single-stranded, non-coding RNA molecules, typically 21–23 nucleotides in length, found in the genomes of a wide range of organisms, including animals and plants. These molecules play a crucial role in RNA silencing and the post-transcriptional regulation of gene expression. miRNAs are now recognized as key regulators of fundamental cellular processes, such as development, proliferation, differentiation, apoptosis, and oncogenic transformation (for review Gomes et al. 2013).

The expression of the HDAC6 gene is post-transcriptionally regulated by distinct miRNAs (see Table 4.4), whose dysregulation often leads to HDAC6 overexpression. This phenomenon, involving HDAC6 overexpression, is commonly linked to enhanced cell proliferation, metastasis, invasion, and tumor progression. For example, miR-548 downregulation is associated with non-Hodgkin lymphoma (Lwin et al. 2013), miR-206 with endometrial cancer cells (Zheng et al. 2020), and downregulation of miR-433 and miR-22 cholangiocarcinoma cells (Mansini et al. 2018). In contrast, in hepatocellular carcinoma, miR-221, regulated by the JNK/c-JUN and NF-κB signaling pathways, suppresses HDAC6 expression. In this context, the depletion of HDAC6 may enhance tumorigenesis (Bae et al. 2015).

Table 4.4 Regulation of HDAC6 and SIRT2 Expression by miRNAs in various biological models

miRNAs	Biological model	miRNA regulation	References
miR-548	Non-Hodgkin lymphoma	Downregulation causes HDAC6 overexpression	Lwin et al. (2013)
miR-206	Endometrial cancer cells	Downregulation causes HDAC6 overexpression	Zheng et al. (2020)
miR-433	Cholangiocarcinoma cells	Downregulation causes HDAC6 overexpression	Mansini et al. (2018)
miR-22	Cholangiocarcinoma cells	Downregulation causes HDAC6 overexpression	Mansini et al. (2018)
miR-221	Hepatocellular carcinoma	Downregulation suppresses HDAC6 expression	Bae et al. (2015)
miR-26a	Embryonic stem cells subjected to hypoxia	Upregulation suppresses HDAC6 expression	Lee et al. (2015)
miR-143	Astrocytes	Upregulation suppresses Sirt2 expression	Wang et al. (2024)
miR-150	Non-small cell lung cancer	Upregulation causes downregulation of Sirt2	Jiang et al. (2018)
miR-212-5p	Parkinson disease C57BL/6 mice; SH-SY5Y neuroblastoma cells	Upregulation causes downregulation of Sirt2	Sun et al. (2018)

miRNAs also regulate HDAC6 expression (Table 4.4) in embryonic stem cells subjected to hypoxia, where an increase in miR-26a that directly targets and suppresses HDAC6 expression leads to expression of myogenic marker genes like Pax7 (Lee et al. 2015).

The SIRT2 is also regulated by specific miRNAs creating complex patterns of regulation involving this deacetylase. Following spinal cord injury in astrocytes, miR-143 is upregulated, targeting SIRT2 and inhibiting transforming growth factor β (TGF-β)-induced proliferation (Wang et al. 2024). This regulation promotes cell apoptosis and reduces the levels of the intermediate filament protein vimentin. Conversely, stimulation with TGF-β suppresses miR-143 expression, leading to an increase in SIRT2 levels (Wang et al. 2024). In non-small cell lung cancer, miR-150 is upregulated, leading to the downregulation of SIRT2, which makes cancer cells more viable (Jiang et al. 2018). Similarly, in Parkinson's disease, increased SIRT2 levels can be counteracted by miR-212-5p, which downregulates SIRT2 expression (Sun et al. 2018).

The fact that HDAC6 and SIRT2 target a wide range of substrates beyond tubulin makes it challenging to directly correlate the studies mentioned above with only MT acetylation status and to determine how these specifically contribute for the described disruption of cell homeostasis.

4.3.2.3 Post-Translational Modifications

PTMs, particularly phosphorylation, play a pivotal role in regulating tubulin, the αTAT1, and the HDAC6 and SIRT2 (See Table 4.5). The autoacetylation of lysine residues has been suggested to enhance the catalytic activity of αTAT1 at MTs

Table 4.5 Post-translational modifications and regulatory mechanisms of αTAT1, HDAC6, and SIRT2

Protein	PTMs	Biological model	Target amino acid residue	Enzymes	Major features	References
αTAT1	Acetylation	Mouse cell lysates	Multiple K residues	αTAT1	Autoacetylation; enhances the catalytic activity	Kalebic et al. (2013)
	Phosphorylation	Fibroblasts	S236	CK2	Activation	You et al. (2022)
		Mouse embryonic endothelial cells (MEFs); COS-7, HeLa, and PANC1	S237	TAK1	Activation	Shah et al. (2018)
		HeLa	n/a	AMPK	Inhibition	Mackeh et al. (2014) and Shah et al. (2018)
		Proplatelets	Multiple sites on its C-terminal tail	PAK1	Inhibition	Dijk et al. (2020)
HDAC6	Phosphorylation	SH-SY5Y cells	S458	CK2	Activation	Watabe and Nakaki (2011)
		CHO, MEFs,293 T, and HeLa S3 cells	S1035	ERK	Activation	Williams et al. (2013)
		CHO, 293 T, and HeLa, A431 cells	Y570	EGFR	Inhibition	Deribe et al. (2009)
		HeLa and HEK-293 cells	S1060, S1062 and S1069	GRK2	Activation	Lafarga et al. (2012)
		RPE-1 cells	n/a	Aurora A kinase	Activation	Pugacheva et al. (2007)
		Hippocampal neurons	S22	GSK3β	Inhibition	Chen et al. (2010)
SIRT2	Phosphorylation	SH-SY5Y cells	S327, S331, and S335	GSK3β	Activation	Liu et al. (2019a)

AMPK AMP-Activated Protein Kinase, *Aurora A kinase* Aurora A Kinase, *CK2* Casein Kinase 2, *EGFR* Epidermal Growth Factor Receptor, *ERK* Extracellular Signal-Regulated Kinase, *GSK3β* Glycogen Synthase Kinase 3 Beta, *GRK2* G Protein-Coupled Receptor Kinase 2, *HDAC6* Histone Deacetylase 6, *PAK1* p21-Activated Kinase 1, *SIRT2* Sirtuin 2, *TAK1* Transforming Growth Factor Beta-Activated Kinase 1, *αTAT1* Alpha-Tubulin Acetyltransferase 1

(Kalebic et al. 2013). However, it seems that tubulin exhibits a stronger binding affinity for αTAT1 than αTAT1 has for itself, which likely accounts for the relatively low levels of αTAT1 autoacetylation observed in cells (Kalebic et al. 2013).

The activation of αTAT1 can also occur through phosphorylation mediated by casein kinase 2 (CK2), which phosphorylates αTAT1 at serine 236 (S236), promoting MT acetylation (You et al. 2022). Interestingly, CK2 not only phosphorylates αTAT1 but can also phosphorylate and activate HDAC6 at the S458 residue. This dual role suggests that CK2 may act as a regulatory switch, modulating the balance between MT acetylation and deacetylation in response to the cellular environment (Watabe and Nakaki 2011; You et al. 2022). In fibroblasts cultured on a soft matrix, this activation is regulated by the TGF-β pathway, whose disruption has been associated with cancer, inflammation, and neurological disorders. Furthermore, TGF-β-activated kinase 1 (TAK1), a key effector of the TGF-β pathway, directly interacts with and phosphorylates αTAT1 at serine 237 (S237), further promoting its activation (Shah et al. 2018). These findings support the existence of a critical interplay between TGF-β pathway and αTAT1 activation.

Another key kinase involved in αTAT1 regulation is Adenosine Monophosphate-Activated Protein Kinase (AMPK), a central enzyme in maintaining cellular energy balance. Under oxidative stress conditions, AMPK phosphorylates αTAT1 (Mackeh et al. 2014; Shah et al. 2018). Additionally, AMPK is also known to downregulate the acetyltransferase p300, which may concurrently repress αTAT1 expression, further influencing tubulin acetylation (Zhang et al. 2011; Mackeh et al. 2014).

The activity of αTAT1 can also be inhibited by phosphorylation in other serine/threonine residues. For example, during platelet maturation the p21-activated kinase 1 (PAK1) phosphorylates αTAT1 at multiple sites on its C-terminal tail inhibiting its activity (Dijk et al. 2020). Moreover, in the absence of PAK1 activity, cells have hyperacetylated MTs and lose their MT network integrity (Dijk et al. 2020).

Tubulin acetylation is also regulated by the phosphorylation of the deacetylase HDAC6. The Extracellular Signal-Regulated Kinase (ERK), a related protein-serine/threonine kinase associated with the MT cytoskeleton in cells (Ojea Ramos et al. 2022), plays a role in this process. ERK binds and phosphorylates HDAC6, stimulating its deacetylase activity and promoting cell migration by facilitating α-tubulin deacetylation (Williams et al. 2013). Specifically, activation of the EGFR-Ras-Raf-MEK-ERK signaling cascade leads to ERK1-mediated phosphorylation of HDAC6 at serine 1035 (S1035), which enhances its enzymatic activity (Williams et al. 2013).

Interestingly, HDAC6 negatively regulates the endocytosis and degradation of the epidermal growth factor receptor (EGFR) by modulating α-tubulin acetylation, which affects receptor trafficking along MTs (Deribe et al. 2009). In turn, EGFR phosphorylates HDAC6 at tyrosine 570 (Y570) upon binding to endosomal receptors, thereby inactivating it and establishing a negative feedback loop (Deribe et al. 2009). This finding revealed the existence of a regulatory interplay between HDAC6 activity, MT acetylation status, and EGFR signaling.

At specific cellular localizations other kinases play key regulatory roles in HDAC6 activity. For example, at the leading edge of migrating cells, the G protein-coupled receptor kinase 2 (GRK2) interacts with HDAC6 enhancing its α-tubulin deacetylase activity and thereby promoting cell migration (Lafarga et al. 2012). This supports the view that MT acetylation is critical in cell migration. Additionally, Aurora A kinase is well-known for its interaction with HDAC6 at the basal body of primary cilia regulating its dynamic disassembly. Phosphorylation of HDAC6 by Aurora A triggers ciliary disassembly and reabsorption, and the inhibition of both Aurora A and HDAC6 has been shown to stabilize primary cilia (Pugacheva et al. 2007).

Similar to HDAC6, the deacetylase SIRT2 activity can also be regulated by phosphorylation with some of the same kinases involved in its regulation. For example, Glycogen synthase kinase-3 beta (GSK3β) has been identified as an upstream regulator of both HDAC6 and SIRT2. In hippocampal neurons, GSK3β co-localizes and phosphorylates HDAC6 in serine residue 22 (S22) and its inhibition leads to increased MT acetylation (Chen et al. 2010). Additionally, GSK3β simultaneously phosphorylates specific serine residues (S327, S331, and S335) of SIRT2, leading to reduced tubulin acetylation and accelerated cell death in neuroblastoma cells (SH-SY5Y) overexpressing GSK3β (Liu et al. 2019a).

SIRT2 is also regulated by phosphorylation during the cell cycle by several cyclin-dependent kinases. However, most of these studies do not provide evidence that this regulation affects tubulin acetylation levels (Kawauchi et al. 2006; North and Verdin 2007a; Pandithage et al. 2008; Zhang et al. 2018).

Taken together, the examples analyzed in this section emphasize the intricate regulation of αTAT1, HDAC6, and SIRT2, which exhibit distinct combinatorial phosphorylation sites that depend on diverse kinase activities and signaling pathways. This regulation may have significant implications for the acetylation patterns of MT cytoskeleton as well as for its organization and dynamics.

4.3.2.4 Protein-Protein Interactions

The correct balance between tubulin acetylase and deacetylase activities also relies on interactions with various proteins that modulate their functions through these associations. For example, the $p27^{kip1}$, a member of the universal cyclin-dependent kinase inhibitor (CDKI) family and a cell cycle regulator, interacts with αTAT1 and stabilizes it by preventing its degradation via the proteasome (Morelli et al. 2018). The complexity of this regulation is further illustrated by the multifunctional roles of $p27^{Kip1}$, which serves as a substrate for both Cdk5-p35 and Cdk2-Cyclin E, acts as an inhibitor of Cdk2, and antagonizes SIRT2 phosphorylation by Cdk5-p35 (Kawauchi et al. 2006; Pandithage et al. 2008). Conversely, in the context of viral replication, the interaction between vimentin and αTAT1 inhibits tubulin acetylation, reducing viral replication of human parainfluenza virus type 3 (HPIV3). This occurs via the degradation of αTAT1, a process that appears crucial for maintaining the virus's replication sites (Liu et al. 2022) and highlights again the importance of the regulatory interplay among distinct components of the cytoskeleton.

Similar to αTAT1, HDAC6 interacts with several proteins that regulate its activity and influence cellular processes. One such protein is tubulin polymerization-promoting protein 1 (TPPP1), which binds to and inhibits HDAC6 activity. This interaction increases MT acetylation, prevents cell cycle transition from G1 to S phase, and inhibits the re-entry in G1 after mitosis (Schofield et al. 2013). Inhibition of TPPP1 by cyclin B/Cdk1 and Rho-associated protein kinase (ROCK)-mediated phosphorylation restores HDAC6 activity and cell cycle progression (Schofield et al. 2013). Another key regulator of HDAC6 is paxillin, a focal adhesion scaffold protein. Paxillin interacts with and inhibits HDAC6, leading to increased MT acetylation, which is associated with Golgi apparatus integrity and supports polarized cell migration (Deakin and Turner 2014). In newly emerging tumoral epithelial cells, paxillin suppression of HDAC6 activity, and consequent acetylation of MTs, is important for apical extrusion of oncogenic cells from the epithelium, acting as an early cancer preventive mechanism (Kasai et al. 2018). Lastly, the MT-associated protein 7 (MAP7) domain-containing protein 3 (Mdp3) contributes to MT stability by binding to MTs and downregulating HDAC6 activity (Tala et al. 2014).

4.3.2.5 Subcellular Localization and the Spatiotemporal Regulation of MT Acetylation

Another level of the regulation of tubulin lysine acetyltransferases/deacetylases activity is by regulating their subcellular localization and availability. Specifically, they must be present in the cytoplasm and colocalize with MTs to exert their effects (Legube and Trouche 2003).

The αTAT1 exhibits dynamic localization throughout the cell cycle, suggesting a direct relationship between the regulation of the dynamic intracellular localization of αTAT1 and its activity in MT acetylation at specific cellular localizations. This regulation is tightly regulated by sequence motifs in its primary structure. Studies by Roy et al. (2022) identified a conserved signal motif in the disordered C-terminus of αTAT1, which is composed of three competing regulatory elements responsible for nuclear export, nuclear import, and cytoplasmic retention (Roy et al. 2022).

During interphase, αTAT1 is detected in centrioles and in the basal bodies of primary cilia. In human fibroblasts at the G0 phase, αTAT1 is found within the nucleus as low-density granular structures that co-localize with a nucleolar marker (Nekooki-Machida et al. 2018). These structures remain visible, exhibiting intense staining during the G1, S, and G2 phases, until the nucleolus disassembles during the M phase. During metaphase and anaphase, αTAT1 localizes to the spindle midzone, spindle poles, peripheral spindle fibers, and centromeres. In telophase, αTAT1 is found in centrioles and within the nuclei of daughter cells. As cells progress to cytokinesis, αTAT1 concentrates at the center of the midbody (Nekooki-Machida et al. 2018).

The regulation of αTAT1 cellular localization is tailored by phosphorylation of αTAT1 by CDK1, cAMP-dependent protein kinase (PKA), and CK2 kinases, as well as dephosphorylation by the PP2A phosphatase. In its unphosphorylated form, αTAT1 is exported from the nucleus via exportin 1 through its nuclear export signal

(NES), rich in hydrophobic residues, and is found in cytoplasm and nucleus (Roy et al. 2022). When phosphorylated at critical regulatory residues of the nuclear localization signal (NLS) αTAT1 binds to 14–3-3 proteins, which integrate and control multiple signaling pathways, and accumulates in the cytoplasm facilitating MT acetylation (Roy et al. 2022). In fact, phosphorylation of αTAT1 at T-322 and/or S-315 by kinases such as CDK1 and CK2 promotes its translocation to the cytoplasm, where it facilitates MT acetylation. On the contrary, dephosphorylation by phosphatase PP2A enhances αTAT1 nuclear localization, inhibiting MT acetylation. Threonine-322 also acts as a binding site for 14–3-3 proteins, which mediate the exclusion of αTAT1 from the nucleus, further promoting MT acetylation. Therefore, the cells may regulate spatiotemporal MT acetylation by sequestering αTAT1 in the nucleus, preventing its interaction with MTs (Roy et al. 2022).

Interestingly, HDAC6 has also a NES1 sequence responsible for its shuttling between the nucleus and the cytoplasm. In proliferating cells, HDAC6 is predominately cytoplasmatic. However, in cell cycle arrest or during differentiation a fraction of HDAC6 migrates to the nucleus (Verdel et al. 2000). Acetylation of HDAC6 by p300 blocks its interaction with importin-α and contributes to its retention in the cytoplasm (Liu et al. 2012).

Likewise, SIRT2 is predominantly cytoplasmatic and contains a chromosome region maintenance 1 (Crm1)-dependent NES in its amino terminus. During mitosis, a small fraction of SIRT2 translocates to the nucleus, where it exhibits dynamic localization: it associates with centrosomes in prophase, the mitotic spindle during metaphase, and the midbody during cytokinesis (North and Verdin 2007b).

4.4 The Tubulin Acetylation: Bridging Metabolism Regulation and Stress Adaptation through Evolution

Given the capacity of PTMs, single or in combinations, to generate true codes for protein functional regulation and cellular localization, it is tempting to think that PTMs may have played important roles in the evolution of eukaryotic cells and metazoan. Analysis of the role of acetylation PTM throughout different organisms of the tree of life has the potential to give us clues about their core roles in modern metazoan and how they evolved for more sophisticated functions probably contributing to multicellularity.

In fact, some PTMs in proteins plausibly appeared early in evolution, being already present in last universal common ancestor (LUCA) (Bradley 2022), since they can be detected across the phylogenetic tree of life, as is the case of phosphorylation, acetylation, and glycosylation (Kim et al. 2006; Yu et al. 2008; Zhang et al. 2009; Choudhary et al. 2009; Wang et al. 2010; Zhao et al. 2010; Huttlin et al. 2010). Large-scale analysis revealed that in comparison to the proteome lysine-acetylated proteins are significantly more conserved through the evolutionary tree which suggests additional selective pressure (Choudhary et al. 2009). Moreover, the acetylome was much more conserved between yeast and prokaryotes than the phosphoproteome (Choudhary et al. 2009).

Several studies showed that lysine-acetylation is an abundant PTM in many bacteria. For example, in *Escherichia coli* this PTM predominantly targets metabolic enzymes involved in metabolic pathways, including most glycolytic and tricarboxylic acid (TCA) cycle enzymes (Yu et al. 2008; Zhang et al. 2009; Nakayasu et al. 2017). These observations led to the proposal that lysine acetylation provides a molecular mechanism to regulate metabolism in bacteria, especially in response to environmental changes.

In vertebrates, global lysine acetylation studies showed that lysine acetyltransferases and deacetylases are predominantly nuclear or mitochondrial which correlates with enrichment of acetylated proteins in these cellular compartments (Choudhary et al. 2009; Kim et al. 2016). Moreover, many of the enzymes participating in distinct metabolic pathways, for example those involved in glucose and fatty acid metabolism, undergo acetylation. Their acetylation status is influenced by the concentrations of glucose, amino acids, and fatty acids (Zhao et al. 2010). This indicates that acetylation levels may change in response to the alterations of extracellular nutrient availability. Since acetylation can also occur non-enzymatically (Baeza et al. 2015; Harmel and Fiedler 2018), it is plausible that one of the earliest conserved roles of protein acetylation was to create a direct link between protein function, nutrient levels and stress conditions. Interestingly, the acetate used for lysin acetylation can be recycled from other acetylation sites through the coordinated activity of acetyltransferases and deacetylases. In yeast and cancer cells, the histone acetylation may use acetate coming from deacetylation of other histone lysine residues preventing the failure of histone acetylation even during periods of oxygen and serum restriction (Mendoza et al. 2022; Bulusu et al. 2017). Therefore, certain lysine acetylation sites may serve as local acetate reservoirs, readily available for rapid transfer to key regulatory sites involved in acetylation and transcriptional activation (Mendoza et al. 2022). On the other hand, the activity of modern Sirtuins depends on NAD^+ (a vital coenzyme that facilitates redox reactions in metabolism by transport and transferring electrons between components of distinct metabolic pathways). This dependency establishes a critical link between the cell's energy status and its homeostasis (Beltrao et al. 2013).

Could Regulation of Tubulin Acetylation Also Be Linked to Metabolism Control?

Although the answer to this question remains complex, the affirmative answer is almost inevitable since tubulin acetylation relies on two important metabolites, CoA and NAD^+ (Beltrao et al. 2013; Siudeja et al. 2011). CoA serves as the primary donor of acetyl groups for tubulin acetylation, while NAD^+ functions as a crucial cofactor for the Sirtuin 2 deacetylase. For example, in neuronal axons loss of ATP citrate lyase activity, an enzyme that regenerates acetyl-CoA from the citrate molecule, led to lower level of tubulin acetylation and consequently inhibition of axonal transport (Even et al. 2021; Jankowska-Kulawy et al. 2022). Also, lower levels of NAD^+ lead to higher tubulin acetylation, on the contrary higher NAD^+ availability leads to less acetylation (Skoge et al. 2014).

Conversely, MTs acetylation may impact on metabolism regulation by affecting intracellular transport and cytoplasmic organelle distribution, namely mitochondria, Golgi apparatus, and endoplasmic reticulum (ER). The impact on intracellular transport may be linked to the observation that MT acetylation promotes the recruitment of molecular motors such as dynein and kinesin-1, which seem to preferentially travel along acetylated MTs (Reed et al. 2006; Dompierre et al. 2007; Friedman et al. 2010). Mitochondria transport impacts on its correct function by affecting the balance between their fission and fusion (Waddell et al. 2021). Inhibition of HDAC6 in Aβ (amyloid beta) hippocampal neurons restored the velocity and motility of the mitochondria in both anterograde and retrograde axonal transports and the length of the mitochondria (Kim et al. 2012). Hyperacetylation of MTs induced by osmotic and oxidative stress conditions contributes to the recruitment of Drp1 a protein that mediates mitochondrial fission and consequently increasing mitochondria fission (Perdiz et al. 2017). The examples above are far from illustrating all the many potential connections between tubulin acetylation and mitochondrial function. As will be discussed further below in another context, MFN2 (mitochondrial fusion protein mitofusin-2), a large GTPase located in the outer mitochondrial membrane, regulates mitochondrial fusion and tethering to the ER, recruits αTAT1 to sites where MTs contact mitochondria (Kumar et al. 2024). These contact points serve as hotspots for tubulin acetylation, which in turn regulates mitochondrial transport.

On the other hand, it has been suggested that MT acetylation may influence the distribution and interactions of ER and mitochondria by restricting their transport and localization to a specific subset of modified MTs (Friedman et al. 2010). This role appears to extend beyond ER and mitochondria interactions, as transport between the ER and Golgi apparatus also requires acetylated MTs, which are abundantly localized around the Golgi apparatus, a MT-organizing center (Chabin-Brion et al. 2001; Rivero et al. 2009). Interestingly, the Golgi-nucleated Mts are rapidly acetylated (Chabin-Brion et al. 2001).

Nutrient availability and metabolism influence the level of acetylation of distinct proteins, which in turn regulate metabolism and the response to nutrient scarcity. This is evident since acetylation controls transcription through histones acetylation and directly regulates several metabolic enzymes. Moreover, given that Mts play a critical role in cytoplasm spatial organization, in signaling transduction and in intracellular transport, directly affecting mitochondria and ER/Golgi membranes, it is plausible that MT acetylation may regulate the ability of cells to cope with nutrient and metabolic alterations. Adding complexity to these distinct regulatory layers is the fact that acetylase and deacetylases may be regulated in coordination with the cell cycle, whose progression is, in turn, dependent on the metabolism. We can also envisage that acetylation modification evolved to integrate not only nutrient-related signals but also other environmental stressors, including chemical or mechanical factors. Again, cell cycle and cytoskeleton regulation will be required to trigger efficient responses to stress conditions, and acetylation modification emerges as a rapid mechanism to regulate the response and adaptation to these stress situations.

4.5 Tubulin Acetylation and Human Diseases

4.5.1 Neurological Disorders

Tubulin acetylation impacts MT-dependent cellular processes such as intracellular trafficking, cell division, cell polarity, and migration. It is therefore not surprising that altered tubulin acetylation has been associated with the etiology of several human diseases. For example, altered tubulin acetylation is associated with multiple neurological disorders like Charcot-Marie-Tooth (peripheral neuropathy), Huntington's, Parkinson's, and Alzheimer's diseases.

Mutations in the HSPB1 (Heat shock protein beta-1) gene cause Charcot-Marie-Tooth disease, and mice expressing mutant HSPB1 in neurons displayed decreased α-tubulin acetylation levels with associated severe axonal defects that could be reversed by inhibiting HDAC6 (d'Ydewalle et al. 2011). Similar observations were made for Huntington's disease, an incurable neurodegenerative disorder (Dompierre et al. 2007; Bobrowska et al. 2011). A subtype of Charcot-Marie-Tooth disease (type 2 A; CMT2A) is caused by mutations in MFN2 a protein that recruits αTAT1 to sites of mitochondrial contact with MTs, which are regions of high α-tubulin acetylation. Axonal mitochondrial motility is dependent on these contacts with MTs and seems to be impaired by defective αTAT1-mediated α-tubulin acetylation in CMT2A (Baloh et al. 2007; Rocha et al. 2018; Kumar et al. 2024).

Joubert syndrome is a rare genetic disease characterized by abnormal brain development caused by affected primary cilia. One of the genes associated with this disorder is KIF7, which codes for the non-motile kinesin 7, a protein that accumulates at the tips of cilia and that regulates Hedgehog (Hh) signaling through MT dynamics. Silencing KIF7 perturbs cilia formation, abnormal centrosome duplication, and fragmentation of the Golgi apparatus, likely due to abnormally elevated tubulin acetylation and affected MT dynamics (Dafinger et al. 2011). In addition to its ciliary role, KIF7 is involved in other MT-dependent processes such as cell division, and its depletion consistently leads to increased tubulin acetylation, and altered MT dynamics and organization in different cell types (Dafinger et al. 2011; Coles et al. 2015).

Parkinson's disease (PD) is a movement neurodegenerative disorder of the central nervous system and is commonly caused by mutations in the LRRK2 (leucine-rich repeat kinase 2) gene. LRRK2 interacts directly with specific isoforms of β-tubulin (TUBB, TUBB4, and TUBB6) and this interaction is perturbed by a pathogenic mutation in LRRK2 (R1441G). Moreover, LRRK2 was shown to regulate tubulin acetylation. Indeed, mouse LRRK2 knockout embryonic fibroblasts display increased levels of tubulin acetylation, and molecular modeling suggested that LRRK2 interacts with the lumen of MT filaments close to the K40 acetylation site in α-tubulin. Finally, LRRK2 accumulates in highly dynamic microtubular structures in dopaminergic cells, and its overexpression or mutation impacts the morphology of growth cones (Law et al. 2014). Indeed, impaired tubulin acetylation regulation by LRRK2 may contribute to the etiology of PD.

Tubulin acetylation levels have also been shown to be altered in Alzheimer's disease (AD) (Zhang et al. 2015; Martínez-Hernández et al. 2022). The analysis of post-mortem brain samples from AD patients with different stages of disease development and age-matched non-demented controls revealed increased levels of tubulin acetylation in patients (Martínez-Hernández et al. 2022). This observation was confirmed in vitro through the analysis of hiPSC-derived neurons carrying an AD-linked mutation (V717I). Compared to control wild-type hiPSC-neurons, the mutant cells which accumulate hyperphosphorylated Tau protein also showed increased levels of tubulin acetylation (Martínez-Hernández et al. 2022). Of note, in the post-mortem samples, there was no significant relationship between the disease stages and the amount of HDAC6, whereas in the mutant hiPSC-neurons there was a big increase in HDAC6 levels while the levels of αTAT1 were unaltered compared to the control cells. These observations indicated that altered tubulin acetylation in AD patients cannot be attributed to changes in the levels of the enzymes that regulate it. Interestingly, tubulin acetylation is correlated with tubulin detyrosination, and the levels of both modifications are increased in the neurons of TTL knockout mice (Martínez-Hernández et al. 2022). TTL null mice present severe defects in neuronal morphology, perturbed neurite outgrowth, and premature axonal differentiation (Erck et al. 2005). These data suggest that TTL loss leads to accumulated detyrosinated tubulin in MTs that become stable possibly allowing αTAT1 more time to hyperacetylate them. Finally, despite the observed elevated levels of tubulin acetylation in AD patients and mouse models, the inhibition of HDAC6 was shown to improve the cognitive deficits and memory in mouse models of AD and a mouse model of Tau deposits (Kilgore et al. 2010; Govindarajan et al. 2013; Selenica et al. 2014). This may be related to tubulin-independent HDAC6 activities.

Amyotrophic lateral sclerosis (ALS), also known as Lou Gehrig's disease, is a neurological disorder characterized by the degeneration of motor neurons and is caused by mutations in several genes like *SOD1* and *TUBA4A*. For example, the ALS-associated *TUBA4A* mutation R320C mutation might block the acetylation of α-tubulin in addition to perturbing tubulin polymerization, MT interaction with motor proteins, and/or create dysfunctional disulfide bridges. Moreover, an in silico analysis suggested that the R320C mutant binds GTP poorly and forms unstable tubulin dimers (Ganne et al. 2023). Perturbed tubulin acetylation may also be an important factor in ASL caused by mutations in different genes. Indeed, in a mouse model of ASL carrying the $SOD1^{G93A}$ mutation, it was shown that the deletion of HDAC6 was beneficial, increasing the survival rate after disease onset, and slowing down the progression of motor neuron degeneration. The significantly increased levels of α-tubulin acetylation in the spinal cord, sciatic nerve, and brain of $SOD1^{G93A}$ mice lacking HDAC6 may lead to improved axonal transport (Taes et al. 2013).

Collectively, these data suggest that altered α-tubulin acetylation is a factor in the development of neurological diseases. Indeed, as this tubulin modification is important for MT stability, the binding of motor proteins, and MT-associated transport, it becomes critical for neuronal development, organization, and maintenance. However, further work will be necessary to understand if abnormal α-tubulin acetylation is the driving force or a consequence of disease development. In addition, more needs to be done in order to understand the role of tubulin-modifying enzymes in disease development and their potential as therapeutic targets. For example, there are several examples showing a benefit of HDAC6 inhibition in several neurological diseases. However, it is not always clear if this benefit comes from an impact on tubulin. Finally, a recent patent also proposes the inhibition of α-tubulin acetylation for the treatment of pain for instance by silencing αTAT1 with antisense oligonucleotides (US Patent n° US11834713B2, 2023) (Heppenstall and Morley 2023), which demonstrates the continued interest in exploring the modulation of tubulin modifications as therapeutic approaches for neurological disorders.

In addition to its involvement in brain development, abnormal tubulin acetylation has also been implicated in left-right (LR) asymmetry establishment during embryonic development, a process that relies on specialized motile cilia in the embryonic node in mammals. For example, loss of function mutations in NEK3, a member of the NIMA (never in mitosis A)-related kinases (Nek1-Nek11) family of Ser-Thr kinases, cause α-tubulin deacetylation, and lead to the downregulation of NUP205 and defective cilia, predisposing individuals to cilia-related abnormal cardiac left-right patterning (Zhang et al. 2020). Of note, NUP205 is a nuclear pore protein with a ciliary role, and whose depletion leads to cilia loss and abnormal cardiac morphology (Marquez et al. 2021). Finally, NEK3 has also been shown to regulate neuronal MT acetylation. Expression of mutant NEK3 in neurons results in perturbed cell morphology and polarity as well as MT deacetylation through HDAC6 (Chang et al. 2009).

4.5.2 Cancer

Tubulin acetylation and the enzymes regulating it have been associated with the development of different types of cancer and resistance to anticancer treatments. For example, in MDA-MB-231 breast cancer cells, tubulin acetylation controls cell invasion by modulating ER stress. Interestingly, when these cells were grown on a stiff matrix, there was an increase in MT acetylation and the downregulation of ER stress markers. Moreover, decreasing tubulin acetylation by silencing or knocking-out αTAT1 leads to elevated expression of ER stress markers, changes in ER morphology, and enhanced tunicamycin-induced unfolded protein response signaling. Also, in the αTAT1 knockout context, several genes associated with cancer progression were found to be downregulated, including genes important for the formation of focal adhesions (Ko et al. 2021). Other studies showed that elevated α-tubulin acetylation is associated with aggressive metastatic behavior in breast cancer, promoting microtentacle formation, adhesion, and invasive migration in metastatic and basal-like breast cancer cells (Boggs et al. 2015). Similarly, tubulin acetylation caused by the induction of αTAT1 activated cell migration and invasion in ameloblastoma cells (Yoshimoto et al. 2022). Conversely, a different study showed that the overexpression of αTAT1 (MEC-17) led to increased cell spreading area, and suppression of pseudopods formation in a three-dimensional (3D) culture system, and it inhibited cancer cell migration and invasion ability and tumor metastasization in an orthotopic lung cancer animal model (Lee et al. 2018). In addition, αTAT1 loss of function resulted in increased pseudopod formation, and aggressive cell migration (Lee et al. 2018). Thus, the acetylated tubulin's role in cell attachment, migration, and metastasis formation seems to be tumor cell-specific.

In lung cancer cells, the levels of α-tubulin acetylation are associated with sensitivity to paclitaxel. Indeed, paclitaxel-resistant H460 lung cancer cells present elevated levels of tubulin acetylation, and cells with elevated tubulin acetylation due to overexpression of αTAT1or HDAC6 inhibition showed significantly reduced paclitaxel-induced apoptosis. In fact, the modulation of tubulin acetylation led to changes in the levels of antiapoptotic Mcl-1, which were increased upon αTAT1 overexpression, and were decreased upon αTAT1 silencing by RNAi. Mcl-1 is a MT binding protein, which binds preferentially to acetylated MTs. Elevated tubulin acetylation stabilizes Mcl-1, protecting it from ubiquitin–proteasome-mediated degradation leading to resistance to paclitaxel-induced cell death (Wattanathamsan et al. 2021). Also, in lung cancer cells (Calu-1), it was shown that the treatment with the anticancer drug tanespimycin, an inhibitor of the molecular chaperone Hsp90, leads to a significant increase in the levels of acetylated tubulin, decreased expression of HDAC6, and apoptosis (Wang and Liu 2020). These results suggest that elevated acetylated tubulin has different impacts in different cell types, even originating from tumors of the same tissue. It is possible that different tumor cells express proteins like Mcl-1, or that tanespimycin blocks the induced stability of Mcl-1 and its anti-apoptosis protection role. Also, in the context of the impact of tubulin acetylation on anticancer treatments, clinical data showed that there is a correlation between the levels of K40 acetylation on α-tubulin with survival of patients with hepatocellular carcinoma (HCC) with better prognosis being associated with lower levels of acetylation. Furthermore, decreasing tubulin acetylation by knocking out αTAT1 or overexpressing HDAC6 in HCC cells increased sensitivity to Eribulin treatment in vitro and in vivo. These results place tubulin acetylation as a prognosis marker and indicator for Eribulin treatment for this patient population (Zhong et al. 2023).

Similarly to neurological diseases, the role of tubulin acetylation in cancer development and therapy remains elusive. It seems clear that it is context-dependent with different impacts in different cell types, tissues, and tumor types with their specific mutation burden. The development of drugs that inhibit tubulin acetylation will bring answers to these open questions. For example, the small molecule inhibitors GM-90257 and GM-90631 (GM compounds) block MT acetylation by interfering with their interaction with αTAT1. These molecules induce apoptosis and mitotic arrest in Triple-Negative Breast Cancer Cells, through the activation of the JNK/AP-1 pathway (Ahn et al. 2023). Furthermore, multiple clinical trials have tested HDAC6 inhibitors (including FDA-approved Panobinostat and Resminostat), alone or in combination with other drugs, for the treatment of different types of cancer (e.g., solid tumors—NCT03008018, NCT05268666; melanoma—NCT02935790; breast cancer—NCT02632071, NCT00788112; myeloma—NCT01323751, NCT01023308; B-cell lymphoma—NCT05526313). Of note, pan HDAC and

HDAC6-specific inhibitors affect not only tubulin acetylation but a plethora of cellular pathways (*e.g.*, cell cycle, apoptosis, autophagy). Thus, it is challenging to dissect the mechanisms underlying the benefits and risks associated with this class of drugs (Eckschlager et al. 2017; Pu et al. 2024). Moreover, further investigation is required to fully grasp how tubulin acetylation, combined with several other tubulin PTMs, as well as the expression of different tubulin isotypes, governs tubulin folding, heterodimer formation and assembly, MT dynamics and interactions with MAPs, and all the MT-dependent processes in the context of health and disease. Finally, ongoing, and future basic and clinical research will help elucidate the potential of this tubulin modification as a therapeutic target.

4.5.3 Cardiac Diseases

Damage and misfolded proteins tend to accumulate with age and as a result of exposure to different forms of stress like osmotic, mechanical, and oxidative stress (Herczenik and Gebbink 2008). In cells continuously exposed to mechanical stress, like cardiomyocytes, cellular dysfunction has been associated with altered acetylated tubulin levels. For example, in cardiomyocytes of a mutated mouse for αβ-crystallin (CryAB^{R120G}) the accumulation of perinuclear aggregates induces desmin-related cardiomyopathy (McLendon et al. 2014). In these heart cells, α-tubulin was found to be hyperacetylated, while other proteins were hypoacetylated. Interestingly, HDAC6 inhibition showed a protective role against these protein aggregates by promoting autophagy (McLendon et al. 2014).

Additionally, atrial fibrillation, a hallmark of clinical tachyarrhythmia, induces activation of HADC6, with concomitant reduction of tubulin acetylation levels (Zhang et al. 2014). The role of tubulin acetylation has been also associated with a dilated cardiomyopathy caused by mutations in lamin A/C gene. In a mouse model, it was shown that a decrease in tubulin acetylation led to the remodeling of gap junctions specifically conexin 43 (Cx43) compromising synchronized action of cardiomyocytes (Macquart et al. 2019).

These findings show the impact of Mts and acetylation modulation in cells with complex actin cytoskeleton. In the heart, MTs are involved in the organization and maintenance of the sarcomere, intercalated disc, protein clearance machinery, and transverse-tubule and sarcoplasmic reticulum membranes (Caporizzo and Prosser 2022).

4.6 Final Remarks on the Role of Microtubule Acetylation in Cellular Response to Environmental Challenges

Despite several controversial aspects, it is now possible to establish a connection between metabolism and the presence of acetylated MTs, particularly through their association with mitochondria and intracellular transport, a relationship that appears more evident in neuronal cells. Acetylated MTs appear to play a critical role in the distribution pattern and transport of mitochondria, as well as their interactions with the ER. These interactions, in turn, influence key mitochondrial functions such as fusion and fission, ultimately impacting cellular metabolism.

This also seems important for the spatial organization and interactions between the ER and Golgi apparatus also with impact in intracellular transport. The acetylation pattern of MTs is regulated by the balance between lysine acetylase and deacetylase activities, with one of the major tubulin deacetylases, SIRT2, being dependent on NAD^+ levels, a key intermediate metabolite. However, it remains unclear how this acetylation pattern is established and how cells "read" it to carry out the associated cellular functions. It has been accepted that acetylation may generate spatial information that is recognized by specific proteins (MAPs and molecular motors), which then perform functions or recruit other factors and organelles, contributing to the dynamic architecture of the cytoplasm in close association with other cytoskeletal filaments. In this scenario, the physical properties of MTs including their ability to cope with mechanical stress, and their self-repair mechanisms that occur along the MT shaft, should also be considered.

The new results show that acetylated MTs are more flexible and more resistant to damage and that damage sites in MTs contribute to the establishment of the acetylation pattern within cells. Since most MT functions rely on their dynamic behavior and interactions with various associated proteins, motor proteins, and organelles, they are constantly subjected to mechanical stress (Fig. 4.3). Additionally, in multicellular organisms, the MT network is also influenced by physical constraints from interactions with other cells and the ECM (Schaedel et al. 2015) through focal adhesions and cell-cell junctions. Do these physical constraints continuously cause damage to specific MT populations, leading to alterations in their acetylation levels and transforming them into more damage-resistant MT populations establishing the acetylation pattern of the cell? Interestingly, Li et al. (2023) observed that the application of compression forces cycles to RPE-1 cells stabilize MTs, making them more resistant to nocodazole, independent of actin filaments, and MT acetylation and detyrosination (Li et al. 2023). They also observed that the knocking-out of the gene encoding CLASP2 plus-tip protein, a regulator of MT plus-end growth, abolished the mechano-stabilization of MTs. Additionally, the authors also found that CLASP2 relocates from the tip to the MT lattice in MTs bent by compressive forces. Consequently, a final answer to the above question is still missing and would be complex.

Additionally, acetylation also occurs through MT ends which may be also linked to regulation of MT dynamics that can be changed by mechanical forces (Trushko et al. 2013). MT dynamics is also governed by a multiplicity of factors, from the critical concentration of tubulin heterodimers competent to polymerize, to distinct MAPs like Tau or even the severing protein katanin. For example, in neurons the sensitivity to MT severing by katanin is regulated by a balance of factors, including the acetylation state of the MTs and the binding of Tau to the MTs (Sudo and Baas 2010). Increased acetylation of MTs by genetic and pharmacological approaches rescues human Tau overexpression-induced toxicity in *Drosophila* muscles and neurons (Mao et al. 2017). Interestingly, *Drosophila* katanin not only can remove tubulin dimers from the ends of MTs but is also targeted to MT defects in vitro (Díaz-Valencia et al. 2011). In plants MT cytoskeleton remodeling in response to different mechanical stresses requires katanin MT severing (Uyttewaal et al. 2012; Sampathkumar et al. 2014; Louveaux et al. 2016). Interestingly, in plant cells, katanin specifically severs microtubules at crossover sites (Wang et al. 2018), which are putative hotspots for damage. In animal cells, katanin-mediated severing can itself create damage sites in microtubule walls, potentially influencing their acetylation state (Andreu-Carbó et al. 2024). Together, the establishment of a MT acetylation pattern emerges as the result of combinatory factors where the specific trigger is still elusive but cannot be reduced to a simple relation between forces applied to cells and MT shaft damage.

The recruitment of αTAT1 to focal adhesions and sites where MTs interact with the outer mitochondrial membrane, such as during fission, is closely linked to the regulation of actin cytoskeleton remodeling and dynamics, likely requiring local force alterations. Furthermore, in both cases, acetylation likely occurs at the MT ends rather than at sites of MT damage. Notably, both processes critically depend on the interplay with MTs and are associated with membranes that constitute a new niche of local information (Fig. 4.3). Supporting this view is the fact that focal adhesions are enriched in cholesterol and phosphatidylcholine with saturated fatty acid chains, compared to the overall cell membrane (Tachibana et al. 2023). For example, integrin activation, by ligand binding, triggers its sequestration into cholesterol-rich membrane regions (Ge et al. 2018), and cholesterol enhances membrane focal adhesion kinase (FAK) activity (Takahashi et al. 2021). At focal adhesions phosphoinositides are required to recruit talin and vinculin the two proteins that connect integrin to actomyosin networks in the cell (Kelley et al. 2020). These findings strongly indicate that the local membrane composition is crucial for regulating the sequential recruitment, activation, and function of proteins within focal adhesions. In this view, it is interesting to remind that αTAT1 depletion did not alter focal adhesion size but affected their number and localization (Seetharaman et al. 2022) strongly suggesting that Mts acetylation, probably in close talk with membrane dynamics triggered by mechanical forces, is critical to define the precise localization of the complex structure of the focal adhesion.

A study on antidepressant therapies highlighted the association of tubulin acetylation with lipid-raft microdomains, showing that tubulin acts as an anchor for Gαs within lipid rafts, where it increases adenylyl cyclase activation and increases cAMP levels in the cell (Singh et al. 2018). Inhibiting HDAC6 enhances tubulin acetylation, facilitating the movement of Gαs out of lipid rafts, which in turn activates

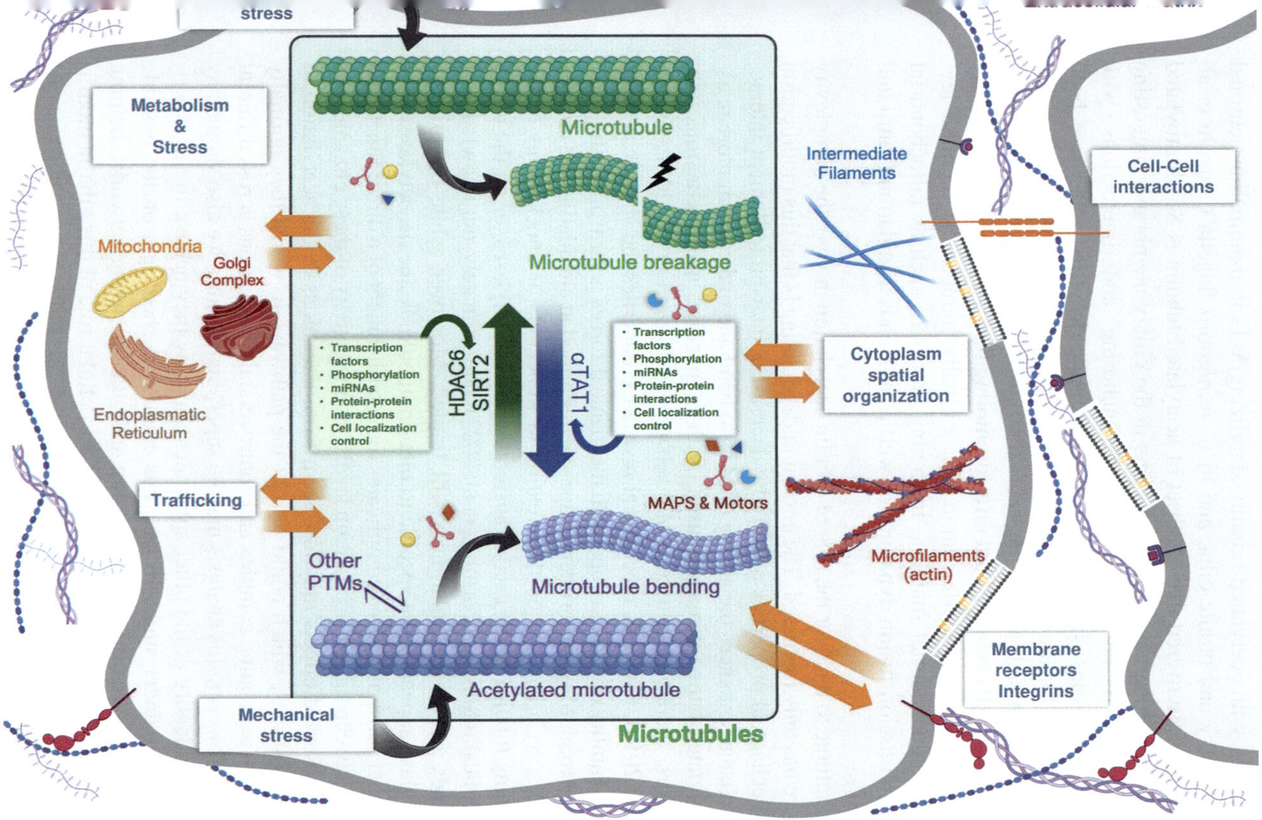

Fig. 4.3 A schematic representation illustrating the major cross-talks between microtubule (MT) acetylation and key regulators of various cellular processes. These include the three-dimensional organization of the cytoplasm (*e.g.*, the dynamic arrangement of actin filaments, intermediate filaments, mitochondria, the Golgi complex, and the endoplasmic reticulum), as well as intracellular trafficking and metabolism. Additionally, cell-cell interactions, cellular adhesion, and the extracellular matrix are highlighted, as they are crucial for tissue architecture and continuously interact with MTs, influencing their acetylation status and vice-versa. Given the complexity and diversity of factors regulating MT acetylation, a simplified central green rectangle represents these interactions. The figure also emphasizes how this intricate system modulates MT mechanical properties in response to cellular stress. The major enzymes involved in tubulin acetylation (acetyltransferases) and deacetylation (deacetylases) are indicated. (Schematic representation created using BioRender)

downstream cAMP signaling. This reduced association of Gαs with tubulin in lipid rafts has been shown to function as an antidepressant mechanism (Singh et al. 2018). Moreover, Tripathi et al. (2021) reported that the sphingolipid ceramide forms a complex with acetylated tubulin, stabilizing MTs in mammalian neuronal processes, primary and motile cilia, and in *C. reinhardtii* flagella (Tripathi et al. 2021). They also discovered that a subset of acetylated tubulin is S-palmitoylated and colocalizes with ceramide-rich platforms in the ciliary membrane, suggesting that these platforms mediate membrane anchoring and interactions with S-palmitoylated proteins. In line with this, ceramide-rich lipid rafts have been shown to be crucial for the transport of over 100 palmitoylated proteins from the Golgi to the plasma membrane (Xiong et al. 2019). Additionally, ceramide has been proposed to structurally modify lipid rafts, promoting their coalescence into signaling platforms (Kinoshita and Matsumori 2022). Finally, intracellular ceramide levels influence mitochondrial function, impacting processes such as mitochondrial biogenesis and fission-fusion dynamics as well as MT ends acetylation (Jamil and Cowart 2023).

As we continue to uncover the role of tubulin acetylation in vivo, the complexity of the picture has only increased. The current understanding is that this modification alters the mechanical properties of MTs by decreasing inter-protofilament interactions. This change enhances the ability of MTs to bend and makes them more resistant to age-related lattice damage caused by their functions and interactions with various factors, thereby facilitating MT self-repair (Portran et al. 2017; Eshun-Wilson et al. 2019). How this mechanical resistance underlies the specific cellular functions and ability of MTs to cope with mechanical forces is far from being completely understood. Most experimental approaches aimed at elucidating the cellular functions of acetylation rely on the creation of knockouts or knockdowns of the genes encoding tubulin lysine acetyltransferases and deacetylases, or on the use of specific chemical inhibitors (Carmona et al. 2023). Unfortunately, our knowledge of these enzymes and their regulation is still limited. Moreover, while the primary tubulin acetylase appears to be specific to tubulin, tubulin lysine deacetylases have numerous other substrates. This complicates the establishment of a direct functional relationship between tubulin acetylation levels and the activity of these enzymes. As a result, the observed effects may also stem from alterations in the acetylation of non-tubulin substrates, which could impact the pathways in which they normally participate often generating complex and controversial observations across different types of cells. However, this analysis may be superficial because we also should pay attention to the identity of the other substrates and how they behave in the experimental conditions under study. In fact, this could also shed light on the possible interconnection between tubulin acetylation and other signaling networks, helping to clarify and complete our understanding of the cellular role of this tubulin modification. For example, it is interesting to note that the other major substrates of HDAC6 are proteins involved in stress response, namely HSP90 and peroxiredoxins (Parmigiani et al. 2008; Su et al. 2016). This strongly supports the idea that MT acetylation is part of a critical regulatory axis of stress response, integrating proteins required for protein folding, trafficking, and response to oxidative stress.

To support MT dynamics, remodeling, and repair of damage sites in their lattice cells should possess a critical pool of free tubulin heterodimers competent to polymerize. Tubulin folding cofactors (TBCA-TBCE) are crucial for controlling those processes by regulating the critical concentration of free tubulin heterodimers. In fact, TBCs have a prominent role in regulating tubulin recycling and degradation (Gonçalves et al. 2010; Serna and Zabala 2016; Nolasco et al. 2021). Therefore, if MTs are subjected to mechanical forces, it is likely that they will require extensive recycling of tubulin heterodimers, making the roles of TBCs crucial in this process. Although no direct evidence currently links MT acetylation to tubulin-binding cofactors (TBCs), it is noteworthy that TBCD overexpression has been shown to cause the disassembly of tight and adherens junctions, leading to cell dissociation from the epithelial monolayer (Shultz et al. 2008). Furthermore, TBCA is highly abundant in mouse heart cells, where MTs must cope with constant mechanical forces (Nolasco et al. 2021). TBCs may play an important role in regulating the composition of tubulin pools, particularly in terms of isotype variation produced by the expression of distinct members of tubulin gene families. It remains unclear the impact of specific isotype abundance within individual MTs. It is not known if different α-tubulin isotypes have varying affinities for lysine acetyltransferases or deacetylases, which could influence the rate and pattern of MT acetylation. Adding further complexity, it is unknown how the differential abundance of tubulin isotypes affects other PTMs beyond acetylation. Notably, the most divergent region among tubulin isotypes is the C-terminal region, a hotspot for various PTMs. Emerging evidence suggests that different PTMs may influence each other. For example, crosstalk between tubulin acetylation and detyrosination has been shown to reduce MT dynamics in Alzheimer's disease (Martínez-Hernández et al. 2022). Although α-tubulin Lys40 is possibly the predominant acetylation site in MTs, little is known about whether acetylation at α-tubulin K40 and K394 can occur simultaneously, or in combination with β-tubulin K252 acetylation, and how these potential patterns impact on MT properties and functions.

The interplay between different PTMs expands the range of their individual roles, potentially creating distinct MT populations within each cell. Given that different cell types, with specialized functions, exhibit, distinct cytoskeletal organizations with specific MT arrays, along with unique tissue architectures and diverse ECM compositions, the range of possible functional combinations of PTMs is expected to increase. This complexity poses challenges in fully understanding the roles of individual PTMs, their associated mechanisms, and their regulation. It also may explain some of the contradictory observations available in the literature reinforcing the need of the next generation of experiments to consider both the specific characteristics of different cell types and their surrounding microenvironments to achieve a more comprehensive understanding. Moreover, it is important to take into account the specific challenges that each type of cell is adapted to face. Thus, in some cases, the specific role of acetylation may only become relevant when the cell is challenged by a specific signal. This raises the possibility that the absence of clear phenotypes in K40 mutant cells may be due to their lack of exposure to the appropriate challenges such as nutrient deprivation, oxidative stress, mechanical forces, or others.

In conclusion, tubulin acetylation has emerged as an important feature of MTs, capable of altering their mechanical properties by integrating signals from both intracellular and extracellular forces (see also chapter Tubulin Acetylation and the Cellular Mechanosensing and Stress Response). This modification enhances the cell's ability to coordinate metabolism with the local and overall 3D organization of the cytoplasm and improves their potential to respond to internal and external stress conditions. These findings highlight the need for further research to fully understand the role of tubulin acetylation in development and tissue homeostasis, which may also provide valuable insights into cancer biology. Several clues have already linked tubulin acetylation to the regulation of cell proliferation and migration. It appears to be essential for establishing specific cellular territories and intrinsic polarity, which impact in Planar cell polarity (PCP) and influence cell-cell and cell-matrix junction dynamics. Furthermore, recent studies showed that tubulin acetylation, by increasing MT flexibility, enhanced a cell's ability to penetrate through cell junctions during radial intercalation. This process facilitates apical movement and integration of cells into an epithelium which is essential for development and tissue homeostasis (Collins et al. 2021), and probably has implications for cancer metastasis.

Acknowledgments This work was funded by Instituto Politécnico de Lisboa IPL/IDI&CA2024/HyperCil_ESTeSL and IPL/IDI&CA2024/ENIGMA_ESTeSL, and also Centro de Química Estrutural is a Research Unit funded by FCT through projects UIDB/00100/2020 (https://doi.org/10.54499/UIDB/00100/2020) and UIDP/00100/2020 (https://doi.org/10.54499/UIDP/00100/2020). Institute of Molecular Sciences is an Associate Laboratory funded by FCT through project LA/P/0056/2020 (https://doi.org/10.54499/LA/P/0056/2020).

References

Aebersold R, Agar JN, Amster IJ et al (2018) How many human proteoforms are there? Nat Chem Biol 14:206–214. https://doi.org/10.1038/nchembio.2576

Ahmad FJ, Pienkowski TP, Baas PW (1993) Regional differences in microtubule dynamics in the axon. J Neurosci 13:856–866. https://doi.org/10.1523/JNEUROSCI.13-02-00856.1993

Ahn S, Kwon A, Oh Y et al (2023) Microtubule acetylation-specific inhibitors induce cell death and mitotic arrest via JNK/AP-1 activation in triple-negative breast cancer cells. Mol Cells 46(6):387–398. https://doi.org/10.14348/molcells.2023.2192

Aillaud C, Bosc C, Saoudi Y et al (2016) Evidence for new C-terminally truncated variants of α- and β-tubulins. Mol Biol Cell 27:640–653. https://doi.org/10.1091/mbc.E15-03-0137

Aillaud C, Bosc C, Peris L et al (2017) Vasohibins/SVBP are tubulin carboxypeptidases (TCPs) that regulate neuron differentiation. Science 358:1448–1453. https://doi.org/10.1126/science.aao4165

Akella JS, Wloga D, Kim J et al (2010) MEC-17 is an alpha-tubulin acetyltransferase. Nature 467:218–222. https://doi.org/10.1038/nature09324

Alper JD, Decker F, Agana B, Howard J (2014) The motility of axonemal dynein is regulated by the tubulin code. Biophys J 107:2872–2880. https://doi.org/10.1016/j.bpj.2014.10.061

Andreu-Carbó M, Egoldt C, Velluz M-C, Aumeier C (2024) Microtubule damage shapes the acetylation gradient. Nat Commun 15:2029. https://doi.org/10.1038/s41467-024-46379-5

Bae HJ, Jung KH, Eun JW et al (2015) MicroRNA-221 governs tumor suppressor HDAC6 to potentiate malignant progression of liver cancer. J Hepatol 63:408–419. https://doi.org/10.1016/j.jhep.2015.03.019

Baeza J, Smallegan MJ, Denu JM (2015) Site-specific reactivity of nonenzymatic lysine acetylation. ACS Chem Biol 10:122–128. https://doi.org/10.1021/cb500848p

Baloh RH, Schmidt RE, Pestronk A, Milbrandt J (2007) Altered axonal mitochondrial transport in the pathogenesis of Charcot-Marie-tooth disease from mitofusin 2 mutations. J Neurosci 27(2):422–430. https://doi.org/10.1523/JNEUROSCI.4798-06.2007

Bance B, Seetharaman S, Leduc C et al (2019) Microtubule acetylation but not detyrosination promotes focal adhesion dynamics and astrocyte migration. J Cell Sci 132:jcs225805. https://doi.org/10.1242/jcs.225805

Barisic M, Silva e Sousa R, Tripathy SK et al (2015) Microtubule detyrosination guides chromosomes during mitosis. Science 348:799–803. https://doi.org/10.1126/science.aaa5175

Barra HS, Arce CA, Rodriguez JA, Caputto R (1973) Incorporation of phenylalanine as a single unit into rat brain protein: reciprocal inhibition by phenylalanine and tyrosine of their respective incorporations. J Neurochem 21:1241–1251. https://doi.org/10.1111/j.1471-4159.1973.tb07578.x

Beltrao P, Bork P, Krogan NJ, Van Noort V (2013) Evolution and functional cross-talk of protein post-translational modifications. Mol Syst Biol 9:1–13. https://doi.org/10.1002/msb.201304521

Bhamidipati A, Lewis SA, Cowan NJ (2000) Adp ribosylation factor-like protein 2 (Arl2) regulates the interaction of tubulin-folding cofactor D with native tubulin. J Cell Biol 149:1087–1096. https://doi.org/10.1083/jcb.149.5.1087

Bobrowska A, Paganetti P, Matthias P, Bates GP (2011) Hdac6 knock-out increases tubulin acetylation but does not modify disease progression in the R6/2 mouse model of Huntington's disease. PLoS One 6:e20696. https://doi.org/10.1371/journal.pone.0020696

Boggs AE, Vitolo MI, Whipple RA et al (2015) A-Tubulin acetylation elevated in metastatic and basal-like breast cancer cells promotes microtentacle formation, adhesion, and invasive migration. Cancer Res 75:203–215. https://doi.org/10.1158/0008-5472.CAN-13-3563

Bradley D (2022) The evolution of post-translational modifications. Curr Opin Genet Dev 76:101956. https://doi.org/10.1016/j.gde.2022.101956

Brown A, Li Y, Slaughter T, Black MM (1993) Composite microtubules of the axon: quantitative analysis of tyrosinated and acetylated tubulin along individual axonal microtubules. J Cell Sci 104:339–352. https://doi.org/10.1242/jcs.104.2.339

Bulusu V, Tumanov S, Michalopoulou E et al (2017) Acetate recapturing by nuclear acetyl-CoA synthetase 2 prevents loss of histone acetylation during oxygen and serum limitation. Cell Rep 18:647–658. https://doi.org/10.1016/j.celrep.2016.12.055

Burbank KS, Mitchison TJ (2000) Quick guides microtubule dynamic instability. Curr Biol 16:516–517

Bursch KL, Goetz CJ, Smith BC (2024) Current trends in Sirtuin activator and inhibitor development. Molecules 29. https://doi.org/10.3390/molecules29051185

Cambray-Deakin MA, Burgoyne RD (1987) Posttranslational modifications of alpha-tubulin: acetylated and detyrosinated forms in axons of rat cerebellum. J Cell Biol 104:1569–1574. https://doi.org/10.1083/jcb.104.6.1569

Camelo C, Peneda C, Carmona B, Soares H (2016) TBCC. In: Choi S (ed) Encyclopedia of signaling molecules. Springer, New York, NY, pp 1–8

Caporizzo MA, Prosser BL (2022) The microtubule cytoskeleton in cardiac mechanics and heart failure. Nat Rev Cardiol 19:364–378. https://doi.org/10.1038/s41569-022-00692-y

Carmona B, Marinho HS, Matos CL et al (2023) Tubulin post-translational modifications: the elusive roles of acetylation. Biology (Basel) 12:561. https://doi.org/10.3390/biology12040561

Caron JM (1997) Posttranslational modification of tubulin by palmitoylation: I. In vivo and cell-free studies. Mol Biol Cell 8:621–636. https://doi.org/10.1091/mbc.8.4.621
Cavalieri V (2021) The expanding constellation of histone post-translational modifications in the epigenetic landscape. Genes 12:1596. https://doi.org/10.3390/genes12101596
Chabin-Brion K, Marceiller J, Perez F et al (2001) The Golgi complex is a microtubule-organizing organelle. MBoC 12:2047–2060. https://doi.org/10.1091/mbc.12.7.2047
Chang J, Baloh RH, Milbrandt J (2009) The NIMA-family kinase Nek3 regulates microtubule acetylation in neurons. J Cell Sci 122:2274–2282. https://doi.org/10.1242/jcs.048975
Chen S, Owens GC, Makarenkova H, Edelman DB (2010) HDAC6 regulates mitochondrial transport in hippocampal neurons. PLoS One 5:e10848. https://doi.org/10.1371/journal.pone.0010848
Chesarone MA, DuPage AG, Goode BL (2010) Unleashing formins to remodel the actin and microtubule cytoskeletons. Nat Rev Mol Cell Biol 11:62–74. https://doi.org/10.1038/nrm2816
Choudhary C, Kumar C, Gnad F et al (2009) Lysine acetylation targets protein complexes and co-regulates major cellular functions. Science 325:834–840. https://doi.org/10.1126/science.1175371
Chu C-W, Hou F, Zhang J et al (2011) A novel acetylation of β-tubulin by san modulates microtubule polymerization via down-regulating tubulin incorporation. Mol Biol Cell 22:448–456. https://doi.org/10.1091/mbc.E10-03-0203
Coles GL, Baglia LA, Ackerman KG (2015) KIF7 controls the proliferation of cells of the respiratory airway through distinct microtubule dependent mechanisms. PLoS Genet 11(10):e1005525. https://doi.org/10.1371/journal.pgen.1005525
Collins C, Kim SK, Ventrella R et al (2021) Tubulin acetylation promotes penetrative capacity of cells undergoing radial intercalation. Cell Rep 36:109556. https://doi.org/10.1016/j.celrep.2021.109556
Conacci-Sorrell M, Ngouenet C, Eisenman RN (2010) Myc-nick: a cytoplasmic cleavage product of Myc that promotes alpha-tubulin acetylation and cell differentiation. Cell 142:480–493. https://doi.org/10.1016/j.cell.2010.06.037
Coombes C, Yamamoto A, McClellan M et al (2016) Mechanism of microtubule lumen entry for the α-tubulin acetyltransferase enzyme αTAT1. Proc Natl Acad Sci USA 113:E7176–E7184. https://doi.org/10.1073/pnas.1605397113
Creppe C, Malinouskaya L, Volvert M-L et al (2009) Elongator controls the migration and differentiation of cortical neurons through acetylation of alpha-tubulin. Cell 136:551–564. https://doi.org/10.1016/j.cell.2008.11.043
Cueva JG, Hsin J, Huang KC, Goodman MB (2012) Posttranslational acetylation of α-tubulin constrains protofilament number in native microtubules. Curr Biol 22:1066–1074. https://doi.org/10.1016/j.cub.2012.05.012
Cuveillier C, Delaroche J, Seggio M et al (2020) MAP6 is an intraluminal protein that induces neuronal microtubules to coil. Sci Adv 6:1–11. https://doi.org/10.1126/sciadv.aaz4344
d'Ydewalle C, Krishnan J, Chiheb DM et al (2011) HDAC6 inhibitors reverse axonal loss in a mouse model of mutant HSPB1-induced Charcot-Marie-tooth disease. Nat Med 17:968–974. https://doi.org/10.1038/nm.2396
Dafinger C, Liebau MC, Elsayed SM et al (2011) Mutations in KIF7 link Joubert syndrome with sonic hedgehog signaling and microtubule dynamics. J Clin Invest Jul 121(7):2662–2667. https://doi.org/10.1172/JCI43639
Dalwadi U, Yip CK (2018) Structural insights into the function of Elongator. Cell Mol Life Sci 75:1613–1622. https://doi.org/10.1007/s00018-018-2747-6
Deakin NO, Turner CE (2014) Paxillin inhibits HDAC6 to regulate microtubule acetylation, Golgi structure, and polarized migration. J Cell Biol 206:395–413. https://doi.org/10.1083/jcb.201403039
Delgado ILS, Gonçalves J, Fernandes R et al (2024) Balancing act: tubulin Glutamylation and microtubule dynamics in toxoplasma gondii. Microorganisms 12. https://doi.org/10.3390/microorganisms12030488

Deribe YL, Wild P, Chandrashaker A et al (2009) Regulation of epidermal growth factor receptor trafficking by lysine deacetylase HDAC6. ScienceSignaling 2:1–12
Deribe YL, Pawson T, Dikic I (2010) Post-translational modifications in signal integration. Nat Struct Mol Biol 17:666–672. https://doi.org/10.1038/nsmb.1842
Díaz-Valencia JD, Morelli MM, Bailey M et al (2011) Drosophila katanin-60 depolymerizes and severs at microtubule defects. Biophys J 100:2440–2449. https://doi.org/10.1016/j.bpj.2011.03.062
Dijk J, Bompard G, Rabeharivelo G et al (2020) PAK1 regulates MEC-17 acetyltransferase activity and microtubule acetylation during proplatelet extension. Int J Mol Sci 21:1–17. https://doi.org/10.3390/ijms21207531
Dompierre JP, Godin JD, Charrin BC et al (2007) Histone deacetylase 6 inhibition compensates for the transport deficit in Huntington's disease by increasing tubulin acetylation. J Neurosci 27:3571–3583. https://doi.org/10.1523/JNEUROSCI.0037-07.2007
Eckschlager T, Plch J, Stiborova M, Hrabeta J (2017) Histone deacetylase inhibitors as anticancer drugs. Int J Mol Sci 18:1414. https://doi.org/10.3390/ijms18071414
Erck C, Peris L, Andrieux A et al (2005) A vital role of tubulin-tyrosine-ligase for neuronal organization. Proc Natl Acad Sci USA 31;102(22):7853–8:7853. https://doi.org/10.1073/pnas.0409626102
Eshun-Wilson L, Zhang R, Portran D et al (2019) Effects of α-tubulin acetylation on microtubule structure and stability. Proc Natl Acad Sci 116:10366–10371. https://doi.org/10.1073/pnas.1900441116
Even A, Morelli G, Turchetto S et al (2021) ATP-citrate lyase promotes axonal transport across species. Nat Commun 12:5878. https://doi.org/10.1038/s41467-021-25786-y
Feng W, Liu R, Xie X et al (2021) SUMOylation of α-tubulin is a novel modification regulating microtubule dynamics. J Mol Cell Biol 13:91–103. https://doi.org/10.1093/jmcb/mjaa076
Fernández-Barrera J, Bernabé-Rubio M, Casares-Arias J et al (2018) The actin-MRTF-SRF transcriptional circuit controls tubulin acetylation via α-TAT1 gene expression. J Cell Biol 217:929–944. https://doi.org/10.1083/jcb.201702157
Fourest-Lieuvin A, Peris L, Gache V et al (2006) Microtubule regulation in mitosis: tubulin phosphorylation by the cyclin-dependent kinase Cdk1. Mol Biol Cell 17:1041–1050. https://doi.org/10.1091/mbc.e05-07-0621
Friedman JR, Webster BM, Mastronarde DN et al (2010) ER sliding dynamics and ER–mitochondrial contacts occur on acetylated microtubules. J Cell Biol 190:363–375. https://doi.org/10.1083/jcb.200911024
Friedmann DR, Aguilar A, Fan J et al (2012) Structure of the α-tubulin acetyltransferase, αTAT1, and implications for tubulin-specific acetylation. Proc Natl Acad Sci 109:19655–19660. https://doi.org/10.1073/pnas.1209357109
Gaertig J, Cruz MA, Bowen J et al (1995) Acetylation of lysine 40 in alpha-tubulin is not essential in Tetrahymena thermophila. J Cell Biol 129:1301–1310. https://doi.org/10.1083/jcb.129.5.1301
Ganne A, Balasubramaniam M, Ayyadevara H et al (2023) In silico analysis of TUBA4A mutations in amyotrophic lateral sclerosis to define mechanisms of microtubule disintegration. Sci Rep 13(1):2096. https://doi.org/10.1038/s41598-023-28381-x
Gao Y, Thomas JO, Chow RL et al (1992) A cytoplasmic chaperonin that catalyzes β-actin folding. Cell 69:1043–1050. https://doi.org/10.1016/0092-8674(92)90622-J
Gao Y, Vainberg I, Chow R, Cowan NJ (1993) Two cofactors and cytoplasmic chaperonin. Mol Cell Biol 13(4):2478–2485
Gardiner J, Barton D, Marc J, Overall R (2007) Potential role of tubulin acetylation and microtubule-based protein trafficking in familial dysautonomia. Traffic 8:1145–1149. https://doi.org/10.1111/j.1600-0854.2007.00605.x
Garnham CP, Roll-Mecak A (2012) The chemical complexity of cellular microtubules: tubulin post-translational modification enzymes and their roles in tuning microtubule functions. Cytoskeleton 69:442–463. https://doi.org/10.1002/cm.21027

Garnham CP, Vemu A, Wilson-Kubalek EM et al (2015) Multivalent microtubule recognition by tubulin tyrosine ligase-like family Glutamylases HHS public access molecular basis for specificity among the enzymes primarily responsible for chemically diversifying cellular microtubules. Cell 161:1112–1123. https://doi.org/10.1016/j.cell.2015.04.003.Multivalent

Ge Y, Gao J, Jordan R, Naumann CA (2018) Changes in cholesterol level Alter integrin sequestration in raft-mimicking lipid mixtures. Biophys J 114:158–167. https://doi.org/10.1016/j.bpj.2017.11.005

Genova M, Grycova L, Puttrich V et al (2023) Tubulin polyglutamylation differentially regulates microtubule-interacting proteins. EMBO J 42:e112101. https://doi.org/10.15252/embj.2022112101

Girdwood D, Bumpass D, Vaughan OA, et al (2003) p300 transcriptional repression is mediated by SUMO modification allele knockouts of p300 and CBP display distinct transcriptional phenotypes indicate that DNA bound transcription factors are likely to be in competition for limiting amounts of p300 and CBP. In: Molecular cell

Gomes AQ, Nolasco S, Soares H (2013) Non-coding RNAs: multi-tasking molecules in the cell. Int J Mol Sci 14:16010–16039. https://doi.org/10.3390/ijms140816010

Gonçalves J, Tavares A, Carvalhal S, Soares H (2010) Revisiting the tubulin folding pathway: new roles in centrosomes and cilia. Biomol Concepts 1:423–434. https://doi.org/10.1515/bmc.2010.033

Govindarajan N, Rao P, Burkhardt S et al (2013) Reducing HDAC6 ameliorates cognitive deficits in a mouse model for Alzheimer's disease. EMBO Mol Med 5:52–63. https://doi.org/10.1002/emmm.201201923

Gray SG, Ekström TJ (2001) The human histone deacetylase family. Exp Cell Res 262:75–83. https://doi.org/10.1006/excr.2000.5080

Greer K, Maruta H, L'hernault SW, Rosenbaum JL (1985) A-tubulin Acetylase activity Chlamydomonas Flagella in isolated. J Cell Biol 101:2081–2084

Gu B, Zhu W-G (2012) Surf the post-translational modification network of p53 regulation. Int J Biol Sci 8:672–684. https://doi.org/10.7150/ijbs.4283

Hallak ME, Rodriguez JA, Barra HS, Caputto R (1977) Release of tyrosine from tyrosinated tubulin. Some common factors that affect this process and the assembly of tubulin. FEBS Lett 73:147–150. https://doi.org/10.1016/0014-5793(77)80968-x

Han Y, Jin Y-H, Kim Y-J et al (2008) Acetylation of Sirt2 by p300 attenuates its deacetylase activity. Biochem Biophys Res Commun 375:576–580. https://doi.org/10.1016/j.bbrc.2008.08.042

Han Y, Jeong HM, Jin Y-H et al (2009) Acetylation of histone deacetylase 6 by p300 attenuates its deacetylase activity. Biochem Biophys Res Commun 383:88–92. https://doi.org/10.1016/j.bbrc.2009.03.147

Hansen BK, Gupta R, Baldus L et al (2019) Analysis of human acetylation stoichiometry defines mechanistic constraints on protein regulation. Nat Commun 10:1055. https://doi.org/10.1038/s41467-019-09024-0

Harmel R, Fiedler D (2018) Features and regulation of non-enzymatic post-translational modifications. Nat Chem Biol 14:244–252. https://doi.org/10.1038/nchembio.2575

Heppenstall P, Morley S (2023) Inhibitors of alpha-tubulin acetylation for the treatment of pain

Herczenik E, Gebbink MFBG (2008) Molecular and cellular aspects of protein misfolding and disease. FASEB J 22:2115–2133. https://doi.org/10.1096/fj.07-099671

Horio T, Hotani H (1986) Visualization of the dynamic instability of individual microtubules by dark-field microscopy. Nature 321:605–607. https://doi.org/10.1038/321605a0

Huang K, Diener DR, Rosenbaum JL (2009) The ubiquitin conjugation system is involved in the disassembly of cilia and flagella. J Cell Biol 186:601–613. https://doi.org/10.1083/jcb.200903066

Hubbert C, Guardiola A, Shao R et al (2002) HDAC6 is a microtubule-associated deacetylase. Nature 417:455–458. https://doi.org/10.1038/417455a

Hübner MR, Eckersley-Maslin MA, Spector DL (2013) Chromatin organization and transcriptional regulation. Curr Opin Genet Dev 23:89–95. https://doi.org/10.1016/j.gde.2012.11.006

Hunter T (2007) The age of crosstalk: phosphorylation, ubiquitination, and beyond. Mol Cell 28:730–738. https://doi.org/10.1016/j.molcel.2007.11.019

Huttlin EL, Jedrychowski MP, Elias JE et al (2010) A tissue-specific atlas of mouse protein phosphorylation and expression. Cell 143:1174–1189. https://doi.org/10.1016/j.cell.2010.12.001

Ibar C, Irvine KD (2020) Integration of hippo-YAP signaling with metabolism. Dev Cell 54:256–267. https://doi.org/10.1016/j.devcel.2020.06.025

Ichikawa M, Liu D, Kastritis PL et al (2017) Subnanometre-resolution structure of the doublet microtubule reveals new classes of microtubuleassociated proteins. Nat Commun 8:1–12. https://doi.org/10.1038/ncomms15035

Iuzzolino A, Pellegrini FR, Rotili D et al (2024) The α-tubulin acetyltransferase ATAT1: structure, cellular functions, and its emerging role in human diseases. Cell Mol Life Sci 81:193. https://doi.org/10.1007/s00018-024-05227-x

Jamil M, Cowart LA (2023) Sphingolipids in mitochondria—from function to disease. Front Cell Dev Biol 11:1302472. https://doi.org/10.3389/fcell.2023.1302472

Janke C (2014) The tubulin code: molecular components, readout mechanisms, and functions. J Cell Biol 206:461–472. https://doi.org/10.1083/jcb.201406055

Janke C, Bulinski JC (2011) Post-translational regulation of the microtubule cytoskeleton: mechanisms and functions. Nat Rev Mol Cell Biol 12:773–786. https://doi.org/10.1038/nrm3227

Janke C, Magiera MM (2020) The tubulin code and its role in controlling microtubule properties and functions. Nat Rev Mol Cell Biol 21:307–326. https://doi.org/10.1038/s41580-020-0214-3

Janke C, Montagnac G (2017) Causes and consequences of microtubule acetylation. Curr Biol 27:R1287. https://doi.org/10.1016/j.cub.2017.10.044

Jankowska-Kulawy A, Klimaszewska-Łata J, Gul-Hinc S et al (2022) Metabolic and cellular compartments of acetyl-CoA in the healthy and diseased brain. Int J Mol Sci 23. https://doi.org/10.3390/ijms231710073

Jiang K, Shen M, Chen Y, Xu W (2018) miR-150 promotes the proliferation and migration of non-small cell lung cancer cells by regulating the SIRT2/JMJD2A signaling pathway. Oncol Rep 40:943–951. https://doi.org/10.3892/or.2018.6487

Kalebic N, Sorrentino S, Perlas E et al (2013) αTAT1 is the major α-tubulin acetyltransferase in mice. Nat Commun 4:1962. https://doi.org/10.1038/ncomms2962

Kasai N, Kadeer A, Kajita M et al (2018) The paxillin-plectin-EPLIN complex promotes apical elimination of RasV12-transformed cells by modulating HDAC6-regulated tubulin acetylation. Sci Rep 8:2097. https://doi.org/10.1038/s41598-018-20146-1

Kawauchi T, Chihama K, Nabeshima YI, Hoshino M (2006) Cdk5 phosphorylates and stabilizes p27kip1 contributing to actin organization and cortical neuronal migration. Nat Cell Biol 8:17–26. https://doi.org/10.1038/ncb1338

Kearns S, Mason FM, Rathmell WK et al (2021) Molecular determinants for α-tubulin methylation by SETD2. J Biol Chem 297:100898. https://doi.org/10.1016/j.jbc.2021.100898

Kelley CF, Litschel T, Schumacher S, et al (2020) Phosphoinositides regulate force-independent interactions between Talin, vinculin, and actin. eLife 9:e56110. https://doi.org/10.7554/eLife.56110

Keryer G, Pineda JR, Liot G et al (2011) Ciliogenesis is regulated by a huntingtin-HAP1-PCM1 pathway and is altered in Huntington disease. J Clin Invest 121:4372–4382. https://doi.org/10.1172/JCI57552

Kilgore M, Miller CA, Fass DM et al (2010) Inhibitors of class 1 histone deacetylases reverse contextual memory deficits in a mouse model of Alzheimer's disease. Epub 2009 Dec 9 PMID: 20010553; PMCID: PMC3055373 Mar; 35(4):870-80:870. https://doi.org/10.1038/npp.2009.197

Kim SC, Sprung R, Chen Y et al (2006) Substrate and functional diversity of lysine acetylation revealed by a proteomics survey. Mol Cell 23:607–618. https://doi.org/10.1016/j.molcel.2006.06.026

Kim C, Choi H, Jung ES et al (2012) HDAC6 inhibitor blocks amyloid Beta-induced impairment of mitochondrial transport in hippocampal neurons. PLoS One 7:e42983. https://doi.org/10.1371/journal.pone.0042983

Kim J-Y, Woo S-Y, Hong YB et al (2016) HDAC6 inhibitors rescued the defective axonal mitochondrial movement in motor neurons derived from the induced pluripotent stem cells of peripheral neuropathy patients with HSPB1 mutation. Stem Cells Int 2016:9475981. https://doi.org/10.1155/2016/9475981

Kinoshita M, Matsumori N (2022) Inimitable impacts of ceramides on lipid rafts formed in artificial and natural cell membranes. Membranes (Basel) 12:727. https://doi.org/10.3390/membranes12080727

Kitamura N, Galligan JJ (2023) A global view of the human post-translational modification landscape. Biochem J 480:1241–1265. https://doi.org/10.1042/BCJ20220251

Ko P, Choi JH, Song S et al (2021) Microtubule acetylation controls MDA-MB-231 breast cancer cell invasion through the modulation of endoplasmic reticulum stress. Int J Mol Sci 22. https://doi.org/10.3390/ijms22116018

Kormendi V, Szyk A, Piszczek G, Roll-Mecak A (2012) Crystal structures of tubulin acetyltransferase reveal a conserved catalytic core and the plasticity of the essential N terminus. J Biol Chem 287:41569–41575. https://doi.org/10.1074/jbc.C112.421222

Kortazar D, Fanarraga ML, Carranza G et al (2007) Role of cofactors B (TBCB) and E (TBCE) in tubulin heterodimer dissociation. Exp Cell Res 313:425–436. https://doi.org/10.1016/j.yexcr.2006.09.002

Kozminski KG, Diener DR, Rosenbaum JL (1993) High level expression of nonacetylatable alpha-tubulin in Chlamydomonas reinhardtii. Cell Motil Cytoskeleton 25:158–170. https://doi.org/10.1002/cm.970250205

Kull FJ, Sloboda RD (2014) A slow dance for microtubule acetylation. Cell 157:1255–1256. https://doi.org/10.1016/j.cell.2014.05.021

Kumar A, Larrea D, Pero ME et al (2024) MFN2 coordinates mitochondria motility with α-tubulin acetylation and this regulation is disrupted in CMT2A. iScience 15;27(6):109994:109994. https://doi.org/10.1016/j.isci.2024.109994

L'Hernault SW, Rosenbaum JL (1985) Chlamydomonas .alpha.-tubulin is posttranslationally modified by acetylation on the. epsilon.-amino group of a lysine. Biochemistry 24:473–478. https://doi.org/10.1021/bi00323a034

Lafarga V, Aymerich I, Tapia O et al (2012) A novel GRK2/HDAC6 interaction modulates cell spreading and motility. EMBO J 31:856–869. https://doi.org/10.1038/emboj.2011.466

Landskron L, Bak J, Adamopoulos A et al (2022) Posttranslational modification of microtubules by the MATCAP detyrosinase. Science 376:eabn6020. https://doi.org/10.1126/science.abn6020

Law BMH, Spain VA, Leinster VHL et al (2014) A direct interaction between leucine-rich repeat kinase 2 and specific β-tubulin isoforms regulates tubulin acetylation. J Biol Chem 289:895–908. https://doi.org/10.1074/jbc.M113.507913

LeDizet M, Piperno G (1987) Identification of an acetylation site of Chlamydomonas alpha-tubulin. Proc Natl Acad Sci 84:5720–5724. https://doi.org/10.1073/pnas.84.16.5720

Lee G, Rook SL (1992) Expression of tau protein in non-neuronal cells: microtubule binding and stabilization. J Cell Sci 102:227–237. https://doi.org/10.1242/jcs.102.2.227

Lee SW, Yang J, Kim SY et al (2015) MicroRNA-26a induced by hypoxia targets HDAC6 in myogenic differentiation of embryonic stem cells. Nucleic Acids Res 43:2057–2073. https://doi.org/10.1093/nar/gkv088

Lee C-C, Cheng Y-C, Chang C-Y et al (2018) Alpha-tubulin acetyltransferase/MEC-17 regulates cancer cell migration and invasion through epithelial-mesenchymal transition suppression and cell polarity disruption. Sci Rep 8:17477. https://doi.org/10.1038/s41598-018-35392-6

Legube G, Trouche D (2003) Regulating histone acetyltransferases and deacetylases. EMBO Rep 4:944–947. https://doi.org/10.1038/sj.embor.embor941

Leo L, Yu W, D'Rozario M et al (2015) Vertebrate Fidgetin restrains axonal growth by severing labile domains of microtubules. Cell Rep 12:1723–1730. https://doi.org/10.1016/j.celrep.2015.08.017

Leutert M, Entwisle SW, Villén J (2021) Decoding post-translational modification crosstalk with proteomics. Mol Cell Proteomics 20:0–11. https://doi.org/10.1016/j.mcpro.2021.100129, 100129

Lewis SA, Tian G, Cowan NJ (1997) The alpha- and beta-tubulin folding pathways. Trends Cell Biol 7:479–484. https://doi.org/10.1016/S0962-8924(97)01168-9

Li L, Liu B, Zhang X, Ye L (2015) The oncoprotein HBXIP promotes migration of breast cancer cells via GCN5-mediated microtubule acetylation. Biochem Biophys Res Commun 458:720–725. https://doi.org/10.1016/j.bbrc.2015.02.036

Li T, Zhang C, Hassan S et al (2018) Histone deacetylase 6 in cancer. J Hematol Oncol 11:111. https://doi.org/10.1186/s13045-018-0654-9

Li C, Zhou Y, Kim JT et al (2021) Regulation of SIRT2 by Wnt/β-catenin signaling pathway in colorectal cancer cells. Biochimica et Biophysica Acta (BBA) - molecular. Cell Res 1868:118966. https://doi.org/10.1016/j.bbamcr.2021.118966

Li Y, Kučera O, Cuvelier D et al (2023) Compressive forces stabilize microtubules in living cells. Nat Mater 22:913–924. https://doi.org/10.1038/s41563-023-01578-1

Liu Y, Peng L, Seto E et al (2012) Modulation of histone deacetylase 6 (HDAC6) nuclear import and tubulin deacetylase activity through acetylation. J Biol Chem 287:29168–29174. https://doi.org/10.1074/jbc.M112.371120

Liu N, Xiong Y, Li S et al (2015a) New HDAC6-mediated deacetylation sites of tubulin in the mouse brain identified by quantitative mass spectrometry. Sci Rep 5:16869. https://doi.org/10.1038/srep16869

Liu N, Xiong Y, Ren Y et al (2015b) Proteomic profiling and functional characterization of multiple post-translational modifications of tubulin. J Proteome Res 14:3292–3304. https://doi.org/10.1021/acs.jproteome.5b00308

Liu S, Zhou Z, Zhang L et al (2019a) Inhibition of SIRT2 by targeting GSK3β-mediated phosphorylation alleviates SIRT2 toxicity in SH-SY5Y cells. Front Cell Neurosci 13. https://doi.org/10.3389/fncel.2019.00148

Liu Y, Tavana O, Gu W (2019b) p53 modifications: exquisite decorations of the powerful guardian. J Mol Cell Biol 11:564–577. https://doi.org/10.1093/jmcb/mjz060

Liu P, Zhang S, Ma J et al (2022) Vimentin inhibits α-tubulin acetylation via enhancing α-TAT1 degradation to suppress the replication of human parainfluenza virus type 3. PLoS Pathog 18:e1010856. https://doi.org/10.1371/journal.ppat.1010856

Loktev AV, Zhang Q, Beck JS et al (2008) A BBSome subunit links ciliogenesis, microtubule stability, and acetylation. Dev Cell 15:854–865. https://doi.org/10.1016/j.devcel.2008.11.001

Lopez-Fanarraga M, Avila J, Guasch A et al (2001) Review: postchaperonin tubulin folding cofactors and their role in microtubule dynamics. J Struct Biol 135:219–229. https://doi.org/10.1006/jsbi.2001.4386

Louveaux M, Rochette S, Beauzamy L et al (2016) The impact of mechanical compression on cortical microtubules in Arabidopsis: a quantitative pipeline. Plant J 88:328–342. https://doi.org/10.1111/tpj.13290

Lundby A, Lage K, Weinert BT et al (2012) Proteomic analysis of lysine acetylation sites in rat tissues reveals organ specificity and subcellular patterns. Cell Rep 2:419–431. https://doi.org/10.1016/j.celrep.2012.07.006

Lwin T, Zhao X, Cheng F et al (2013) A microenvironment-mediated c-Myc/miR-548m/HDAC6 amplification loop in non-Hodgkin B cell lymphomas. J Clin Invest 123:4612–4626. https://doi.org/10.1172/JCI64210

Ma M, Stoyanova M, Rademacher G et al (2019) Structure of the decorated ciliary doublet microtubule. Cell 179:909–922.e12. https://doi.org/10.1016/j.cell.2019.09.030

Mackeh R, Lorin S, Ratier A et al (2014) Reactive oxygen species, AMP-activated protein kinase, and the transcription cofactor p300 regulate α-tubulin acetyltransferase-1 (αTAT-1/MEC-17)-

dependent microtubule hyperacetylation during cell stress. J Biol Chem 289:11816–11828. https://doi.org/10.1074/jbc.M113.507400
Macquart C, Jüttner R, Morales Rodriguez B et al (2019) Microtubule cytoskeleton regulates Connexin 43 localization and cardiac conduction in cardiomyopathy caused by mutation in A-type lamins gene. Hum Mol Genet 28:4043–4052. https://doi.org/10.1093/hmg/ddy227
MacTaggart B, Shimogawa M, Lougee M et al (2023) Global analysis of post-translational side-chain Arginylation using pan-Arginylation antibodies. Mol Cell Proteomics 22:100664. https://doi.org/10.1016/j.mcpro.2023.100664
Mansini AP, Lorenzo Pisarello MJ, Thelen KM et al (2018) MicroRNA (miR)-433 and miR-22 dysregulations induce histone-deacetylase-6 overexpression and ciliary loss in cholangiocarcinoma. Hepatology 68:561–573. https://doi.org/10.1002/hep.29832
Mao C-X, Xiong Y, Xiong Z et al (2014) Microtubule-severing protein Katanin regulates neuromuscular junction development and dendritic elaboration in drosophila. Development 141:1064–1074. https://doi.org/10.1242/dev.097774
Mao C-X, Wen X, Jin S, Zhang YQ (2017) Increased acetylation of microtubules rescues human tau-induced microtubule defects and neuromuscular junction abnormalities in drosophila. Dis Model Mech 10:1245–1252. https://doi.org/10.1242/dmm.028316
Marquez J, Bhattacharya D, Lusk CP, Khokha MK (2021) Nucleoporin NUP205 plays a critical role in cilia and congenital disease. Dev Biol 469:46. https://doi.org/10.1016/j.ydbio.2020.10.001
Martínez-Hernández J, Parato J, Sharma A et al (2022) Crosstalk between acetylation and the tyrosination/detyrosination cycle of α-tubulin in Alzheimer's disease. Front Cell Dev Biol 10:926914. https://doi.org/10.3389/fcell.2022.926914
Maruta H, Greer K, Rosenbaum JL (1986) The acetylation of alpha-tubulin and its relationship to the assembly and disassembly of microtubules. J Cell Biol 103:571–579. https://doi.org/10.1083/jcb.103.2.571
Matsuyama A, Shimazu T, Sumida Y et al (2002) In vivo destabilization of dynamic microtubules by HDAC6-mediated deacetylation. EMBO J 21:6820–6831. https://doi.org/10.1093/emboj/cdf682
McKenna ED, Sarbanes SL, Cummings SW, Roll-Mecak A (2023) The tubulin code, from molecules to health and disease. Annu Rev Cell Dev Biol 39:331–361. https://doi.org/10.1146/annurev-cellbio-030123-032748
McLendon PM, Ferguson BS, Osinska H et al (2014) Tubulin hyperacetylation is adaptive in cardiac proteotoxicity by promoting autophagy. Proc Natl Acad Sci USA 111:E5178–E5186. https://doi.org/10.1073/pnas.1415589111
Mendoza M, Egervari G, Sidoli S et al (2022) Enzymatic transfer of acetate on histones from lysine reservoir sites to lysine activating sites. Sci Adv 8:eabj5688. https://doi.org/10.1126/sciadv.abj5688
Millán-Zambrano G, Burton A, Bannister AJ, Schneider R (2022) Histone post-translational modifications - cause and consequence of genome function. Nat Rev Genet 23:563–580. https://doi.org/10.1038/s41576-022-00468-7
Mitchison T, Kirschner M (1984) Dynamic instability of microtubule growth. Nature 312:237–242. https://doi.org/10.1038/312237a0
Miyake Y, Keusch JJ, Wang L et al (2016) Structural insights into HDAC6 tubulin deacetylation and its selective inhibition. Nat Chem Biol 12:748–754. https://doi.org/10.1038/nchembio.2140
Morelli G, Even A, Gladwyn-Ng I et al (2018) p27Kip1 modulates axonal transport by regulating α-tubulin acetyltransferase 1 stability. Cell Rep 23:2429–2442. https://doi.org/10.1016/j.celrep.2018.04.083
Nakayasu ES, Burnet MC, Walukiewicz HE et al (2017) Ancient regulatory role of lysine acetylation in central metabolism. mBio 8:e01894–17. https://doi.org/10.1128/mBio.01894-17
Nekooki-Machida Y, Nakakura T, Nishijima Y et al (2018) Dynamic localization of α-tubulin acetyltransferase ATAT1 through the cell cycle in human fibroblastic KD cells. Med Mol Morphol 51:217–226. https://doi.org/10.1007/s00795-018-0195-x

Neumann B, Hilliard MA (2014) Loss of MEC-17 leads to microtubule instability and axonal degeneration. Cell Rep 6:93–103. https://doi.org/10.1016/j.celrep.2013.12.004
Nieuwenhuis J, Adamopoulos A, Bleijerveld OB et al (2017) Vasohibins encode tubulin detyrosinating activity. Science 358:1453–1456. https://doi.org/10.1126/science.aao5676
Nolasco S, Bellido J, Serna M et al (2021) Colchicine blocks tubulin heterodimer recycling by tubulin cofactors TBCA, TBCB, and TBCE. Front Cell Dev Biol 9:656273. https://doi.org/10.3389/fcell.2021.656273
North BJ, Verdin E (2007a) Mitotic regulation of SIRT2 by cyclin-dependent kinase 1-dependent phosphorylation. J Biol Chem 282:19546–19555. https://doi.org/10.1074/jbc.M702990200
North BJ, Verdin E (2007b) Interphase nucleo-cytoplasmic shuttling and localization of SIRT2 during mitosis. PLoS One 2:e784. https://doi.org/10.1371/journal.pone.0000784
North BJ, Marshall BL, Borra MT et al (2003) The human Sir2 ortholog, SIRT2, is an NAD+-dependent tubulin deacetylase. Mol Cell 11:437–444. https://doi.org/10.1016/s1097-2765(03)00038-8
Ohkawa N, Sugisaki S, Tokunaga E et al (2008) N-acetyltransferase ARD1-NAT1 regulates neuronal dendritic development. Genes Cells 13:1171–1183. https://doi.org/10.1111/j.1365-2443.2008.01235.x
Ojea Ramos S, Feld M, Fustiñana MS (2022) Contributions of extracellular-signal regulated kinase 1/2 activity to the memory trace. Front Mol Neurosci 15:988790. https://doi.org/10.3389/fnmol.2022.988790
Orbach R, Howard J (2019) The dynamic and structural properties of axonemal tubulins support the high length stability of cilia. Nat Commun 10:1838. https://doi.org/10.1038/s41467-019-09779-6
Palazzo A, Ackerman B, Gundersen GG (2003) Tubulin acetylation and cell motility. Nature 421:230. https://doi.org/10.1038/421230a
Pandithage R, Lilischkis R, Harting K et al (2008) The regulation of SIRT2 function by cyclin-dependent kinases affects cell motility. J Cell Biol 180:915–929. https://doi.org/10.1083/jcb.200707126
Park S-Y, Kim J-S (2020) A short guide to histone deacetylases including recent progress on class II enzymes. Exp Mol Med 52:204–212. https://doi.org/10.1038/s12276-020-0382-4
Park IY, Powell RT, Tripathi DN et al (2016) Dual chromatin and cytoskeletal remodeling by SETD2. Cell 166:950–962. https://doi.org/10.1016/j.cell.2016.07.005
Parmigiani RB, Xu WS, Venta-Perez G et al (2008) HDAC6 is a specific deacetylase of peroxiredoxins and is involved in redox regulation. Proc Natl Acad Sci USA 105:9633–9638. https://doi.org/10.1073/pnas.0803749105
Paturle-Lafanechère L, Eddé B, Denoulet P et al (1991) Characterization of a major brain tubulin variant which cannot be tyrosinated. Biochemistry 30:10523–10528. https://doi.org/10.1021/bi00107a022
Perdiz D, Lorin S, Leroy-Gori I, Poüs C (2017) Stress-induced hyperacetylation of microtubule enhances mitochondrial fission and modulates the phosphorylation of Drp1 at 616Ser. Cell Signal 39:32–43. https://doi.org/10.1016/j.cellsig.2017.07.020
Piperno G, Fuller MT (1985) Monoclonal antibodies specific for an acetylated form of alpha-tubulin recognize the antigen in cilia and flagella from a variety of organisms. J Cell Biol 101:2085–2094. https://doi.org/10.1083/jcb.101.6.2085
Poirier K, Saillour Y, Bahi-Buisson N et al (2010) Mutations in the neuronal β-tubulin subunit TUBB3 result in malformation of cortical development and neuronal migration defects. Hum Mol Genet 19:4462–4473. https://doi.org/10.1093/hmg/ddq377
Portran D, Schaedel L, Xu Z et al (2017) Tubulin acetylation protects long-lived microtubules against mechanical ageing. Nat Cell Biol 19:391–398. https://doi.org/10.1038/ncb3481
Prota AE, Magiera MM, Kuijpers M et al (2013) Structural basis of tubulin tyrosination by tubulin tyrosine ligase. J Cell Biol 200:259–270. https://doi.org/10.1083/jcb.201211017

Pu J, Liu T, Wang X et al (2024) Exploring the role of histone deacetylase and histone deacetylase inhibitors in the context of multiple myeloma: mechanisms, therapeutic implications, and future perspectives. Exp Hematol Oncol 13:45. https://doi.org/10.1186/s40164-024-00507-5

Pugacheva EN, Jablonski SA, Hartman TR et al (2007) HEF1-dependent Aurora a activation induces disassembly of the primary cilium. Cell 129:1351–1363. https://doi.org/10.1016/j.cell.2007.04.035

Raybin D, Flavin M (1975) An enzyme tyrosylating α-tubulin and its role in microtubule assembly. Biochem Biophys Res Commun 65:1088–1095. https://doi.org/10.1016/S0006-291X(75)80497-9

Redeker V, Levilliers N, Vinolo E et al (2005) Mutations of tubulin Glycylation sites reveal cross-talk between the C termini of α- and β-tubulin and affect the ciliary matrix in Tetrahymena. J Biol Chem 280:596–606. https://doi.org/10.1074/jbc.M408324200

Reed NA, Cai D, Blasius TL et al (2006) Microtubule acetylation promotes kinesin-1 binding and transport. Curr Biol 16:2166–2172. https://doi.org/10.1016/j.cub.2006.09.014

Rivero S, Cardenas J, Bornens M, Rios RM (2009) Microtubule nucleation at the cis-side of the Golgi apparatus requires AKAP450 and GM130. EMBO J 28:1016–1028. https://doi.org/10.1038/emboj.2009.47

Rocha AG, Franco A, Krezel AM et al (2018) MFN2 agonists reverse mitochondrial defects in preclinical models of Charcot-Marie-tooth disease type 2A. Science 360:336–341. https://doi.org/10.1126/science.aao1785

Rogowski K, Juge F, van Dijk J et al (2009) Evolutionary divergence of enzymatic mechanisms for posttranslational polyglycylation. Cell 137:1076–1087. https://doi.org/10.1016/j.cell.2009.05.020

Roll-Mecak A (2019) How cells exploit tubulin diversity to build functional cellular microtubule mosaics. Curr Opin Cell Biol 56:102–108. https://doi.org/10.1016/j.ceb.2018.10.009

Roll-Mecak A (2020) The tubulin code in microtubule dynamics and information encoding. Dev Cell 54:7–20. https://doi.org/10.1016/j.devcel.2020.06.008

Roux KJ, Kim DI, Raida M, Burke B (2012) A promiscuous biotin ligase fusion protein identifies proximal and interacting proteins in mammalian cells. J Cell Biol 196:801–810. https://doi.org/10.1083/jcb.201112098

Roy AD, Gross EG, Pillai GS et al (2022) Non-catalytic allostery in α-TAT1 by a phospho-switch drives dynamic microtubule acetylation. J Cell Biol 221. https://doi.org/10.1083/jcb.202202100

Saji S, Kawakami M, Hayashi S et al (2005) Significance of HDAC6 regulation via estrogen signaling for cell motility and prognosis in estrogen receptor-positive breast cancer. Oncogene 24:4531–4539. https://doi.org/10.1038/sj.onc.1208646

Sampathkumar A, Krupinski P, Wightman R et al (2014) Subcellular and supracellular mechanical stress prescribes cytoskeleton behavior in Arabidopsis cotyledon pavement cells. eLife 3:e01967. https://doi.org/10.7554/eLife.01967

Saragoni L, Hernández P, Maccioni RB (2000) Differential association of tau with subsets of microtubules containing posttranslationally-modified tubulin variants in neuroblastoma cells. Neurochem Res 25:59–70. https://doi.org/10.1023/a:1007587315630

Schaedel L, John K, Gaillard J et al (2015) Microtubules self-repair in response to mechanical stress. Nature Mater 14:1156–1163. https://doi.org/10.1038/nmat4396

Schofield AV, Gamell C, Suryadinata R et al (2013) Tubulin polymerization promoting protein 1 (Tppp1) phosphorylation by rho-associated coiled-coil kinase (Rock) and cyclin-dependent kinase 1 (Cdk1) inhibits microtubule dynamics to increase cell proliferation. J Biol Chem 288:7907–7917. https://doi.org/10.1074/jbc.M112.441048

Schulze E, Kirschner M (1987) Dynamic and stable populations of microtubules in cells. J Cell Biol 104:277–288. https://doi.org/10.1083/jcb.104.2.277

Schwartz CL, Heumann JM, Dawson SC, Hoenger A (2012) A detailed, hierarchical study of Giardia lamblia's ventral disc reveals novel microtubule-associated protein complexes. PLoS One 7:e43783. https://doi.org/10.1371/journal.pone.0043783

Seetharaman S, Vianay B, Roca V et al (2022) Microtubules tune mechanosensitive cell responses. Nat Mater 21:366–377. https://doi.org/10.1038/s41563-021-01108-x
Selenica M-L, Benner L, Housley SB et al (2014) Histone deacetylase 6 inhibition improves memory and reduces total tau levels in a mouse model of tau deposition. Alzheimers Res Ther 6:12. https://doi.org/10.1186/alzrt241
Serna M, Zabala JC (2016) Tubulin folding and degradation. In: eLS. Wiley, pp 1–9
Shah N, Kumar S, Zaman N et al (2018) TAK1 activation of alpha-TAT1 and microtubule hyperacetylation control AKT signaling and cell growth. Nat Commun 9:1696. https://doi.org/10.1038/s41467-018-04121-y
Shen Q, Zheng X, McNutt MA et al (2009) NAT10, a nucleolar protein, localizes to the midbody and regulates cytokinesis and acetylation of microtubules. Exp Cell Res 315:1653–1667. https://doi.org/10.1016/j.yexcr.2009.03.007
Shida T, Cueva JG, Xu Z et al (2010) The major α-tubulin K40 acetyltransferase αTAT1 promotes rapid ciliogenesis and efficient mechanosensation. Proc Natl Acad Sci 107:21517–21522. https://doi.org/10.1073/pnas.1013728107
Shultz T, Shmuel M, Hyman T, Altschuler Y (2008) B-Tubulin cofactor D and ARL2 take part in apical junctional complex disassembly and abrogate epithelial structure. FASEB J 22:168–182. https://doi.org/10.1096/fj.06-7786com
Singh L, Rana S, Mehan S (2018) Role of adenylyl cyclase activator in controlling experimental diabetic nephropathy in rats. Int J Physiol Pathophysiol Pharmacol 10:144–153
Siudeja K, Srinivasan B, Xu L et al (2011) Impaired coenzyme a metabolism affects histone and tubulin acetylation in drosophila and human cell models of pantothenate kinase associated neurodegeneration. EMBO Mol Med 3:755–766. https://doi.org/10.1002/emmm.201100180
Skoge RH, Dölle C, Ziegler M (2014) Regulation of SIRT2-dependent α-tubulin deacetylation by cellular NAD levels. DNA Repair 23:33–38. https://doi.org/10.1016/j.dnarep.2014.04.011
Skultetyova L, Ustinova K, Kutil Z et al (2017) Human histone deacetylase 6 shows strong preference for tubulin dimers over assembled microtubules. Sci Rep 7:1–13. https://doi.org/10.1038/s41598-017-11739-3
Solinger JA, Paolinelli R, Klöss H et al (2010) The Caenorhabditis elegans Elongator complex regulates neuronal alpha-tubulin acetylation. PLoS Genet 6:e1000820. https://doi.org/10.1371/journal.pgen.1000820
Song Y, Kirkpatrick LL, Schilling AB et al (2013) Transglutaminase and polyamination of tubulin: posttranslational modification for stabilizing axonal microtubules. Neuron 78:109–123. https://doi.org/10.1016/j.neuron.2013.01.036
Soppina V, Herbstman JF, Skiniotis G, Verhey KJ (2012) Luminal localization of α-tubulin K40 acetylation by cryo-EM analysis of fab-labeled microtubules. PLoS One 7:e48204. https://doi.org/10.1371/journal.pone.0048204
Steczkiewicz K, Kinch L, Grishin NV et al (2006) Eukaryotic domain of unknown function DUF738 belongs to Gcn5-related N-acetyltransferase superfamily. Cell Cycle 5:2927–2930. https://doi.org/10.4161/cc.5.24.3572
Su M, Guan H, Zhang F et al (2016) HDAC6 regulates the chaperone-mediated autophagy to prevent oxidative damage in injured neurons after experimental spinal cord injury. Oxidative Med Cell Longev 2016:7263736. https://doi.org/10.1155/2016/7263736
Sudo H, Baas PW (2010) Acetylation of microtubules influences their sensitivity to severing by Katanin in neurons and fibroblasts. J Neurosci 30:7215–7226. https://doi.org/10.1523/JNEUROSCI.0048-10.2010
Sun S, Han X, Li X et al (2018) MicroRNA-212-5p prevents dopaminergic neuron death by inhibiting SIRT2 in MPTP-induced mouse model of Parkinson's disease. Front Mol Neurosci 11. https://doi.org/10.3389/fnmol.2018.00381
Szczesna E, Zehr EA, Cummings SW et al (2022) Combinatorial and antagonistic effects of tubulin glutamylation and glycylation on katanin microtubule severing. Dev Cell 57:2497–2513.e6. https://doi.org/10.1016/j.devcel.2022.10.003

Szyk A, Deaconescu AM, Spector J et al (2014) Molecular basis for age-dependent microtubule acetylation by tubulin acetyltransferase. Cell 157:1405–1415. https://doi.org/10.1016/j.cell.2014.03.061

Tachibana H, Minoura K, Omachi T et al (2023) The plasma membrane of focal adhesions has a high content of cholesterol and phosphatidylcholine with saturated acyl chains. J Cell Sci 136:jcs260763. https://doi.org/10.1242/jcs.260763

Taes I, Timmers M, Hersmus N et al (2013) Hdac6 deletion delays disease progression in the SOD1G93A mouse model of ALS. Hum Mol Genet 22:1783–1790. https://doi.org/10.1093/hmg/ddt028

Takahashi H, Nomura H, Iriki H et al (2021) Cholesterol 25-hydroxylase is a metabolic switch to constrain T cell-mediated inflammation in the skin. Sci Immunol 6:eabb6444. https://doi.org/10.1126/sciimmunol.abb6444

Takemura R, Okabe S, Umeyama T et al (1992) Increased microtubule stability and alpha tubulin acetylation in cells transfected with microtubule-associated proteins MAP1B, MAP2 or tau. J Cell Sci 103(Pt 4):953–964. https://doi.org/10.1242/jcs.103.4.953

Tala, Sun X, Chen J, Zhang L et al (2014) Microtubule stabilization by Mdp3 is partially attributed to its modulation of HDAC6 in addition to its association with tubulin and microtubules. PLoS One 9:e90932. https://doi.org/10.1371/journal.pone.0090932

Talukdar PD, Chatterji U (2023) Transcriptional co-activators: emerging roles in signaling pathways and potential therapeutic targets for diseases. Sig Transduct Target Ther 8:1–41. https://doi.org/10.1038/s41392-023-01651-w

Tan Y, Zheng J, Liu X et al (2018) Loss of nucleolar localization of NAT10 promotes cell migration and invasion in hepatocellular carcinoma. Biochem Biophys Res Commun 499:1032–1038. https://doi.org/10.1016/j.bbrc.2018.04.047

Tas RP, Chazeau A, Cloin BMC et al (2017) Differentiation between oppositely oriented microtubules controls polarized neuronal transport. Neuron 96:1264–1271.e5. https://doi.org/10.1016/j.neuron.2017.11.018

Taschner M, Vetter M, Lorentzen E (2012) Atomic resolution structure of human α-tubulin acetyltransferase bound to acetyl-CoA. Proc Natl Acad Sci USA 109:19649–19654. https://doi.org/10.1073/pnas.1209343109

Tischfield MA, Baris HN, Wu C et al (2010) Human TUBB3 mutations perturb microtubule dynamics, kinesin interactions, and axon guidance. Cell 140:74–87. https://doi.org/10.1016/j.cell.2009.12.011

Topalidou I, Keller C, Kalebic N et al (2012) Genetically separable functions of the MEC-17 tubulin acetyltransferase affect microtubule organization. Curr Biol 22:1057–1065. https://doi.org/10.1016/j.cub.2012.03.066

Tripathi P, Zhu Z, Qin H et al (2021) Palmitoylation of acetylated tubulin and association with ceramide-rich platforms is critical for ciliogenesis. J Lipid Res 62:100021. https://doi.org/10.1194/jlr.RA120001190

Trushko A, Schäffer E, Howard J (2013) The growth speed of microtubules with XMAP215-coated beads coupled to their ends is increased by tensile force. Proc Natl Acad Sci 110:14670–14675. https://doi.org/10.1073/pnas.1218053110

Uyttewaal M, Burian A, Alim K et al (2012) Mechanical stress acts via katanin to amplify differences in growth rate between adjacent cells in Arabidopsis. Cell 149:439–451. https://doi.org/10.1016/j.cell.2012.02.048

Verdel A, Curtet S, Brocard M-P, et al (2000) Active maintenance of mHDA2/mHDAC6 histone-deacetylase in the cytoplasm results and discussion. In: Current biology

Verhey KJ, Gaertig J (2007) The tubulin code. Cell Cycle 6:2152–2160. https://doi.org/10.4161/cc.6.17.4633

Voloshin O, Gocheva Y, Gutnick M et al (2010) Tubulin chaperone E binds microtubules and proteasomes and protects against misfolded protein stress. Cell Mol Life Sci 67:2025–2038. https://doi.org/10.1007/s00018-010-0308-8

Waddell J, Banerjee A, Kristian T (2021) Acetylation in mitochondria dynamics and neurodegeneration. Cells 10:3031. https://doi.org/10.3390/cells10113031

Walsh CT, Garneau-Tsodikova S, Gatto GJ Jr (2005) Protein posttranslational modifications: the chemistry of proteome diversifications. Angew Chem Int Ed 44:7342–7372. https://doi.org/10.1002/anie.200501023

Wang Q, Liu X (2020) The dual functions of α-tubulin acetylation in cellular apoptosis and autophage induced by tanespimycin in lung cancer cells. Cancer Cell Int 20:369. https://doi.org/10.1186/s12935-020-01453-y

Wang Q, Zhang Y, Yang C et al (2010) Acetylation of metabolic enzymes coordinates carbon source utilization and metabolic flux. Science 327:1004–1007. https://doi.org/10.1126/science.1179687

Wang G, Wang C, Liu W et al (2018) Augmin antagonizes Katanin at microtubule crossovers to control the dynamic organization of plant cortical arrays. Curr Biol 28:1311–1317.e3. https://doi.org/10.1016/j.cub.2018.03.007

Wang X, Fu Y, Beatty WL et al (2021) Cryo-EM structure of cortical microtubules from human parasite toxoplasma gondii identifies their microtubule inner proteins. Nat Commun 12:1–14. https://doi.org/10.1038/s41467-021-23351-1

Wang C, Zhu Y, Chen R et al (2024) Targets SIRT2 to mediate the histone acetylation of PLAUR and modulates functions of astrocytes in spinal cord injury. Chem Biol Interact 390:143. https://doi.org/10.1016/j.cbi.2023.110854

Watabe M, Nakaki T (2011) Protein kinase CK2 regulates the formation and clearance of aggresomes in response to stress. J Cell Sci 124:1519–1532. https://doi.org/10.1242/jcs.081778

Wattanathamsan O, Thararattanobon R, Rodsiri R et al (2021) Tubulin acetylation enhances lung cancer resistance to paclitaxel-induced cell death through Mcl-1 stabilization. Cell Death Discov 7(1):67. https://doi.org/10.1038/s41420-021-00453-9

Weinert BT, Wagner SA, Horn H et al (2011) Proteome-wide mapping of the drosophila acetylome demonstrates a high degree of conservation of lysine acetylation. Sci Signal 4:ra48. https://doi.org/10.1126/scisignal.2001902

Williams KA, Zhang M, Xiang S et al (2013) Extracellular signal-regulated kinase (ERK) phosphorylates histone deacetylase 6 (HDAC6) at serine 1035 to stimulate cell migration. J Biol Chem 288:33156–33170. https://doi.org/10.1074/jbc.M113.472506

Wilson PJ, Forer A (1989) Acetylated α-tubulin in spermatogenic cells of the crane fly Nephrotoma suturalis: kinetochore microtubules are selectively acetylated. Cell Motil 14:237–250. https://doi.org/10.1002/cm.970140210

Wloga D, Gaertig J (2010) Post-translational modifications of microtubules. J Cell Sci 123:3447–3455. https://doi.org/10.1242/jcs.063727

Wloga D, Webster DM, Rogowski K et al (2009) TTLL3 is a tubulin glycine ligase that regulates the assembly of cilia. Dev Cell 16:867–876. https://doi.org/10.1016/j.devcel.2009.04.008

Wloga D, Joachimiak E, Fabczak H (2017) Tubulin post-translational modifications and microtubule dynamics. Int J Mol Sci 18:2207. https://doi.org/10.3390/ijms18102207

Wu H, Coskun V, Tao J et al (2010) Dnmt3a-dependent nonpromoter DNA methylation facilitates transcription of neurogenic genes. Science 329:444–447. https://doi.org/10.1126/science.1190485

Wynshaw-Boris A (2009) Elongator bridges tubulin acetylation and neuronal migration. Cell 136:393–394. https://doi.org/10.1016/j.cell.2009.01.024

Xiong X, Lee C-F, Li W et al (2019) Acid sphingomyelinase regulates the localization and trafficking of palmitoylated proteins. Biol Open 8:bio040311. https://doi.org/10.1242/bio.040311

Xu Z, Schaedel L, Portran D et al (2017) Microtubules acquire resistance from mechanical breakage through intralumenal acetylation. Science 356:328–332. https://doi.org/10.1126/science.aai8764

Yin H, Ju Z, Zheng M et al (2023) Loss of the m6A methyltransferase METTL3 in monocyte-derived macrophages ameliorates Alzheimer's disease pathology in mice. PLoS Biol 21:3002017. https://doi.org/10.1371/journal.pbio.3002017

Yoshimoto S, Morita H, Okamura K, et al (2022) αTAT1-induced tubulin acetylation promotes ameloblastoma migration and invasion. 102(1):80–89. Epub 2021 Sep 10 PMID: 34508164; PMCID: PMC8695380. https://doi.org/10.1038/s41374-021-00671-w
You E, Jeong J, Lee J et al (2022) Casein kinase 2 promotes the TGF-β-induced activation of α-tubulin acetyltransferase 1 in fibroblasts cultured on a soft matrix. BMB Rep 55:192–197. https://doi.org/10.5483/BMBRep.2022.55.4.021
Yu BJ, Kim JA, Moon JH et al (2008) The diversity of lysine-acetylated proteins in Escherichia coli. J Microbiol Biotechnol 18:1529–1536
Zhang Y, Li N, Caron C et al (2003) HDAC-6 interacts with and deacetylates tubulin and microtubules in vivo. EMBO J 22:1168–1179. https://doi.org/10.1093/emboj/cdg115
Zhang J, Sprung R, Pei J et al (2009) Lysine acetylation is a highly abundant and evolutionarily conserved modification in Escherichia coli. Mol Cell Proteomics 8:215–225. https://doi.org/10.1074/mcp.M800187-MCP200
Zhang Y, Qiu J, Wang X et al (2011) AMP-activated protein kinase suppresses endothelial cell inflammation through phosphorylation of transcriptional coactivator p300. Arterioscler Thromb Vasc Biol 31:2897–2908. https://doi.org/10.1161/ATVBAHA.111.237453
Zhang D, Wu C-T, Qi X et al (2014) Activation of histone deacetylase-6 induces contractile dysfunction through derailment of α-tubulin proteostasis in experimental and human atrial fibrillation. Circulation 129:346–358. https://doi.org/10.1161/circulationaha.113.005300
Zhang F, Su B, Wang C et al (2015) Posttranslational modifications of α-tubulin in alzheimer disease. Translat Neurodegen 4:9. https://doi.org/10.1186/s40035-015-0030-4
Zhang Z, Zhang P, Qi GJ et al (2018) CDK5-mediated phosphorylation of Sirt2 contributes to depressive-like behavior induced by social defeat stress. Biochim Biophys Acta Mol basis Dis 1864:533–541. https://doi.org/10.1016/j.bbadis.2017.11.012
Zhang Y, Chen W, Zeng W et al (2020) Biallelic loss of function NEK3 mutations deacetylate α-tubulin and downregulate NUP205 that predispose individuals to cilia-related abnormal cardiac left-right patterning. Cell Death Dis 11:1005. https://doi.org/10.1038/s41419-020-03214-1
Zhao S, Xu W, Jiang W et al (2010) Regulation of cellular metabolism by protein lysine acetylation. Science 327:1000–1004. https://doi.org/10.1126/science.1179689
Zheng Y, Yang X, Wang C et al (2020) HDAC6, modulated by miR-206, promotes endometrial cancer progression through the PTEN/AKT/mTOR pathway. Sci Rep 10:3576. https://doi.org/10.1038/s41598-020-60271-4
Zhong Y, Wang C, Wang Y et al (2023) Suppression of alpha-tubulin acetylation potentiates therapeutic efficacy of Eribulin in liver cancer. Am J Cancer Res 13:5698–5718

Chapter 5
Tubulin Acetylation and the Cellular Mechanosensing and Stress Response

Bruno Carmona, Inês L. S. Delgado, Sofia Nolasco, Rita Marques, João Gonçalves, and Helena Soares

Abstract Microtubule (MT) acetylation has emerged as a critical regulator of cellular stress responses, integrating mechanical and oxidative stimuli to support cellular adaptability and survival. This post-translational modification (PTM) enhances MT flexibility and resilience, enabling cells to withstand mechanical challenges such as changes in extracellular matrix stiffness and applied forces. Through its impact on MT physical properties, acetylation minimizes cytoskeletal breakage, reducing the need for constant remodeling and supporting cellular integrity under

B. Carmona · H. Soares (✉)
Centro de Química Estrutural, Institute of Molecular Sciences, Faculdade de Ciências, Universidade de Lisboa, Lisbon, Portugal

Escola Superior de Tecnologia da Saúde de Lisboa, Instituto Politécnico de Lisboa, Lisbon, Portugal
e-mail: mhsoares@ciencias.ulisboa.pt; helena.soares@estesl.ipl.pt

I. L. S. Delgado
Laboratório Associado para Ciência Animal e Veterinária (AL4AnimalS), Lisbon, Portugal

CIISA—Centro de Investigação Interdisciplinar em Sanidade Animal, Faculdade de Medicina Veterinária, Universidade de Lisboa, Lisbon, Portugal

Faculdade de Medicina Veterinária, Universidade Lusófona-Centro Universitário de Lisboa, Lisbon, Portugal

S. Nolasco
Escola Superior de Tecnologia da Saúde de Lisboa, Instituto Politécnico de Lisboa, Lisbon, Portugal

Laboratório Associado para Ciência Animal e Veterinária (AL4AnimalS), Lisbon, Portugal

CIISA—Centro de Investigação Interdisciplinar em Sanidade Animal, Faculdade de Medicina Veterinária, Universidade de Lisboa, Lisbon, Portugal

R. Marques
Centro de Química Estrutural, Institute of Molecular Sciences, Faculdade de Ciências, Universidade de Lisboa, Lisbon, Portugal

J. Gonçalves
Evotec, Toulouse, France

M. Halasa, A. Wawruszak (eds.), *Histone and Non-Histone Reversible Acetylation in Development, Aging and Disease*, Results and Problems in Cell Differentiation 75,
https://doi.org/10.1007/978-3-031-91459-1_5

mechanical stress. Furthermore, tubulin acetylation regulates intracellular trafficking by modulating interactions with molecular motors, allowing for efficient cargo transport and precise spatial organization without disrupting the MT network. In the context of oxidative stress, tubulin acetylation responds to redox imbalances by stabilizing MTs and influencing cellular pathways that regulate reactive oxygen species (ROS). This modification is linked to enhanced antioxidant responses, autophagy regulation, and mitochondrial dynamics, highlighting its role in maintaining cellular homeostasis under oxidative conditions. The dual function of tubulin acetylation, responding to and integrating signals from mechanical and oxidative stress, acts as a bridging mechanism between physical and chemical signaling pathways. Consequently, it has the potential to be a therapeutic target in diseases characterized by dysregulated stress responses, including neurodegenerative disorders, cancer, and cardiovascular conditions. Despite significant progress has been made, unanswered questions persist, particularly regarding the molecular mechanisms by which acetylated MTs encode spatial and functional information and their interplay with other tubulin PTMs.

5.1 Microtubule Acetylation

The cytoskeleton, a fundamental component of eukaryotic cells, is essential for maintaining cellular structure, organization, and function. Composed of actin filaments, intermediate filaments, and microtubules (MTs), this intricate network supports a wide range of processes, including cell spatial organization, establishment of cellular asymmetries and polarity, intracellular transport, cell division, migration, morphogenesis, and signal transduction including mechanosensing (Dogterom and Koenderink 2019). Among these components, MTs are hollow dynamic tube-like structures assembled from highly conserved α/β-tubulin heterodimers. In cells, these MTs are also the main components of complex structures such as the mitotic and meiotic spindles, centrioles/basal bodies, and the ciliary axoneme.

In the last years, cytoskeleton has emerged as an essential cellular module to sense mechanical forces (mechanosensing) and cues from their environment in animal cells. Through mechanotransduction, cells translate these signals into biochemical cues, influencing gene programs and triggering responses like growth, proliferation, polarization, migration, and differentiation (Jansen et al. 2015; Wolfenson et al. 2019).

MTs are approximately 100 times more resistant to bending compared to actin filaments (Gittes et al. 1993). This inherent strength allows MTs to withstand substantial compressive forces from various physiological sources. Research by Brangwynne et al. (2006) demonstrated that the mechanical coupling of MTs with the elastic components of the cytoskeleton, including actin and intermediate filaments, enhances their ability to endure these forces playing a significant structural role in the cell's mechanical behavior (Brangwynne et al. 2006).

The MT scaffold is structurally consistent yet functionally diverse. This diversity arises from different tubulin isotypes, encoded by distinct tubulin gene family members, which can be incorporated into MTs in varying proportions (Poirier et al. 2010; Tischfield et al. 2010). In addition to their structural and mechanical roles, MTs are dynamically regulated through various post-translational modifications (PTMs), which fine-tune their function and adapt them to the cell's physiological needs (Verhey and Gaertig 2007). Acetylation is one of tubulin's PTMs and involves the addition of an acetyl group (CH_3CO) to specific amino acid residues, primarily lysines (K). In tubulin, this reversible modification can occur at multiple K residues, with lysine 40 (K40) of α-tubulin being the most studied site (L'Hernault and Rosenbaum 1985; LeDizet and Piperno 1987; Choudhary et al. 2009). Interestingly, unlike other tubulin PTMs, this modification is uniquely located within the MT lumen (for details see Chapter Tubulin Acetylation: A Critical Regulator of Microtubule Function). Tubulin acetylation is catalyzed by tubulin acetyltransferase (αTAT1), which exhibits a preference for polymerized MTs over soluble tubulin (Akella et al. 2010; Shida et al. 2010). Conversely, the removal of the acetyl group is mediated by two primary deacetylases: histone deacetylase 6 (HDAC6), which favors soluble tubulin, and sirtuin 2 (SIRT2), which can modify both polymerized and soluble forms of tubulin (Hubbert et al. 2002; North et al. 2003). These enzymes enable the reversible and dynamic nature of acetylation, making it a crucial mechanism for regulating MT properties in response to cellular and environmental cues. Research has revealed that tubulin acetylation is enriched in stable MT populations, such as those found in axons, ciliary structures, and mitotic spindles (Orbach and Howard 2019). In recent years, studies have revealed that a major effect of tubulin acetylation is its impact on flexibility (Portran et al. 2017; Eshun-Wilson et al. 2019). This increased flexibility enhances MT resilience, reducing brittleness under mechanical stress (see Fig. 5.1) (Portran et al. 2017; Xu et al. 2017). Acetylated MTs are also more resistant to damage and exhibit improved self-repair mechanisms, which include the incorporation of new GTP-tubulin heterodimers (Fig. 5.1) (Schaedel et al. 2015; Portran et al. 2017; Janke and Montagnac 2017; Eshun-Wilson et al. 2019). This mechanical stability is thought to arise from altered interactions between protofilaments within the MT lattice.

While early studies debated the role of acetylation in MT stability and dynamics, recent evidence suggests that it also influences the activity of molecular motors, including dynein and kinesin, as well as MT-associated proteins (MAPs) such as *tau*. These interactions affect intracellular transport and cytoskeletal organization. Acetylation also regulates MT-severing proteins like katanin and fidgetin, which exhibit altered activity depending on the acetylation state of MTs. Additionally, the balance between acetylation and other PTMs, such as methylation at K40, is influenced by the competition between acetyltransferases and methyltransferases, further increasing the complexity of the regulation of MT dynamics (Andreu-Carbó et al. 2024).

Dysregulated tubulin acetylation is implicated in several diseases, including neurodegenerative disorders, cancer, cardiac conditions, and ciliopathies, highlighting its potential as a therapeutic target (McKenna et al. 2023). For example, altered

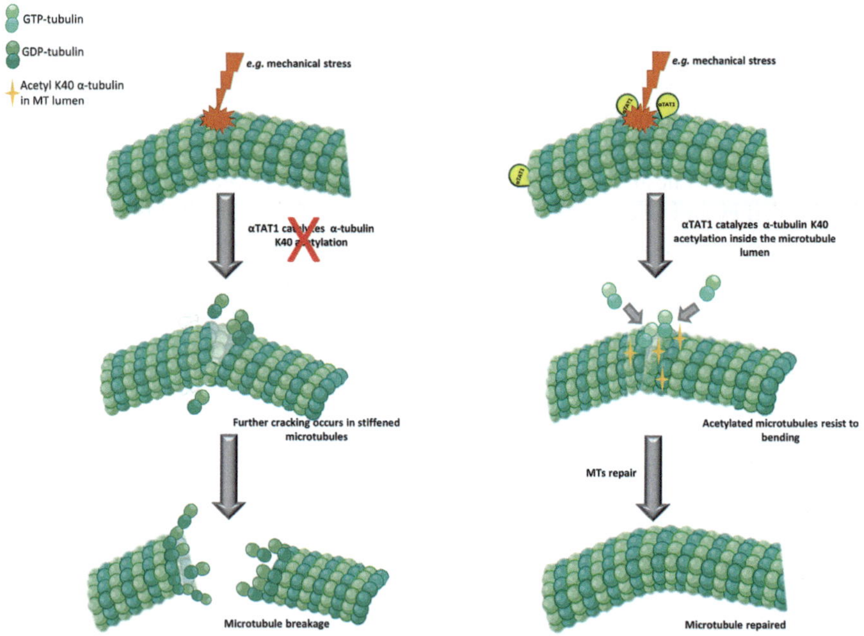

Fig. 5.1 The role of tubulin K40 acetylation in MT damage by mechanical stress. Mechanical stress induces MT bending, which can lead to cracks in the MT wall and, ultimately, MT breakage. Acetylation of K40 on α-tubulin, located in the MT lumen, protects the MT wall from damage by increasing its flexibility. The lysine acetyltransferases αTAT1 can access the MT lumen through MT ends or cracks in the MT wall. Once inside, αTAT1 localizes to damaged sites, increasing MT acetylation locally. This protects MTs from further bending damage and facilitates repair by promoting the incorporation of new GTP-tubulin heterodimers into the cracks in the MT wall. (Adapted from Janke and Montagnac, 2017)

acetylation patterns are associated with ciliary dysfunction in ciliopathies and defects in axonal transport in neurological diseases (McKenna et al. 2023). Beyond structural and functional regulation, tubulin acetylation plays a significant role in cellular stress responses, particularly mechanical and oxidative stress. In this chapter, we will explore how tubulin acetylation acts in the interplay between MTs, mechanical forces, and oxidative stress, highlighting its importance in cellular adaptation and its implications in various physiological and pathological conditions.

5.2 Microtubule Acetylation in Response to Mechanical and Oxidative Stress

Mechanical forces apply stress to cells, leading to alterations in cytoskeletal organization and dynamics, intracellular transport, and gene expression. These forces cause changes in the cytoskeleton, allowing cells to respond to their mechanical surroundings (Di et al. 2023). In addition, oxidative stress, caused by increased levels of reactive oxygen species (ROS), adds further complexity to cytoskeleton regulation. ROS can disturb the stability of cytoskeletal components and trigger cellular reactions that change gene expression patterns (Tee et al. 2009; Wilson and González-Billault 2015). Therefore, both mechanical and oxidative stresses work together to impact the cytoskeleton's role in maintaining cell structure and coordinating essential cellular functions like intracellular transport and signaling. This double impact highlights how crucial it is for cytoskeletal adaptability to guarantee cell survival and optimal performance in stressful situations (Wang and Li 2010). The connection between these stress conditions and the cytoskeleton demonstrates the intricate regulatory processes that regulate how cells respond to their environmental challenges.

5.2.1 Cellular Response to Mechanical Forces

In recent years, the study of cell physics has gained significant traction, driven by mounting evidence that mechanical forces play a crucial role in cellular functions. Research has shown that external forces applied to cells and the resistance from extracellular matrices to cell-generated forces can produce signals similar to those of biochemical signaling. These mechanical cues are now recognized as critical factors for various cellular processes, including cell growth, differentiation, gene and protein expression, and cell movement (Huang and Ingber 2005; Ainsworth 2008).

The mechanical signals prevalent in living systems are diverse and include hydrostatic pressure, fluid shear stress, tensile force, and extracellular matrix (ECM) rigidity (Di et al. 2023). The effects of these signals are not uniform but are influenced by their magnitude, direction, and spatial and temporal aspects, enabling them to modulate their impacts in complex ways (Swaminathan and Gloerich 2021). These forces are particularly evident in tissues that experience constant physical activity, such as the cardiovascular system and skeletal muscles (Di et al. 2023).

In addition to these external forces, cells exhibit internal mechanical forces generated by their cytoskeleton dynamics (Sanger et al. 1983; Kolega et al. 1991; Wang and Li 2010). These forces act on the ECM via focal adhesions, controlling cell migration, cell shape, and ECM structure and composition (Katsuta et al. 2024). Moreover, internal mechanical forces influence cellular metabolism, proliferation, and differentiation (Lavagnino and Arnoczky 2005; Stooke-Vaughan and Campàs 2018).

MTs, actin filaments, and intermediate filaments have a crucial role in mechanosensing, by playing a concerted action to sense and respond to mechanical forces (Fletcher and Mullins 2010; Uray and Uray 2021; Janshoff 2021).

The cellular tensegrity model, introduced by Ingber et al. (2014), explains how cells maintain their shape and function under mechanical stress by balancing tensile and compressive forces (Ingber et al. 2014). This balance distributes mechanical stresses throughout the cytoskeleton, preserving cellular integrity despite the constant dynamics of the cytoskeleton polymers. Pre-stress, generated by contractile elements acting against compressive structures like MTs, is key to this model. Additionally, other components, such as the ECM, organelles, and nuclei, contribute to this compressive role (Ingber et al. 2014; Butler 2016). The model requires cytoskeletal connections to cellular and nuclear membranes, which involves focal adhesion complexes (i.e., integrins), and the Linker of Nucleoskeleton and Cytoskeleton (LINC) complex, respectively.

Integrins serve as key adhesive receptors, linking the ECM to the intracellular cytoskeleton. Clustered integrins are components of focal adhesions, large protein complexes, connecting them to actin filaments. These structures function as mechanical hubs that relay signals, in both senses, between the ECM and the cytoskeleton (Mui et al. 2016). At the nuclear envelope, the LINC complex, composed of nesprins, connects the cytoskeleton to the nuclear interior by interacting with actin filaments, MTs, and intermediate filaments. Inside the nucleus, it links to the nucleoskeleton, including lamins (Hieda 2019). This system allows external mechanical signals to be transmitted from the periphery throughout the cell into the nucleus, facilitating cellular adaptation and response to the mechanical environment, which is critical for many cellular and tissue functions.

5.2.1.1 Microtubules as Key Players in Cellular Mechanosensing

Among the cytoskeletal elements, MTs stand out for their unique physical properties and crucial role in cellular mechanics and mechanosensing. These structures are characterized by their rigid yet flexible nature, which allows them to act as compression-resistant elements within cells. MTs bear the brunt of external compressive forces and play a critical role in stabilizing cellular structure during mechanical stress (Brangwynne et al. 2006; Fletcher and Mullins 2010; Uray and Uray 2021). Despite their rigidity, MTs exhibit a highly dynamic structure, that coupled with their physical properties, enables MTs to serve multiple crucial functions in cellular mechanics.

One of the key roles of MTs is to provide attachment points for actin and myosin contractile system, enabling the generation of tensile forces during processes like cell migration (Hawkins et al. 2010). Their robustness allows them to withstand the movement of cargo-carrying motor proteins like kinesins and dyneins, ensuring efficient intracellular trafficking of organelles, vesicles, and proteins throughout the cytoplasm (Hawkins et al. 2010). MTs have shown a remarkable ability to respond to ECM changes and alter cellular interactions with ECM. Research by Rafiq et al. (2019) demonstrated that MTs can reorganize focal adhesions, thereby modifying how cells interact with the ECM (Rafiq et al. 2019). In a breast cancer model, Torrino et al. (2021) observed that MTs enhanced their stability in response to matrix stiffening by remodeling their post-translational glutamylation status, which in turn promoted increased cell invasion (Torrino et al. 2021). Earlier works by Putnam et al. (2001) and Myers et al. (2011) found that MT dynamics were dependent on ECM stiffness, with softer matrices associated with more dynamic MTs

(Putnam et al. 2001; Myers et al. 2011). Interestingly, Kaverina et al. (2002) observed that MT polymerization and guidance are directed toward adhesion sites under increased stress, possibly due to interaction with actin filaments (Kaverina et al. 2002).

MTs also show a remarkable ability to reorganize in response to external mechanical forces such as stretching and compression. In plant cells, it has been observed that the division plane is organized under the influence of mechanical forces (Lintilhac and Vesecky 1984; Lynch and Lintilhac 1997). Similar phenomena have been observed in animal cells, where Théry et al. (2007) and Fink et al. (2011) showed that spindle orientation in dividing cells is dependent on cortical force generators and aligns parallel to applied forces (Théry et al. 2007; Fink et al. 2011). In different plant tissues, cortical MTs have been found to align parallel to tensile forces applied to these cells (Hejnowicz et al. 2000; Jacques et al. 2013; Robinson and Kuhlemeier 2018; Hamant et al. 2019). The organization of the MT cytoskeleton appears to be self-organized and dependent on interactions with MT-associated proteins, such as the severing protein katanin (Wasteneys and Ambrose 2009; Sampathkumar et al. 2014). External forces regulate not only MT orientation but also their bundling. Studies in *Arabidopsis* plant tissues have shown that compression forces lead to excessive bundling of MTs (Jacques et al. 2013; Louveaux et al. 2016). Louveaux et al. (2016) further demonstrated that this bundling is enhanced by the MT severing activity of katanin, highlighting the complex interplay between mechanical forces and MAPs in shaping cellular responses (Louveaux et al. 2016).

The dynamic stability of MTs is another aspect regulated by mechanical forces. Celik et al. (2013) observed, in Chinese hamster ovary cells, a rapid depolymerization of MTs at an indentation point where external force was applied, followed by slow polymerization in the surrounding area (Celik et al. 2013). Similar results were observed in plant cells by Hardham et al. (2008) (Hardham et al. 2008). Generally, MT dynamics favor depolymerization under compressive forces and polymerization under tensile forces (Zheng et al. 1993; Putnam et al. 1998; Landrein and Hamant 2013). This feature is particularly relevant in processes such as axon extension promoted by the growth cone, where oriented tubulin polymerization plays a crucial role (Tamariz and Varela-Echavarría 2015).

MTs play a multifaceted role in cellular mechanics and mechanosensing, responding to various mechanical cues through changes in organization, bundling, and dynamic stability. However, we are still far from understanding how factors that regulate MTs dynamic functions, like MAPs, motors, tubulin isoforms, and PTMs, participate in mechano-signaling pathways mediated by MTs.

5.2.1.2 Tubulin Acetylation in Response to Mechanical Stress

Mechanical forces not only induce structural and organizational changes in MTs but also trigger responses that often result in tubulin K40 acetylation, a key PTM that stabilizes the MT network. As previously described, tubulin acetylation enhances the resilience of MTs, making them more adaptable to mechanical stress (Portran et al. 2017; Xu et al. 2017). This section explores how mechanotransduction pathways are linked to tubulin acetylation and their physiological significance across various cellular contexts.

Recent research has unveiled the intricate relationship between mechanical forces, tubulin acetylation, and cellular behavior (summarized in Fig. 5.2). In fibrotic skin tissues, Wen et al. (2023) observed high levels of tubulin acetylation, which not only responded to increased substrate stiffness but also proved critical for fibroblast activation, migration, and ECM modification (Wen et al. 2023). This finding highlights the role of tubulin acetylation in pathological conditions associated with altered tissue mechanics. Interestingly, the cellular response to mechanical cues through tubulin acetylation is not uniform across all cell types and contexts. Marchant et al. (2022) demonstrated that migrating clusters of cephalic neural crest cells from *Xenopus* decreased their stiffness when in contact with a stiff surface, which triggers collective cell migration (Marchant et al. 2022). This adaptation, which seems contradictory to some previous observations (Putnam et al. 2001; Myers et al. 2011), was found to be due to MT deacetylation mediated by a mechanosensitive pathway controlling the Piezo1 ion channel (Marchant et al. 2022).

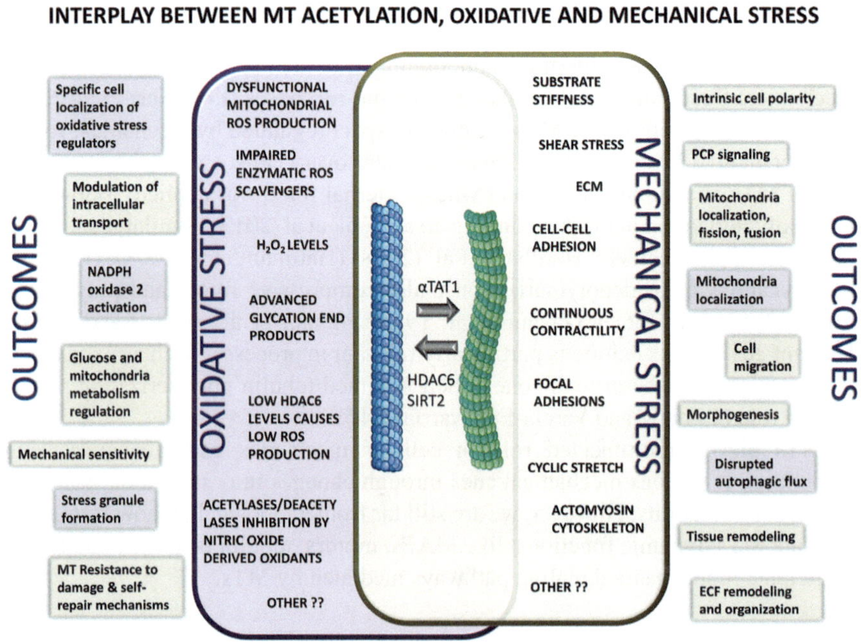

Fig. 5.2 Interplay between MT acetylation, oxidative, and mechanical stress. This schematic highlights the major known mechanical and oxidative factors that influence tubulin acetylation (blue—deacetylated MTs, green—acetylated MTs) and illustrates their downstream consequences for cellular and tissue homeostasis, as well as development (Outcomes). (Abbreviations: *MTs* microtubules)

The impact of MT acetylation on cellular behavior extends to intercellular interactions. Collins et al. (2021) showed that *Xenopus* embryonic skin cells with high levels of MT acetylation exhibited an enhanced ability to penetrate and intercalate among other cells (Collins et al. 2021). This finding suggests a role for tubulin acetylation in tissue remodeling and morphogenesis. In endothelial cells, McCue et al. (2006) observed that shear stress promotes tubulin acetylation. Notably, the distribution of acetylated MTs was highly asymmetric throughout the cell, since they were predominantly found downstream of the nucleus in relation to the shear forces (McCue et al. 2006). This pattern may contribute to the establishment of intrinsic cell polarity with impact in planar cell polarity (PCP), highlighting the potential role of tubulin acetylation in cellular orientation and directional responses to mechanical stimuli.

The influence of mechanical forces on tubulin acetylation is particularly relevant in neuronal systems. Falconieri et al. (2023) demonstrated that axons subjected to stretch forces presented a higher ratio of acetylated versus tyrosinated tubulin compared to control neurons, leading to modulation of axonal transport (Falconieri et al. 2023). These results are especially significant considering that mechanical forces drive axon extension (Tamariz and Varela-Echavarría 2015). The relationship between mechanical stress and tubulin acetylation has also been observed in vascular tissues. Using a rat model for hypertension, Mozzicato et al. (2024) found that mesenteric arteries from hypertensive rats showed higher levels of tubulin acetylation compared to normotensive rats, although this increase did not change the cell's Young's modulus (Mozzicato et al. 2024). In contrast, Hirase and Node (2012) observed that endothelial cells under cyclic stretch exhibited tubulin deacetylation at the cell periphery (Hirase and Node 2012). Wang et al. (2010) further elucidated that endothelial cells modulate their acetylated tubulin levels in response to normal shear stress and interactions with vascular smooth muscle cells. They observed that stimuli resulted in decreased tubulin acetylation associated with increased expression of the tubulin deacetylase HDAC6 (Wang et al. 2010). This process was shown to be the basis for endothelial cell migration. The impact of mechanical forces on tubulin acetylation extends to other cell types as well. In lung cells, Geiger et al. (2009) found that cyclic stretch dramatically increased tubulin acetylation levels due to decreased HDAC6 activity (Geiger et al. 2009). Similarly, Bae et al. (2024) observed increased levels of acetylated tubulin in periodontal ligament cells subjected to cyclic stretch (Bae et al. 2024).

The functional significance of tubulin acetylation is particularly evident in contractile cells. Coleman et al. (2021) demonstrated that tubulin acetylation is critical for the functioning of cells that exhibit contraction, such as cardiomyocytes or skeletal myofibers (Coleman et al. 2021). Increased tubulin acetylation led to enhanced stiffness and viscoelastic resistance of the cells, causing a decrease in contraction kinetics. However, in intact muscle, changes in tubulin acetylation showed a low impact on muscle function, suggesting complex tissue-level compensatory mechanisms (Coleman et al. 2021).

It's important to note that the relationship between mechanical forces and tubulin modifications is not always straightforward. Li et al. (2023) observed increased MT stabilization in retinal pigment epithelial cells (RPE-1) subjected to stretch and compression cycles, but they did not detect differences in either tubulin acetylation or detyrosination (Li et al. 2023). This finding emphasizes the complexity of the mechanisms underlying MT organization and response to mechanical stimuli, suggesting that factors beyond PTMs play also important roles.

Intriguingly, tubulin acetylation appears to be not only a response to mechanical stimuli but also an orchestrator of these stimuli. Studies in various model organisms have highlighted the importance of tubulin acetylation in mechanosensation. Shida et al. (2010) described tubulin acetylation as fundamental for mechanical sensitivity in the touch receptor neurons of *C. elegans* (Shida et al. 2010). Subsequently, Morley et al. (2016) and Yan et al. (2018) reported similar findings in sensory neurons of mice and *Drosophila*, respectively (Morley et al. 2016; Yan et al. 2018). Mutations in the αTAT1 or in the tubulin K40 impaired several behavioral responses to mechanical stimuli in these model organisms, emphasizing the conserved role of tubulin acetylation in mechanosensation across species (Morley et al. 2016; Yan et al. 2018).

The interplay between tubulin acetylation and internal mechanical forces, resulting from MT interaction with other cytoskeletal components, adds another layer of complexity to this process. Seetharaman et al. (2022) demonstrated that MT acetylation was promoted by the mechanosensing of substrate stiffness through actomyosin (Seetharaman et al. 2022). The acetylation resulted from the recruitment of αTAT1 to focal adhesions. Interestingly, tubulin acetylation itself triggered further effects, activating RhoA, actomyosin contractility, and traction forces. This study built upon earlier work by Bance et al. (2019), which showed that tubulin acetylation was important for focal adhesion turnover and impacted cell migration, highlighting the interdependence of tubulin interactions with the actin cytoskeleton and cellular adhesion sites at the membrane (Bance et al. 2019). The connection between actin dynamics and tubulin acetylation is further illustrated by the work of Fernández-Barrera et al. (2018) (Fernández-Barrera et al. 2018). They showed that in RPE-1 cells, depletion of the formin INF2 resulted in a significant decrease in tubulin acetylation accompanied by an increase in the ratio between G-actin and F-actin. This decrease in tubulin acetylation was attributed to an almost complete inactivation of αTAT1 expression, induced by impaired myocardin-related transcription factor (MRTF)/serum response factor (SRF). Formins, known primarily for their role in nucleating and polymerizing actin filaments (Goode and Eck 2007), have also been shown to bind to MTs, modulating their stability, acetylation, and alignment with actin fibers (Thurston et al. 2012; Fernández-Barrera et al. 2018).

It is plausible that an unexplored regulatory pathway, mediated by cytoskeleton, links mechanical signals to the regulation of metabolism. In fact, the cytoskeleton remodeling requires for example an activation and increased levels of specific enzymes and available levels of GTP and ATP for tubulin heterodimers and actin monomers polymerization, which demands a metabolic response (Dawson et al. 2023). Indeed, it is estimated that 50% of the total ATP consumed by cells is used by the actin cytoskeleton dynamics (Daniel et al. 1986; Bernstein and Bamburg 2003). For example, epithelial cells modulate their response to mechanical stress by increasing glucose uptake, recruiting glucose transporter 1 (GLUT1) to sites of E-cadherin-mediated force transmission. This process is facilitated by non-muscle myosin IIA-driven contractility and the protein Ankyrin G, which links E-cadherin to G GLUT1 (Salvi et al. 2021). Another example is available from the work of Park et al. (2020) that showed that glycolysis is coupled to mechanical properties of the microenvironment in human bronchial epithelial cells (HBECs) (Park et al. 2020). When these cells are transferred from stiff to soft substrates, glycolysis is downregulated through the proteasomal degradation of the rate-limiting enzyme phosphofructokinase (PFK). This process is triggered by the disassembly of actin stress fibers, leading to changes in the actomyosin cytoskeleton. Not much is known in these processes about the role of MTs, but it is inevitable that these polymers will play a role since there is a close interplay with actin networks. Mechanical stress can also affect mitochondria biogenesis, structure and function and consequently cell metabolism (Dawson et al. 2023).

5.2.2 *Impact of Oxidative Stress on Cellular Processes*

Oxidative stress is a critical cellular phenomenon defined as an imbalance between the production of oxidants and antioxidants within cells. This imbalance can lead to the damage of biological molecules, and cell structure potentially resulting in the development of various pathologies (Sies 2015; Forman and Zhang 2021). The process involves several molecular species derived from oxygen and nitrogen, with some playing particularly significant roles in the development of oxidative stress.

Key players in oxidative stress include superoxide, hydrogen peroxide (H_2O_2), hydroxyl radicals, and peroxynitrite, each with distinct roles in cellular damage (Sies 2015; Forman and Zhang 2021). Superoxide is formed primarily through the mitochondrial respiratory chain and NADPH oxidases, and while not a potent oxidant itself, it generates H_2O_2 via superoxide dismutase (Wilson and González-Billault 2015; Forman and Zhang 2021). Hydrogen peroxide, though weak as an oxidant, can diffuse through membranes and produce highly reactive hydroxyl radicals via Fenton reactions. Peroxynitrite, formed by the reaction between superoxide and nitric oxide, along with hydroxyl radicals, is among the most potent oxidants, driving oxidative stress through rapid oxidation of cellular molecules (Wilson and González-Billault 2015; Forman and Zhang 2021).

5.2.2.1 The Interplay Between Oxidative Stress and Tubulin Acetylation

In recent years, the MT cytoskeleton's response to oxidative stress has garnered significant attention. Accumulating evidence, both direct and indirect, suggests that MTs modify their structure in response to cellular redox imbalance (summarized in Fig. 5.2). This chapter delves into research demonstrating how oxidative stress can modulate the MT cytoskeleton through acetylation.

The relationship between MT acetylation and oxidative stress was first described in 2009 when Giustiniani et al. discovered that the localization of endothelial nitric oxide synthase near the Golgi apparatus depended on stable acetylated MTs (Giustiniani et al. 2009). This study provided evidence that acetylated MTs are not only targets of oxidative stress modulation but also active participants in redox regulation. Subsequent research has revealed a complex interplay between oxidative stress, MT acetylation, and various cellular processes. For instance, Wu and coworkers suggested in 2010 that excessive ROS produced by defective mitochondria in chronic progressive external ophthalmoplegia patients led to increased tubulin acetylation through decreased SIRT2 levels (Wu et al. 2010). Interestingly, they also observed that mitochondria in this scenario were distributed in the perinuclear region, an area typically associated with acetylated MTs (Wu et al. 2010).

The role of tubulin acetylation in neurodegenerative diseases has been a subject of particular interest. In 2013, Taes et al. demonstrated that in a mouse model for amyotrophic lateral sclerosis (ALS) with mutant superoxide dismutase 1 (SOD1), deletion of the HDAC6 gene significantly extended survival and increased tubulin acetylation (Taes et al. 2013). Although not directly evaluated in this study, other researchers have shown increased oxidative stress in this ALS mouse model (Méndez-López et al. 2021). Notably, the deletion of HDAC6 also maintained motor axon integrity, while Sirt2 gene deletion did not affect disease course or tubulin acetylation levels, suggesting HDAC6 as the primary tubulin deacetylase in the nervous system (Taes et al. 2013). Further studies using SOD1 mutants have provided insights into tubulin acetylation in the context of ALS. Gal et al. (2013) showed that HDAC6 knockdown increased the aggregation of mutant SOD1 protein, a phenomenon related to increased tubulin acetylation (Gal et al. 2013). This finding aligns with the importance of tubulin acetylation in axonal transport.

The response of tubulin acetylation to oxidative stress has been observed across various cell types and conditions. Mackeh et al. (2014) demonstrated in HeLa cells that tubulin acetylation responds to H_2O_2 levels in a dose-dependent manner, with αTAT1 identified as the responsible enzyme (Mackeh et al. 2014). However, the precise mechanism triggering this response remains elusive, as treatment with the antioxidant N-acetyl-cysteine only partially prevented MT acetylation. In rat cardiomyocytes, Yuan et al. (2015) showed that advanced glycation end products resulting from diabetes were responsible for increased tubulin acetylation levels due to inhibition of the tubulin deacetylase SIRT2 (Yuan et al. 2015). This finding highlights the potential role of tubulin acetylation in diabetes-related cellular changes.

Tubulin acetylation also plays a crucial role in regulating the autophagic process triggered by oxidative stress. Bonet-Ponce et al. (2016) found that exposure to rotenone and the consequent increase in ROS production led to an accumulation of 4-hydroxynonenal inclusions in ARPE-19 cells, which are known as oxidative stress markers (Bonet-Ponce et al. 2016). This process was accompanied by increased tubulin acetylation, which disrupted the autophagic flux and blocked the clearance of these inclusions. This study further emphasizes the role of tubulin acetylation in controlling intracellular transport.

The relationship between oxidative stress and tubulin acetylation is not always straightforward. Borgas et al. (2016) observed a decrease in tubulin acetylation when exposing cultured lung tissue to tobacco smoke, associated with increased HDAC6 activation (Borgas et al. 2016). This effect was reversed by incubation with the antioxidant N-acetyl cysteine, linking the observed phenotypes to increased oxidative stress.

The regulation of tubulin deacetylases, particularly SIRT2 and HDAC6, appears to be intricately linked to oxidative stress responses. Wang et al. (2007) found that SIRT2 expression was higher in the white adipose tissue and kidney of calorie-restricted mice and increased upon exposure to H_2O_2 (Wang et al. 2007). SIRT2 was shown to regulate FOXO3a, a transcription factor involved in metabolism and oxidative stress response. In mouse embryos, SIRT2 expression seems to be regulated by increased oxidative stress. Yu et al. (2016) showed that SIRT2 expression was reduced in embryos of a pre-gestational maternal diabetes model, an effect reversed by SOD1 overexpression (Yu et al. 2016). This reduction was related to the development of diabetes-induced neural tube defects, potentially linking tubulin acetylation to nervous system development.

Recent studies have shed light on the mechanisms by which oxidative stress can directly affect tubulin deacetylases. Kalous et al. (2020) demonstrated that SIRT2 could be inhibited by S-nitrosation of cysteines caused by nitric oxide-derived oxidants (Kalous et al. 2020). Similarly, Bohl et al. (2024) showed that peroxynitrite can also inhibit SIRT2 activity (Bohl et al. 2024). Interestingly, HDAC6 can also be inhibited by nitric oxide-derived oxidants, leading to increased tubulin acetylation (Okuda et al. 2015).

The role of HDAC6 in oxidative stress responses has been extensively studied. Youn et al. (2016) observed that macrophages overexpressing HDAC6 presented higher levels of ROS and pro-inflammatory cytokine expression (Youn et al. 2016). Inhibition of HDAC6 with tubastatin A decreased ROS production and pro-inflammatory cytokine expression while increasing acetylated tubulin levels. HDAC6 also appears to play a role in hypoxia-ischemia-induced oxidative stress. Su et al. (2016) showed that HDAC6 inhibition accelerated ROS production in response to hypoxia-ischemia in a mouse model, accompanied by increased HSP90 acetylation and cell apoptosis (Su et al. 2016). Conversely, a more recent study demonstrated that HDAC6 knockout mice were resistant to oxidative stress upon intracerebral hemorrhage and presented reduced neuronal apoptosis (Wang et al. 2022).

The role of HDAC6 and tubulin acetylation seems particularly relevant in modulating retinal disorders, especially in response to oxidative stress derived from photoreceptor activity. Leyk and colleagues showed that HDAC6 inhibition by Tubastatin A in a cone-like photoreceptor cell line increased tubulin acetylation and extended cell survival after hydrogen peroxide exposure (Leyk et al. 2017). In an in vivo model for inherited blindness using zebrafish, Tubastatin A treatment was sufficient to rescue visual function and retinal morphology, with tubulin hyperacetylation observed in the eyes. The impact of tubulin acetylation on stress responses extends to stress granule formation, a cellular defense mechanism against ROS. In

HeLa cells after H_2O_2 exposure HDAC6 inhibition with Tubastatin A resulted in decreased stress granule formation (Feng et al. 2020). In the context of myocardial ischemia/reperfusion injury, particularly crucial for diabetic patients, HDAC6 inhibition has shown promising results. Leng et al. (2018) demonstrated that selective HDAC6 inhibition improved cardiac function and lowered ROS production in a rat model of myocardial ischemia/reperfusion injury (Leng et al. 2018). Similarly, Abouhish et al. (2020) found that HDAC6 inhibition prevented retinal microvascular hyperpermeability, decreased senescence markers, and lowered free radical levels in a rat model of type 1 diabetes (Abouhish et al. 2020).

Altogether these data strongly suggest that oxidative stress can regulate tubulin deacetylases. However, in many studies, MT acetylation levels were not directly assessed. Given that key tubulin deacetylases, SIRT2 and HDAC6, target multiple substrates beyond tubulin, many of which are involved in stress regulation, it remains challenging to establish a direct link between tubulin acetylation and the observed changes in enzyme activity. Nonetheless, several affected molecules are part of pathways influenced by tubulin acetylation (Fig. 5.2). For instance, HSP90's binding to MTs is acetylation-dependent, and tubulin acetylation enhances the signaling and binding of proteins like the kinase AKT/PKB and the transcription factor p53, both of which are HSP90 substrates (Giustiniani et al. 2009). This shared deacetylase activity between tubulin and stress-related molecules suggests that tubulin acetylation may play a broader role in cellular stress responses.

5.2.3 *Crosstalk Between Mechanical and Oxidative Stress*

Mechanical and oxidative stress, though often studied independently, are emerging as intricately connected in their impact on cellular functions, particularly through tubulin acetylation. Recent research has highlighted the complex interplay between mechanical forces, oxidative stress, and MT regulation (Fig. 5.2). For example, Coleman et al. (2021) found that changes in tubulin acetylation in striated muscle not only affected muscle contraction rates but also increased the activation of NADPH oxidase 2, an enzyme responsible for generating superoxide radicals (Coleman et al. 2021). This shows how tubulin modifications can influence both mechanical processes (contraction) and oxidative stress responses (NADPH oxidase activation) within the same cell type. Similarly, Fu et al. (2019) explored the effects of mechanical loading on cartilage explants, revealing that mechanical stress reduced nitric oxide release, activated HDAC6, and decreased tubulin acetylation (Fu et al. 2019). Epithelial cells cultured on a soft ECM exhibit increased perimitochondrial F-actin and enhanced mitochondrial fission (Romani et al. 2022), a process linked to MTs acetylation (Waddell et al. 2021). This is accompanied by elevated mitochondrial ROS production and activation of the NRF2-dependent antioxidant response, which includes increased cystine uptake and glutathione metabolism. This direct link between mechanical forces, tubulin acetylation, and oxidative stress markers underscores the interconnected nature of these processes.

5.3 Concluding Remarks

In recent decades, MTs have emerged as dynamic regulators of cellular integrity and adaptability. These roles are enabled by their remarkable dynamic properties and capacity to withstand mechanical forces, largely attributed to post-translational modifications such as acetylation. These modifications allow MTs to mediate essential processes, including mechanosensing, intracellular trafficking, and cytoskeletal remodeling, which collectively sustain cellular homeostasis. Tubulin acetylation enhances MT flexibility and resilience, underpinning its pivotal role in enabling cells to adapt to environmental changes. By modulating interactions with molecular motors, MT-associated proteins, and other cytoskeletal elements, acetylation facilitates key cellular responses, including the reorganization of adhesion sites, the modulation of intracellular transport, the regulation of autophagy, the orchestration of antioxidant responses, and the fine-tuning of signaling pathways critical for cellular survival.

However, several questions remain unanswered, requiring further research to achieve a complete understanding of how MT acetylation facilitates these stress responses. Perhaps the most intriguing question is the precise mechanism by which PTM influences the role of MTs in responding to environmental challenges. This can be envisioned as a two-fold response.

On one hand, by altering the physical properties of MTs, tubulin acetylation may enhance the cellular response to physical challenges imposed by the environment, such as changes in the extracellular matrix (ECM). By providing MTs with a more resilient and flexible structure, tubulin acetylation likely reduces the frequency of MT breakage which is accompanied by improved damage repair of the MT wall. This, in turn, indirectly affects MT dynamics by minimizing the need for constant cytoskeletal remodeling, thereby supporting cellular integrity under stress.

On the other hand, MT acetylation has been shown to alter interactions with other molecules associated with MTs. By modifying these interactions, tubulin acetylation can influence several critical processes. One such process is intracellular trafficking, where acetylation seems to affect the affinity of molecular motors like dynein and kinesin. This enables the cell to finely regulate the transport of molecules and organelles without altering the physical "railways" on which these cargos travel. Such a mechanism likely allows cells to adapt more rapidly to specific needs without compromising the overall structure of the MT cytoskeleton. Additionally, by establishing specific MT acetylation patterns, cells can direct and concentrate components to specific regions of the cytoplasm without the need for physical barriers. This capability facilitates localized responses rather than global changes in the cell's behavior, ensuring precise and efficient cellular adaptation to environmental demands.

The interplay between mechanical and oxidative stress highlights the multifaceted nature of MT acetylation. It shows how this PTM of tubulin responds to physical and chemical stimuli giving a concerted answer. This highlights the central role of MT, and specifically MT acetylation, as a crucial hub that not only acts as a responsive modification but also mediates the integration and distribution of diverse cellular signals. This dual functionality emphasizes its potential significance in pathological conditions where cellular stress responses are dysregulated, such as neurodegenerative disorders, cancer, and cardiovascular diseases.

Future research should delve into unresolved questions, such as the precise molecular mechanisms that enable acetylated MTs to encode spatial and functional information. The interconnections between acetylation, distinct tubulin isotypes, and other PTMs also warrant further investigation, particularly in the context of cellular metabolism and its coupling to cytoskeletal dynamics. Finally, it has been completely unexplored how proteins involved in tubulin dimer assembly, recycling, and degradation are affected and/or recognize PTM tubulin dimers and regulate MT dynamics in these different processes. By advancing our understanding of these processes, new therapeutic avenues may emerge, targeting tubulin acetylation to mitigate disease and enhance cellular resilience.

Acknowledgments This work was funded by Instituto Politécnico de Lisboa IPL/IDI&CA2024/HyperCil_ESTeSL and IPL/IDI&CA2024/ENIGMA_ESTeSL, and also Centro de Química Estrutural is a Research Unit funded by FCT through projects UIDB/00100/2020 (https://doi.org/10.54499/UIDB/00100/2020) and UIDP/00100/2020 (https://doi.org/10.54499/UIDP/00100/2020). Institute of Molecular Sciences is an Associate Laboratory funded by FCT through project LA/P/0056/2020 (https://doi.org/10.54499/LA/P/0056/2020).

References

Abouhish H, Thounaojam MC, Jadeja RN et al (2020) Inhibition of HDAC6 attenuates diabetes-induced retinal redox imbalance and microangiopathy. Antioxidants (Basel) 9:599. https://doi.org/10.3390/antiox9070599

Ainsworth C (2008) Cell biology: stretching the imagination. Nature 456:696–699. https://doi.org/10.1038/456696a

Akella JS, Wloga D, Kim J et al (2010) MEC-17 is an alpha-tubulin acetyltransferase. Nature 467:218–222. https://doi.org/10.1038/nature09324

Andreu-Carbó M, Egoldt C, Velluz M-C, Aumeier C (2024) Microtubule damage shapes the acetylation gradient. Nat Commun 15:2029. https://doi.org/10.1038/s41467-024-46379-5

Bae H-J, Shin S-J, Jo SB et al (2024) Cyclic stretch induced epigenetic activation of periodontal ligament cells. Mater Today Bio 26:101050. https://doi.org/10.1016/j.mtbio.2024.101050

Bance B, Seetharaman S, Leduc C et al (2019) Microtubule acetylation but not detyrosination promotes focal adhesion dynamics and astrocyte migration. J Cell Sci 132:jcs225805. https://doi.org/10.1242/jcs.225805

Bernstein BW, Bamburg JR (2003) Actin-ATP hydrolysis is a major energy drain for neurons. J Neurosci 23:1–6. https://doi.org/10.1523/JNEUROSCI.23-01-00002.2003

Bohl K, Wynia-Smith SL, Jones Lipinski RA, Smith BC (2024) Inhibition of sirtuin deacylase activity by peroxynitrite. Biochemistry 63:2463. https://doi.org/10.1021/acs.biochem.4c00257

Bonet-Ponce L, Saez-Atienzar S, da Casa C et al (2016) Rotenone induces the formation of 4-hydroxynonenal aggresomes. Role of ROS-mediated tubulin hyperacetylation and autophagic flux disruption. Mol Neurobiol 53:6194–6208. https://doi.org/10.1007/s12035-015-9509-3

Borgas D, Chambers E, Newton J et al (2016) Cigarette smoke disrupted lung endothelial barrier integrity and increased susceptibility to acute lung injury via histone deacetylase 6. Am J Respir Cell Mol Biol 54:683–696. https://doi.org/10.1165/rcmb.2015-0149OC

Brangwynne CP, MacKintosh FC, Kumar S et al (2006) Microtubules can bear enhanced compressive loads in living cells because of lateral reinforcement. J Cell Biol 173:733–741. https://doi.org/10.1083/jcb.200601060

Butler PJ (2016) Mechanotransduction of shear stress by the endothelium. In: Tanishita K, Yamamoto K (eds) Vascular engineering: new prospects of vascular medicine and biology with a multidiscipline approach. Springer Japan, Tokyo, pp 159–197

Celik E, Abdulreda MH, Maiguel D et al (2013) Rearrangement of microtubule network under biochemical and mechanical stimulations. Methods 60:195–201. https://doi.org/10.1016/j.ymeth.2013.02.014

Choudhary C, Kumar C, Gnad F et al (2009) Lysine Acetylation Targets Protein Complexes and Co-Regulates Major Cellular Functions. Science 325:834–840. https://doi.org/10.1126/science.1175371

Coleman AK, Joca HC, Shi G et al (2021) Tubulin acetylation increases cytoskeletal stiffness to regulate mechanotransduction in striated muscle. J Gen Physiol 153:e202012743. https://doi.org/10.1085/jgp.202012743

Collins C, Kim SK, Ventrella R et al (2021) Tubulin acetylation promotes penetrative capacity of cells undergoing radial intercalation. Cell Rep 36:109556. https://doi.org/10.1016/j.celrep.2021.109556

Daniel JL, Molish IR, Robkin L, Holmsen H (1986) Nucleotide exchange between cytosolic ATP and F-actin-bound ADP may be a major energy-utilizing process in unstimulated platelets. Eur J Biochem 156:677–683. https://doi.org/10.1111/j.1432-1033.1986.tb09631.x

Dawson LW, Cronin NM, DeMali KA (2023) Mechanotransduction: Forcing a change in metabolism. Curr Opin Cell Biol 84:102219. https://doi.org/10.1016/j.ceb.2023.102219

Di X, Gao X, Peng L et al (2023) Cellular mechanotransduction in health and diseases: from molecular mechanism to therapeutic targets. Sig Transduct Target Ther 8:282. https://doi.org/10.1038/s41392-023-01501-9

Dogterom M, Koenderink GH (2019) Actin-microtubule crosstalk in cell biology. Nat Rev Mol Cell Biol 20:38–54. https://doi.org/10.1038/s41580-018-0067-1

Eshun-Wilson L, Zhang R, Portran D et al (2019) Effects of α-tubulin acetylation on microtubule structure and stability. Proc Natl Acad Sci 116:10366–10371. https://doi.org/10.1073/pnas.1900441116

Falconieri A, De Vincentiis S, Cappello V et al (2023) Axonal plasticity in response to active forces generated through magnetic nano-pulling. Cell Rep 42:111912. https://doi.org/10.1016/j.celrep.2022.111912

Feng S, Daw JN, Chen QM (2020) Histone deacetylase inhibitors prevent H2O2 from inducing stress granule formation. Curr Res Toxicol 1:141–148. https://doi.org/10.1016/j.crtox.2020.10.004

Fernández-Barrera J, Bernabé-Rubio M, Casares-Arias J et al (2018) The actin-MRTF-SRF transcriptional circuit controls tubulin acetylation via α-TAT1 gene expression. J Cell Biol 217:929–944. https://doi.org/10.1083/jcb.201702157

Fink J, Carpi N, Betz T et al (2011) External forces control mitotic spindle positioning. Nat Cell Biol 13:771–778. https://doi.org/10.1038/ncb2269

Fletcher DA, Mullins RD (2010) Cell mechanics and the cytoskeleton. Nature 463:485–492. https://doi.org/10.1038/nature08908

Forman HJ, Zhang H (2021) Targeting oxidative stress in disease: promise and limitations of antioxidant therapy. Nat Rev Drug Discov 20:689–709. https://doi.org/10.1038/s41573-021-00233-1

Fu S, Thompson CL, Ali A et al (2019) Mechanical loading inhibits cartilage inflammatory signalling via an HDAC6 and IFT-dependent mechanism regulating primary cilia elongation. Osteoarthr Cartil 27:1064–1074. https://doi.org/10.1016/j.joca.2019.03.003

Gal J, Chen J, Barnett KR et al (2013) HDAC6 regulates mutant SOD1 aggregation through two SMIR motifs and tubulin acetylation. J Biol Chem 288:15035–15045. https://doi.org/10.1074/jbc.M112.431957

Geiger RC, Kaufman CD, Lam AP et al (2009) Tubulin Acetylation and Histone Deacetylase 6 Activity in the Lung under Cyclic Load. Am J Respir Cell Mol Biol 40:76–82. https://doi.org/10.1165/rcmb.2007-0307OC

Gittes F, Mickey B, Nettleton J, Howard J (1993) Flexural rigidity of microtubules and actin filaments measured from thermal fluctuations in shape. J Cell Biol 120:923–934. https://doi.org/10.1083/jcb.120.4.923

Giustiniani J, Couloubaly S, Baillet A et al (2009) Basal endothelial nitric oxide synthase (eNOS) phosphorylation on Ser(1177) occurs in a stable microtubule- and tubulin acetylation-dependent manner. Exp Cell Res 315:3509–3520. https://doi.org/10.1016/j.yexcr.2009.07.018

Goode BL, Eck MJ (2007) Mechanism and function of formins in the control of actin assembly. Annu Rev Biochem 76:593–627. https://doi.org/10.1146/annurev.biochem.75.103004.142647

Hamant O, Inoue D, Bouchez D et al (2019) Are microtubules tension sensors? Nat Commun 10:2360. https://doi.org/10.1038/s41467-019-10207-y

Hardham AR, Takemoto D, White RG (2008) Rapid and dynamic subcellular reorganization following mechanical stimulation of Arabidopsisepidermal cells mimics responses to fungal and oomycete attack. BMC Plant Biol 8:63. https://doi.org/10.1186/1471-2229-8-63

Hawkins T, Mirigian M, Selcuk Yasar M, Ross JL (2010) Mechanics of microtubules. J Biomech 43:23–30. https://doi.org/10.1016/j.jbiomech.2009.09.005

Hejnowicz Z, Rusin A, Rusin T (2000) Tensile tissue stress affects the orientation of cortical microtubules in the epidermis of sunflower hypocotyl. J Plant Growth Regul 19:31–44. https://doi.org/10.1007/s003440000005

Hieda M (2019) Signal transduction across the nuclear envelope: role of the LINC complex in bidirectional signaling. Cells 8:124. https://doi.org/10.3390/cells8020124

Hirase T, Node K (2012) Endothelial dysfunction as a cellular mechanism for vascular failure. Am J Phys Heart Circ Phys 302:H499–H505. https://doi.org/10.1152/ajpheart.00325.2011

Huang S, Ingber DE (2005) Cell tension, matrix mechanics, and cancer development. Cancer Cell 8:175–176. https://doi.org/10.1016/j.ccr.2005.08.009

Hubbert C, Guardiola A, Shao R et al (2002) HDAC6 is a microtubule-associated deacetylase. Nature 417:455–458. https://doi.org/10.1038/417455a

Ingber DE, Wang N, Stamenovic D (2014) Tensegrity, cellular biophysics, and the mechanics of living systems. Rep Prog Phys 77:046603. https://doi.org/10.1088/0034-4885/77/4/046603

Jacques E, Verbelen J-P, Vissenberg K (2013) Mechanical stress in Arabidopsis leaves orients microtubules in a 'continuous' supracellular pattern. BMC Plant Biol 13:163. https://doi.org/10.1186/1471-2229-13-163

Janke C, Montagnac G (2017) Causes and Consequences of Microtubule Acetylation. Curr Biol 27:R1287. https://doi.org/10.1016/j.cub.2017.10.044

Jansen KA, Donato DM, Balcioglu HE et al (2015) A guide to mechanobiology: where biology and physics meet. Biochimica et Biophysica Acta (BBA) Mol Cell Res 1853:3043–3052. https://doi.org/10.1016/j.bbamcr.2015.05.007

Janshoff A (2021) Viscoelastic properties of epithelial cells. Biochem Soc Trans 49:2687–2695. https://doi.org/10.1042/BST20210476

Kalous KS, Wynia-Smith SL, Summers SB, Smith BC (2020) Human sirtuins are differentially sensitive to inhibition by nitrosating agents and other cysteine oxidants. J Biol Chem 295:8524–8536. https://doi.org/10.1074/jbc.RA119.011988

Katsuta H, Sokabe M, Hirata H (2024) From stress fiber to focal adhesion: a role of actin cross linkers in force transmission. Front Cell Dev Biol 12:1444827. https://doi.org/10.3389/fcell.2024.1444827

Kaverina I, Krylyshkina O, Beningo K et al (2002) Tensile stress stimulates microtubule outgrowth in living cells. J Cell Sci 115:2283–2291. https://doi.org/10.1242/jcs.115.11.2283

Kolega J, Janson LW, Taylor DL (1991) The role of solation-contraction coupling in regulating stress fiber dynamics in nonmuscle cells. J Cell Biol 114:993–1003. https://doi.org/10.1083/jcb.114.5.993
L'Hernault SW, Rosenbaum JL (1985) Chlamydomonas alpha.-tubulin is post translationally modified by acetylation on the .epsilon.-amino group of a lysine. Biochemistry 24:473–478. https://doi.org/10.1021/bi00323a034
Landrein B, Hamant O (2013) How mechanical stress controls microtubule behavior and morphogenesis in plants: history, experiments and revisited theories. Plant J 75:324–338. https://doi.org/10.1111/tpj.12188
Lavagnino M, Arnoczky SP (2005) In vitro alterations in cytoskeletal tensional homeostasis control gene expression in tendon cells. J Orthop Res 23:1211–1218. https://doi.org/10.1016/j.orthres.2005.04.001
LeDizet M, Piperno G (1987) Identification of an acetylation site of Chlamydomonas alpha-tubulin. Proc Natl Acad Sci 84:5720–5724. https://doi.org/10.1073/pnas.84.16.5720
Leng Y, Wu Y, Lei S et al (2018) Inhibition of HDAC6 activity alleviates myocardial ischemia/reperfusion injury in diabetic rats: potential role of peroxiredoxin 1 acetylation and redox regulation. Oxidative Med Cell Longev 2018:9494052. https://doi.org/10.1155/2018/9494052
Leyk J, Daly C, Janssen-Bienhold U et al (2017) HDAC6 inhibition by tubastatin A is protective against oxidative stress in a photoreceptor cell line and restores visual function in a zebrafish model of inherited blindness. Cell Death Dis 8:e3028. https://doi.org/10.1038/cddis.2017.415
Li Y, Kučera O, Cuvelier D et al (2023) Compressive forces stabilize microtubules in living cells. Nat Mater 22:913–924. https://doi.org/10.1038/s41563-023-01578-1
Lintilhac PM, Vesecky TB (1984) Stress-induced alignment of division plane in plant tissues grown in vitro. Nature 307:363–364. https://doi.org/10.1038/307363a0
Louveaux M, Rochette S, Beauzamy L et al (2016) The impact of mechanical compression on cortical microtubules in Arabidopsis: a quantitative pipeline. Plant J 88:328–342. https://doi.org/10.1111/tpj.13290
Lynch TM, Lintilhac PM (1997) Mechanical signals in plant development: a new method for single cell studies. Dev Biol 181:246–256. https://doi.org/10.1006/dbio.1996.8462
Mackeh R, Lorin S, Ratier A et al (2014) Reactive oxygen species, AMP-activated protein kinase, and the transcription cofactor p300 regulate α-tubulin acetyltransferase-1 (αTAT-1/MEC-17)-dependent microtubule hyperacetylation during cell stress. J Biol Chem 289:11816–11828. https://doi.org/10.1074/jbc.M113.507400
Marchant CL, Malmi-Kakkada AN, Espina JA, Barriga EH (2022) Cell clusters softening triggers collective cell migration in vivo. Nat Mater 21:1314–1323. https://doi.org/10.1038/s41563-022-01323-0
McCue S, Dajnowiec D, Xu F et al (2006) Shear stress regulates forward and reverse planar cell polarity of vascular endothelium in vivo and in vitro. Circ Res 98:939–946. https://doi.org/10.1161/01.RES.0000216595.15868.55
McKenna ED, Sarbanes SL, Cummings SW, Roll-Mecak A (2023) The tubulin code, from molecules to health and disease. Annu Rev Cell Dev Biol 39:331–361. https://doi.org/10.1146/annurev-cellbio-030123-032748
Méndez-López I, Sancho-Bielsa FJ, Engel T et al (2021) Progressive mitochondrial SOD1G93A accumulation causes severe structural, metabolic and functional aberrations through OPA1 down-regulation in a mouse model of amyotrophic lateral sclerosis. Int J Mol Sci 22:8194. https://doi.org/10.3390/ijms22158194
Morley SJ, Qi Y, Iovino L et al (2016) Acetylated tubulin is essential for touch sensation in mice. eLife 5:e20813. https://doi.org/10.7554/eLife.20813
Mozzicato AM, Bastrup JA, Sanchez-Alonso JL et al (2024) Mesenteric artery smooth muscle cells from hypertensive rats have increased microtubule acetylation. Biochem J 481:387–403. https://doi.org/10.1042/BCJ20230420

Mui KL, Chen CS, Assoian RK (2016) The mechanical regulation of integrin–cadherin crosstalk organizes cells, signaling and forces. J Cell Sci 129:1093–1100. https://doi.org/10.1242/jcs.183699

Myers KA, Applegate KT, Danuser G et al (2011) Distinct ECM mechanosensing pathways regulate microtubule dynamics to control endothelial cell branching morphogenesis. J Cell Biol 192:321–334. https://doi.org/10.1083/jcb.201006009

North BJ, Marshall BL, Borra MT et al (2003) The human Sir2 ortholog, SIRT2, is an NAD+-dependent tubulin deacetylase. Mol Cell 11:437–444. https://doi.org/10.1016/S1097-2765(03)00038-8

Okuda K, Ito A, Uehara T (2015) Regulation of histone deacetylase 6 activity via S-nitrosylation. Biol Pharm Bull 38:1434–1437. https://doi.org/10.1248/bpb.b15-00364

Orbach R, Howard J (2019) The dynamic and structural properties of axonemal tubulins support the high length stability of cilia. Nat Commun 10:1838. https://doi.org/10.1038/s41467-019-09779-6

Park JS, Burckhardt CJ, Lazcano R et al (2020) Mechanical regulation of glycolysis via cytoskeleton architecture. Nature 578:621–626. https://doi.org/10.1038/s41586-020-1998-1

Poirier K, Saillour Y, Bahi-Buisson N et al (2010) Mutations in the neuronal β-tubulin subunit TUBB3 result in malformation of cortical development and neuronal migration defects. Hum Mol Genet 19:4462–4473. https://doi.org/10.1093/hmg/ddq377

Portran D, Schaedel L, Xu Z et al (2017) Tubulin acetylation protects long-lived microtubules against mechanical ageing. Nat Cell Biol 19:391–398. https://doi.org/10.1038/ncb3481

Putnam AJ, Cunningham JJ, Dennis RG et al (1998) Microtubule assembly is regulated by externally applied strain in cultured smooth muscle cells. J Cell Sci 111(Pt 22):3379–3387. https://doi.org/10.1242/jcs.111.22.3379

Putnam AJ, Schultz K, Mooney DJ (2001) Control of microtubule assembly by extracellular matrix and externally applied strain. Am J Phys Cell Phys 280:C556–C564. https://doi.org/10.1152/ajpcell.2001.280.3.C556

Rafiq NBM, Nishimura Y, Plotnikov SV et al (2019) A mechano-signalling network linking microtubules, myosin IIA filaments and integrin-based adhesions. Nat Mater 18:638–649. https://doi.org/10.1038/s41563-019-0371-y

Robinson S, Kuhlemeier C (2018) Global compression reorients cortical microtubules in arabidopsis hypocotyl epidermis and promotes growth. Curr Biol 28:1794–1802.e2. https://doi.org/10.1016/j.cub.2018.04.028

Romani P, Nirchio N, Arboit M et al (2022) Mitochondrial fission links ECM mechanotransduction to metabolic redox homeostasis and metastatic chemotherapy resistance. Nat Cell Biol 24:168–180. https://doi.org/10.1038/s41556-022-00843-w

Salvi AM, Bays JL, Mackin SR et al (2021) Ankyrin G organizes membrane components to promote coupling of cell mechanics and glucose uptake. Nat Cell Biol 23:457–466. https://doi.org/10.1038/s41556-021-00677-y

Sampathkumar A, Krupinski P, Wightman R et al (2014) Subcellular and supracellular mechanical stress prescribes cytoskeleton behavior in Arabidopsis cotyledon pavement cells. eLife 3:e01967. https://doi.org/10.7554/eLife.01967

Sanger JW, Sanger JM, Jockusch BM (1983) Differences in the stress fibers between fibroblasts and epithelial cells. J Cell Biol 96:961–969. https://doi.org/10.1083/jcb.96.4.961

Schaedel L, John K, Gaillard J et al (2015) Microtubules self-repair in response to mechanical stress. Nature Mater 14:1156–1163. https://doi.org/10.1038/nmat4396

Seetharaman S, Vianay B, Roca V et al (2022) Microtubules tune mechanosensitive cell responses. Nat Mater 21:366–377. https://doi.org/10.1038/s41563-021-01108-x

Shida T, Cueva JG, Xu Z et al (2010) The major α-tubulin K40 acetyltransferase αTAT1 promotes rapid ciliogenesis and efficient mechanosensation. Proc Natl Acad Sci 107:21517–21522. https://doi.org/10.1073/pnas.1013728107

Sies H (2015) Oxidative stress: a concept in redox biology and medicine. Redox Biol 4:180–183. https://doi.org/10.1016/j.redox.2015.01.002

Stooke-Vaughan GA, Campàs O (2018) Physical control of tissue morphogenesis across scales. Curr Opin Genet Dev 51:111–119. https://doi.org/10.1016/j.gde.2018.09.002
Su M, Guan H, Zhang F et al (2016) HDAC6 regulates the chaperone-mediated autophagy to prevent oxidative damage in injured neurons after experimental spinal cord injury. Oxidative Med Cell Longev 2016:7263736. https://doi.org/10.1155/2016/7263736
Swaminathan V, Gloerich M (2021) Decoding mechanical cues by molecular mechanotransduction. Curr Opin Cell Biol 72:72–80. https://doi.org/10.1016/j.ceb.2021.05.006
Taes I, Timmers M, Hersmus N et al (2013) Hdac6 deletion delays disease progression in the SOD1G93A mouse model of ALS. Hum Mol Genet 22:1783–1790. https://doi.org/10.1093/hmg/ddt028
Tamariz E, Varela-Echavarría A (2015) The discovery of the growth cone and its influence on the study of axon guidance. Front Neuroanat 9:51. https://doi.org/10.3389/fnana.2015.00051
Tee S-Y, Bausch AR, Janmey PA (2009) The mechanical cell. Curr Biol 19:R745–R748. https://doi.org/10.1016/j.cub.2009.06.034
Théry M, Jiménez-Dalmaroni A, Racine V et al (2007) Experimental and theoretical study of mitotic spindle orientation. Nature 447:493–496. https://doi.org/10.1038/nature05786
Thurston SF, Kulacz WA, Shaikh S et al (2012) The ability to induce microtubule acetylation is a general feature of formin proteins. PLoS One 7:e48041. https://doi.org/10.1371/journal.pone.0048041
Tischfield MA, Baris HN, Wu C et al (2010) Human TUBB3 mutations perturb microtubule dynamics, kinesin interactions, and axon guidance. Cell 140:74–87. https://doi.org/10.1016/j.cell.2009.12.011
Torrino S, Grasset EM, Audebert S et al (2021) Mechano-induced cell metabolism promotes microtubule glutamylation to force metastasis. Cell Metab 33:1342–1357.e10. https://doi.org/10.1016/j.cmet.2021.05.009
Uray IP, Uray K (2021) Mechanotransduction at the plasma membrane-cytoskeleton interface. Int J Mol Sci 22:11566. https://doi.org/10.3390/ijms222111566
Verhey KJ, Gaertig J (2007) The tubulin code. Cell Cycle 6:2152–2160
Waddell J, Banerjee A, Kristian T (2021) Acetylation in mitochondria dynamics and neurodegeneration. Cells 10:3031. https://doi.org/10.3390/cells10113031
Wang JH, Li B (2010) Mechanics rules cell biology. BMC Sports Sci Med Rehabil 2:16. https://doi.org/10.1186/1758-2555-2-16
Wang F, Nguyen M, Qin FX-F, Tong Q (2007) SIRT2 deacetylates FOXO3a in response to oxidative stress and caloric restriction. Aging Cell 6:505–514. https://doi.org/10.1111/j.1474-9726.2007.00304.x
Wang Y-H, Yan Z-Q, Qi Y-X et al (2010) Normal shear stress and vascular smooth muscle cells modulate migration of endothelial cells through histone deacetylase 6 activation and tubulin acetylation. Ann Biomed Eng 38:729–737. https://doi.org/10.1007/s10439-009-9896-6
Wang M, Zhou C, Yu L et al (2022) Upregulation of MDH1 acetylation by HDAC6 inhibition protects against oxidative stress-derived neuronal apoptosis following intracerebral hemorrhage. Cell Mol Life Sci 79:356. https://doi.org/10.1007/s00018-022-04341-y
Wasteneys GO, Ambrose JC (2009) Spatial organization of plant cortical microtubules: close encounters of the 2D kind. Trends Cell Biol 19:62–71. https://doi.org/10.1016/j.tcb.2008.11.004
Wen D, Gao Y, Liu Y et al (2023) Matrix stiffness-induced α-tubulin acetylation is required for skin fibrosis formation through activation of Yes-associated protein. Med Comm 4:e319. https://doi.org/10.1002/mco2.319
Wilson C, González-Billault C (2015) Regulation of cytoskeletal dynamics by redox signaling and oxidative stress: implications for neuronal development and trafficking. Front Cell Neurosci 9:381. https://doi.org/10.3389/fncel.2015.00381
Wolfenson H, Yang B, Sheetz MP (2019) Steps in Mechanotransduction Pathways that Control Cell Morphology. Annu Rev Physiol 81:585–605. https://doi.org/10.1146/annurev-physiol-021317-121245

Wu Y-T, Wu S-B, Lee W-Y, Wei Y-H (2010) Mitochondrial respiratory dysfunction-elicited oxidative stress and posttranslational protein modification in mitochondrial diseases. Ann N Y Acad Sci 1201:147–156. https://doi.org/10.1111/j.1749-6632.2010.05631.x

Xu Z, Schaedel L, Portran D et al (2017) Microtubules acquire resistance from mechanical breakage through intralumenal acetylation. Science 356:328–332. https://doi.org/10.1126/science.aai8764

Yan C, Wang F, Peng Y et al (2018) Microtubule acetylation is required for mechanosensation in Drosophila. Cell Rep 25:1051–1065.e6. https://doi.org/10.1016/j.celrep.2018.09.075

Youn GS, Lee KW, Choi SY, Park J (2016) Overexpression of HDAC6 induces pro-inflammatory responses by regulating ROS-MAPK-NF-κB/AP-1 signaling pathways in macrophages. Free Radic Biol Med 97:14–23. https://doi.org/10.1016/j.freeradbiomed.2016.05.014

Yu J, Wu Y, Yang P (2016) High glucose-induced oxidative stress represses sirtuin deacetylase expression and increases histone acetylation leading to neural tube defects. J Neurochem 137:371–383. https://doi.org/10.1111/jnc.13587

Yuan Q, Zhan L, Zhou Q-Y et al (2015) SIRT2 regulates microtubule stabilization in diabetic cardiomyopathy. Eur J Pharmacol 764:554–561. https://doi.org/10.1016/j.ejphar.2015.07.045

Zheng J, Buxbaum RE, Heidemann SR (1993) Investigation of microtubule assembly and organization accompanying tension-induced neurite initiation. J Cell Sci 104(Pt 4):1239–1250. https://doi.org/10.1242/jcs.104.4.1239

Part III
Acetylation in Development

Chapter 6
Reversible Histone Acetylation During Preimplantation Embryo Development in Mammals

Giulia Musmeci, Fernanda Fagali Franchi, Francesca Mossa, Alberto Maria Luciano, Valentina Lodde, and Federica Franciosi

Abstract Histone acetylation is an epigenetic modification responsible for changes in chromatin architecture, accessibility, and ultimately gene expression. At the onset of a new life, when the fully differentiated parental genomes fuse together to generate a new totipotent cell, the gametes' epigenetic program must be erased, and new ones are progressively installed. Together with other epigenetic modifications, histone acetylation participates in the early events of embryogenesis, undergoing dynamic changes that involve several amino acid residues on different histone proteins. By analyzing studies that followed these changes during the preimplantation development in different mammals, we identified critical windows of acetylation/deacetylation in relation to the oocyte-to-zygote transition, the activation of the embryonic genome, and the specification of cell lineages, all crucial events for early embryo development, the establishment of pluripotent embryonic tissue, and ultimately of a multicellular organism.

Finally, this survey points out the possibility that while contributing to the necessary plasticity of the embryonic stem cells, the reversibility of histone acetylation/deacetylation patterns renders this mechanism prone to be hijacked by environmental conditions, such as maternal diet or pollutants, leading to the alterations of epigenetic marks that can be potentially transmitted to the daughter cells and up to adulthood.

G. Musmeci · F. F. Franchi · A. M. Luciano · V. Lodde · F. Franciosi (✉)
Reproductive and Developmental Biology Laboratory, Department of Veterinary Medicine and Animal Sciences, University of Milan, Lodi, LO, Italy
e-mail: federica.franciosi1@unimi.it

F. Mossa
Department of Veterinary Medicine, University of Sassari, Sassari, SS, Italy

M. Halasa, A. Wawruszak (eds.), *Histone and Non-Histone Reversible Acetylation in Development, Aging and Disease*, Results and Problems in Cell Differentiation 75,
https://doi.org/10.1007/978-3-031-91459-1_6

6.1 Reversible Histone Acetylation During Preimplantation Embryo Development in Mammals

Multicellular organisms originate from a single totipotent cell, the zygote, that in turn derives from fully differentiated gametes. Therefore, the maternal and paternal genomes contributing to the zygote must undergo complete nuclear and epigenetic reprogramming to establish a totipotent state and progressively install the required changes. This program of "molecular erasure and writing" ultimately drives the differentiation of the pluripotent embryonic tissues.

Histone acetylation, like many other epigenetic modifications, endures drastic remodeling, both at the global and site-specific levels, which are necessary to modify the chromatin conformation and accessibility, hence controlling the ability to interact with transcription factors and, consequently, having a role in gene expression (Shirvaliloo 2022).

The reversible nature of histone acetylation, regulated by enzymes that introduce or remove the acetylated groups from specific amino acid residues (Lodde et al. 2017; Peserico and Simone 2011), is particularly evident at the oocyte-to-zygote transition, entailing the removal of the gamete-specific marks. However, reports of changes in the histone acetylation patterns also exist for later crucial developmental passages, as the embryo genome activation (EGA) and the first cell lineages commitments, indicating the participation of histone acetylation/deacetylation events in the progressive fine-tuning of the functions of the embryonic genome.

However, this exact reversible nature also exposes the program of histone acetylation to the influence of external factors that, by acting on the availability of substrates and/or on the activity of the enzymes, might interfere with the normal patterns of histone acetylation, affecting the survival of the embryo, the dynamics of development, or even setting the stage for future anomalies, as postulated by the hypothesis of developmental programming of adult disease (Huber et al. 2020).

After a brief recapitulation of the main events in mammalian embryo development, with this review we will analyze the key histone acetylation changes of the prominent amino acid residues during preimplantation development and finally explore the scenarios that may compromise their establishment.

Most of the knowledge on histone acetylation during early embryo development comes from studies on humans, mice, pigs, and cattle, while research on other mammals, such as dromedaries, rabbits, buffalos, and primates, is relatively rare. This review will emphasize findings from the primary species of interest. Furthermore, due to the capacity of the embryo to develop outside the maternal body for several days, many of the studies considered rely on experimental models of in vitro embryo culture.

6.2 Main Events of Preimplantation Embryo Development in Mammals

The early stages of embryo development comprehend a series of events starting immediately after fertilization and resulting in the formation of a multicellular organism. This process is crucial for successful implantation and subsequent fetal growth (Alberio 2020). After fertilization, the zygote is formed through the fusion of the male and female pronuclei, resulting in a single cell that contains a complete set of chromosomes inherited from both parents. The zygote then undergoes a series of mitotic divisions, with the initial stages being regulated by stored mRNAs and proteins of maternal origin before a switch to embryo-encoded transcripts is made (Coticchio et al. 2023a). In mammals, the early divisions lead to the progression from the 2-cell stage to a more complex structure, the morula, which typically consists of around 32–64 blastomeres. These first cellular divisions mark a critical phase in the embryo's development as they anticipate and prepare the cellular and molecular machinery for successive differentiation and specialization. Upon reaching the morula stage, the embryo begins to form a fluid-filled cavity called the blastocoel. This event marks the differentiation of the first two distinct cell populations: the inner cell mass (ICM), which will give rise to the fetus, and the trophectoderm (TE), which will form the embryonic annexes. The ICM will further differentiate into an epithelial monolayer named primitive endoderm (PrE) or hypoblast, which contributes to the yolk sac and the remaining enclosed epiblast cells (Epi), which give rise to all the fetal tissues (Cui and Mager 2018; Rossant 2018; Wilkinson et al. 2023). At the beginning of blastocoel formation, the early blastocyst remains enveloped within the zona pellucida, an oocyte-derived glycoprotein shell. Later, with the enlargement of the cavity and continuous cell proliferation, the blastocyst expands and the zona pellucida thins, leading to embryo hatching (Seshagiri et al. 2016).

Once free, the hatched blastocyst continues its expansion, increasing in size. In species such as cattle and pigs, this phase involves flattening out and transitioning to an elongated shape, significantly increasing its surface area and contact with the maternal endometrium (Davenport et al. 2023; Miles et al. 2023). This elongation and flattening are typical of non-invasive implantation, where the increased surface contact with the uterine lining facilitates nutrient exchange and placental development. In contrast, in mice and humans implantation is invasive and the blastocyst does not elongate before implantation, but it keeps a cylindrical or spherical shape (Kagawa et al. 2022; Weberling and Zernicka-Goetz 2021).

The preimplantation stages of development are characterized by cell proliferation and differentiation, setting the stage for successful implantation and the subsequent development of the embryo within the maternal environment (Y. Zhang et al. 2007). Various factors, including hormonal levels, uterine receptivity, and the health and quality of the embryo itself influence the success of early embryogenesis. If implantation is successful, the embryo begins to establish a connection with the maternal blood supply, ensuring proper nourishment and support for subsequent development (Muter et al. 2023).

The preimplantation stages of embryogenesis can be replicated in vitro using standardized protocols, except for the stage of blastocyst elongation, which is still experimental (Ramos-Ibeas et al. 2020). Thanks to in vitro embryo production (IVP) protocols, embryos of several mammalian species are cultured in specialized media for up to 7–8 days until the blastocyst stage. Bovine IVP is one of the best-established protocols, allowing to obtain 20–40% blastocyst development starting from prophase I-arrested, fully-grown immature oocytes (Lonergan et al. 2001). Upon isolation from the follicular environment, such oocytes resume meiosis and reach metaphase II (MII), corresponding to the stage at which they are when ovulation occurs. After this step, called in vitro maturation (IVM), they can be fertilized in vitro (IVF), and the resulting zygotes are cultured for several days. When reaching the morula-early blastocyst, the embryos can be transferred into recipients or cryopreserved for future, timed transfer. However, only high-quality embryos can be successfully used for cryopreservation, as they bear the potential for higher pregnancy rates (Layek et al. 2022).

During the early stages of development, the embryo utilizes stored maternal mRNAs and proteins before the embryonic genome becomes activated. The EGA occurs with species-specific timing: around the 4–8-cell stage in humans, the 8–16-cell stage in cattle, the 4-cell stage in pigs, and the 2-cell stage in mice (L. Li et al. 2013), with large mammals exhibiting comparable timelines during the preimplantation stages up to the hatching, compared to mice. Similarly, after fertilization, human embryos progress to the blastocyst stage in approximately 7 days (Coticchio et al. 2023b), as cows (Mamo et al. 2012). Conversely, the development of murine embryos is faster, and the blastocyst stage is already reached at 4 days after fertilization. Indeed, in mice, compaction begins earlier, at the 8–16 cell stage, and blastocoel cavity formation occurs between the 16–32 cells (Kidder and Watson 2005). In contrast, bovine embryos compact near the 32-cell stage, with cavitation generally observed after the 64 cells (Van Soom et al. 1997).

Besides the timeline, the mechanisms regulating early cell lineage specification in the mouse differ from those of large mammals, as evidenced by gene expression profiling (Berg et al. 2011; Blakeley et al. 2015; Fogarty et al. 2017; Gerri et al. 2020; Niakan and Eggan 2013). In mice, the expression of genes such as the POU domain, class 5, transcription factor 1 (*Pou5f1*, also known as octamer-binding transcription factor 4—*Oct4*), and SRY-box 2 (*Sox2*) is critical for the maintenance of the pluripotency in the undifferentiated cells of the ICM. At the same time, caudal type homeobox 2 (*Cdx2*) promotes the differentiation of the TE by repressing the expression of the pluripotency factors (Kim et al. 2008; Niwa et al. 2005). The ICM is further separated in Epi and PrE, in which the Epi marker, homeobox protein NANOG (*Nanog*), and the PrE marker, GATA-binding factor 6 (*Gata6*), repress each other during the second lineage segregation (Thompson et al. 2022).

In higher-order mammals, the expression of *POU5F1* is not limited to the ICM as in mice, but it is widespread, extending to all cells in the Day 8 blastocyst (Aguila et al. 2022; Fogarty et al. 2017), while *SOX2* is detected in the ICM nuclei of bovine, mouse, and human embryos (Aguila et al. 2022; De Paepe et al. 2013; Yao et al. 2019). Overall, the transcription factors *POU5F1* and *CDX2* exhibit comparable temporal and spatial expression patterns in human and bovine embryos (Berg et al. 2011; A. E. Chen et al. 2009; Niakan and Eggan 2013), while mice display notable differences from these species since the proteins POU5F1 and CDX2 start to be detected from the 8- and the 16-cell stages, respectively (Niakan and Eggan 2013; Niwa et al. 2005). Overall, both bovine and murine models offer distinct advantages and limitations for studying early embryonic development. The features of each model highlight the necessity for a complementary approach to enhance the ability to investigate biological processes and refine the strategies to study human development (Carreiro et al. 2021).

6.2.1 Epigenetic Remodeling in Preimplantation Embryo Development

As anticipated above, during the early stages of embryogenesis, the embryo undergoes significant epigenetic remodeling. Epigenetic modifications are heritable changes that alter chromatin structure and gene expression without modifying the DNA sequence (Hamilton 2011). Thanks to these mechanisms, a single genotype expresses multiple phenotypes, such as the cells composing different tissues and organs in an organism or during the commitment and determination of cell fate. They can be passed down to daughter cells and even from parents to offspring through multiple generations via transgenerational epigenetic inheritance (Moelling 2024). Epigenetic changes include an array of different mechanisms, such as DNA methylation at specific cytosine residues, covalent modifications of histone proteins, and expression of non-coding RNAs (ncRNAs) (Kiefer 2007; Mondal and Kanduri 2013).

The embryo epigenetic reprogramming begins after fertilization. For instance, the DNA is globally demethylated and DNA methylation is then re-established according to sex-specific patterns, with the notable exception of imprinted genes which retain their methylation status throughout these processes (Chen et al. 2023). Genomic imprinting is an epigenetic process that leads to monoallelic expression based on parental origin (Crespi 2019). Imprinted genes are crucial in mediating fetal and childhood growth, significantly influencing early developmental patterns and metabolic programming (Plasschaert and Bartolomei 2014). Disruptions in imprinting can result in abnormal phenotypes linked to human diseases, such as Prader–Willi syndrome, an epigenetic variation at the H19 gene, which encodes an imprinted maternally expressed non-coding transcript, and at insulin-like growth factor 2 (*Igf2*), both of which influence birth weight and long-term health risks like obesity and type 2 diabetes (Peters 2014).

In the embryo, epigenetic mechanisms ensure that a single genotype—the one formed by the syngamy of paternal and maternal genomes in the zygote—is progressively committed to express different phenotypes, therefore allowing proper development, influencing processes such as lineage commitment, blastocyst formation, and implantation.

However, epigenetic patterns can be altered by environmental factors, such as nutrition, stress, and pollutants, during critical periods of development (Peral-Sanchez et al. 2021). Consequently, epigenetic modifications are considered key mechanisms through which early-life environmental exposures impact long-term health outcomes. Recent research has provided valuable insights into how specific environmental exposures can affect epigenetic mechanisms and predispose individuals to various chronic diseases (Ho et al. 2012; Vaiserman 2015). For example, maternal consumption of different nutrients during the first trimester of pregnancy has been associated with epigenetic modifications that alter gene expression and may result in increased susceptibility to developing obesity and related metabolic changes, as proposed by the Developmental Origins of Health and Disease (DOHaD) (Barker 2007), also known as fetal programming of adult disease. This theory highlights how environmental conditions, like maternal malnutrition, during early development can influence susceptibility to chronic diseases in adulthood (Bokor et al. 2024). Similarly, exposure to stress or toxicants during critical periods of development can have significant consequences on epigenetic mechanisms and increase the risk for altered physiological and psychiatric outcomes in the offspring (de Magalhães-Barbosa et al. 2022; Dieckmann and Czamara 2024). Despite the intensive research on this topic, the critical window(s) of exposure is not entirely defined, and a possible involvement of the first stages of embryo development has not received full attention. Furthermore, DNA methylation has been mainly considered thus far. However, the interplay of various epigenetic mechanisms is likely to occur.

In the next session, we will describe the dynamics of histone acetylation during the normal preimplantation embryo development in the main studied mammalian species and examine the putative influence that the environment can exert on such dynamics.

6.3 Histone Acetylation Changes During Early Embryo Development

The amino-terminal tails of histone proteins extend outward from the nucleosome and undergo various post-translational covalent modifications, including acetylation, methylation, phosphorylation, biotinylation, sumoylation, crotonylation, and lactylation (Gao et al. 2024; Jenuwein and Allis 2001; Strahl and Allis 2000; Yang et al. 2024). These modifications influence chromatin architecture and DNA accessibility, affecting transcriptional activity and representing important epigenetic mechanisms for regulating gene expression.

Histone acetylation is one of the better-understood modifications and it is mediated by histone acetyltransferases (HATs), which add acetyl groups, and histone deacetylases (HDACs), which remove them. Acetylation occurs predominantly on lysine residues (K) within the N-terminal tail, neutralizing their positive charge and weakening the interaction between the histone proteins and the negatively charged DNA (Bannister and Kouzarides 2011; Millán-Zambrano et al. 2022). Importantly, histone acetylation directly increases the rate of transcription in vitro (Durrin et al. 1991; Protacio et al. 2000). Histone acetylation occurs on several core histones, each contributing uniquely to chromatin regulation and gene expression.

6.3.1 Histone H3

Histone H3 is the most studied and significant target for acetylation, particularly at the K residues H3K9, H3K14, and H3K27, which play essential roles in transcriptional regulation and chromatin structure. Additional acetylation sites, including H3K18, H3K23, H3K56, H3K64, and H3K122, have been noted, although they have been less extensively investigated. A graphical representation of the acetylation dynamics of key residues on histone H3 during early embryo development is provided in Fig. 6.1.

The acetylation of **lysine 9** on histone H3 (H3K9ac) has been a focal point in early embryonic development studies. Research in mice using immunofluorescence or Chip-seq techniques has not demonstrated significant fluctuations in H3K9 acetylation levels during early embryogenesis up to the blastocyst stage (Bošković et al. 2012; Guo et al. 2012; J. Li et al. 2022; Stein et al. 1997).

The bovine model appears to follow a similar pattern, as observed in some immunofluorescence studies (Bošković et al. 2012; Salehi et al. 2020; M. Wang et al. 2017). However, other studies using the same technique have reported a reduction in H3K9ac levels at the 8-cell stage (Wu et al. 2011, 2020), which coincides with the EGA in cattle. This observation suggests a putative role of this residue in epigenetic reprogramming at this critical stage. Notably, the same study also considered H3K9ac starting from the oocytes, observing a decrease in acetylation at the MII stage, prior to fertilization, followed by a rapid increase post-fertilization, pointing out that H3K9ac is erased after meiosis resumption (Wu et al. 2020). A similar pattern was also reported for H3K18ac in the same study (Wu et al. 2020).

Only one study has investigated the 4-cell stage in porcine embryos, which corresponds to the EGA in pigs. This study reported a reduction in H3K9ac levels, similar to what was observed in cattle oocytes. However, conclusions should be drawn carefully since the antibody used also stains H3K14ac, implying that the observed signal may overlap with the detection of other residues (Nguyen et al. 2011).

Research on histone H3 **lysine 14** acetylation (H3K14ac) has revealed fluorescence signal reduction at the 4-cell stage in porcine embryos, coinciding with the EGA (Liu et al. 2012; Sun et al. 2020), such as observed for H3K9 in the cattle, again suggesting a dynamic epigenetic modulation specific to this developmental transition. The observation of high global histone acetylation levels before the EGA, at a time of transcriptional silencing, could be reconducted to this modification's

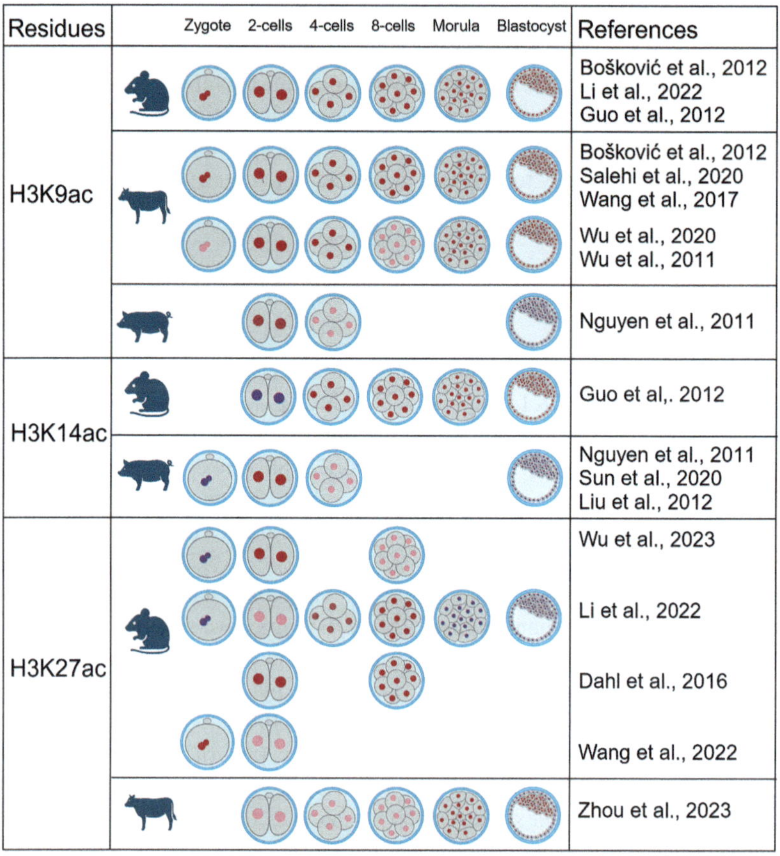

Fig. 6.1 Dynamic changes in the acetylation of histone H3 during early embryo development. A summary of acetylation patterns at different lysine residues on histone H3 (H3K9ac, H3K14ac, and H3K27ac) across various stages of preimplantation development in mouse, bovine, and porcine embryos is provided. The chart illustrates the presence and intensity of acetylation, represented by shades ranging from pink (least acetylated) to red and purple (most acetylated) at each residue from the zygote to the blastocyst stage. The stages analyzed include zygote, 2-cell, 4-cell, 8-cell, morula, and blastocyst. References for each residue are indicated on the right. (Created with BioRender)

multifaceted functions. Indeed, histone acetylation not only facilitates gene transcription by promoting the accessibility of genetic information to transcription factors, but it also has a significant role in regulating chromatin structure (Shahbazian and Grunstein 2007). An increase in the fluorescent signal was then detected at the blastocyst stage in porcine oocytes (Liu et al. 2012; Nguyen et al. 2011; Sun et al. 2020).

This residue doesn't reveal significant fluctuations in studies conducted on mouse and bovine embryos (Carey et al. 2015; Stein et al. 1997; P. Zhao et al. 2019). In contrast, a study in mice examining stages from the 2-cells to the blastocyst reported a substantially stronger acetylation signal at the 2-cell stage that slightly decreased thereafter and was stably maintained up to the blastocyst stage (Guo et al. 2012). Unfortunately, many studies have focused on a single developmental stage, failing to highlight dynamic changes that might occur throughout the preimplantation period. Further investigation is therefore required to resolve these discrepancies and better understand the regulation of H3K14ac during early development.

Patterns of acetylation of **lysine 27** (H3K27ac) differ across the same species. In fact, in mice, conflicting findings have been reported: one study which used ChIP-seq technique observed reduced acetylation at the 8-cell stage (Wu et al. 2023), while another study, also using ChIP-seq, reported a decrease as early as the 2-cell stage (Li et al. 2022), corresponding to EGA in the mice. A yet additional study using μChIP-seq reported a marked increase in the genomic coverage of H3K27ac from oocytes to the 2-cell stage, but no significant changes were observed between the 2-cell and 8-cell stages (Dahl et al. 2016). Conversely, a different study using immunofluorescence revealed a dramatic decrease in H3K27ac levels from the prophase I stage to the MII in mice (Wang et al. 2022). These levels were rapidly restored in zygotes to values comparable to those observed in prophase I oocytes, suggesting that de novo acetylation occurs shortly after fertilization. Notably, acetylation levels decreased again at the early 2-cell stage. These findings were further corroborated using the CUT&RUN technique, which demonstrated noncanonical broad H3K27ac domains in prophase I oocytes and in zygotes, indicating chromatin hyperacetylation during these stages (M. Wang et al. 2022). Notably, in a study in humans using ChIP-seq on patients undergoing intracytoplasmic sperm injection (ICSI) to overcome infertility, genomic regions marked by broad H3K27ac domains are narrowed down during development. At the 8-cell stage and onward, most of the broad H3K27ac domains (>10 kb) detected in 2-cell embryos disappear or transit to narrow H3K27ac peaks (Wu et al. 2023).

A recent study (Zhou et al. 2023) reported increased H3K27ac levels in cattle from the morula stage onward, with pronounced acetylation in both the ICM and TE at the blastocyst stage, suggesting its role in lineage specification.

Other lysine residues, such as H3K18, H3K23, H3K56, H3K64, and H3K122, remain understudied, with limited and even more inconsistent data reported across species and developmental stages. An immunofluorescence signal of histone H3 acetylation at **lysine 18** (H3K18ac) was erased after re-entry into the meiotic cell cycle in bovine oocytes and then newly established in the zygotes. Then, it significantly decreased at the 8-cell stage, suggesting a transient reduction in acetylation during EGA (Wu et al. 2011, 2020). This trend reverses by the morula stage, with a significant increase in staining intensity. Interestingly, at the blastocyst stage, H3K18ac displays different intensities according to the localization, with the acetylation signal being less intense in the ICM compared to the TE cells (Wu et al. 2011). This difference likely reflects the contrasting states of these two lineages: the TE is already differentiated and committed to forming the extraembryonic structures, whereas the ICM remains in a pluripotent state, capable of giving rise to all embryonic tissues. While these observations suggest a potential link between H3K18ac and lineage specification, the underlying mechanisms and functional implications remain unclear. A similar pattern has also been observed for some residues of histone H4, as reported below.

The acetylation of **lysine 23** on histone H3 (H3K23ac) has shown contrasting patterns in different species, reflecting variability in the epigenetic regulation during preimplantation development. In mouse embryos, H3K23ac was observed to be distributed uniformly throughout the nucleus, with no apparent stage-specific changes (Stein et al. 1997). Conversely, H3K23ac is absent in porcine embryos during the 1-, 2-, 4-, and 8-cell stages but becomes enriched at the blastocyst stage (Lin et al. 2020). These discrepancies highlight potential species-specific regulatory roles for H3K23ac in embryogenesis. Notably, its enrichment at the blastocyst stage in porcine embryos suggests a potential link to the emergence of differentiated lineages. Still, in this case, more studies are needed to clarify its function and relevance to lineage-specific processes.

The acetylation patterns of **lysine 56, 64, and 122** on histone H3 (H3K56ac, H3K64ac, and H3K122ac) remain poorly characterized, with limited studies reporting conflicting findings. In a mouse model, H3K56ac and H3K64ac were consistently present throughout preimplantation, whereas H3K122ac, although detected from prophase I to 8-cell stage, was observed at low levels before implantation. These findings suggest a more prominent role for H3K56ac and H3K64ac than H3K122ac in mouse preimplantation development (Ziegler-Birling et al. 2016). In contrast, a study in human embryos demonstrated that all three modifications are detectable throughout the preimplantation stages, albeit at varying levels (Wang et al. 2020). H3K64ac showed relatively stronger signals, while H3K122ac levels were much weaker. Interestingly, H3K56ac exhibited a peak at the 2-cell stage. These interspecies differences underscore the need for standardized methodologies to elucidate the specific roles of these acetylation marks.

6.3.2 Histone H4

Histone H4 acetylation at **lysine 5, 8, 12, and 16** (H4K5ac, H4K8ac, H4K12ac, and H4K16ac) plays a crucial role in chromatin remodeling and transcriptional regulation during the oocyte-to-zygote transition and the early embryo development. Despite their importance, studies on these residues remain limited, with findings revealing complex and stage-specific patterns influenced by environmental stimuli and experimental conditions, similar to those emerged from the survey of H3 acetylation patterns. A graphical representation of acetylation dynamics at key residues on histone H4 during early embryo development is provided in Fig. 6.2.

A study on bovine embryos has shown similar dynamics for H4K5ac and H4K8ac during early development (Wu et al. 2020). Specifically, the fluorescent signal of H4K5ac and H4K8ac is erased after meiosis resumption and during fertilization, while they are re-established on the male and female chromatin at 4 and 8 hrs post-fertilization, respectively (Ma and Schultz 2008; Wu et al. 2020). Notably, in the same study, histone methylations exhibit no significant changes during meiosis. As

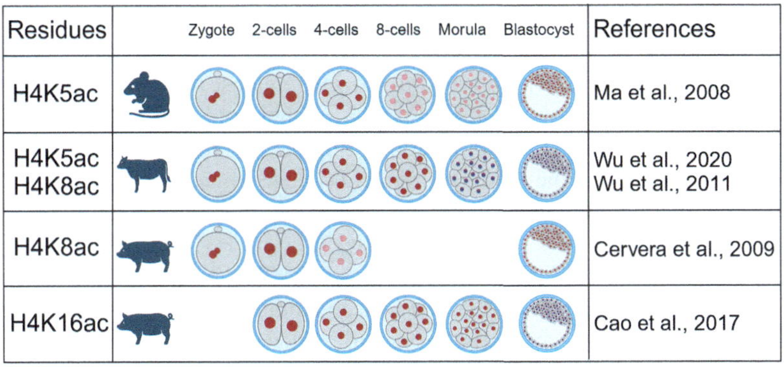

Fig. 6.2 Dynamic changes in the acetylation of histone H4 during early embryo development. A summary of acetylation patterns at different lysine residues on histone H4 (H4K5ac, H4K8ac, and H4K16ac) across various stages of preimplantation development in mouse, bovine, and porcine embryos is provided. The chart illustrates the presence and intensity of acetylation, represented by shades ranging from pink (least acetylated) to red and purple (most acetylated) at each residue from the zygote to the blastocyst stage. The stages analyzed include zygote, 2-cell, 4-cell, 8-cell, morula, and blastocyst. References for each residue are indicated on the right. (Created with BioRender)

preimplantation development progresses, the signal of H4K5ac and H4K8ac increases significantly after the morula stage (Wu et al. 2011), as observed in another study (Wu et al. 2020). An independent study has also confirmed the deacetylation of MII bovine oocytes (Lodde et al. 2021).

In mice, a decrease in the H4K5ac fluorescence signal was observed in MII oocytes, followed by a re-establishment in the zygote and a gradual decrease at the 8-cell and morula stages, followed by a resurgence at the blastocyst stage (Ma and Schultz 2008). The H4K5ac and H4K8ac levels of the blastomeres during interphase seem to be more pronounced at the nuclear periphery from the 2-cell to the 8-cell stages (Stein et al. 1997; Wu et al. 2020). Interestingly, during mitosis, the H4K8ac signal becomes particularly intense at the chromosome periphery (Wu et al. 2020). These observations highlight the potential regulatory roles of H4K5ac and H4K8ac in key transitions, such as EGA and chromatin organization. However, the specific functional implications of their localization at the nuclear and chromosome peripheries remain to be elucidated.

A study conducted in 2007 on mouse embryos reported a homogeneous immunofluorescence staining pattern of acetylated H4 (AcH4) from the zygote to the blastocyst stage in both in vivo and IVF-derived embryos (Huang et al. 2007). Conversely, in 2004, another study that included the oocytes' prophase I and MII stages described distinct patterns of hyperacetylated H4 (Sarmento et al. 2004). In immature prophase I oocytes, hyperacetylation was confined to condensing chromatin surrounding the nucleolar-like body, while at metaphase, global levels of acetylation dramatically declined, with almost no staining on egg chromatin (Sarmento et al. 2004). However, strong AcH4 signals were observed in interphase blastomeres. At the blastocyst stage, AcH4 staining was less intense in the ICM compared to the TE. We observed similar patterns in bovine blastocysts obtained by IVP and immunostained with antibodies directed against the acetylated forms of H4K8 and, to a lower extent, H4K12 (Fig. 6.3, unpublished data). The difference in acetylated signal intensity may reflect the commitment of the TE toward a differentiated state compared to the ICM's pluripotent state (Sarmento et al. 2004).

Notably, the residues seem particularly sensitive to environmental stimuli. In mouse zygotes, the signals of H4K8ac, H4K12ac, and H4K16ac are significantly reduced 3 h after somatic cell nuclear transfer (SCNT) compared to 3 h after intracytoplasmic sperm injection (ICSI), where acetylation levels are higher (Wang et al. 2007). This discrepancy may arise from different requirements between somatic cells, bearing histones. and the spermatozoa, bearing protamines, to reorganize the nucleosomes. Additionally, it may suggest that external factors, such as the fertilization mode, influence these residues' acetylation patterns. In line with this hypothesis, we repeatedly observed that the oocytes' acetylation at specific residues is affected by the mode of maturation. Indeed, acetylation at the H4K16 is retained in

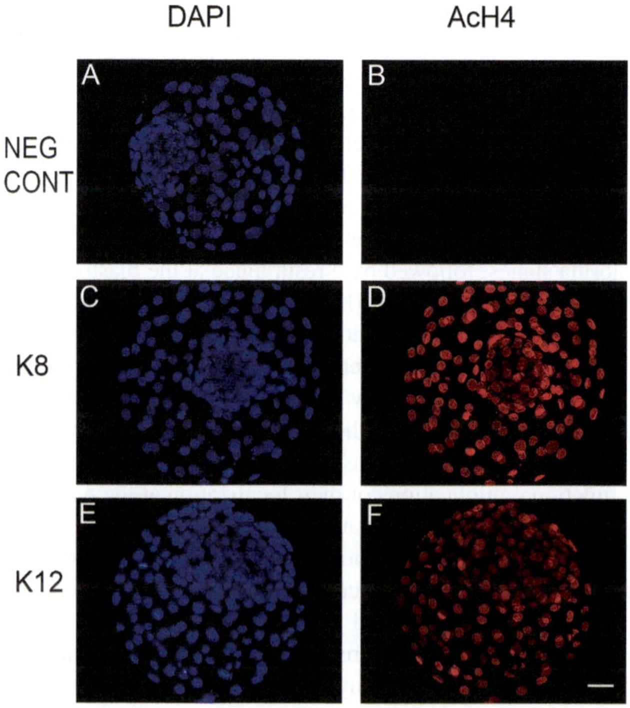

Fig. 6.3 Acetylation of H4K8 and H4K12 in bovine blastocysts. Representative images of hatched/hatching blastocysts stained with DAPI (blue, A, C, E), negative control (B), AcH4K8 (D), and AcH4K12 (F) in red. Bar = 50 μm. Embryos were fixed and stained using specific antibodies, as described in (Franciosi et al. 2012; Lodde et al. 2021) and show a lower immunostaining intensity at the ICM

in vivo ovulated horse oocytes, but it becomes undetectable in in vitro matured ones (Franciosi et al. 2017a, b). Notably, other H4 residues, K8 and K12, were not affected (Franciosi et al. 2012), probably indicating a different sensibility of residue-specific histone acetyl-transferases/deacetylases to environmental conditions.

In porcine embryos, at the time of EGA, which occurs at the 4-cell stage, a marked reduction in the H4K8ac signal has been observed (Cervera et al. 2009). However, this acetylation level recovers at the blastocyst stage, aligning with developmental requirements for chromatin remodeling and lineage specification.

Studies on H4K12ac and H4K16ac conducted in mouse, bovine, and porcine embryos did not identify significant differences at key developmental stages (Cao et al. 2019; Huang et al. 2016; Stein et al. 1997; Xiao et al. 2022). Interestingly, an increase in H4K16ac fluorescence was only observed in one study conducted in the porcine blastocyst (Cao et al. 2017). Nonetheless, the limited focus on these residues highlights the need for more comprehensive research to uncover their potential regulatory roles during embryogenesis.

Interestingly, among the lysine residues of histone H4, K5 and K8 emerge as the most dynamically regulated and consistently implicated in acetylation patterning during early embryo development. Their pronounced stage-specific acetylation patterns, particularly during critical transitions, such as the MII stage and EGA, suggest a central role in orchestrating chromatin remodeling and transcriptional activation. However, future studies are essential to clarify whether the observed acetylation dynamics reflect universal mechanisms or species-specific variations and elucidate the interplay between different acetylation marks in shaping the embryonic epigenetic landscape.

6.3.3 *Histones H2A and H2B*

While much attention has been given to histones H3 and H4 modifications, fewer studies have focused on histone H2A and H2B during early embryonic development. Among the limited studies available, an early investigation conducted in 1997 on mouse embryos revealed that histone H2B was uniformly distributed in the nuclei at the 2-cell stage. In contrast, the single acetylated form of H2A (H2A.Ac5) localization showed enrichment at the nuclear periphery in 2-cell embryos (Stein et al. 1997).

More recently, another study compared the dynamics of H2A.Z acetylation (H2A.Zac) in bovine and mouse embryos, revealing differences between these species (Bošković et al. 2012). In mice, H2A.Zac was undetectable in embryonic chromatin at the early 2-cell stage and began to re-incorporate it at very low levels only in the late 2-cell stage. In contrast, in bovine embryos, H2A.Zac was strongly enriched on the maternal chromatin of zygotes and subsequently exhibited a uniform distribution across all nuclei during later stages (Bošković et al. 2012).

These findings suggest that the dynamics of H2A.Z acetylation may not be conserved across mammals and may reflect species-specific regulatory mechanisms during early development. However, the scarcity of studies on H2A and H2B acetylation underscores the need for further research to elucidate their precise roles in chromatin remodeling and EGA.

6.4 Reversible Histone Acetylation/Deacetylation and the Adaptation to Environmental Conditions

During preimplantation embryo development, cells use the energetic substrates required for cell division, growth, and differentiation that are provided through the maternal reproductive organs (ovarian follicle for the oocyte, oviduct, and uterus for the embryo) or in the culture medium. The embryo metabolism is an intricate and dynamic process that varies according to the stages of oocyte/embryo growth and it is crucial for successful development (Bruna de Lima et al. 2023). For instance, during EGA, the embryo switches from anaerobic to aerobic metabolism, consequently changing substrate requirement for energy production (Leese 1995; Martin 2000).

A recapitulation of key biochemistry concepts is provided to better understand the mechanisms behind embryo metabolism and how the maternal/cultural environment may affect it.

6.4.1 Embryo Metabolism

Glycolysis breaks down glucose to produce 2 molecules of ATP, pyruvate and lactate. Pyruvate produced during glycolysis can be oxidized into acetyl-CoA and enter the tricarboxylic cycle (TCA), also known as the citric acid or Krebs cycle. The TCA cycle is a mitochondrial pathway essential for energy production and biosynthetic processes, also during embryonic development (Chakrabarty and Chandel 2022; Sharpley et al. 2021). It takes place in the mitochondrial matrix and comprises 8 enzymes, except the outlier succinate dehydrogenase, which is related to the respiratory chain on the inner mitochondrial membrane. The TCA also plays a role in replenishing precursors for the storage form of fuels such as amino acids and cholesterol (Cavalcanti et al. 2014). The TCA cycle does not require oxygen directly; however, oxygen is needed for the final phase of aerobic respiration, the oxidative phosphorylation, which produces most of the ATP. The presence of coenzymes like Nicotinamide adenine dinucleotide (oxidized, NAD+) and Flavin adenine dinucleotide (FAD) controls and regulates the Krebs cycle, while high concentrations of Nicotinamide adenine dinucleotide (reduced, NADH) inhibit it (Alabduladhem and Bordoni 2024).

Before entering the TCA cycle, organic molecules, like carbohydrates, lipids, and amino acids, are transformed directly or through a further transformation in acetyl-CoA, a molecule formed by an acetyl group (CH3CO-) and an acyl transporter called coenzyme A. The acetyl group is then oxidized, and the energy obtained is stored as ATP in association with oxidative phosphorylation (Martínez-Reyes and Chandel 2020). However, the primary source of acetyl-CoA remains glycolysis. In conclusion, the TCA cycle integrates glucose, pyruvate, lactate, and amino acids into a cohesive energy production system vital for embryonic development (Fig. 6.4). Its ability to maintain energy homeostasis and provide biosynthetic precursors is essential for the rapid proliferation and differentiation of embryonic cells (Arnold and Finley 2023).

During the first cleavages up to the morula, the embryo uses mainly pyruvate and partially lactate to generate energy in most species, including humans (Lee and Rinaudo 2024). Glycolysis is also present during these stages, but less active (Lee and Rinaudo 2024). Initially the energy requirements are low, and the metabolism is mainly sustained by transcripts and proteins produced and stored during oocyte

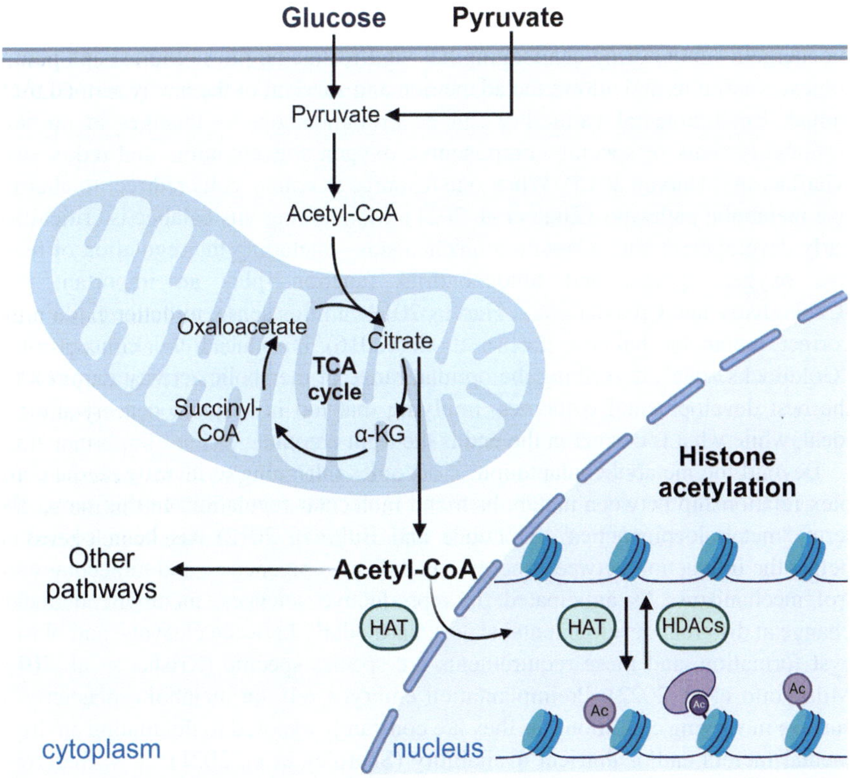

Fig. 6.4 Proposed model of embryo metabolism affecting histone acetylation. Glucose metabolism generates pyruvate, which enters the mitochondria and is converted into acetyl-CoA via the tricarboxylic acid (TCA) cycle. Acetyl-CoA is transported to the nucleus where it is used as a substrate for histone acetylation mediated by histone acetyltransferases (HATs). Histone acetylation levels are dynamically regulated by the opposing actions of HATs and histone deacetylases (HDACs), influencing chromatin structure and gene expression. Created with BioRender

growth and maturation. At this point, the high ATP:ADP ratio maintains the inhibition of phosphofructokinase, limiting glucose oxidation. As the embryo develops, glycolysis increases, and glucose is transported into the blastomeres by the facilitated glucose transporters GLUT1 and GLUT3 (de Lima et al. 2020; Pantaleon et al. 1997). Hence, as the embryo reaches the blastocyst stage, the glucose becomes the dominant substrate, most likely to face the energy demands of the Na^+/K^+ ATPase pump necessary to form the blastocoel cavity (Barrie 2020; Leese et al. 2016). In addition to the increase in glycolysis, there is also an increase in oxygen consumption. In fact, the TE relies on oxidative phosphorylation, while ICM remains glycolytic in mammals (Lee and Rinaudo 2024).

6.4.2 Environmental Conditions and Histone Acetylation

The embryo metabolism is highly sensitive to environmental changes, which makes bioenergetics monitoring challenging but vital for facing physiological and pathological conditions and allows the adaptation and survival of the newly formed individual. Environmental variability can be brought about by changes in nutrient availability (bulk or specific components), oxygen concentration, and redox state (Gardner and Harvey 2015). When one resource is scarce, cells redirect to alternative metabolic pathways (Zhao et al. 2021). Maintaining this balance is critical for early development, and adaptation mechanisms—including the regulation of reactive oxygen species and mitochondrial function—play an important role (Chakrabarty and Chandel 2022; Harvey 2019). In this sense, to better explain the correct metabolic balance, Leese et al. (2016) introduced the concept of a "Goldilocks zone", describing the optimal range of metabolic activity required for the best developmental outcomes, implying that too little and too much are not ideal, while what is correct in the end is the right amount.

Beyond the metabolic adaptation, it becomes intriguing to investigate the complex relationship between metabolism and molecular regulation. In this sense, the term "metaboloepigenetics" (Donohoe and Bultman 2012) has been coined to define the interaction between energy metabolism, epigenetics, and molecular control mechanisms. As anticipated, in reproductive sciences, metabolic demands change at different developmental stages, particularly between cleavage and blastocyst formation, and these requirements are species-specific (Krisher et al. 2015; Milazzotto et al. 2022). Preimplantation embryos rely on metabolic plasticity to survive in varying conditions, as they are constantly exposed to fluctuating environmental factors and/or nutrient availability (Sharpley et al. 2021). It is, therefore, conceivable that changes in energetic supplies, through the maternal diet or the culture medium, determine alterations of the ideal TCA substrate/product concentration that, in turn, can be "sensed" by the cell and used to install adaptive mechanism(s) (Zhang et al. 2021). Given that acetyl groups are the natural substrate of histone acetylation/deacetylation enzymes, the addition/removal of such groups in response to increased or decreased availability may represent a direct way to introduce nutrient-related epigenetic changes (Fig. 6.4) (Charidemou and Kirmizis 2024). In this scenario, establishing new acetylation patterns dependent on nutrient availability would serve the double purpose of operating changes in the chromatin landscape that allow the immediate adaptation and survival of the developing embryo while also introducing an epigenetic mark that can be passed down to the daughter cells. Despite histone acetylation being reversible, the epigenetic signature might become fixated in the genome through additional mechanisms, especially when cell plasticity is progressively lost. A similar mode of action can be hypothesized for other histone covalent modifications, such as lactylation (Yang et al. 2024).

This metabolic flexibility allows embryos to adjust their energy production and use according to the available resources, enabling them to maintain cellular functions and development under stress. However, while this adaptability is essential for immediate survival and proper development, it may have long-term consequences, potentially influencing the health and viability of the embryo as it progresses through later stages of development and into adulthood (Sharpley et al. 2021). The ability to shift metabolic pathways and adjust cellular functions during early development can leave lasting imprints on cellular behavior, potentially affecting key developmental outcomes, including gene expression and cell fate decisions. The change can be passed to daughter cells, and the altered phenotype can be transmitted transgenerationally through epigenetic inheritance (Perez and Lehner 2019).

6.5 Conclusions

This review article summarized the main patterns of histone acetylation during the early stages of embryo development, a time of intense epigenetic remodeling (Wilkinson et al. 2023). Our survey highlighted that, despite a relatively conserved program of totipotency and pluripotency among mammals, few reports agreed on common dynamics of histone acetylation. At the same time, several discrepancies were found between the different reports, also when the same species was considered. While variations of the experimental setting and methodologies might partially be held accountable for lack of reproducibility, an emerging theme of our analysis was the capacity of the early embryo to fine-tune its metabolism to adapt to changing environments. Notably, histone acetylation participates in such regulation by being the acceptor/donor of acetyl groups—that ultimately are a function of nutrients availability—and as an epigenetic modifier (Kumar et al. 2021).

In this scenario, the lack of a strict consensus is less appalling, as variations in nutrient availability from the maternal diet or the culture conditions are expected to introduce changes at which the developing embryos are ready to react in light of the inherent plasticity.

If this hypothesis is confirmed, it will be essential to ascertain how permanent these modifications can be and if there are safe ranges in which the embryo tolerates exposure to nutrient defect or excess without impacting the long-term epigenetic reprogramming.

Acknowledgments This work was supported by: 'Programma di Ricerca Scientifica di Rilevante Interesse Nazionale-PRIN 2017' by Ministero dell'Istruzione dell'Università e della Ricerca-MIUR, grant number 20172N2WL3_002; Piano di Sostegno alla Ricerca UNIMI 2022: Linea 2—Azione A. F.F.F. was supported by fellowships from 'Programma di Ricerca Scientifica di Rilevante Interesse Nazionale-PRIN 2017' by Ministero dell'Istruzione dell'Università e della Ricerca-MIUR, grant number 20172N2WL3_002 and Italian Ministry of University and Research (MUR), PRIN2020, No. 20209L8BN4 (InfinitEGG).

Conflict of Interest Statement The authors declare no conflicts of interest.

Author Contributions Conceptualization: G.M. and F.F.; Methodology: G.M.; Writing—Original Draft: G.M. and F.F.; Writing—Review & Editing: F.F.F., F.M., A.M.L., and V.L.; Visualization: G.M.; Supervision: F.M., A.M.L., and V.L..; Funding Acquisition: F.M., A.M.L., and F.F.

References

Aguila L, Osycka-Salut C, Treulen F, Felmer R (2022) Pluripotent core in bovine embryos: a review. Animals (Basel) 12(8). https://doi.org/10.3390/ani12081010

Alabduladhem TO, Bordoni B (2024) Physiology, krebs cycle. In: StatPearls

Alberio R (2020) Regulation of cell fate decisions in early mammalian embryos. Annu Rev Anim Biosci 8:377–393. https://doi.org/10.1146/annurev-animal-021419-083841

Arnold PK, Finley LWS (2023) Regulation and function of the mammalian tricarboxylic acid cycle. J Biol Chem 299(2):102838. https://doi.org/10.1016/j.jbc.2022.102838

Bannister AJ, Kouzarides T (2011) Regulation of chromatin by histone modifications. Cell Res 21(3):381–395. https://doi.org/10.1038/cr.2011.22

Barker DJ (2007) The origins of the developmental origins theory. J Intern Med 261(5):412–417. https://doi.org/10.1111/j.1365-2796.2007.01809.x

Barrie A (2020) Embryo development: from zygote to blastocyst. In: Allahbadia GN, Ata B, Lindheim SR, Woodward BJ, Bhagavath B (eds) Textbook of assisted reproduction. Springer, Singapore

Berg DK, Smith CS, Pearton DJ, Wells DN, Broadhurst R, Donnison M, Pfeffer PL (2011) Trophectoderm lineage determination in cattle. Dev Cell 20(2):244–255. https://doi.org/10.1016/j.devcel.2011.01.003

Blakeley P, Fogarty NM, Del Valle I, Wamaitha SE, Hu TX, Elder K et al (2015) Defining the three cell lineages of the human blastocyst by single-cell RNA-seq. Development 142(20):3613. https://doi.org/10.1242/dev.131235

Bokor S, Csölle I, Felső R, Vass RA, Funke S, Ertl T, Molnár D (2024) Dietary nutrients during gestation cause obesity and related metabolic changes by altering DNA methylation in the offspring. Front Endocrinol (Lausanne) 15:1287255. https://doi.org/10.3389/fendo.2024.1287255

Bošković A, Bender A, Gall L, Ziegler-Birling C, Beaujean N, Torres-Padilla ME (2012) Analysis of active chromatin modifications in early mammalian embryos reveals uncoupling of H2A.Z acetylation and H3K36 trimethylation from embryonic genome activation. Epigenetics 7(7):747–757. https://doi.org/10.4161/epi.20584

Bruna de Lima C, Cristina Dos Santos É, Sirard MA (2023) DOHaD: a menagerie of adaptations and perspectives: the interplay between early embryo metabolism and mitoepigenetic programming of development. Reproduction 166(1):F15–F26. https://doi.org/10.1530/REP-22-0424

Cao Z, Wu R, Gao D, Xu T, Luo L, Li Y et al (2017) Maternal histone acetyltransferase KAT8 is required for porcine preimplantation embryo development. Oncotarget 8(52):90250–90261. https://doi.org/10.18632/oncotarget.21657

Cao Z, Xu T, Tong X, Zhang D, Liu C, Wang Y et al (2019) HASPIN kinase mediates histone deacetylation to regulate oocyte meiotic maturation in pigs. Reproduction 157(6):501–510. https://doi.org/10.1530/REP-18-0447

Carey TS, Cao Z, Choi I, Ganguly A, Wilson CA, Paul S, Knott JG (2015) BRG1 governs nanog transcription in early mouse embryos and embryonic stem cells via antagonism of histone H3 lysine 9/14 acetylation. Mol Cell Biol 35(24):4158–4169. https://doi.org/10.1128/MCB.00546-15

Carreiro LE, Dos Santos GS, Luedke FE, Goissis MD (2021) Cell differentiation events in pre-implantation mouse and bovine embryos. Anim Reprod 18(4):e20210054. https://doi.org/10.1590/1984-3143-AR2021-0054

Cavalcanti JH, Esteves-Ferreira AA, Quinhones CG, Pereira-Lima IA, Nunes-Nesi A, Fernie AR, Araújo WL (2014) Evolution and functional implications of the tricarboxylic acid cycle as revealed by phylogenetic analysis. Genome Biol Evol 6(10):2830–2848. https://doi.org/10.1093/gbe/evu221

Cervera RP, Martí-Gutiérrez N, Escorihuela E, Moreno R, Stojkovic M (2009) Trichostatin A affects histone acetylation and gene expression in porcine somatic cell nucleus transfer embryos. Theriogenology 72(8):1097–1110. https://doi.org/10.1016/j.theriogenology.2009.06.030

Chakrabarty RP, Chandel NS (2022) Beyond ATP, new roles of mitochondria. Biochem (Lond) 44(4):2–8. https://doi.org/10.1042/bio_2022_119

Charidemou E, Kirmizis A (2024) A two-way relationship between histone acetylation and metabolism. Trends Biochem Sci 49(12):1046–1062. https://doi.org/10.1016/j.tibs.2024.10.005

Chen AE, Egli D, Niakan K, Deng J, Akutsu H, Yamaki M et al (2009) Optimal timing of inner cell mass isolation increases the efficiency of human embryonic stem cell derivation and allows generation of sibling cell lines. Cell Stem Cell 4(2):103–106. https://doi.org/10.1016/j.stem.2008.12.001

Chen Y, Wang L, Guo F, Dai X, Zhang X (2023) Epigenetic reprogramming during the maternal-to-zygotic transition. MedComm *4*(4):e331. https://doi.org/10.1002/mco2.331

Coticchio G, Cimadomo D, Cermisoni GC, Rienzi L, Papaleo E, Ubaldi FM et al (2023a) The first mitotic division: a perilous bridge connecting the zygote and the early embryo. Hum Reprod 38(6):1019–1027. https://doi.org/10.1093/humrep/dead067

Coticchio G, Ezoe K, Lagalla C, Zacà C, Borini A, Kato K (2023b) The destinies of human embryos reaching blastocyst stage between DAY 4 and day 7 diverge as early as fertilization. Hum Reprod 38(9):1690–1699. https://doi.org/10.1093/humrep/dead136

Crespi BJ (2019) Why and how imprinted genes drive fetal programming. Front Endocrinol (Lausanne) 10:940. https://doi.org/10.3389/fendo.2019.00940

Cui W, Mager J (2018) Transcriptional regulation and genes involved in first lineage specification during preimplantation development. Adv Anat Embryol Cell Biol 229:31–46. https://doi.org/10.1007/978-3-319-63187-5_4

Dahl JA, Jung I, Aanes H, Greggains GD, Manaf A, Lerdrup M et al (2016) Broad histone H3K4me3 domains in mouse oocytes modulate maternal-to-zygotic transition. Nature 537(7621):548–552. https://doi.org/10.1038/nature19360

Davenport KM, Ortega MS, Johnson GA, Seo H, Spencer TE (2023) Review: implantation and placentation in ruminants. Animal 17(Suppl 1):100796. https://doi.org/10.1016/j.animal.2023.100796

de Lima CB, Dos Santos ÉC, Ispada J, Fontes PK, Nogueira MFG, Dos Santos CMD, Milazzotto MP (2020) The dynamics between in vitro culture and metabolism: embryonic adaptation to environmental changes. Sci Rep 10(1):15672. https://doi.org/10.1038/s41598-020-72221-1

de Magalhães-Barbosa MC, Prata-Barbosa A, da Cunha AJLA (2022) Toxic stress, epigenetics and child development. J Pediatr 98(Suppl 1):S13–S18. https://doi.org/10.1016/j.jped.2021.09.007

De Paepe C, Cauffman G, Verloes A, Sterckx J, Devroey P, Tournaye H et al (2013) Human trophectoderm cells are not yet committed. Hum Reprod 28(3):740–749. https://doi.org/10.1093/humrep/des432

Dieckmann L, Czamara D (2024) Epigenetics of prenatal stress in humans: the current research landscape. Clin Epigenetics 16(1):20. https://doi.org/10.1186/s13148-024-01635-9
Donohoe DR, Bultman SJ (2012) Metaboloepigenetics: interrelationships between energy metabolism and epigenetic control of gene expression. J Cell Physiol 227(9):3169–3177. https://doi.org/10.1002/jcp.24054
Durrin LK, Mann RK, Kayne PS, Grunstein M (1991) Yeast histone H4 N-terminal sequence is required for promoter activation in vivo. Cell 65(6):1023–1031. https://doi.org/10.1016/0092-8674(91)90554-c
Fogarty NME, McCarthy A, Snijders KE, Powell BE, Kubikova N, Blakeley P et al (2017) Genome editing reveals a role for OCT4 in human embryogenesis. Nature 550(7674):67–73. https://doi.org/10.1038/nature24033
Franciosi F, Lodde V, Goudet G, Duchamp G, Deleuze S, Douet C et al (2012) Changes in histone H4 acetylation during in vivo versus in vitro maturation of equine oocytes. Mol Hum Reprod 18(5):243–252. https://doi.org/10.1093/molehr/gar077
Franciosi F, Goudet G, Tessaro I, Papillier P, Dalbies-Tran R, Reigner F et al (2017a) In vitro maturation affects chromosome segregation, spindle morphology and acetylation of lysine 16 on histone H4 in horse oocytes. Reprod Fertil Dev 29(4):721–730. https://doi.org/10.1071/RD15350
Franciosi F, Tessaro I, Dalbies-Tran R, Douet C, Reigner F, Deleuze S et al (2017b) Analysis of chromosome segregation, histone acetylation, and spindle morphology in horse oocytes. J Vis Exp 123. https://doi.org/10.3791/55242
Gao D, Li C, Liu SY, Xu TT, Lin XT, Tan YP et al (2024) P300 regulates histone crotonylation and preimplantation embryo development. Nat Commun 15(1):6418. https://doi.org/10.1038/s41467-024-50731-0
Gardner DK, Harvey AJ (2015) Blastocyst metabolism. Reprod Fertil Dev 27(4):638–654. https://doi.org/10.1071/RD14421
Gerri C, McCarthy A, Alanis-Lobato G, Demtschenko A, Bruneau A, Loubersac S et al (2020) Initiation of a conserved trophectoderm program in human, cow and mouse embryos. Nature 587(7834):443–447. https://doi.org/10.1038/s41586-020-2759-x
Guo L, Qi ST, Miao DQ, Liang XW, Li H, Ou XH et al (2012) The roles of parathyroid hormone-like hormone during mouse preimplantation embryonic development. PLoS One 7(7):e40528. https://doi.org/10.1371/journal.pone.0040528
Hamilton JP (2011) Epigenetics: principles and practice. Dig Dis 29(2):130–135. https://doi.org/10.1159/000323874
Harvey AJ (2019) Mitochondria in early development: linking the microenvironment, metabolism and the epigenome. Reproduction 157(5):R159–R179. https://doi.org/10.1530/REP-18-0431
Ho SM, Johnson A, Tarapore P, Janakiram V, Zhang X, Leung YK (2012) Environmental epigenetics and its implication on disease risk and health outcomes. ILAR J 53(3–4):289–305. https://doi.org/10.1093/ilar.53.3-4.289
Huang JC, Lei ZL, Shi LH, Miao YL, Yang JW, Ouyang YC et al (2007) Comparison of histone modifications in in vivo and in vitro fertilization mouse embryos. Biochem Biophys Res Commun 354(1):77–83. https://doi.org/10.1016/j.bbrc.2006.12.163
Huang J, Zhang H, Yao J, Qin G, Wang F, Wang X et al (2016) BIX-01294 increases pig cloning efficiency by improving epigenetic reprogramming of somatic cell nuclei. Reproduction 151(1):39–49. https://doi.org/10.1530/REP-15-0460
Huber E, Notaro US, Recce S, Rodríguez FM, Ortega HH, Salvetti NR, Rey F (2020) Fetal programming in dairy cows: effect of heat stress on progeny fertility and associations with the hypothalamic-pituitary-adrenal axis functions. Anim Reprod Sci 216:106348. https://doi.org/10.1016/j.anireprosci.2020.106348
Jenuwein T, Allis CD (2001) Translating the histone code. Science 293(5532):1074–1080. https://doi.org/10.1126/science.1063127
Kagawa H, Javali A, Khoei HH, Sommer TM, Sestini G, Novatchkova M et al (2022) Human blastoids model blastocyst development and implantation. Nature 601(7894):600–605. https://doi.org/10.1038/s41586-021-04267-8

Kidder GM, Watson AJ (2005) Roles of Na,K-ATPase in early development and trophectoderm differentiation. Semin Nephrol 25(5):352–355. https://doi.org/10.1016/j.semnephrol.2005.03.011

Kiefer JC (2007) Epigenetics in development. Dev Dyn 236(4):1144–1156. https://doi.org/10.1002/dvdy.21094

Kim J, Chu J, Shen X, Wang J, Orkin SH (2008) An extended transcriptional network for pluripotency of embryonic stem cells. Cell 132(6):1049–1061. https://doi.org/10.1016/j.cell.2008.02.039

Krisher RL, Heuberger AL, Paczkowski M, Stevens J, Pospisil C, Prather RS et al (2015) Applying metabolomic analyses to the practice of embryology: physiology, development and assisted reproductive technology. Reprod Fertil Dev 27(4):602–620. https://doi.org/10.1071/RD14359

Kumar V, Thakur JK, Prasad M (2021) Histone acetylation dynamics regulating plant development and stress responses. Cell Mol Life Sci 78(10):4467–4486. https://doi.org/10.1007/s00018-021-03794-x

Layek SS, Patil SP, Gorani S, Karuppanasamy K, Kishore G, Gupta RO (2022) Ovum pick-up and in vitro embryo production in bovine. In: Kumaresan A, Srivastava AK (eds) Frontier technologies in bovine reproduction. Springer, Singapore

Lee SH, Rinaudo PF (2024) Metabolic regulation of preimplantation embryo development in vivo and in vitro: Molecular mechanisms and insights. Biochem Biophys Res Commun 726:150256. https://doi.org/10.1016/j.bbrc.2024.150256

Leese HJ (1995) Metabolic control during preimplantation mammalian development. Hum Reprod Update 1(1):63–72. https://doi.org/10.1093/humupd/1.1.63

Leese HJ, Guerif F, Allgar V, Brison DR, Lundin K, Sturmey RG (2016) Biological optimization, the Goldilocks principle, and how much is lagom in the preimplantation embryo. Mol Reprod Dev 83(9):748–754. https://doi.org/10.1002/mrd.22684

Li L, Lu X, Dean J (2013) The maternal to zygotic transition in mammals. Mol Asp Med 34(5):919–938. https://doi.org/10.1016/j.mam.2013.01.003

Li J, Zhang J, Hou W, Yang X, Liu X, Zhang Y et al (2022) Metabolic control of histone acetylation for precise and timely regulation of minor ZGA in early mammalian embryos. Cell Discov 8(1):96. https://doi.org/10.1038/s41421-022-00440-z

Lin T, Sun L, Lee JE, Lee JB, Kim SY, Jin DI (2020) Changes of histone H3 lysine 23 acetylation and methylation in porcine somatic cells, oocytes and preimplantation embryos. Theriogenology 148:162–173. https://doi.org/10.1016/j.theriogenology.2020.03.006

Liu L, Liu Y, Gao F, Song G, Wen J, Guan J et al (2012) Embryonic development and gene expression of porcine SCNT embryos treated with sodium butyrate. J Exp Zool B Mol Dev Evol 318(3):224–234. https://doi.org/10.1002/jez.b.22440

Lodde V, Luciano AM, Franciosi F, Labrecque R, Sirard MA (2017) Accumulation of chromatin remodelling enzyme and histone transcripts in bovine oocytes. Results Probl Cell Differ 63:223–255. https://doi.org/10.1007/978-3-319-60855-6_11

Lodde V, Luciano AM, Musmeci G, Miclea I, Tessaro I, Aru M et al (2021) A nuclear and cytoplasmic characterization of bovine oocytes reveals that cysteamine partially rescues the embryo development in a model of low ovarian reserve. Animals (Basel) 11(7). https://doi.org/10.3390/ani11071936

Lonergan P, Rizos D, Ward F, Boland MP (2001) Factors influencing oocyte and embryo quality in cattle. Reprod Nutr Dev 41(5):427–437. https://doi.org/10.1051/rnd:2001142

Ma P, Schultz RM (2008) Histone deacetylase 1 (HDAC1) regulates histone acetylation, development, and gene expression in preimplantation mouse embryos. Dev Biol 319(1):110–120. https://doi.org/10.1016/j.ydbio.2008.04.011

Mamo S, Rizos D, Lonergan P (2012) Transcriptomic changes in the bovine conceptus between the blastocyst stage and initiation of implantation. Anim Reprod Sci 134(1–2):56–63. https://doi.org/10.1016/j.anireprosci.2012.08.011

Martin KL (2000) Nutritional and metabolic requirements of early cleavage stage embryos and blastocysts. Hum Fertil (Camb) 3(4):247–254. https://doi.org/10.1080/1464727002000199071

Martínez-Reyes I, Chandel NS (2020) Mitochondrial TCA cycle metabolites control physiology and disease. Nat Commun 11(1):102. https://doi.org/10.1038/s41467-019-13668-3

Milazzotto MP, Noonan MJ, de Almeida Monteiro Melo Ferraz, M. (2022) Mining RNAseq data reveals dynamic metaboloepigenetic profiles in human, mouse and bovine pre-implantation embryos. iScience 25(3):103904. https://doi.org/10.1016/j.isci.2022.103904

Miles JR, Walsh SC, Rempel LA, Pannier AK (2023) Mechanisms regulating the initiation of porcine conceptus elongation. Mol Reprod Dev 90(7):646–657. https://doi.org/10.1002/mrd.23623

Millán-Zambrano G, Burton A, Bannister AJ, Schneider R (2022) Histone post-translational modifications - cause and consequence of genome function. Nat Rev Genet 23(9):563–580. https://doi.org/10.1038/s41576-022-00468-7

Moelling K (2024) Epigenetics and transgenerational inheritance. J Physiol 602(11):2537–2545. https://doi.org/10.1113/JP284424

Mondal T, Kanduri C (2013) Maintenance of epigenetic information: a noncoding RNA perspective. Chromosom Res 21(6–7):615–625. https://doi.org/10.1007/s10577-013-9385-5

Muter J, Lynch VJ, McCoy RC, Brosens JJ (2023) Human embryo implantation. Development 150(10). https://doi.org/10.1242/dev.201507

Nguyen NT, Lo NW, Chuang SP, Jian YL, Ju JC (2011) Sonic hedgehog supplementation of oocyte and embryo culture media enhances development of IVF porcine embryos. Reproduction 142(1):87–97. https://doi.org/10.1530/REP-11-0049

Niakan KK, Eggan K (2013) Analysis of human embryos from zygote to blastocyst reveals distinct gene expression patterns relative to the mouse. Dev Biol 375(1):54–64. https://doi.org/10.1016/j.ydbio.2012.12.008

Niwa H, Toyooka Y, Shimosato D, Strumpf D, Takahashi K, Yagi R, Rossant J (2005) Interaction between Oct3/4 and Cdx2 determines trophectoderm differentiation. Cell 123(5):917–929. https://doi.org/10.1016/j.cell.2005.08.040

Pantaleon M, Harvey MB, Pascoe WS, James DE, Kaye PL (1997) Glucose transporter GLUT3: ontogeny, targeting, and role in the mouse blastocyst. Proc Natl Acad Sci USA 94(8):3795–3800. https://doi.org/10.1073/pnas.94.8.3795

Peral-Sanchez I, Hojeij B, Ojeda DA, Steegers-Theunissen RPM, Willaime-Morawek S (2021) Epigenetics in the uterine environment: how maternal diet and ART may influence the epigenome in the offspring with long-term health consequences. Genes (Basel) 13(1). https://doi.org/10.3390/genes13010031

Perez MF, Lehner B (2019) Intergenerational and transgenerational epigenetic inheritance in animals. Nat Cell Biol 21(2):143–151. https://doi.org/10.1038/s41556-018-0242-9

Peserico A, Simone C (2011) Physical and functional HAT/HDAC interplay regulates protein acetylation balance. J Biomed Biotechnol 2011:371832. https://doi.org/10.1155/2011/371832

Peters J (2014) The role of genomic imprinting in biology and disease: an expanding view. Nat Rev Genet 15(8):517–530. https://doi.org/10.1038/nrg3766

Plasschaert RN, Bartolomei MS (2014) Genomic imprinting in development, growth, behavior and stem cells. Development 141(9):1805–1813. https://doi.org/10.1242/dev.101428

Protacio RU, Li G, Lowary PT, Widom J (2000) Effects of histone tail domains on the rate of transcriptional elongation through a nucleosome. Mol Cell Biol 20(23):8866–8878. https://doi.org/10.1128/MCB.20.23.8866-8878.2000

Ramos-Ibeas P, Lamas-Toranzo I, Martínez-Moro Á, de Frutos C, Quiroga AC, Zurita E, Bermejo-Álvarez P (2020) Embryonic disc formation following post-hatching bovine embryo development in vitro. Reproduction 160(4):579–589. https://doi.org/10.1530/REP-20-0243

Rossant J (2018) Genetic control of early cell lineages in the mammalian embryo. Annu Rev Genet 52:185–201. https://doi.org/10.1146/annurev-genet-120116-024544

Salehi M, Abouhamzeh B, Hosseini A, Zare Z, Bakhtari A (2020) Comparison of epigenetic modifier genes in bovine adipose tissue-derived stem cell based embryos, as donors, with in vitro and parthenogenesis embryos. Cell J 22(2):149–157. https://doi.org/10.22074/cellj.2020.6714

Sarmento OF, Digilio LC, Wang Y, Perlin J, Herr JC, Allis CD, Coonrod SA (2004) Dynamic alterations of specific histone modifications during early murine development. J Cell Sci 117(Pt 19):4449–4459. https://doi.org/10.1242/jcs.01328

Seshagiri PB, Vani V, Madhulika P (2016) Cytokines and blastocyst hatching. Am J Reprod Immunol 75(3):208–217. https://doi.org/10.1111/aji.12464

Shahbazian MD, Grunstein M (2007) Functions of site-specific histone acetylation and deacetylation. Annu Rev Biochem 76:75–100. https://doi.org/10.1146/annurev.biochem.76.052705.162114

Sharpley MS, Chi F, Hoeve JT, Banerjee U (2021) Metabolic plasticity drives development during mammalian embryogenesis. Dev Cell 56(16):2329–2347.e2326. https://doi.org/10.1016/j.devcel.2021.07.020

Shirvaliloo M (2022) The landscape of histone modifications in epigenomics since 2020. Epigenomics 14(23):1465–1477. https://doi.org/10.2217/epi-2022-0437

Stein P, Worrad DM, Belyaev ND, Turner BM, Schultz RM (1997) Stage-dependent redistributions of acetylated histones in nuclei of the early preimplantation mouse embryo. Mol Reprod Dev 47(4):421–429. https://doi.org/10.1002/(SICI)1098-2795(199708)47:4<421::AID-MRD8>3.0.CO;2-M

Strahl BD, Allis CD (2000) The language of covalent histone modifications. Nature 403(6765):41–45. https://doi.org/10.1038/47412

Sun J, Cui K, Li Z, Gao B, Jiang J, Liu Q et al (2020) Histone hyperacetylation may improve the preimplantation development and epigenetic status of cloned embryos. Reprod Biol 20(2):237–246. https://doi.org/10.1016/j.repbio.2020.02.005

Thompson JJ, Lee DJ, Mitra A, Frail S, Dale RK, Rocha PP (2022) Extensive co-binding and rapid redistribution of NANOG and GATA6 during emergence of divergent lineages. Nat Commun 13(1):4257. https://doi.org/10.1038/s41467-022-31938-5

Vaiserman A (2015) Epidemiologic evidence for association between adverse environmental exposures in early life and epigenetic variation: a potential link to disease susceptibility? Clin Epigenetics 7(1):96. https://doi.org/10.1186/s13148-015-0130-0

Van Soom A, Boerjan ML, Bols PE, Vanroose G, Lein A, Coryn M, de Kruif A (1997) Timing of compaction and inner cell allocation in bovine embryos produced in vivo after superovulation. Biol Reprod 57(5):1041–1049. https://doi.org/10.1095/biolreprod57.5.1041

Wang F, Kou Z, Zhang Y, Gao S (2007) Dynamic reprogramming of histone acetylation and methylation in the first cell cycle of cloned mouse embryos. Biol Reprod 77(6):1007–1016. https://doi.org/10.1095/biolreprod.107.063149

Wang M, Gao Y, Qu P, Qing S, Qiao F, Zhang Y et al (2017) Sperm-borne miR-449b influences cleavage, epigenetic reprogramming and apoptosis of SCNT embryos in bovine. Sci Rep 7(1):13403. https://doi.org/10.1038/s41598-017-13899-8

Wang XF, Xie SM, Guo SM, Su P, Zhou LQ (2020) Dynamic pattern of histone H3 core acetylation in human early embryos. Cell Cycle 19(17):2226–2234. https://doi.org/10.1080/15384101.2020.1806433

Wang M, Chen Z, Zhang Y (2022) CBP/p300 and HDAC activities regulate H3K27 acetylation dynamics and zygotic genome activation in mouse preimplantation embryos. EMBO J 41(22):e112012. https://doi.org/10.15252/embj.2022112012

Weberling A, Zernicka-Goetz M (2021) Trophectoderm mechanics direct epiblast shape upon embryo implantation. Cell Rep 34(3):108655. https://doi.org/10.1016/j.celrep.2020.108655

Wilkinson AL, Zorzan I, Rugg-Gunn PJ (2023) Epigenetic regulation of early human embryo development. Cell Stem Cell 30(12):1569–1584. https://doi.org/10.1016/j.stem.2023.09.010

Wu X, Li Y, Xue L, Wang L, Yue Y, Li K et al (2011) Multiple histone site epigenetic modifications in nuclear transfer and in vitro fertilized bovine embryos. Zygote 19(1):31–45. https://doi.org/10.1017/S0967199410000328

Wu X, Hu S, Wang L, Li Y, Yu H (2020) Dynamic changes of histone acetylation and methylation in bovine oocytes, zygotes, and preimplantation embryos. J Exp Zool B Mol Dev Evol 334(4):245–256. https://doi.org/10.1002/jez.b.22943

Wu K, Fan D, Zhao H, Liu Z, Hou Z, Tao W et al (2023) Dynamics of histone acetylation during human early embryogenesis. Cell Discov 9(1):29. https://doi.org/10.1038/s41421-022-00514-y
Xiao L, Dang Y, Hu B, Luo L, Zhao P, Wang S, Zhang K (2022) Overlapping functions of RBBP4 and RBBP7 in regulating cell proliferation and histone H3.3 deposition during mouse preimplantation development. Epigenetics 17(10):1205–1218. https://doi.org/10.1080/15592294.2021.1999006
Yang D, Zheng H, Lu W, Tian X, Sun Y, Peng H (2024) Histone lactylation is involved in mouse oocyte maturation and embryo development. Int J Mol Sci 25(9). https://doi.org/10.3390/ijms25094821
Yao C, Zhang W, Shuai L (2019) The first cell fate decision in pre-implantation mouse embryos. Cell Regen 8(2):51–57. https://doi.org/10.1016/j.cr.2019.10.001
Zhang Y, Yang Z, Wu J (2007) Signaling pathways and preimplantation development of mammalian embryos. FEBS J 274(17):4349–4359. https://doi.org/10.1111/j.1742-4658.2007.05980.x
Zhang H, Yan K, Sui L, Li P, Du Y, Hu J et al (2021) Low-level pyruvate inhibits early embryonic development and maternal mRNA clearance in mice. Theriogenology 166:104–111. https://doi.org/10.1016/j.theriogenology.2021.02.022
Zhao P, Li S, Wang H, Dang Y, Wang L, Liu T et al (2019) Sin3a regulates the developmental progression through morula-to-blastocyst transition. FASEB J 33(11):12541–12553. https://doi.org/10.1096/fj.201901213R
Zhao J, Yao K, Yu H, Zhang L, Xu Y, Chen L et al (2021) Metabolic remodelling during early mouse embryo development. Nat Metab 3(10):1372–1384. https://doi.org/10.1038/s42255-021-00464-x
Zhou C, Halstead MM, Bonnet-Garnier A, Schultz RM, Ross PJ (2023) Histone remodeling reflects conserved mechanisms of bovine and human preimplantation development. EMBO Rep 24(3):e55726. https://doi.org/10.15252/embr.202255726
Ziegler-Birling C, Daujat S, Schneider R, Torres-Padilla ME (2016) Dynamics of histone H3 acetylation in the nucleosome core during mouse pre-implantation development. Epigenetics 11(8):553–562. https://doi.org/10.1080/15592294.2015.1103424

Chapter 7
Epigenetic Echoes: Decoding the Acetylation Journey from Neural Crest to Melanocyte

Ayesha Nasreen, Sribas Chowdhury, Dharani Alagar Selvam, and Vivek T. Natarajan

Abstract Epigenetic mechanisms influence early developmental events, shaping gene expression in exciting ways that go beyond the DNA blueprint. The state of chromatin is governed by an interplay between various histone modifications, variants, nucleosome remodeling complexes, and other chromatin modifiers that work in sync to prime the chromatin for specific biological outcomes. In this chapter, we explore neural crest cells (NCCs), a critical progenitor population that retains the extensive developmental potential of their blastula origins. The formation and differentiation of NCCs into diverse cell types are influenced by the regulation of their acetylation state through various epigenetic factors. This chapter delves into the intricate interplay between histone acetylases (HATs) and deacetylases (HDACs), highlighting how these enzymes modify chromatin to create a permissive environment for the induction of NCCs and steer their fate toward the melanocytic lineage. The shift in acetylation profiles during the transition from melanocytes to melanoma suggests that the transcriptional machinery may override normal regulatory mechanisms, promoting a neural crest-like state in melanoma development. Epigenetic regulation, particularly through histone acetylation, plays a pivotal role in neural crest cell development and melanoma initiation offering potential therapeutic targets.

Authors Ayesha Nasreen and Sribas Chowdhury have equally contributed to this chapter.

A. Nasreen · D. A. Selvam · V. T. Natarajan (✉)
CSIR-Institute of Genomics and Integrative Biology, New Delhi, India

Academy of Scientific and Innovative Research (AcSIR), Ghaziabad, India
e-mail: tnvivek@igib.in

S. Chowdhury
CSIR-Institute of Genomics and Integrative Biology, New Delhi, India

M. Halasa, A. Wawruszak (eds.), *Histone and Non-Histone Reversible Acetylation in Development, Aging and Disease*, Results and Problems in Cell Differentiation 75,
https://doi.org/10.1007/978-3-031-91459-1_7

7.1 Introduction

Epigenetic reprogramming during early development is crucial for establishing cell identity and fate as undifferentiated cells transition into specialized lineages. Factors such as histone modifications and chromatin remodeling work together to create a flexible and responsive chromatin environment, enabling precise regulation of developmental pathways. Among these modifications, acetylation is the most extensively studied. Its significant impact on gene regulation and chromatin dynamics has made it a focal point in epigenetic research (Kurdistani and Grunstein 2003). This is particularly relevant in the context of neural crest cells (NCCs), which are a multipotent embryonic cell type that emerges at the end of gastrulation.

NCCs play a crucial role in development, requiring precise regulation of gene expression for their differentiation. The ectoderm comprises both neural and non-neural regions, giving rise to the central nervous system (CNS) and the epidermis, respectively. At their junction, the initial precursors to NCCs are located in the dorsal part of the neural tube. Here, these cells undergo an epithelial-to-mesenchymal transition, which allows them to become migratory (Lowery and Sive 2004). Neural crest cells being multipotent have the ability to differentiate into a diverse number of cell types including cartilage and bone cells, neurons, and melanocytes (Simões-Costa and Bronner 2015; Martik and Bronner 2017). Thus, it is important to understand how these diverse types of cells are produced from the same set of genetic information encoded in the genome that is identical in all cells.

The commitment of neural crest cells to specific fates has long been a topic of debate, with various models proposed to explain the process. The two extreme hypotheses are: one proposes early fate restriction, where NCCs are considered inherently heterogeneous and pre-determined to adopt specific fates; the other suggests that NCCs are initially homogeneous and progressively restricted to particular fates during migration as they encounter environmental cues (Weston 1982; LaBonne and Bronner-Fraser 1998; Le Douarin and Kalcheim 1999; Kelsh et al. 2021). Recent studies using single-cell RNA sequencing strongly support the progressive fate restriction model, which suggests that sequential binary fate decisions guide cells toward adopting a specific fate (Soldatov et al. 2019). A classic example of this is the development of melanocytes, which arise from an upstream neural/glia/melanocyte precursor population (Raible and Eisen 1994).

Melanocytes are pigment-producing cells responsible for synthesizing melanin, the pigment that gives color to the skin, hair, and eyes (White and Zon 2008). Besides contributing to camouflage and cosmetic variations, melanin primarily protects the skin from UV-induced DNA damage, serving as a crucial defense mechanism against harmful factors (Natarajan et al. 2014). Among other lineages arising out of NCCs, melanocytes are exclusively derived from these cells, making them an ideal model for studying fate decisions. Abnormal functioning of melanocytes is linked to diseases like vitiligo and melanoma, underpinning the necessity to understand the factors underlying melanocyte biology.

Although extensive research has been conducted on deciphering the role of transcription factors and regulatory circuits in neural crest and melanocyte development, there are still additional layers of regulatory complexity to consider. Epigenetic regulation plays a significant role, involving mechanisms such as chromatin remodelers, histone-modifying enzymes, and DNA methylation and acetylation events. Acetylation is an interesting epigenetic mechanism where an acetyl (-$COCH_3$) is added to specific conserved lysine residues in histones by enzymes such as histone acetylases. This neutralizes the positive charge on lysine, reducing its affinity for the negatively charged DNA. Furthermore, due to the bulky nature of acetyl group, the chromatin structure becomes more relaxed, transitioning from a condensed, transcriptionally inactive state (heterochromatin) to a more "open", transcriptionally active conformation known as euchromatin. In this euchromatin state, transcription factors and other regulatory proteins can access gene promoters and enhancers more easily, leading to the activation of gene expression. This is followed by transcriptional machinery binding to their respective sites and mediating transcription of necessary genes. Similarly, deacetylation by histone deacetylases enables chromatin to reattain its compact state and become "closed". Hence, epigenetic mechanisms dictate the changes in spatio-temporal architecture and dynamics of chromatin, guiding NCCs to differentiate into specific cell types. In this chapter, we focus on the collective role of histone and non-histone acetylation in NCCs and how it subsequently shapes the epigenetic landscape to adopt a melanocyte fate and its implications in melanocyte-related disorders.

7.2 Role of Acetylation During Neural Crest Induction

The neural plate border is established as gastrulation concludes. This region gives rise to future NCCs, and the signaling events during the initial induction are similar across vertebrates, as shown in studies involving frogs, chicks, and zebrafish (*Danio rerio*) (Sauka-Spengler and Bronner-Fraser 2008). Fibroblast growth factor-2 (FGF2), along with Bone Morphogenic Protein (BMP) antagonists, help promote neural crest induction, as demonstrated in *Xenopus laevis*. Intermediate levels of BMP signaling may be necessary to maintain this newly established population. Additionally, Notch signaling, in conjunction with BMP and Wingless-related Integration Site (WNT) signaling, helps restrict the neural crest territory and establishes intermediate levels of these signals. This gradient influences the neural plate border region and induces the expression of key neural crest border specifier genes such as *tfap2, zic1, and pax3/7* (LaBonne and Bronner-Fraser 1998; Sato et al. 2005; Sauka-Spengler et al. 2007; Sauka-Spengler and Bronner-Fraser 2008). These factors in turn synergistically activate the expression of neural crest specifiers like *snail2* (Snail Family Transcriptional Repressor 2) and *foxd3* (forkhead box D3) that now make them bonafide NC cells (Dutton et al. 2001).

The precise spatiotemporal control of gene expression during embryonic development requires chromatin regulation and alterations in its accessibility, which is governed primarily by histone-modifying enzymes. The core nucleosomal unit undergoes modifications like methylation, acetylation, deacetylation, and phosphorylation. Among them, methylation and acetylation/deacetylation are known to play essential roles during NCC development. DNA methylation, primarily involving the addition of a methyl group to cytosine bases in DNA by DNA methyltransferases such as DNMT3A/B, plays a crucial role in tissue-specific regulation during early development (Cedar and Bergman 2010). Deletion of these methyltransferases is linked to embryonic lethality and growth impairment (Okano et al. 1999). In chick embryos, loss of function studies of *dnmt3A* results in the downregulation of neural crest specifier genes, highlighting its critical role as a molecular switch, repressing neural to favor neural crest cell fate (Hu et al. 2012). Another interesting study talks about JMJD2A, a histone demethylase that removes methylation marks from histones occupying the promoter elements of key neural crest specifier genes (Strobl-Mazzulla et al. 2010). Another important histone modification involves acetylation and deacetylation of histones, mediated by HATs and histone deacetylases HDACs (Shahbazian and Grunstein 2007). This process regulates the genome by activating or repressing the chromatin. Acetylation allows the chromatin structure to open up, making it increasingly accessible to subsequent transcription factor binding.

HDACs are a well-known group of enzymes that remove acetylation marks from histones and regulate the acetylation state of the chromatin. There are reportedly 11 HDAC enzymes encoded by the mammalian genome and that are then grouped into four different classes based on similarity and function (Seto and Yoshida 2014). Surprisingly, deletion studies involving these HDACs have revealed highly specific and non-redundant roles for these enzymes in various cellular and in-vivo systems. Among them class I HDACs are known to play a crucial role during early development, with HDAC1 and HDAC2 regulating stem cell maintenance by controlling the expression of essential pluripotent transcription factors (Jamaladdin et al. 2014). An insightful study demonstrates that treating *Xenopus* embryos with HDAC inhibitors, such as Trichostatin A, results in a total loss of neural crest cells, as evidenced by the analysis of key neural crest marker genes like s*nail2* and *foxd3* (Murko et al. 2013). In fact, in zebrafish harboring *co1b382* mutation that encodes for HDAC1, there is severe effect on the melanocytes being specified from NCCs (Ignatius et al. 2008). In *hdac1col* mutants, *foxd3* expression is extended for a prolonged time repressing melanophore specification by directly inhibiting *mitfa* expression. Loss of HDAC3 in cardiac NCCs results in cardiovascular abnormalities by affecting smooth muscle differentiation (Singh et al. 2011). Global deletion of another histone deacetylase HDAC8 has been shown to cause aberrant patterning of the skull by specifically repressing homeobox transcription factors like OTX2 and LHX1 (Haberland et al. 2009).

Overall, HDACs and HATs maintain the acetylation switch of the chromatin state by modulating gene expression either by deacetylating the histone tails or the target transcription factors or by both and thereby altering the chromatin structure of the cellular landscape (Fig. 7.1).

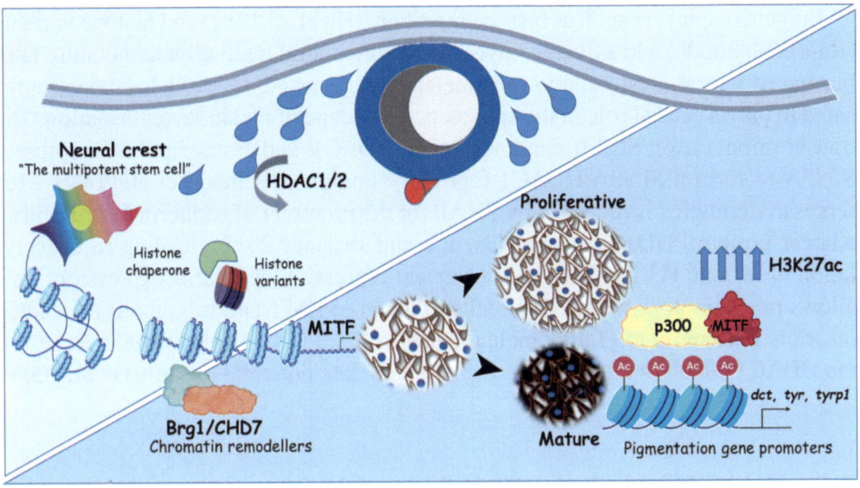

Fig. 7.1 Interplay of epigenetic forces in the neural crest to melanocyte transition. The transition from neural crest cells to melanocytes is influenced by a variety of epigenetic factors that prepare the chromatin for lineage specification. This schematic illustrates the crucial role of acetylation during this process and how this epigenetic modification shapes the evolving dynamics of cell specification

7.3 Histone Deacetylase: The Navigator of Neural Crest Cell Migration

Neural crest cells migrate from the dorsal neural tube after going through an epithelial-to-mesenchymal transition (Bronner 2012) under the influence of FOXD3 and SNAIL transcription factors. *Snail* proteins are central players in driving the repression of cell-cell adhesion molecules like E-cadherin, thus enabling the cells to break free from the neural epithelium and migrate. FOXD3 further modulates this process by promoting the multipotency of neural crest cells and ensuring their survival during migration (Leathers and Rogers 2022). At this stage, they segregate into two populations having distinct migratory properties. One of them is the neurogenic population, which migrates ventrally toward the developing somites and peripheral tissues under the influence of FOXD3. FOXD3 functions as a repressor of the melanogenic program and an activator of neural and glial lineages (Thomas and Erickson 2009), ensuring that the ventrally migrating cells give rise to peripheral neurons and Schwann cells, which are key components of the peripheral nervous system. In contrast, the melanogenic population follows the dorsolateral pathway and localizes between the surface epidermis and the somites (Luciani et al. 2011). The migration and specification of the melanogenic population is driven by SOX10 (Sry-related high-mobility-group box 10) and PAX3 (paired box gene 3), both of which are essential for the early onset of melanocyte specification and survival (Bondurand et al. 2000).

Epigenetic players such as histone modifiers (Hu et al. 2012) and histone variants (Raja et al. 2020b) add a distinct layer of spatiotemporal regulation in dictating fate-biasing of neural crest cells toward melanocytic lineage. HDAC1 has been implicated to play a central role in the epigenetic landscape of melanocyte formation. The transcription factor SNAIL interacts with HDAC1 and transcriptional repressor SIN3A to form a SIN3A/HDAC1 repressive complex (Peinado et al. 2004). This repressive complex is recruited by SNAIL to the promoter of cadherin Cad6B via an adapter protein PHD12 (Strobl-Mazzulla and Bronner 2012), leading to deacetylation of histone H3 and H4 and subsequent repression of Cad6B expression. This allows pre-migratory neural crest cells to undergo EMT and migrate to their fated location. Further, in migrating melanogenic NCCs, FOXD3 expression is repressed via HDAC1 which maintains their melanogenic fate potential (Ignatius et al. 2008).

7.4 HDAC1/2 Mediates Switching of Roles in Differentiating Melanocytes

During their migration along the dorsolateral pathway, melanogenic NCCs differentiate into two distinct precursor cell types: Schwann cell precursors and melanoblasts (Vandamme and Berx 2019). Schwann cell precursors are initially multipotent and have the potential to differentiate into both Schwann cells and melanocytes. Melanoblasts, on the other hand, are committed progenitors of the melanocyte lineage and express Microphthalmia-associated Transcription Factor (MITF) (Bronner-Fraser 1993). SOX10 interacts with chromatin remodeler BRG1 (Brahma-related gene 1) to establish melanocyte lineage via increased expression of MITF (Marathe et al. 2017).

MITF is the master transcription factor that governs the specification, differentiation, and survival of melanoblasts and melanocytes (Capowski et al. 2014; Goding and Arnheiter 2019). It is one of the earliest markers of the melanocyte lineage and plays a central role in activating the melanogenic gene program. MITF regulates the expression of numerous genes involved in melanin biosynthesis, including dopachrome tautomerase (DCT) and pre-melanosome protein (PMEL), both of which are key players in the formation and function of melanosomes, the organelles responsible for melanin production (Adema et al. 1994; Kobayashi et al. 1994; Costin et al. 2005). The expression of MITF is itself regulated by multiple upstream transcription factors, including SOX10 (Elworthy et al. 2003) and PAX3 (Bondurand et al. 2000), which work in tandem to drive the development of melanoblasts. FOXD3, on the other hand, negatively regulates MITF expression, biasing migratory neural crest cells toward neurogenic fate and limiting the number of melanoblasts forming. In Zebrafish, an interesting case of role switching arises between MITF and SOX10, with HDAC1/2 being the major molecular player. In-situ hybridization studies have reported a decrease in SOX10 expression levels in specified melanocytes, which arises as a result of MITF's interaction with HDAC1/2 (Greenhill et al. 2011). The paper suggested a model where SOX10 drives MITF expression in migrating NCC and early melanoblasts, after which MITF takes over and represses SOX10 expression via HDAC1/2. In conflict with this, studies carried out in mouse embryos and cell lines describe SOX10 and MITF as concomitant partners driving the expression of downstream pigmentation targets (Murisier et al. 2007). It highlights the differential role of acetylation as an epigenetic mechanism and how it affects the multipotency of NCCs.

7.5 The Acetylation Profile of Melanocytes and Its Role in Cellular Function

Across the developmental landscape of cells, particularly in melanocytes, epigenetic programming plays a pivotal role in regulating gene expression, differentiation, and response to environmental stimuli (Dilshat et al. 2021). One of the most dynamic and essential modifications in this context is the acetylation of histones, a process that is tightly controlled by HATs and HDACs. In melanocytes, these modifications are fine-tuned to modulate processes such as pigmentation, proliferation, and the maintenance of cellular identity. Among other major acetylation programs, the acetylation of H3 and H4 by EP300 (E1A-binding protein p300, also known as p300) in melanocytes is critical for the expression of pigmentation-related genes such as *dct*, *tyr*, and *tyrp1*, which are essential for melanin production. The specificity of histone acetylation in melanocytes is crucial for dictating the cell state at a particular time. For example, p300-mediated acetylation of histone H3 at lysine 27 (H3K27ac) is enriched at the enhancer and promoter regions of *dct*, *tyr*, and *tyrp1*, thereby driving the transcriptional program that leads to melanin synthesis (Creyghton et al. 2010).

7.6 Increase in H3K27ac Marks Onset of Pigmentation

MITF exudes diverse roles in melanocytes including pigmentation, proliferation, cell maintenance, and homeostasis. To facilitate these diverse effector functions, MITF target genes are selectively activated during distinct transitions in melanocytes. This necessitates the existence of several interconnected regulatory loops that allow for the selective activation of specific genes. For instance, the onset of pigmentation is mediated by the expression of *dct, tyr*, and *tyrp1* genes which are under the regulation of MITF. They are characterized by a district H3K27 acetylation (H3K27ac) mark, which is associated with active transcription. The enzyme EP300 is a major acetyltransferase responsible for adding acetyl groups to histones (Dancy and Cole 2015), thereby facilitating a more open chromatin structure conducive to transcription. EP300 is recruited to enhancer and promoter regions of *dct*, *tyr,* and *tyrp1* by dynamically interacting with MITF and SOX10, leading to increased H3K27 acetylation at these sites and enhancing the transcription of these pigmentation-related genes.

Chip-seq-based studies have indicated that during the pigmentation process, the distribution of H3K27ac undergoes dynamic changes (Gorkin et al. 2012). Initially, the mark is sparse, but it progressively accumulates as pigmentation develops, reflecting an increase in gene expression related to pigmentation. A study using cellular model systems shows the coexistence of a mature (high pigmentation; HP) and a proliferative (low pigmentation; LP) state in cultured melanocytes which potentially aids their response to external cues such as UV (Aggarwal et al. 2024). Hyperacetylation of H3K27 is predominantly associated with the mature (HP) state having double the amount of H3K27ac especially on MITF and other pigmentation-associated genes. This suggests that the mature pigmenting state is primarily driven by histone acetylation while the proliferative state is maintained by alternative modes.

Besides acetylation, histones undergo a myriad of epigenetic changes such as methylation, phosphorylation, ubiquitination, and replacement by histone variants. Histone variants have been recognized as crucial regulators of chromatin structure, introducing an additional dimension to the epigenetic regulation of the 3D chromatin state. Numerous studies have concentrated on H2A and H3 variants, demonstrating their functional significance during development (Sivasubbu et al. 2006; Cox et al. 2012; Pünzeler et al. 2017; Greenberg et al. 2019). It has been long known that histone variant H3.3 is linked to active chromatin and is enriched at transcriptionally active genes in mouse embryonic stem cells (mESCs) (Goldberg et al. 2010). A recent study (Martire et al. 2019) showed that phosphorylation S31 on H3.3 stimulates p300 and its acetylation activity on H3 histone in trans. Its depletion reduces enhancer acetylation and reprogramming capacity, highlighting its importance in genome regulation. Another exciting study establishes the role of an epigenetic modifier, histone variant H2A.Z, in cell fate decisions during melanocyte development from the multipotent neural crest cells (NCCs) (Raja et al. 2020b). It illustrates that the mode of action of H2A.Z is via its occupancy at key melanocyte regulatory genes like *mitf* and *sox9*. The occupancy of H2A.Z at these promoters is positively correlated with their acetylation status. This points to the possibility of dual epigenetic regulation of selective genes resulting in a more open and activated state of the chromatin.

7.7 Extrinsic Factors Influence the Acetylation State of Melanocytes

Among the core processes in melanocytes is melanogenesis, a dynamic process that uses tyrosine as a substrate to produce melanin. The process involves an intricate balance of intercellular crosstalk and intracellular signaling. Melanogenesis starts when the keratinocytes in the vicinity of melanocytes secrete the hormone α-MSH (alpha Melanocyte Stimulating Hormone) in response to UV radiation (Chakraborty et al. 1996; Cui et al. 2007). α-MSH serves as the ligand for melanocortin 1 receptor

(MC1R) (Yang et al. 1997), a GPCR present on the surface of melanocytes. Activation of MC1R leads to increased levels of secondary messenger cAMP (Im et al. 1998) in melanocytes which activates the protein kinase A (PKA) pathway. Activation of PKA leads to phosphorylation of cAMP-response element binding protein (CREB) (Naikoo et al. 2023) which then recruits CBP/p300 to MITF activation domain and results in increased levels of MITF (Brown et al. 2023a). MITF recognizes H3K27ac mark on the enhancer and promoter regions of *dct, tyr,* and *tyrp1* and recruits BRG and p300 to these regions, leading to their expression and subsequent pigmentation (Laurette et al. 2015; Brown et al. 2023a, b). p300 has a histone-acetyltransferase (HAT) domain which acetylates H3K27 under a dynamic regulation of intracellular pH (pH_i) and metabolic reprogramming (Raja et al. 2020a; Sultan et al. 2022).

A study by Raja et al. has demonstrated the role of a MITF-regulated carbonic anhydrase 14 (CA14) involved in a feed-forward loop and mediates the activation of pigmentation genes by aiding the deposition of H3K27ac marks at their promoters (Raja et al. 2020a). Activation of CA14 increases the pH_i and creates an alkaline environment in the cell, making it conducive for p300 HAT activity. The HAT domain of p300 has a "Theorell Chance" mechanism in which the ε amino group of lysine in H3 undergoes deprotonation when the pH is 7.8 to 8.0 (Liu et al. 2008; Zhang et al. 2014). Deprotonation of the ε amino group is coupled with addition of acetyl from acetyl coenzyme A (CoA), leading to increased H3K27ac marks in pigmenting melanocytes.

Acetylation of histones is subject to the availability of acetyl CoA in the nuclear and cytosolic compartment (Takahashi et al. 2006; Galdieri and Vancura 2012) and arises as a result of metabolic reprogramming in melanocytes during melanogenesis. A study exploring how fatty acid metabolism affects melanogenic programs of melanocytes delineated increased fatty acid production and accumulation of lipid droplets (Sultan et al. 2022). Transient decoupling of glycolysis and TCA cycle leads to increased efflux of citrate from mitochondria (Martínez-Reyes and Chandel 2020) into the cytosol and subsequent conversion of citrate to acetyl CoA by ATP citrate lyase (ACL) (Wellen et al. 2009). Signaling mediated by α-MSH activates Sterol regulatory element-binding protein 1 (SREBF1) via MITF which upregulates genes involved in the fatty acid synthesis pathway. This results in increased availability of acetyl CoA in the cytosolic environment which leads to de-novo synthesis of triacylglycerol (TAGs) and subsequent accumulation of lipid droplets. The lipid droplets subsequently undergo beta-oxidation resulting in the formation of acetyl-CoA which gets then incorporated into the TCA cycle, thereby restoring coupling of glycolysis and TCA. Meanwhile, increase in pH_i due to activation of CA14 leads to influx of available acetyl-CoA into the nucleosomal compartment. Acetyl-CoA serves as substrate for p300 which transfers acetyl group to H3K27, ultimately increasing levels of H3K27ac at enhancers of *dct, tyr,* and *tyrp1*. MITF and SOX10 recognize this acetylation mark and bind to the promoter regions of the genes, resulting in increased expression and subsequent melanogenesis. Hence, pigmentation in melanocytes due to acetylation of H3K27 is a result of a highly dynamic interplay of metabolic factors and p300 which concomitantly work to maintain homeostasis and melanogenesis in maturing melanocytes (Fig. 7.2).

7.8 Acetylation at K206 Switches MITF Targets

MITF being a central factor governs dynamic aspects of melanocyte development, requiring the selective activation of target genes during cell state transitions which in turn affect specific biological outcomes. MITF stimulates both proliferation and differentiation, yet how it distinguishes between targets associated with these two processes is still a matter of debate.

To tackle this issue, a rheostat model was proposed to explain the conflicting functions of MITF (Carreira et al. 2006). According to this model, MITF activity varies across processes: it is low during invasion, intermediate during proliferation, and high during differentiation. Numerous studies are currently focusing on post-translational modifications of non-histone regulatory proteins as crucial factors in the regulation of cellular gene transcription. An interesting study by (Louphrasitthiphol et al. 2023) reveals MITF K206 acetylation to be an important factor for shifting the equilibrium of MITF binding away from differentiation-associated genes. They show using DNA binding affinity assays that CATGTG M-box motifs linked to the promoters of differentiation genes are lower affinity binding sites for MITF, while CACGTG E-boxes are higher affinity sites and can be recognized even at lower MITF levels. However, aside from protein levels, they show that MITF's ability to regulate its target genes is also influenced by post-translational modifications. A critical residue acetylation at K206 significantly reduces MITF binding to CATGTG M-box motifs. The ability of K206 acetylation to distinguish between CATGTG and CACGTG elements implies that the regulation of MITF acetylation could play a crucial role in shaping the range of target genes that MITF can influence. This suggests that acetylation modifications at K206 may serve as a key mechanism in fine-tuning MITF's interactions with specific genetic elements, ultimately affecting which genes are activated or repressed in response to various cellular cues.

7.9 Acetylation in Melanoma—A Double-Edged Sword

Melanoma arises when melanocytes undergo uncontrolled cell division and proliferation, which results in tumor formation. Ultraviolet ray is the leading agent that results in melanoma which is characterized by vast genetic alterations and epigenetic modifications (Besaratinia and Tommasi 2014). Along with the mutations that have been reported for the genes involved in cell checkpoint, DNA damage response and repair pathways, tumor suppressors, and oncogenes, mutations that are harbored in MITF gene regulates melanoma in different aspects like cell proliferation,

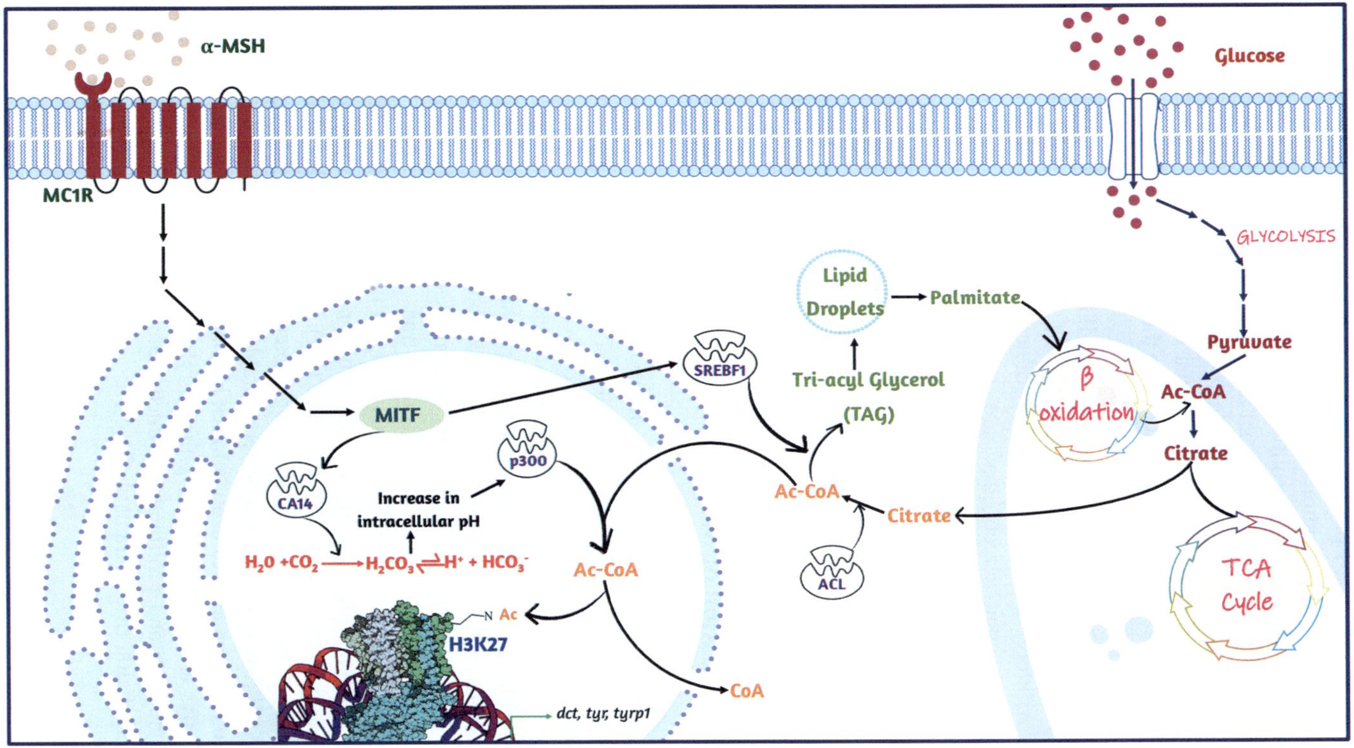

Fig. 7.2 pH-dependent metabolic reprogramming supports HAT activity of p300, facilitating acetylation of H3K27. Citrate efflux from mitochondria increases cytosolic acetyl CoA, which is made available at the nucleus as a substrate for p300 for H3K27 acetylation. This leads to *Dct, Tyr*, and *Tyrp1* expression and subsequent pigmentation. MITF activates the enzyme CA14 to raise intracellular pH, enhancing p300 HAT activity. Under the influence of α-MSH, MITF induces SREBF1 promoting fatty acid synthesis from acetyl CoA which is stored in lipid droplet as accumulated TAGs. Subsequent β-oxidation of these stored fatty acids restores acetyl CoA levels and cellular homeostasis. The schematic illustrates how metabolic programming mediated by extracellular crosstalk influences acetylation state in pigmenting melanocytes

invasion, and metastasis. Hence MITF is considered a lineage-specific oncogene (Kawakami and Fisher 2017). Epigenetic modifications such as phosphorylation, methylation, and acetylation play immense roles in melanoma pathogenesis at various stages such as progression and proliferation, cell dedifferentiation, invasiveness, and metastasis. These modifications act as a molecular switch by turning on and off the genes that promote melanoma. Such DNA modification results in the mutations that drive and contribute to melanoma development crucially. Driver mutations like BRAFV600E, NRAF, and KRAS are the key mutations in melanoma by activating the mitogen-activated protein kinase (MAPK) signaling pathway which in turn promotes the phosphorylation of MITF by extracellular signa-regulated kinase (ERK) pathway (Paluncic et al. 2016), that results in acetylation of MITF at K206. As aforementioned, the inability of acetylated MITF K206 not only results in developmental defects that are seen in Waardenburg's disease (Louphrasitthiphol et al. 2023) but also regulates the expression of essential genes involved in melanoma. The later part of this chapter discusses about how acetylation and deacetylation on histone and non-histone targets regulate melanoma at its various stages of pathogenesis.

7.10 Role of Acetylation in Cell Proliferation and Progression of Melanoma

Acetylation plays a crucial role in the regulation of cell proliferation and progression of melanoma. Based on the target gene and position, acetylation regulates cell proliferation and progression of melanoma both positively and negatively is as same as the two sides of the coin. Acetylation of genes which are involved in cell proliferation promotes melanoma proliferation by enabling the access for target specific transcription factors to mediate transcription (Mendelson et al. 2024). It has been reported that H2AZ chaperone enhances cell cycle progression in melanoma by H4 acetylation at promoter region *e2f*, which encodes a transcription factor that targets genes involved in proliferation by P400-TIP60 complex thereby it drives the melanoma cell proliferation which implies the role of histone and non-histone acetylation in melanoma cell proliferation (Jostes et al. 2024). Whereas another study tells that BRM (Brahma—bromodomain-containing protein), one of the catalytic subunits of SWI/SNF (Switch/Sucrose non-fermentable) complexes which regulates chromatin remodeling undergoes acetylation at histone H4 when MAPK and ERK signaling pathways are inhibited thereby the acetylated state of BRM inhibits the melanoma progression and proliferation which indicates the negative regulation mediated by acetylation (Mehrotra et al. 2014).

7.11 Role of Acetylation in Melanoma Dedifferentiation, Invasiveness, and Metastasis.

Dedifferentiation is a crucial step for the cancerous population wherein cells behave less differentiated and mimic stem cell character which renders them to switch from proliferation state to invasion and ultimately results in metastasis of a tumor (Richard et al. 2021). Acetylation and deacetylation benefit melanoma cells by shutting down the genes involved in differentiation thereby promoting dedifferentiation and turning on the genes involved in cell proliferation and cell self-renewal thereby rendering them stem cell-like property thereby it enhances the tumor heterogeneity. Additionally, in melanoma, histone acetylation plays a key role in switching from a proliferative state to an invasive state. Transcriptome analysis of the melanoma model suggests that acetylation at histone 3 lysine 27 (H3K27) of super-enhancers results in the activation of genes (such as ITGA3) involved in cell migration, invasion, and metastasis (Mendelson et al. 2024). Acetylation is not only regulated by HATs and HDAC but also by the substrate availability within the cell. The study (Lu et al. 2024) found that ALDH1A3-high melanoma cells use acetaldehyde metabolism with increased expression of ALDH1A3 (Aldehyde Dehydrogenase 1A3), a key enzyme in the detoxification of aldehydes, including acetaldehyde, a by-product of cellular metabolism thereby it fuels histone acetylation at H3K27, and alters the transcription of key genes involved in cell survival, proliferation, and lineage differentiation of melanoma. The protein–protein interaction between HDACs and their histone or non-histone targets regulates the genes involved in invasiveness. HDAC-8 deacetylases belonging to class-I HDACs are known to regulate the genes involved in cancer progression, epithelial-mesenchymal transition (EMT), and apoptosis suppression by targeting both histone and non-histone substrates such as transcription factors, kinases, cytoskeletal proteins, and tumor suppressors (Kim et al. 2022). It has been revealed in a study that HDAC-8 promotes the invasiveness and metastasis of brain melanoma by downregulating EP300 (histone acetyltransferase) thereby shutting down the melanocyte-specific genes such as *MLANA, PMEL, TYR, DCT,* and *MITF* (Emmons et al. 2023). In contrast, HDAC-2 interacts with SALL4, a zinc finger transcription factor which maintains the pluripotency and self-renewal property of embryonic stem cells (Diener et al. 2021) to negatively regulate the invasiveness by suppressing the key genes involved in it and promotes primary melanoma. Hence, association with different types of HATs and HDACs, histones and non-histone substrates undergo various regulation that determines the fate of melanoma.

7.12 Conclusions and Perspectives

The findings presented here enhance our current understanding of how the acetylation state of chromatin influences the neural crest program, mapping its role in development from the initial formation at the neural plate border to its differentiation into various cell types, particularly melanocytes. Development is a complex and highly regulated process that relies on the collaboration of signaling molecules, transcriptional and translational controls, and epigenetic factors that work together seamlessly. Although much research has focused on the roles of transcription factors and regulatory circuits in neural crest development, it is becoming increasingly clear that the process involves multiple layers of regulatory complexity.

Epigenetic regulation is a key component of this complexity, utilizing mechanisms that influence gene expression without changing the DNA sequence. Chromatin remodeling through histone variants or modifiers adjusts nucleosome positioning, affecting gene accessibility for transcription. Histone modifications can either enhance or suppress gene expression by altering chromatin structure, which is essential for defining specific neural crest cell identities. Together, these epigenetic changes facilitate a coordinated spatiotemporal rearrangement of chromatin during neural crest development, ensuring the proper establishment of each cell's identity and function in response to developmental cues.

Although the molecular intricacies involving gene regulatory loops during the early induction, migration, and overt differentiation of NCCs into diverse cell types are well documented, the epigenetic signatures appear to be convoluted, likely due to increased complexity. While many aspects of the chromatin landscape are understood, the specific connections, direct interactions, and organizational structures still need to be clarified. In this regard identifying histone codes that mediate epigenetic regulation of the involved gene regulatory loop is important. This will ultimately help us delineate the multipotent behavior of NCCs and their subsequent fate biasing to distinct cells.

This chapter examines the establishment of the melanocytic lineage from neural crest cells (NCCs), highlighting the dynamics of MITF and the essential role of H3K27 acetylation as a key signature influencing these states. As MITF is central to driving a wide range of changes in melanocytes—from proliferation to pigmentation—it is essential to understand how a single transcription factor can fulfill such diverse functions. While various theories suggest that this diversity may arise from differential binding to specific target sequence motifs, the underlying mechanisms appear to be more intricate than this. It is intriguing to consider whether the master transcription factor can independently regulate these opposing functions or if epigenetic inputs are involved in shaping its effects. For example, it has been found that H3K27 acetylation mark is heavily enriched on pigmentation-associated genes but not on cell cycle genes pointing to the possibility that apart from MITF activity, this chromatin signature is essential for switching on the expression of pigmentation genes (Malcov-Brog et al. 2018). What also is interesting to postulate is the functional interaction of CA14 and MITF and how this could hasten the influence of MITF on its target repertoires.

Interaction of MITF with CA14 and its subsequent activation leads to increase in pH_i and provides the necessary conditions for the HAT activity of p300. Activity of p300 is metabolically determined by acetyl CoA levels and indicates how cellular processes and systems are dynamically interconnected. The transient uncoupling of glycolysis from the TCA cycle leads to a metabolic shift characterized by an increase in β-oxidation. Interestingly, this entire metabolic reprogramming is temporally regulated, unfolding in distinct time-restricted phases. The process of melanogenesis occurs over a certain period of time and during its onset and peak, the HAT activity of p300 on H3K27 is heightened and increase in fatty acid levels and lipid droplet accumulation is observed, reminiscent of acidosis usually seen in cancer cells. As the process nears its culmination, the accumulated lipid droplets undergo β-oxidation, restoring cellular homeostasis.

Furthermore, a study by Henry et al. (2015) describes the selectivity of p300 being dependent on acetyl CoA levels present in the nucleus, wherein different lysine residues are acetylated in the presence of different concentrations of acetyl CoA. Regulation of acetyl CoA levels through metabolic reprogramming induced by extracellular cues, such as α-MSH, prompts critical questions about the variability of acetylation states across histones and their variants and how extracellular cues play a role in it. It necessitates gaining a more dynamic overview of how perturbation in epigenetic changes, metabolism, and extracellular crosstalk might implicate the onset and progression of melanocyte-associated diseases such as vitiligo and melanoma.

Melanoma is the fifth common most cancer globally with 77% of survival rate when it is localized and 5% when it has metastasized. Apart from the driver mutations reported, melanoma undergoes extensive genetic alteration and epigenetic modification which leads to tumor heterogeneity that promotes malignant melanoma (Ferlay et al. 2021). Modern endocrine medicines target a multitude of molecular events and pathways through the use of kinases, HATs, KATs, HDACs, phosphatase inhibitors, etc. However, melanoma develops resistance through genetic alterations and epigenetic modification; therefore, it has become imperative need to investigate these mechanisms in depth as they lead to aggressive melanoma.

The role of acetylation and deacetylation in melanoma is still being investigated, despite the fact that epigenetic changes in the disease are being well-studied. Histone and non-histone acetylation in melanogenesis and their dualistic nature in melanoma emphasize the importance of acetylation coupled gene expression, necessitating investigation into potential interplay or crosstalk between them. For example, it has been revealed in two independent studies that SOX10 undergoes both histone and non-histone acetylation which promotes melanoma invasion (Mendelson et al. 2024; Waddell et al. 2024). Hence deciphering their relationship with each other might lead to comprehend the underlying molecular mechanism and also will be helpful to develop drugs that can be used as combinational therapy to combat melanoma.

According to the recent report (Mendelson et al. 2024), it is evident that the H3K27 acetylation plays a vital role not only in the regulation of genes with respect to pigmentation but also in melanoma. The acetylation status of important genes declines with the progression of melanogenesis, when neural crest stem cells commit to becoming mature melanocytes, and increases with the transformation of primary melanoma into malignant melanoma (Fig. 7.3). Therefore, it is comprehensible that the acetylation at H3K27 functions as a gatekeeper of gene expression, influencing both melanogenesis and melanoma.

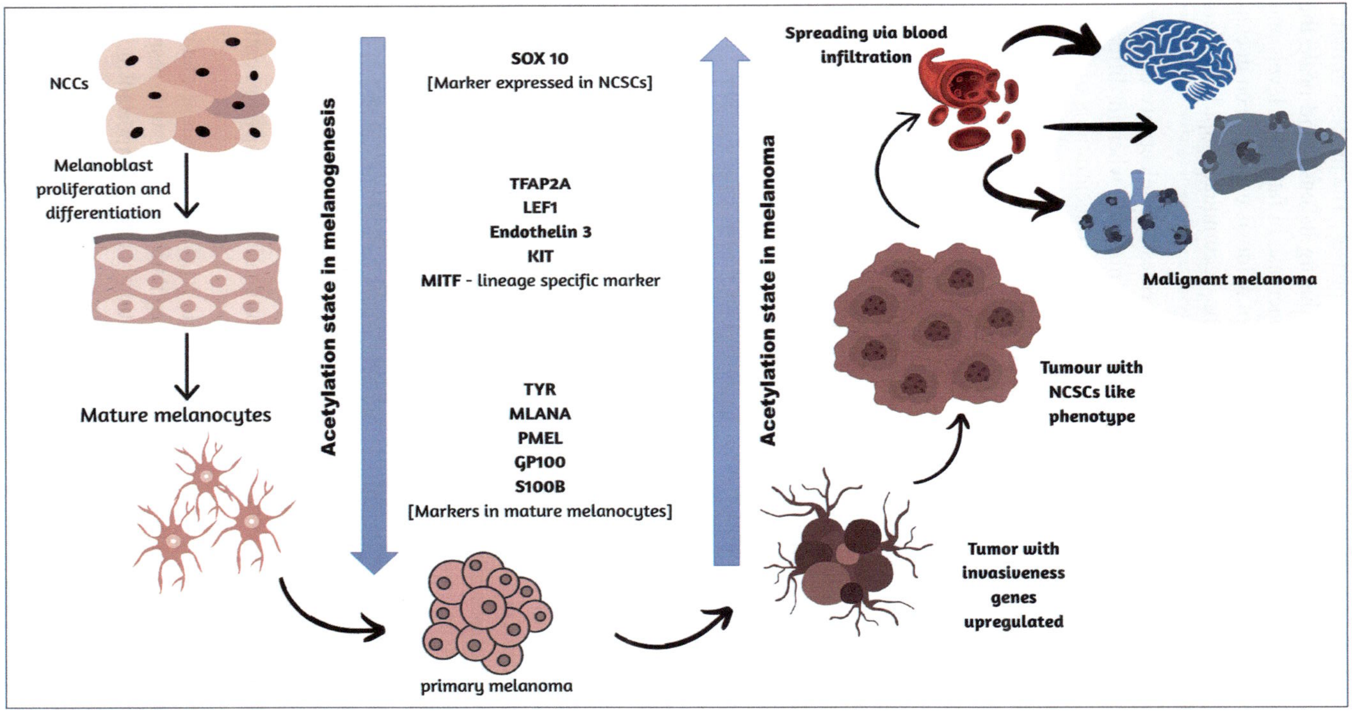

Fig. 7.3 From normal to malignant: The transformative acetylation dynamics in melanocytes. The schematic demonstrates the shifting acetylation profile that occurs during the development of melanocytes to initiation of melanoma, highlighting the re-establishment of a dedifferentiated state reminiscent of neural crest cells, which supports melanoma development

References

Adema GJ, de Boer AJ, Vogel AM, Loenen W, Figdor CG (1994) Molecular characterization of the melanocyte lineage-specific antigen gp100 269(31):20126–20133

Aggarwal A, Nasreen A, Sharma B, Sahoo S, Aswin K, Faruq M, Pandey R, Jolly MK, Singh A, Gokhale RS (2024) Distinct melanocyte subpopulations defined by stochastic expression of proliferation or maturation programs enable a rapid and sustainable pigmentation response 22(8):e3002776

Besaratinia A, Tommasi SJ (2014) Epigenetics of human melanoma: promises and challenges 6(5):356–367

Bondurand N, Pingault V, Goerich DE, Lemort N, Sock E, Caignec CL, Wegner M, Goossens M (2000) Interaction among SOX10, PAX3 and MITF, three genes altered in Waardenburg syndrome 9(13):1907–1917

Bronner ME (2012) Formation and migration of neural crest cells in the vertebrate embryo 138:179–186

Bronner-Fraser M (1993) Mechanisms of neural crest cell migration 15(4):221–230

Brown AD, Lynch K, Langelaan DN (2023a) The C-terminal transactivation domain of MITF interacts promiscuously with co-activator CBP/p300 13(1):16094

Brown AD, Vergunst KL, Branch M, Blair CM, Dupré DJ, Baillie GS, Langelaan DN (2023b) Structural basis of CBP/p300 recruitment by the microphthalmia-associated transcription factor 1870(7):119520

Capowski EE, Simonett JM, Clark EM, Wright LS, Howden SE, Wallace KA, Petelinsek AM, Pinilla I, Phillips MJ, Meyer JS (2014) Loss of MITF expression during human embryonic stem cell differentiation disrupts retinal pigment epithelium development and optic vesicle cell proliferation 23(23):6332–6344

Carreira S, Goodall J, Denat L, Rodriguez M, Nuciforo P, Hoek KS, Testori A, Larue L, Goding CR (2006) Mitf regulation of Dia1 controls melanoma proliferation and invasiveness 20(24):3426–3439

Cedar H, Bergman Y (2010) Epigenetic silencing during early lineage commitment. StemBook

Chakraborty AK, Funasaka Y, Slominski A, Ermak G, Hwang J, Pawelek JM, Ichihashi M (1996) Production and release of proopiomelanocortin (POMC) derived peptides by human melanocytes and keratinocytes in culture: regulation by ultraviolet B 1313(2):130–138

Costin G-E, Valencia JC, Wakamatsu K, Ito S, Solano F, Milac AL, Vieira WD, Yamaguchi Y, Rouzaud F, Petrescu A-J (2005) Mutations in dopachrome tautomerase (Dct) affect eumelanin/pheomelanin synthesis, but do not affect intracellular trafficking of the mutant protein 391(2):249–259

Cox SG, Kim H, Garnett AT, Medeiros DM, An W, Crump JG (2012) An essential role of variant histone H3.3 for ectomesenchyme potential of the cranial neural crest. PLoS Genetics 8(9):e1002938.

Creyghton MP, Cheng AW, Welstead GG, Kooistra T, Carey BW, Steine EJ, Hanna J, Lodato MA, Frampton GM, Sharp PA (2010) Histone H3K27ac separates active from poised enhancers and predicts developmental state 107(50):21931–21936

Cui R, Widlund HR, Feige E, Lin JY, Wilensky DL, Igras VE, D'Orazio J, Fung CY, Schanbacher CF, Granter SR (2007) Central role of p53 in the suntan response and pathologic hyperpigmentation 128(5):853–864

Dancy BM, Cole PA (2015) Protein lysine acetylation by p300/CBP 115(6):2419–2452

Diener J, Baggiolini A, Pernebrink M, Dalcher D, Lerra L, Cheng PF, Varum S, Häusel J, Stierli S, Treier M (2021) Epigenetic control of melanoma cell invasiveness by the stem cell factor SALL4 12(1):5056

Dilshat R, Vu HN, Steingrímsson E (2021) Epigenetic regulation during melanocyte development and homeostasis 30(8):1033–1050

Dutton KA, Pauliny A, Lopes SS, Elworthy S, Carney TJ, Rauch JR, Geisler R, Haffter P, Kelsh RN (2001) Zebrafish colourless encodes sox10 and specifies non-ectomesenchymal neural crest fates. Development. 128(21):4113-25

Elworthy S, Lister JA, Carney TJ, Raible DW, Kelsh RN (2003) Transcriptional regulation of mitfa accounts for the sox10 requirement in zebrafish melanophore development Development 130(12):2809-18

Emmons MF, Bennett RL, Riva A, Gupta K, Carvalho LADC, Zhang C, Macaulay R, Dupéré-Richér D, Fang B, Seto E (2023) HDAC8-mediated inhibition of EP300 drives a transcriptional state that increases melanoma brain metastasis 14(1):7759

Ferlay J, Colombet M, Soerjomataram I, Parkin DM, Piñeros M, Znaor A, Bray F (2021) Cancer statistics for the year 2020. An overview 149(4):778–789

Galdieri L, Vancura A (2012) Acetyl-CoA carboxylase regulates global histone acetylation 287(28):23865–23876

Goding CR, Arnheiter H (2019) MITF—the first 25 years. Genes & Development. 33(15–16):983–1007

Goldberg AD, Banaszynski LA, Noh K-M, Lewis PW, Elsaesser SJ, Stadler S, Dewell S, Law M, Guo X, Li X (2010) Distinct factors control histone variant H3 3 localization at specific genomic regions 140(5):678–691

Gorkin DU, Lee D, Reed X, Fletez-Brant C, Bessling SL, Loftus SK, Beer MA, Pavan WJ, McCallion AS (2012) Integration of ChIP-seq and machine learning reveals enhancers and a predictive regulatory sequence vocabulary in melanocytes 22(11):2290–2301

Greenberg RS, Long HK, Swigut T, Wysocka J (2019) Single amino acid change underlies distinct roles of H2A. Z subtypes in human syndrome. Cell. 178(6):1421–1436. e1424

Greenhill ER, Rocco A, Vibert L, Nikaido M, Kelsh RN (2011) An iterative genetic and dynamical modelling approach identifies novel features of the gene regulatory network underlying melanocyte development 7(9):e1002265

Haberland M, Mokalled MH, Montgomery RL, Olson EN (2009) Epigenetic control of skull morphogenesis by histone deacetylase 8 23(14):1625–1630

Henry RA, Kuo Y-M, Bhattacharjee V, Yen TJ, Andrews AJ (2015) Changing the selectivity of p300 by acetyl-CoA modulation of histone acetylation 10(1):146–156

Hu N, Strobl-Mazzulla P, Sauka-Spengler T, Bronner ME (2012) DNA methyltransferase3A as a molecular switch mediating the neural tube-to-neural crest fate transition 26(21):2380–2385

Ignatius MS, Moose HE, El-Hodiri HM, Henion PD (2008) colgate/hdac1 Repression of foxd3 expression is required to permit mitfa-dependent melanogenesis 313(2):568–583

Im S, Moro O, Peng F, Medrano EE, Cornelius J, Babcock G, Nordlund JJ, Abdel-Malek ZA (1998) Activation of the cyclic AMP pathway by α-melanotropin mediates the response of human melanocytes to ultraviolet B radiation 58(1):47–54

Jamaladdin S, Kelly RD, O'Regan L, Dovey OM, Hodson GE, Millard CJ, Portolano N, Fry AM, Schwabe JW, Cowley SM (2014) Histone deacetylase (HDAC) 1 and 2 are essential for accurate cell division and the pluripotency of embryonic stem cells 111(27):9840–9845

Jostes S, Vardabasso C, Dong J, Carcamo S, Singh R, Phelps R, Meadows A, Grossi E, Hasson D, Bernstein E (2024) H2A. Z chaperones converge on E2F target genes for melanoma cell proliferation. Denes & Development. 38(7–8):336–353

Kawakami A, Fisher DE (2017) The master role of microphthalmia-associated transcription factor in melanocyte and melanoma biology 97(6):649–656

Kelsh RN, Camargo Sosa K, Farjami S, Makeev V, Dawes JH, Rocco A (2021) Cyclical fate restriction: a new view of neural crest cell fate specification 148(22):dev176057

Kim JY, Cho H, Yoo J, Kim GW, Jeon YH, Lee SW, Kwon SHJC (2022) Pathological role of HDAC8: cancer and beyond 11(19):3161

Kobayashi T, Urabe K, Orlow SJ, Higashi K, Imokawa G, Kwon BS, Potterf B, Hearing VJ (1994) The Pmel 17/silver locus protein. Characterization and investigation of its melanogenic function 269(46):29198–29205

Kurdistani SK, Grunstein M (2003) Histone acetylation and deacetylation in yeast 4(4):276–284

LaBonne C, Bronner-Fraser M (1998) Neural crest induction in Xenopus: evidence for a two-signal model 125(13):2403–2414

Laurette P, Strub T, Koludrovic D, Keime C, Le Gras S, Seberg H, Van Otterloo E, Imrichova H, Siddaway R, Aerts S (2015) Transcription factor MITF and remodeller BRG1 define chromatin organisation at regulatory elements in melanoma cells 4:e06857

Le Douarin N, Kalcheim C (1999) The neural crest. Development. 140(11):2247-51

Leathers TA, Rogers CD (2022) Time to go: neural crest cell epithelial-to-mesenchymal transition 149(15):dev200712

Liu X, Wang L, Zhao K, Thompson PR, Hwang Y, Marmorstein R, Cole PA (2008) The structural basis of protein acetylation by the p300/CBP transcriptional coactivator 451(7180):846–850

Louphrasitthiphol P, Loffreda A, Pogenberg V, Picaud S, Schepsky A, Friedrichsen H, Zeng Z, Lashgari A, Thomas B, Patton EE (2023) Acetylation reprograms MITF target selectivity and residence time 14(1):6051

Lowery LA, Sive H (2004) Strategies of vertebrate neurulation and a re-evaluation of teleost neural tube formation 121(10):1189–1197

Lu Y, Travnickova J, Badonyi M, Rambow F, Coates A, Khan Z, Marques J, Murphy LC, Garcia-Martinez P, Marais R (2024) ALDH1A3-acetaldehyde metabolism potentiates transcriptional heterogeneity in melanoma 43(7):114406

Luciani F, Champeval D, Herbette A, Denat L, Aylaj B, Martinozzi S, Ballotti R, Kemler R, Goding CR, De Vuyst F (2011) Biological and mathematical modeling of melanocyte development 138(18):3943–3954

Malcov-Brog H, Alpert A, Golan T, Parikh S, Nordlinger A, Netti F, Sheinboim D, Dror I, Thomas L, Cosson C (2018) UV-protection timer controls linkage between stress and pigmentation skin protection systems 72(3):444–456. e447

Marathe HG, Watkins-Chow DE, Weider M, Hoffmann A, Mehta G, Trivedi A, Aras S, Basuroy T, Mehrotra A, Bennett DC (2017) BRG1 interacts with SOX10 to establish the melanocyte lineage and to promote differentiation 45(11):6442–6458

Martik ML, Bronner ME (2017) Regulatory logic underlying diversification of the neural crest 33(10):715–727

Martínez-Reyes I, Chandel NS (2020) Mitochondrial TCA cycle metabolites control physiology and disease 11(1):102

Martire S, Gogate AA, Whitmill A, Tafessu A, Nguyen J, Teng Y-C, Tastemel M, Banaszynski LA (2019) Phosphorylation of histone H3 3 at serine 31 promotes p300 activity and enhancer acetylation 51(6):941–946

Mehrotra A, Saladi SV, Trivedi AR, Aras S, Qi H, Jayanthy A, Setaluri V, de la Serna IL (2014) Modulation of BRAHMA expression by the mitogen-activated protein kinase/extracellular signal regulated kinase pathway is associated with changes in melanoma proliferation 563:125–135

Mendelson K, Martin TC, Nguyen CB, Hsu M, Xu J, Lang C, Dummer R, Saenger Y, Messina JL, Sondak VK (2024) Differential histone acetylation and super-enhancer regulation underlie melanoma cell dedifferentiation. JCI Insight. 9(6):e166611

Murisier F, Guichard S, Beermann F (2007) The tyrosinase enhancer is activated by Sox10 and Mitf in mouse melanocytes 20(3):173–184

Murko C, Lagger S, Steiner M, Seiser C, Schoefer C, Pusch OJD (2013) Histone deacetylase inhibitor Trichostatin A induces neural tube defects and promotes neural crest specification in the chicken neural tube 85(1–2):55–66

Naikoo SH, Rashid H, Kumar S, Archoo S, Sheikh UA, Lone NA, Singh PP, Tasduq SA (2023) Inhibition of melanogenesis by 3-(1′-methyltetrahydropyridinyl)-2, 4–6-trihydroxy acetophenone via suppressing the activity of cAMP response element-binding protein (CREB) and nuclear exclusion of CREB-regulated transcription coactivator 1 (CRTC1) 952:175734

Natarajan VT, Ganju P, Singh A, Vijayan V, Kirty K, Yadav S, Puntambekar S, Bajaj S, Dani PP, Kar HK (2014) IFN-γ signaling maintains skin pigmentation homeostasis through regulation of melanosome maturation 111(6):2301–2306

Okano M, Bell DW, Haber DA, Li E (1999) DNA methyltransferases Dnmt3a and Dnmt3b are essential for de novo methylation and mammalian development 99(3):247–257

Paluncic J, Kovacevic Z, Jansson PJ, Kalinowski D, Merlot AM, Huang ML-H, Lok HC, Sahni S, Lane DJ, Richardson DR (2016) Roads to melanoma: key pathways and emerging players in melanoma progression and oncogenic signaling 1863(4):770–784

Peinado H, Ballestar E, Esteller M, Cano A (2004) Snail mediates E-cadherin repression by the recruitment of the Sin3A/histone deacetylase 1 (HDAC1)/HDAC2 complex. Molecular and Cellular Biology. 24(1):306-19

Pünzeler S, Link S, Wagner G, Keilhauer EC, Kronbeck N, Spitzer RM, Leidescher S, Markaki Y, Mentele E, Regnard C (2017) Multivalent binding of PWWP2A to H2A. Z regulates mitosis and neural crest differentiation. EMBO J. 36(15):2263–2279

Raible DW, Eisen JS (1994) Restriction of neural crest cell fate in the trunk of the embryonic zebrafish 120(3):495–503

Raja DA, Gotherwal V, Burse SA, Subramaniam YJ, Sultan F, Vats A, Gautam H, Sharma B, Sharma S, Singh A (2020a) pH-controlled histone acetylation amplifies melanocyte differentiation downstream of MITF 21(1):e48333

Raja DA, Subramaniam Y, Aggarwal A, Gotherwal V, Babu A, Tanwar J, Motiani RK, Sivasubbu S, Gokhale RS, Natarajan VT (2020b) Histone variant dictates fate biasing of neural crest cells to melanocyte lineage 147(5):dev182576

Richard V, Kumar TS, Pillai RM (2021) Transitional dynamics of cancer stem cells in invasion and metastasis 14(1):100909

Sato T, Sasai N, Sasai Y (2005) Neural crest determination by co-activation of Pax3 and Zic1 genes in Xenopus ectoderm Development 132(10):2355-63

Sauka-Spengler T, Bronner-Fraser M (2008) A gene regulatory network orchestrates neural crest formation 9(7):557–568

Sauka-Spengler T, Meulemans D, Jones M, Bronner-Fraser M (2007) Ancient evolutionary origin of the neural crest gene regulatory network 13(3):405–420

Seto E, Yoshida M (2014) Erasers of histone acetylation: the histone deacetylase enzymes 6(4):a018713

Shahbazian MD, Grunstein M (2007) Functions of site-specific histone acetylation and deacetylation 76(1):75–100

Simões-Costa M, Bronner MEJD (2015) Establishing neural crest identity: a gene regulatory recipe 142(2):242–257

Singh N, Trivedi CM, Lu M, Mullican SE, Lazar MA, Epstein JA (2011) Histone deacetylase 3 regulates smooth muscle differentiation in neural crest cells and development of the cardiac outflow tract 109(11):1240–1249

Sivasubbu S, Balciunas D, Davidson AE, Pickart MA, Hermanson SB, Wangensteen KJ, Wolbrink DC, Ekker SC (2006) Gene-breaking transposon mutagenesis reveals an essential role for histone H2afza in zebrafish larval development 123(7):513–529

Soldatov R, Kaucka M, Kastriti ME, Petersen J, Chontorotzea T, Englmaier L, Akkuratova N, Yang Y, Häring M, Dyachuk V (2019) Spatiotemporal structure of cell fate decisions in murine neural crest 364(6444):eaas9536

Strobl-Mazzulla PH, Bronner ME (2012) A PHD12–Snail2 repressive complex epigenetically mediates neural crest epithelial-to-mesenchymal transition 198(6):999–1010

Strobl-Mazzulla PH, Sauka-Spengler T, Bronner-Fraser M (2010) Histone demethylase JmjD2A regulates neural crest specification 19(3):460–468

Sultan F, Basu R, Murthy D, Kochar M, Attri KS, Aggarwal A, Kumari P, Dnyane P, Tanwar J, Motiani RK (2022) Temporal analysis of melanogenesis identifies fatty acid metabolism as key skin pigment regulator 20(5):e3001634

Takahashi H, McCaffery JM, Irizarry RA, Boeke JD (2006) Nucleocytosolic acetyl-coenzyme a synthetase is required for histone acetylation and global transcription 23(2):207–217

Thomas AJ, Erickson CA (2009) FOXD3 regulates the lineage switch between neural crest-derived glial cells and pigment cells by repressing MITF through a non-canonical mechanism Development 136(11):1849-58

Vandamme N, Berx G (2019) From neural crest cells to melanocytes: cellular plasticity during development and beyond 76:1919–1934

Waddell A, Grbic N, Leibowitz K, Wyant WA, Choudhury S, Park K, Collard M, Cole PA, Alani RM (2024) p300 KAT regulates SOX10 stability and function in human melanoma 4(8):1894–1907

Wellen KE, Hatzivassiliou G, Sachdeva UM, Bui TV, Cross JR, Thompson CB (2009) ATP-citrate lyase links cellular metabolism to histone acetylation 324(5930):1076–1080

Weston JA (1982) Neural crest cell development 85:359–379

White RM, Zon LI (2008) Melanocytes in development, regeneration, and cancer 3(3):242–252

Yang YK, Dickinson C, Haskell-Luevano C, Gantz IJT (1997) Molecular basis for the interaction of [Nle4, D-Phe7] melanocyte stimulating hormone with the human melanocortin-1 receptor 272(37):23000–23010

Zhang X, Ouyang S, Kong X, Liang Z, Lu J, Zhu K, Zhao D, Zheng M, Jiang H, Liu X (2014) Catalytic mechanism of histone acetyltransferase p300: from the proton transfer to acetylation reaction 118(8):2009–2019

Part IV
Acetylation in Physiological and Pathological Processes

Chapter 8
Impact of Histone Acetyltransferases and Histone Deacetylases on Adult Brain Myelin Plasticity

Alessandra Dominicis, Tommaso Fabiano, Simone Peria, Aland Ibrahim Ahmed Al Jaf, and Antonella Ragnini-Wilson

Abstract Myelin plasticity is a key process for acquiring new motor skills and preventing neurodegeneration during ageing. Neural precursor cells (NPCs) and parenchymal oligodendrocyte precursor cells (OPCs) play a key role in myelin plasticity in the central nervous system (CNS), being specialized in reconstituting the myelin sheath upon damage. Reversible acetylation, regulated by histone acetyltransferases (HATs) and histone deacetylases (HDACs) activity, controls these stem cells' differentiation in myelinating oligodendrocytes (mOLs) during their proliferation and remyelination processes. By modulating cytosolic protein activity and precisely orchestrating the spatial and timely regulated activity of the transcription factors participating in the NPC and OPC differentiation process, these enzymes play a vital role in preserving the adult brain's cognitive capacity during ageing. This review highlights the role of reversible acetylation in the regulation of stem cell differentiation during remyelination, as disruptions in this process contribute to severe neurodegenerative impairments and accelerated ageing.

Abbreviations

Acetyl-CoA	Acetyl coenzyme A
AD	Alzheimer's disease
AGC1	Mitochondrial aspartate-glutamate carrier isoform 1

A. Dominicis · S. Peria · A. I. A. Al Jaf · A. Ragnini-Wilson (✉)
Department of Biology, University of Rome "Tor Vergata", Rome, Italy
e-mail: alessandra.dominicis.01@alumni.uniroma2.eu;
ALANDIBRAHIMAHMED.ALJAF01@icatt.it; antonella.ragnini@uniroma2.eu

T. Fabiano
Department of Biology, University of Rome "Tor Vergata", Rome, Italy

Department of Life Sciences, University of Trieste, Trieste, Italy

M. Halasa, A. Wawruszak (eds.), *Histone and Non-Histone Reversible Acetylation in Development, Aging and Disease*, Results and Problems in Cell Differentiation 75,
https://doi.org/10.1007/978-3-031-91459-1_8

APP	Amyloid precursor protein
AR	Androgen receptor
ASPA	Aspartoacylase
aSyn	α-synuclein
BBB	Blood-brain barrier
Bcas1	Brain enriched myelin-associated protein 1
BMPs	Bone morphogenetic proteins
CBP	CREB-binding protein
CC	Corpus callosum
CD	Canavan disease
ChIP	Chromatin immunoprecipitation
CNPase	2′,3′-Cyclic nucleotide 3′-phosphodiesterase
CNS	Central nervous system
eEF1A1	Eukaryotic translation elongation factor 1 alpha 1
ER	Oestrogen receptor
FAS	Fatty acid synthase
FGF	Fibroblast growth factor
GalC	Galactocerebroside
GCN5	General control nonderepressible 5
Gli	Glioma-Associated Oncogene Family Zinc Finger
GNATs	General control nonderepressible 5-related N-acetyl transferases
GPR17	G protein-coupled receptor 17
HATs	Histone acetyltransferases
HD	Huntington's disease
HDACi	Histone deacetylase inhibitor
HDACs	Histone deacetylases
Id	Inhibitor of DNA binding
Klf	Krüppel-like factor
KO	Knockout
Mash1	Mammalian achaete-scute homolog 1
MBP	Myelin basic protein
MOG	Myelin oligodendrocyte glycoprotein
mOLs	Myelinating oligodendrocytes
MS	Multiple sclerosis
MyRF	Myelin regulatory factor
Myt1	Myelin transcription factor 1
NAA	N-acetylaspartate
NAD^+	Nicotinamide adenine dinucleotide
NCOA	Nuclear receptor coactivator
NCOR	Nuclear receptor corepressor
NFOs	Newly formed oligodendrocytes
NG2	Neural/glial antigen 2
Nkx2.2	NK2 Homeobox 2
NPCs	Neural precursor cells
Olig	Oligodendrocyte transcription factor

OLs	Oligodendrocytes
OPCs	Oligodendrocyte precursor cells
OTX2	Orthodenticle homeobox 2
PAX6	Paired Box 6
PD	Parkinson's disease
PDGFRα	Platelet-derived growth factor receptor alpha
PDH	Pyruvate dehydrogenase
PLP	Proteolipid protein
PPARγ	Peroxisome proliferator-activated receptor gamma
RA	Retinoic acid
SB	Sodium butyrate
SGZ	Subgranular zone
Shh	Sonic Hedgehog
SIRTs	Sirtuins
Smo	Smoothened
SMRT	Silencing mediator of retinoic acid and thyroid hormone receptors
SOX	Sex-determining region Y-box
SREBPs	Sterol regulatory element-binding proteins
STAT3	Signal transducer and activator of transcription 3
SVZ	Subventricular zone
TCA	Tricarboxylic acid cycle
TCF	Transcription factor
TSA	Trichostatin A
VPA	Valproic acid
WM	White matter

8.1 Introduction

Epigenetic signatures and transcriptional changes can act as potent factors altering the course of ageing and neurodegenerative diseases. Epigenetic mechanisms mediate transient or stable hereditary changes in cell transcription programs in response to external and environmental signals, thus constituting an essential pillar of brain plasticity. Acetylation and deacetylation are among the epigenetic signatures regulating the adaptation and plasticity of the central nervous system (CNS). A correct balance between acetylation and deacetylation is essential for maintaining cognitive function during ageing. Indeed, increasing evidence links the imbalance of histone acetylation/deacetylation to different pathologies affecting the CNS (Conway et al. 2012; Egawa et al. 2019; Tapias and Wang 2017).

Myelin plasticity depends on the capacity of oligodendrocyte precursor cells (OPCs) to regenerate the myelin sheath that covers the axons. The regulatory role of acetylation and deacetylation in OPC differentiation is therefore fundamental for the maintenance of brain function during ageing. Myelin is continuously generated over life and the process of axon remyelination is required to perform novel cognitive tasks, including motor skill learning and recalling fear spatial memory (Hill et al. 2018; Hughes et al. 2018; McKenzie et al. 2017; Pan and Chan 2021; Steadman et al. 2020; F. Wang et al. 2020). The importance of remyelination, at any stage of life, is highlighted by the neurodegenerative pathologies that arise from the defective remyelination process. Among the most studied is multiple sclerosis (MS), an autoimmune inflammatory pathology characterized by the inability of OPCs to differentiate into mature oligodendrocytes (OLs), with severe consequences on the motor and cognitive functions of affected patients (Bebo et al. 2022). Demyelination has also been observed as an early event in a variety of age-related neurodegenerative diseases among which Alzheimer's disease (AD) and Parkinson's disease (PD), suggesting that also in these pathologies alteration in OPC differentiation might be at least in part at the basis of the neurodegenerative processes observed (Chen et al. 2021; Dean et al. 2016; Papuć and Rejdak 2018; Xie et al. 2022; Yang et al. 2023). Recently, genetically associated psychiatric syndromes such as schizophrenia (Valdés-Tovar et al. 2022), and some neurodevelopmental disorders that involve deficits in interpersonal communication, social interaction, and repetitive behaviours of autism spectrum disorder (ASD) were featured as potential consequences of defective remyelination (Galvez-Contreras et al. 2020; Graciarena et al. 2019).

To understand how acetylation and deacetylation can regulate myelin plasticity during ageing and myelin injury recovery, it is necessary to describe how new myelin sheath is generated during remyelination processes. Here, we will first examine how oligodendrocytes participate in acetyl coenzyme A (Acetyl-CoA) generation and myelin formation. In the following paragraphs, we will describe how histone acetyltransferases (HATs) and histone deacetylases (HDACs) participate in the process that allows myelin plasticity by describing how oligodendrogenesis is regulated during remyelination and how different classes of HATs and HDACs participate in this process. Finally, we will describe how histone modification participates in some pathologies affecting myelin regeneration.

8.2 Reversible Acetylation in New Myelin Formation: The Metabolic Pathway

Histone acetylation occurs on positively charged lysine residues and thus weakens DNA-histone binding, opening areas of the nucleosome to make them more readable. HATs are responsible for histone acetylation, by neutralizing the positive load of lysine residues, causing weaker interactions between histone proteins and DNA, eventually resulting in an "open" chromatin structure, and therefore chromatin relaxation and transcription increase. HATs catalyse the acetylation by transferring an acetyl group of the Acetyl-CoA to the ε-amino group of lysine residues of target proteins (Roth et al. 2001). On the other hand, the removal of the acetyl group is exerted by the HDACs (Berry et al. 2020).

To better understand the role of Acetyl-CoA in oligodendrocyte development we have first to briefly summarize the metabolic pathways that can lead to the synthesis of this molecule, which is crucial for lipid synthesis and histone/protein acetylation (Douvaras et al. 2016; Magistretti and Allaman 2018; Sajad et al. 2024; Yan and Rivkees 2006). Acetyl-CoA is a thioester, mainly derived from glucose and anabolized by glycolysis or the pentose phosphate pathway (PPP) and finally transformed into pyruvate. The latter can be used for Acetyl-CoA formation in the mitochondria, by the enzyme pyruvate dehydrogenase (PDH). During the transition between OPCs to myelinating oligodendrocytes (mOLs) there is high protein and lipid synthesis, which results in increased mitochondrial metabolism, high oxidative stress and iron accumulation (Amaral et al. 2016; Sajad et al. 2024). Indeed, PDH decreases during cuprizone-induced myelin damage allowing the proliferation and migration of OPCs to the site of the myelin lesion, while an increased PDH kinase (PDH inhibitor) lowers the production rate of myelin basic protein (MBP) (Sajad et al. 2024).

Sajad et al. (2024) demonstrated how Acetyl-CoA is involved in the acetylation of the oligodendrocyte transcription factor (Olig) 1 that determines its translocation from the nucleus towards the cytoplasm (Sajad et al. 2024). In addition, it is incorporated into the tricarboxylic acid cycle (TCA), producing 10% of the total glutamate produced by the Krebs cycle (Amaral et al. 2016).

The main source of Acetyl-CoA for oligodendrocytes comes from N-acetylaspartate (NAA), supplied by neurons and metabolized by the aspartoacylase (ASPA) into aspartate and acetate in OLs (Balboni et al. 2024; Moffett et al. 2007). The acetate can be used for Acetyl-CoA production, contributing to myelin lipid synthesis in the CNS and supporting the oxidative metabolism during myelination in addition to being oxidized in mitochondria, taking part in the TCA cycle (Ariyannur et al. 2010; Amaral et al. 2016; Chakraborty et al. 2001; Francis et al. 2016). Potentially, the aspartate released by NAA hydrolysis in oligodendrocytes could be sent back to neurons, thus avoiding the depletion of anaplerotic substrates. However, Amaral et al. (2017), using a labelled precursor ([1,6-13C] glucose), demonstrated that aspartate is not released to the medium by oligodendrocytes in culture, but it accumulates inside the cell and, probably, takes part in the malate-aspartate shuttle (Amaral et al. 2016; Amaral et al. 2017). The malate–aspartate shuttle is tightly coupled to glycolysis as it is used to carry reducing equivalents across the inner membrane of the mitochondria, impermeable to nicotinamide adenine dinucleotide phosphate (NADPH). Among the carriers involved in the malate-aspartate shuttle, the mitochondrial aspartate-glutamate carrier isoform 1 (AGC1), encoded by the solute carrier family 25-member 12 (*SLC25A12*) gene, has been deeply studied as a mutation on this gene causes severe diseases characterized by hypotonia and hypomyelination (Balboni et al. 2024). A reduced AGC1 activity has been associated with epigenetic/transcriptional changes that affect the OPCs pool maintenance and their differentiation towards OLs. Interestingly, studies using neural precursor cells (NPCs) neurospheres from the subventricular zone (SVZ) have revealed that post-translational modifications are affected by *AGC1* downregulation: histone H3 acetylation appeared reduced as well as p300/CREB-binding protein (CBP), the most important protein of the HATs family involved in brain development (Poeta et al. 2021). Furthermore, *AGC1* inhibition reduces proliferation by inhibiting platelet-derived growth factor receptor alpha (PDGFRα) and simultaneously promotes OPC differentiation (Fang et al. 2019; Petralla et al. 2019). Moreover, *AGC1* silencing led to important readouts in the brain related to the decrease of NAA, TCA products, aspartate, and glutamate and reduced levels of transcriptional factors and enzymes required for lipid synthesis, namely the sterol regulatory element-binding proteins (SREBPs), retinoic acid (RA), and fatty acid synthase (FAS) (Balboni et al. 2024; Falk et al. 2014).

Both acetate and aspartate derived from NAA can act as a source for histone acetylation, although recent studies have shown that the increase in mitochondrial metabolism triggered by NAA leads to a decrease in overall histone acetylation levels (Singhal et al. 2017). Singhal et al. (2017) demonstrated that NAA increases the deacetylation of histone H3, which is essential for OPC differentiation (Singhal et al. 2017). These data are supported also by our recent in vitro study that highlights a strong correlation between NAA, *HDACs* expression and OPC differentiation. We showed that the external administration of NAA to oligodendroglia mouse cell line Oli-neuM promotes their differentiation till the stimulation of synthetic axon engagement as well as a decrease in the total acetylation levels by promoting an increase in specific *HDACs* expression (*HDAC1*, *6*, and *11*). Since the NAA promyelinating effect in combination with the ACY-738 HDAC inhibitor (HDACi) is reduced, we concluded that HDAC activity is required upon NAA treatment for oligodendroglia differentiation in vitro (Dominicis et al. 2023).

In summary, growing pieces of evidence highlight the crucial role of acetyl metabolism not only for its metabolic functions but also for its role in regulating OPC differentiation in the adult brain.

8.3 HATs and HDACs Families

HATs are enzymes that can be classified based on their localization, either in the cytoplasm or in the nucleus. In the cytoplasm, HATs acetylate newly synthesized histone and non-histone proteins, which can then be transported into the nucleus for further modification or to interact with downstream factors. Nuclear HATs can directly acetylate nuclear histones, resulting in altered chromatin compaction and a change in gene expression (Roth et al. 2001). Up to date, about 30 human HATs are known and classified into four different families based on the functional and structural similarity of their catalytic domains (Dekker and Haisma 2009). HATs are distinct independent multi-subunit complexes that can be finely regulated, ensuring them a high substrate specificity and selectivity. This behaviour allows HATs to target specific locations while avoiding non-target proteins, ensuring precise acetylation without modifying accidental substrates (Berndsen and Denu 2008; Wapenaar and Dekker 2016).

Functionally, HATs can be classified into four major families: the general control nonderepressible 5 (GCN5)-related N-Acetyl transferases (GNATs) family, the

MYST family, the EP300/CREBBP (CBP) family, and the Nuclear receptor coactivator (NCOA) family (Table 8.1) (Kamei et al. 1996; Roth et al. 2001; Sterner and Berger 2000; Wapenaar and Dekker 2016).

The GNATs family name derives from the structural similarity to yeast HAT GCN5, as well as to some distantly related HATs like HAT1, elongator complex protein 3 (ELP3), and heparinase 2 (Hpa2), which share a highly conserved acetylation-related structural motif (Choi et al. 2023; Roth et al. 2001).

CBP is a key member of the EP300/CREBBP family, shown to mediate the differentiation of NPCs to OPCs (González-Martínez et al. 2022). Notably, Dai et al.

Table 8.1 Categorization of HATs and their interaction in oligodendrocyte maturation. N.A.: Not Available

Family	Protein	Cell localization	Interactors/targets	Inhibitors
GNAT	KAT2A	Nucleus (Arede and Pina 2021)	SAGA, ATAC (Arede and Pina 2021; Fournier et al. 2016)	CPTH2, MB-3 (Zhang et al. 2023)
	KAT2B	Nucleus (Arede and Pina 2021)	SAGA, ATAC (Arede and Pina 2021; Fournier et al. 2016)	L-Moses (Pavlou et al. 2023)
MYST	KAT5	Nucleus (He et al. 2013)	Androgen receptor (AR) (He et al. 2013)	NU9056 (Xu et al. 2022)
	KAT6A (MYST3)	Nucleus (Ullah et al. 2008)	H3, H4, p53, Runx1 (Marmorstein and Trievel 2009; Voss and Thomas 2009)	PF-07248144 (Mukohara et al. 2024)
	KAT6B (MYST4)	Nucleus (He et al. 2013)	Brahma (BRM), ING5 (He et al. 2013)	PF-07248144 (Mukohara et al. 2024)
	KAT7 (MYST2)	Nucleus (Yan et al. 2018)	VEGFR-2, VEGFR-1 (Yan et al. 2018)	WM-3835 (Mi et al. 2023)
	KAT8 (MYST1)	Nucleus/cytosol (Burrell and Stephens 2021)	H4K16, H4K5, H4K8, LSD1, DBC1 (Burrell and Stephens 2021)	KAT8 inhibitors 19 and 34 (Fiorentino et al. 2023)
EP300/ CREBBP	CBP (KAT3A)	Nucleus/cytosol (Dai et al. 2015)	Olig1 (Dai et al. 2015)	C646 (Dai et al. 2015)
	p300 (KAT3B)	Nucleus (Zeng et al. 2023)	AR, Oestrogen receptor (ER), Maml1, YTHDF2 (Zeng et al. 2023)	A485, C646 (Dai et al. 2015; Zeng et al. 2023)
NCOA	NCOA1	Nucleus (Sun and Xu 2020)	M-CSF1 (Qin et al. 2014)	Bufalin, gossypol (Wang et al. 2011; Wang et al. 2014)
	NCOA2	Nucleus (Sun and Xu 2020)	KAT2B, NF-κB, IL-1B, IL-6, TNFA, ICAM1, VCAM1, CD38, CD40, CD80, NOS2 (Zhou et al. 2022)	N.A.
	NCOA3	Nucleus (Sun and Xu 2020)	ER, XBP1-S (Gupta et al. 2016)	Bufalin, gossypol (Wang et al. 2011; Wang et al. 2014)

(2015) demonstrate that CBP acetyltransferase is responsible for Olig1 acetylation and the subsequent translocation in the cytoplasm, which signs the passage from PDGFR+ to CC1+ cells, and from immature to mature (not myelinating) OLs, as CBP inhibition by HAT p300 inhibitor C646 blocks OPC differentiation (Fig. 8.1) (Dai et al. 2015; Montani 2021). Moreover, genetic and RNA knockdown of the *CBP* in the embryonic forebrain resulted in decreased expression of oligodendrocyte markers in cortical NPCs, and *CBP* haploinsufficiency leads to aberrant development of the corpus callosum (CC) (J. Wang et al. 2010).

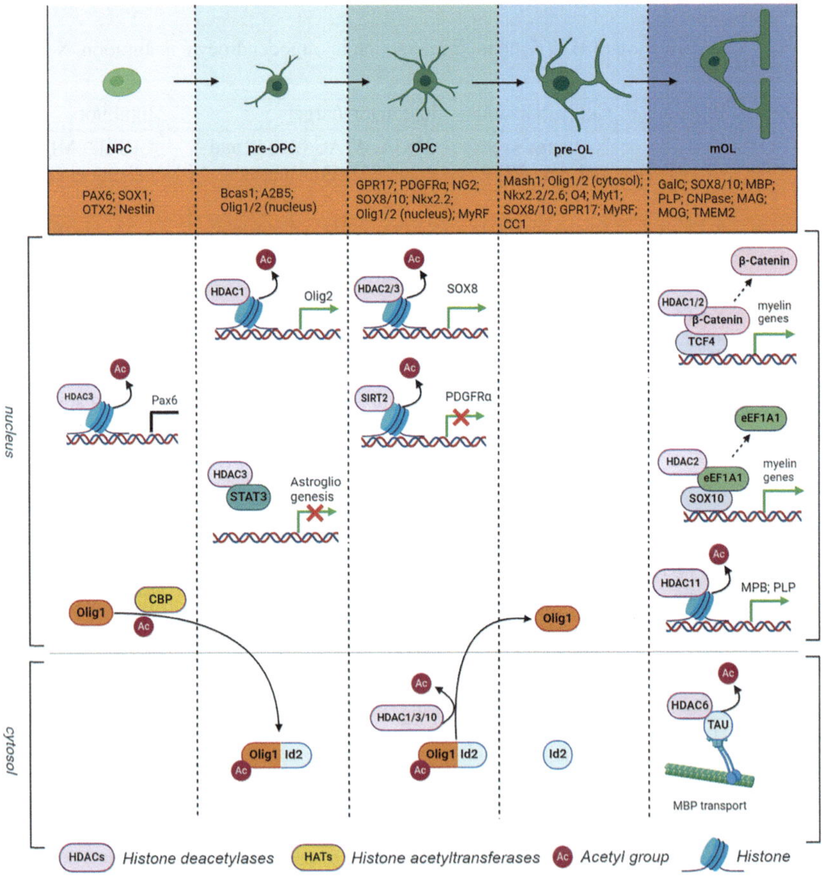

Fig. 8.1 Influence of acetylation and deacetylation on oligodendrocyte lineage progression. The differentiation of NPC to oligodendrocytes in the adult brain can be described in 5 steps: NPC, pre-OPC, OPC, pre-OL, and mOL (the markers reported in the orange box have been added taking into consideration Gregath and Lu 2018; Hughes and Stockton 2021; Juarez et al. 2016; Long et al. 2021). Within each stage, different HDAC isoforms have an important impact on the progression of OLs maturation, both at a histonic and protein level. A detailed description of the various molecular mechanisms represented in the figure can be found in the text. (Created in BioRender. Ragnini, A. (2025) https://BioRender.com/g76z235)

On the contrary, HDACs are enzymes that remove acetyl groups from N-acetyl lysine amino acids in histones. The deacetylation of histones, exerted by HDACs, compacts chromatin and prevents transcriptional processes. Based on their structural characteristics, mammalian HDACs can be classified into four major classes. Class I includes HDAC1, HDAC2, HDAC3, and HDAC8; class II comprehends HDAC4, HDAC5, HDAC6, HDAC7, HDAC9, and HDAC10; class III is represented by sirtuins (SIRTs) 1–7; and class IV includes HDAC11 (Table 8.2) (Koreman et al. 2018).

Among the HDACs, SIRTs family mammal proteins are characterized by a catalytic area that shares the nicotinamide adenine dinucleotide (NAD+) binding, which can act on different substrates depending on biological processes (Zhao et al. 2020).

Individual *HDACs* exhibit a different expression pattern during OL development (Table 8.3) (Gregath and Lu 2018; L. Zhang et al. 2016; Y. Zhang et al. 2014). In 2014, Zhang et al. purified oligodendrocytes and other cell types from the cortex of the mouse and generated a wide transcriptome database of more than 22.000 genes (Zhang et al. 2014; see brainrnaseq.org). Specifically, they distinguished the maturation process of OPCs in mOLs in three stages: OPCs, newly formed oligodendrocytes (NFOs) and mOLs. OPCs were selected by using an anti-PDGFRα coated plate; mature OLs were harvested by using an anti-myelin oligodendrocyte glycoprotein (MOG) coated plate and NFOs were isolated using an anti-galactocerebroside (GalC) hybridoma-coated plate (Zhang et al. 2014).

In Table 8.3, we report the different expressions of *HDAC* genes at these three stages in the mouse brain. It is possible to observe that most of the *HDACs* decrease their expression during the OL maturation stages. Notably, *HDAC5* and *HDAC11* expression exhibits an opposite trend, by increasing during OL maturation. This could be related to their specific role in this process (see below, Sect. 8.4.3). Additional confirmation that specific HDACs are active in different limited time windows comes from the fact that HDAC inhibitors suppress myelination only during the early phase of OPCs differentiation and not after the onset of myelination (Shen et al. 2005). In 2019, this result was confirmed by Egawa et al. (2019) through the observation that *HDACs* are highly expressed in early OPC stages in comparison to other OL maturation steps (Egawa et al. 2019).

Notably, HDACs work in complexes together with some interactors. The HDAC-associated DNA binding transcription factors are a class of proteins regulating neural proliferation and differentiation during brain development. This class of proteins comprises the neuron-restrictive silencer factor (NRSF), the nuclear receptor corepressor (NCOR/Ncor1), and the silencing mediator of RA and thyroid hormone receptors (SMRT/Ncor2). NCOR and/or SMRT-HDAC complexes modulate histone modifications arising at single lysines of histone tails seeming to display progenitor-specific signatures. NCOR plays a critical role in a repression checkpoint of OLs differentiation in embryonic multipotent NPCs and its absence starts specification into OLs cell types (Castelo-Branco et al. 2014).

Table 8.2 Classification of HDACs in classes, localization, and their main interactors. Agonists and inhibitors affecting oligodendrocyte development are listed

CLASS	Mechanism	Protein	Cell localization	Interactors/ targets in oligodendrocytes development	Inhibitors	Agonists
I	Zn^{2+}-dependent, arginase-like	HDAC1	Nucleus	YY1, Transcription Factor (TCF) 4; Olig2, Olig1 (Berry et al. 2020; Gregath and Lu 2018)	CI994 (Wheeler et al. 2015)	Autotaxin (Wheeler et al. 2015)
		HDAC2	Nucleus	TCF4; eukaryotic translation elongation factor 1 alpha 1 (eEF1A1) (Duman et al. 2020; Gregath and Lu 2018)	CI994 (Wheeler et al. 2015), MI192 (Al-Griw et al. 2021)	Theophylline; autotaxin (Duman et al. 2020; Wheeler et al. 2015)
		HDAC3	Nucleus	Signal transducer and activator of transcription 3 (STAT3), p300; Olig1; PPARγ; FAS (Balboni et al. 2024; Gregath and Lu 2018; Kanakasabai et al. 2012)	CI994 (Wheeler et al. 2015), VPA (Ding et al. 2020); MI192 (Al-Griw et al. 2021), RGFP966 (Liu et al. 2025)	N.A.
		HDAC8	Nucleus/ cytoplasm	EST1B; SMC3; p53 (Zhao et al. 2024)	PCI-34051 (Luo et al. 2025)	N.A.
II	Zn^{2+}-dependent, arginase-like. Work only associated with I class HDACs (Fischle et al. 2002)	HDAC4	Nucleus/ cytoplasm	N.A.	LMK-235 (Huang et al. 2024)	N.A.
		HDAC5	Nucleus/ cytoplasm	N.A.	LMK-235 (Huang et al. 2024)	N.A.
		HDAC6	Cytoplasm	Tau (Noack et al. 2014)	ACY-738; Tubastatin A (Wheeler et al. 2015; Phipps et al. 2024)	N.A.
		HDAC7	Nucleus/ cytoplasm	N.A.	N.A.	N.A.
		HDAC9	Nucleus/ cytoplasm	N.A.	N.A.	N.A.
		HDAC10	Cytoplasm	N.A.	N.A.	N.A.

(continued)

Table 8.2 (continued)

CLASS	Mechanism	Protein	Cell localization	Interactors/ targets in oligodendrocytes development	Inhibitors	Agonists
III	NAD^+-dependent deacetylation of the target protein	SIRT1-7	Nucleus/ cytoplasm	H4K16 (Vaquero et al. 2007)	Selisistat (SIRT1; Qu et al. 2024), Sirtinol (SIRT1 and SIRT2; Wu et al. 2023), 3-TYP (SIRT3; Zhou et al. 2025)	Resveratrol-like compounds (SIRT1; Ciccone et al. 2022)
IV	Zn^{2+}-dependent, arginase-like	HDAC11	Nucleus/ cytoplasm	H3K9, H3K14 residues (Berry et al. 2020)	SIS17 (Xie et al. 2025)	N.A.

N.A. Not Available

8.4 Acetylation and Deacetylation in Myelin Formation

OPCs and NPCs are essential for adult brain myelin repair. In neurovascular niches like the SVZ and subgranular zone (SGZ), NPCs differentiate into oligodendrocytes under the control of HATs and HDACs that regulate cell identity and the timing of oligodendroglial development (Berry et al. 2020; Llorente et al. 2022). This paragraph examines how HATs and HDACs influence NPC differentiation into OLs, offering insights into potential treatments for neurodegenerative diseases.

8.4.1 Regulation of NPC and OPC Differentiation During Remyelination

The main neurovascular niches of the brain include the SVZ and the SGZ (Llorente et al. 2022). Here, NPCs can give rise to astrocytes, oligodendrocytes, and neurons. The regulatory events leading to NPC-fating to oligodendroglia lineage are controlled by HATs and HDACs that safeguard their identity and prevent them from becoming neurons or astrocytes (Berry et al. 2020).

The NPC-fating into the oligodendroglia lineage is regulated by factors present in the neurovascular niches. Upon demyelination, chemotactic factors stimulate NPCs to become OPCs; meanwhile, parenchymal OPCs, dispersed in the white matter (WM), get stimulated to proliferate and migrate towards the lesion, where they differentiate into mature mOLs (Fig. 8.1; Tiane et al. 2019).

Table 8.3 Differential expressions of the *HDACs* along the three different stages of oligodendrocytes maturation in mice

CLASS	GENE	Mouse			Human
		OPC	NFO	mOL	OL
I	HDAC1	**1.49 ± 0.18**	1.38 ± 0.12	1.07 ± 0.05	6.00 ± 0.92
	HDAC2	**38.30 ± 5.25**	30.63 ± 6.00	10.72 ± 0.51	3.29 ± 0.64
	HDAC3	**38.74 ± 3.99**	24.52 ± 4.47	14.55 ± 1.24	2.09 ± 0.39
	HDAC8	**8.92 ± 0.09**	5.09 ± 0.05	3.33 ± 0.17	2.80 ± 0.44
II	HDAC4	**2.86 ± 0.21**	1.35 ± 0.03	0.90 ± 0.30	0.76 ± 0.13
	HDAC5	18.88 ± 5.44	30.03 ± 1.55	**39.66 ± 0.36**	0.34 ± 0.07
	HDAC6	**16.36 ± 1.60**	5.70 ± 1.27	1.96 ± 0.46	1.08 ± 0.16
	HDAC7	**4.54 ± 0.07**	3.19 ± 0.24	1.91 ± 0.60	0.09 ± 0.01
	HDAC9	**2.35 ± 0.35**	1.71 ± 0.63	1.71 ± 0.17	2.57 ± 0.39
	HDAC10	**5.04 ± 0.14**	3.68 ± 0.13	2.87 ± 0.26	0.090 ± 0.01
III	SIRT1	5.18 ± 1.52	**5.59 ± 1.18**	0.82 ± 0.25	3.92 ± 0.59
	SIRT2	N.A.	N.A.	N.A.	N.A.
	SIRT3	11.21 ± 2.45	8.33 ± 1.07	**13.60 ± 1.34**	1.15 ± 0.22
	SIRT4	**6.75 ± 2.49**	4.17 ± 0.62	2.19 ± 0.16	0.16 ± 0.03
	SIRT5	**4.78 ± 0.77**	1.86 ± 0.22	0.69 ± 0.30	1.01 ± 0.15
	SIRT6	**5.67 ± 0.37**	3.96 ± 0.27	2.00 ± 0.33	0.09 ± 0.01
	SIRT7	13.32 ± 0.58	**17.28 ± 0.86**	13.55 ± 0.19	0.17 ± 0.03
IV	HDAC11	17.51 ± 4.65	25.61 ± 2.97	**57.2 ± 0.45**	2.68 ± 0.50

Data from Zhang et al. (2014). Values are reported by Fragments Per Kilobase of transcript per Million mapped reads (FPKM). The highest level of expression for each *HDAC* is written in bold. Source: https://brainrnaseq.org/ https://www.ncbi.nlm.nih.gov/pmc/articles/PMC5858588/ (Keil et al. 2018). *N.A.* Not Available

The proliferation behaviour of quiescent NPCs is regulated by the fibroblast growth factor (FGF) 2, and the bone morphogenetic proteins (BMPs) 4 and 7 (BMP4 and BMP7, respectively) (Douet et al. 2013; Llorente et al. 2022). In the neurogenetic niches, Sonic Hedgehog (Shh) signalling promotes NPCs proliferation in part through HDAC1 and HDAC2, while treatment with BMP4 decreases HDACs activity and hampers differentiation (Koreman et al. 2018). Chromatin modifications progressively restrict the expression of pluripotency and multipotency genes of NPCs and facilitate the initiation of the differentiation program, which is linked to a rise in NPCs markers such as Paired Box 6 (PAX6), sex-determining region Y-box (SOX) 1, orthodenticle homeobox 2 (OTX2), and Nestin (Long et al. 2021; Qiao et al. 2016). Conversely, BMPs and transcription factors SOX5/6 prevent OPC differentiation (Fig. 8.1) (Miron et al. 2011).

NPC-fated to the oligodendroglia lineage are specified into pre-OPC. Pre-OPCs are characterized by the expression of high levels of A2B5+, Olig1+, and Olig2+ (Xiao et al. 2016). The OPC stage is defined by the expression of *PDGFRα* and neural/glial antigen 2 (*NG2)* (Gregath and Lu 2018; Hughes and Stockton 2021).

Upon demyelination, chemotactic factors stimulate parenchymal OPCs proliferation and their migration at the lesion, where they differentiate into mature mOLs

(Tiane et al. 2019). The parenchymal OPCs are the first cell type responding to a demyelination stimulus, as shown by a CreER-loxP fate mapping of NPCs and OPCs used to directly assess the relative contribution of these distinct precursor populations to oligodendrogenesis and remyelination after cuprizone-induced demyelination of the CC. This study indicates that demyelination induces the migration of large numbers of NPCs into the CC, where they adopt a predominantly oligodendroglial fate and contribute substantially to remyelination (Gregath and Lu 2018; Hughes and Stockton 2021; Xing et al. 2014). A recent study confirmed that while parenchymal OPCs are the main remyelinating cells in the CC, NPC-derived cells are also activated and recruited at demyelinating lesions. Indeed, blocking parenchymal OPC differentiation genetically resulted in a significant increase in NPC-derived cell recruitment in the demyelinated CC and their differentiation into OLs. These results support the hypothesis that parenchymal OPCs and NPCs-progeny compete to repair WM lesions (Moyon et al. 2023).

In the cortical niches, where the proliferation of the NPCs initiates, the Glioma-Associated Oncogene Family Zinc Finger (*Gli*) expression is upregulated due to the presence of the Shh morphogen that activates the smoothened (Smo) receptor signalling. The parallel upregulation of *FGF*, epidermal growth factor receptor (*EGFR*), and brain-derived neurotrophic factor *(BDNF*) leads to NPCs proliferation. Smo is a morphogenetic factor that regulates both proliferation and differentiation (Namchaiw et al. 2019). Gli1 is the main effector of Smo and is regulated by Gli2 and Gli3 proteins. The balance between Gli1 and Gli2 is important for the fating of NPCs to the oligodendroglial lineage (Radecki et al. 2020). Gene silencing experiments performed in mouse model for demyelination and remyelination indicated that *Gli1* downregulation is essential to fate NPCs in the glial lineage while Smo is not (Nocera et al. 2024; Radecki et al. 2020; Samanta et al. 2015). The transcription factor Gli2 is endogenously acetylated at a single conserved lysine (K757): this modification decreases the transcriptional activity by preventing its promoter recruitment. Gli2 acetylation is induced by p300 and mutation of K757 abrogates Gli2 acetylation. Gli2 WT, K757R, and K757Q mutants do not exhibit differences in their cellular localization, which emerges to be predominantly nuclear, suggesting that Gli2 inhibition by acetylation is a nuclear process. Therefore, acetylation seems to prevent Gli2 recruitment to target chromatin, and since the residue K757 maps outside the zinc finger domain (DNA binding domain) it is unlikely that acetylation affects Gli2/DNA affinity. One possible explanation is that K757 acetylation prevents the association with chromatin-bound cofactors, which are essential for the promoter entry/transactivation of Gli2. Thus, this data holds promising therapeutic potential for diseases that are linked to aberrant Hedgehog (Hh) pathway activation (Coni et al. 2013).

OPCs are lifelong residents of the CNS and consist of a heterogenous population present in both grey and WM (Boshans et al. 2020; Dimou and Simons 2017; Foerster et al. 2019; Jäkel et al. 2019). The process of differentiation of OPCs to OLs passes through four phases: OPCs, pre-mOLs, new oligodendrocytes, and mOLs (Hughes and Stockton 2021). Each of the four OPC differentiation stages is characterized by a specific gene expression pattern and requires epigenetic remodelling of the main transcription factors involved in differentiation (Hughes and Stockton 2021; Liu et al. 2015; Liu and Casaccia 2010). Activation of OPCs is associated with an upregulation of *Olig2* and NK2 Homeobox 2 (*Nkx2.2*) transcription factors, while the differentiation along the oligodendroglial lineage to pre-OLs and mOLs is signed by the expression of the surface antigen O4 and GalC, respectively. Pre-OLs are characterized by the expression of high levels of the G protein-coupled receptor 17 (GPR17) together with brain enriched myelin-associated protein 1 (Bcas1) and start to express ectonucleotide pyrophosphatase/phosphodiesterase 6 (Enpp6) (Hughes and Stockton 2021; Xiao et al. 2016).

To understand the dynamics of global gene expression during oligodendrocyte differentiation, a microarray dataset that included expression values of all transcripts during a detailed time course of rat primary OPCs differentiation has been analysed. Results identified several important transcriptional regulators, such as inhibitor of DNA binding (*Id*) 2, Krüppel-like factor (*Klf*) 4, *Klf10,* and *SOX11*, among the genes downregulated during the early stages of differentiation. Then, microarray analysis on OPCs differentiated for one day in the presence or absence of the inhibitor of HDAC activity trichostatin A (TSA) showed that the increased expression of genes coding for several transcriptional regulators, such as *Klf4*, *Klf10*, and *SOX11* depends on HDACs (Lyssiotis et al. 2007; Shen et al. 2008; Swiss et al. 2011). Histone H3 at the transcriptional start site of *SOX11* was hyperacetylated on lysine 9 residue in TSA-treated cells, compared to untreated controls. In addition, *SOX11* overexpression in primary OPCs was characterized by lower transcript levels for myelin regulatory factor (*MyRF*) compared to controls, identifying SOX11 as a potential upstream regulator of *MyRF* expression (Swiss et al. 2011). *MyRF* is an evolutionarily conserved nuclear protein containing a DNA binding domain conserved from yeast to man. Knockdown of *MyRF* in OPCs blocks myelin gene expression and results in the downregulation of selected OPC markers *NG2* and *PDGFRα*, while proliferation during differentiation is not affected. MyRF induces myelin gene expression in vivo as well as in cultured OPCs by cooperating with SOX10 (Emery et al. 2009). *MyRF*-conditional knockout (KO) mice are impaired at learning a new motor skill and the recall of long-term (28-day) fear memory, though not for fear motor skill learning and short-term recall (McKenzie et al. 2017). In addition, Steadman et al. (2020) found that long-term spatial memory was impaired although spatial learning per se was relatively unaffected. Together, these studies suggest that *MyRF* gene expression is required for the passage from OPCs to pre-OLs and new myelin is required to stabilize task-relevant circuits, with short and longer-term behavioural consequences that might be context-dependent (Huang et al., 2021; Steadman et al. 2020). All together these data indicate that acetylation/deacetylation is fundamental for the passage of OPC in the premyelinating stage by controlling key elements regulating this passage, namely *SOX10, SOX11,* and *MyRF* gene expression.

Finally, OLs reach the myelination stage that is characterized by the expression of myelin genes: *MBP*, proteolipid protein (*PLP*), 2′,3′-cyclic nucleotide 3′-phosphodiesterase (*CNPase*), and myelin-associated glycoprotein (*MAG*). Upon *MBP* gene expression the actin cytoskeleton starts to depolymerize due to the association of MBP with phosphatidylinositol 4,5-bisphosphate (PIP2) at the membrane and the release of cofilin and gelsolin (Zuchero et al. 2015). This causes OL membrane to wrap and envelope multiple axons. OLs are also characterized by the expression of transmembrane protein 2 (*TMEM2*), Prominin 1 (*CD133*), and Connexin 47 (*Cx47*) (Jäkel et al. 2019; Marques et al. 2016). Mature OLs are characterized by expression of MOG, oligodendrocyte myelin glycoprotein (OMgp), and CNPase and by down-regulation of *GPR17* expression (Gregath and Lu 2018; Hughes and Stockton 2021; Lu et al. 2018; Miron et al. 2011).

In conclusion, the differentiation of NPCs and OPCs to myelinating OLs is a stepwise process intricately regulated by chromatin modifications guided by HATs and HDACs. A deeper investigation of HDACs in the epigenetic control of NPC and OPC differentiation into mOLs will open new therapeutic possibilities for promoting brain repair in neurodegenerative diseases.

8.4.2 Oligodendrogenesis in the Adult Brain Requires HDACs Activity

Histone acetylation has been linked to the control of oligodendrocyte formation through studies with pan or selective HDAC inhibitors (Table 8.2) (Conway et al. 2012). Given the crucial role of proper OPC differentiation in maintaining the brain regenerative capacity throughout life, it is essential to understand the extent to which HATs and HDACs regulate adult oligodendrogenesis. In this section, we explore this by highlighting how oligodendrogenesis is impaired or altered when HDACs are inhibited.

In general, the inhibition of HDACs reduces the NPCs differentiation into OLs and promotes the development of neurons (Berry et al. 2020). Among the most used HDACi are TSA, valproic acid (VPA), and sodium butyrate (SB) which are pan-HDACi able to block the activity of all the isoforms (Gregath and Lu 2018; Lyssiotis et al. 2007). It is also possible to selectively inhibit specific HDACs: for example, HDAC6 is inhibited by Tubastatin A (Leyk et al. 2015). The treatment of NPCs with the HDACi VPA shows inhibition of the specification of OPCs. Through the pharmacological treatment with VPA on rat NPCs the differentiation towards the OPCs fate is blocked in favour of a neuronal determination probably due, at least in part, to the effect of HDACs activity on the neurogenic transcription factor NeuroD. The inhibition of HDACs increases the expression of *NeuroD* which is a transcription factor that promotes neurogenesis and inhibits oligodendrogenesis (Hsieh et al. 2004). Even though VPA displayed inhibition of myelination when used in OLs early development, it has no effect after the onset of myelination, suggesting a stage-dependent switch in HDACs regulation (Koreman et al. 2018).

The treatment with TSA suppresses OPCs differentiation both under the developmental stage and under disease conditions in adult CNS (Lyssiotis et al. 2007; Shen et al. 2008). Co-treatment with TSA and SB decreases differentiation of OPCs and stops inhibition of the expression of *Id2*, early growth response 1 (*EGR-1*), and *SOX11* in rats and *Id4* and *SOX2* in humans, indicating a role of class I HDACs in promoting OPCs differentiation (Swiss et al. 2011).

A study with the pan-HDACi Quisinostat demonstrated the role of HDACs on the mechano-transduction of strain to chromatin. Jagielska et al. (2017) analysed the effect of a sustained tensile strain on OPC differentiation, to mimic pathological conditions such as tumour, glial scaring, axon swelling and general trauma or physiological processes such as spinal cord blending or cerebrospinal fluid pulsation, all phenomena that lead to mechanical strain in vivo (Jagielska et al. 2017). It has been shown that applied mechanical strain induces changes in OPCs' nuclear shape, and chromatin organization, and results in enhanced histone deacetylation increasing oligodendrocyte differentiation. RNA sequencing data showed an increased gene expression for *HDAC9* and *HDAC11* in strained cell populations, measured at 24 h of strain duration. HDACs inhibition by Quisinostat decreased OPCs differentiation in both strained and unstrained conditions, suggesting that HDACs are involved in the mechano-transduction of strain in the OPCs nucleus (Jagielska et al. 2017).

In summary, studies involving HDAC inhibitors suggest that modifications in histone acetylation can selectively hamper early OPCs differentiation (Li et al. 2022b). The impact of HDAC activity appears to be different in the diverse stages of OPC development, and its dysregulation significantly affects oligodendroglial maturation. These results suggest that HATs and HDACs can be potential therapeutic targets to treat demyelinating diseases (Jaworska et al. 2019; Kabir et al. 2023; Lu et al. 2018; Nishiyama et al. 2021; Sherafat et al. 2021).

In the following paragraph, we discuss how the different HDACs play fundamental roles for proper remyelination in the adult brain.

8.4.3 *Different HDACs Regulate Specific Steps of the Remyelination Process*

A broad study by Shen et al. (2005) postulated that HDACs aid in the myelination process in OLs by suppressing the transcription of myelination inhibitors. HDACs work both at a protein and chromatin level to regulate the passage from OPC to OLs. Since histone deacetylation is associated with chromatin compaction, two models have been proposed to explain HDACs possible role in this context. According to the first model, the genes that code for differentiation markers, like MBP, have regulatory sites on their promoters, and the myelin compaction would hamper the binding of repressors. An alternative model proposes that chromatin compaction may occur in the promoter region of genes coding for inhibitors of the differentiation site. These models are not mutually exclusive and can both occur within the same cell at a given time (Marin-Husstege et al. 2002; Shen et al. 2005). The general role of the different HDACs in oligodendrogenesis in the adult brain can be broadly analysed following their classification in the four HDAC classes reported in Table 8.2.

Class I HDACs expression levels and functions are essential for oligodendrocyte fate determination and differentiation and show redundancy. *HDAC1/2* double KO in rat pups show a significant impairment in OPCs proliferation or differentiation, meanwhile, *HDAC1* and *HDAC2* single KO do not appear to affect OPCs production, proliferation, nor differentiation, which suggests that HDAC1 and HDAC2 can functionally compensate for each other's loss in oligodendrocyte lineage determination (Ye et al. 2009). This indicates a functional redundancy of HDAC1/2 in mammalian OPCs development (Koreman et al. 2018).

HDACs of class I are reported to be active in brain tissue development: a study on zebrafish elucidated the importance of HDAC1 in OPCs specification, by promoting *Olig2* expression (Fig. 8.1) (Cunliffe and Casaccia-Bonnefil 2006; Montgomery et al. 2009). A later study also showed the role of HDAC1 and HDAC2 in preventing the repression of myelin genes expression of TCF4/TCF7L2, by antagonizing the binding of β-catenin to the promoter, in mammalian OPCs (Fig. 8.1) (Ye et al. 2009).

HDAC2 is a specific repressor checkpoint in control of OL differentiation, and it has been found on the *SOX8* promoter (Fig. 8.1) (Castelo-Branco et al. 2014). *HDAC2* ablation in NPCs, together with the treatment with thyroid hormone Triiodothyronine (T3), ends up derepressing *SOX8* and *SOX10* expression, factors promoting OLs lineage, suggesting that the function of HDACs in differentiation/specification in OLs development is stage-dependent (Koreman et al. 2018). Moreover, the use of theophylline, a known activator of HDAC2, deacetylate eEF1A1. eEF1A1 is a known inhibitor of SOX10 that can bind this transcription factor only upon acetylation. HDAC2-dependent deacetylation of eEF1A1 leads to its translocation in the cytoplasm and to the expression of SOX10 target genes. HDAC2, by decreasing the levels of acetylated eEF1A1, enhances the remyelination process in the CNS after a demyelinating lesion in young and old mice (Fig. 8.1) (Duman et al. 2020).

The role of the HDACs (1,3,10) in promoting NPCs differentiation comprehends the modulation of the nucleus to cytoplasm relocation of Olig1, the helix-loop-helix (HLH) transcription factor crucial for oligodendrocyte specification and differentiation. As oligodendrocytes mature, human Olig1 gets acetylated on Lys150 by the HAT CBP (Fig. 8.1). This acetylation increases its affinity with Id2, decreases its association with chromatin, and eventually leads to Olig1 nuclear to cytoplasmic relocation. The deacetylation of Olig1 by HDAC1 prevents its physical interaction with the Id2 protein promoting Olig1 nuclear transportation. The mechanism by which Olig1 acetylation modulates its cytoplasmic translocation is critical for its function in promoting oligodendrocyte lineage progression and morphological differentiation (Dai et al. 2015).

Concerning *HDAC3* and/or *HDAC8,* their genetic ablation was not able to impair OPCs differentiation, showing a different role of the various isoforms during oligodendrogenesis (Shen et al. 2008; Wheeler et al. 2015).

HDAC3 was found on the promoters of genes that encode for transcription factors associated with neuronal and oligodendrocyte differentiation of NPCs, such as *PAX6* and *SOX8* (Fig. 8.1). Chromatin immunoprecipitation (ChIP) sequencing and consequent single-gene ChIP analysis in NPCs derived from embryonic rats' cortices showed that, at embryonic day 15, HDAC3 is linked to the regulation of oligodendrocyte lineage, in a different mechanism from the ones noted in HDAC1 (Castelo-Branco et al. 2014). *HDAC3* deletion causes an oligodendrocyte to astrocyte destiny transition at the same stages as HDAC1 and HDAC2, indicating that HDAC3 controls primitive OPCs' decision between astrocytic and oligodendrocytes cell lineages. HDAC3 has been shown to inhibit STAT3 acetylation impeding the formation of the Janus kinase (JAK)-STAT3 complex, fundamental for astrogliogenesis. In this sense, HDAC3 plays a role in the specification of NPC towards the OL lineage instead of the astrocyte's development (Fig. 8.1) (Zhang et al., 2014).

Another way HDAC3 can regulate oligodendrogenesis is by regulating lipid metabolism, which is essential for oligodendrocyte differentiation. Once peroxisome proliferator-activated receptor gamma (PPARγ) gets acetylated, it forms heterodimers with retinoid X receptors (RXRs) which regulate the transcription of various genes mainly involved in lipid synthesis and cell differentiation (Kanakasabai et al. 2012). When HDAC3 deacetylate PPARγ, it hampers its activity thereby promoting OPCs proliferation (Ding et al. 2020). VPA has been proposed as a therapeutical target for demyelinating diseases as it inhibits HDAC3 thereby increasing PPARγ activity, leading to an increase in the proliferation of OPCs near the lesion and their differentiation (Ding et al. 2020).

On the other hand, HDAC3 is responsible for FAS deacetylation. FAS is the main responsible for fatty acid synthesis related to oligodendrocyte differentiation and its stability is dependent on the post-translation process of acetylation/deacetylation. In this context, HDAC3 promotes lipid synthesis and OPCs differentiation (Balboni et al. 2024; Poeta et al. 2021).

Among class II HDACs, HDAC6 is involved in OL myelination by deacetylating Tau, a microtubule-associated protein that is necessary for the correct transportation of MBP on microtubules (Fig. 8.1) (Noack et al. 2014; Richter-Landsberg 2016).

SIRTs are class III HDACs that deacetylate K16 of histone 4 (H4K16) and play crucial roles in OLs development (Koreman et al. 2018). However, the role of SIRT in remyelination is not fully clear: this class has been proven to both induce and hamper oligodendrocyte differentiation in different cell lines. SIRT1 is upregulated in NG2+ and PDGFRα+ OPCs in demyelinated Experimental autoimmune encephalomyelitis (EAE) lesions, and blockade of SIRT1 activity in cultured mouse A2B5+ OPCs, by the small molecule compound Ex527, enhanced OPC mitotic activity without affecting their differentiation (Prozorovski et al. 2019). These data are consistent with the observation that a significant increase in oligodendroglial SIRT1 expression has been reported in plaques and normal-appearing white matter (NAWM) in the brain of MS patients (Tegla et al. 2014).

Experiments investigating the effects of *SIRT2* depletion in OPCs were conducted using *SIRT2*$^{-/-}$ mice. In primary OPC cultures, the KO of *SIRT2* led to impairments in both OPC proliferation and differentiation (Jablonska et al. 2022).

During OPC specification, SIRT1 and SIRT2 exhibit overlapping functions that involve nicotinamide phosphoribosyltransferase (NAMPT), an enzyme responsible for catalysing NAD+ synthesis in mammals (Stein and Imai 2014). Indeed, it has been demonstrated that the depletion of *SIRT2* in the aged OPCs is followed by NAD+ reduction. These events are both correlated with declined remyelination during ageing (Ma et al. 2022).

It has also been proven that SIRT2 regulates α-tubulin acetylation, which affects microtubule stability and cell structure. However, its effect on α-tubulin acetylation and *MBP* expression, crucial for oligodendrocyte maturation, is still debated. As OPCs differentiate into OLs, *SIRT2* expression increases along with α-tubulin acetylation; it is also true that SIRT2 may decrease α-tubulin acetylation in OLs, thereby hindering their differentiation (Koreman et al. 2018; Li et al. 2007; Tang and Chua 2008). The role of SIRT2 is conflictual: *SIRT2* silencing has been shown to increase α-tubulin acetylation and *MBP* expression leading to an extension of cellular processes (Hisahara et al. 2021); on the other hand, the overexpression of *SIRT2* leads to a decrease of tubulin acetylation and increased *MBP* expression. The involvement of SIRT2 in OL differentiation has been recently clarified by Fang et al. (2019). SIRT2 downregulates *PDGFRα* by deacetylating its promoter on the CpG islands and facilitating their methylation, hampering *PDGFRα* transcription (Fig. 8.1). This leads to an inhibition of proliferation and facilitates the maturation from OPC to pre-OLs (Fang et al. 2019). Additionally, Nkx2.2 together with HDAC1, binds to the promoter region of SIRT2, inhibiting its transcription and thus blocking OLs differentiation (Ji et al. 2011). These studies underscore the critical role of SIRT2 in the oligodendrocyte differentiation process, as its downregulation appears to hinder proper OLs development. Moreover, SIRT2 has many roles in the regulation of lipogenesis and lipid synthesis, essential for oligodendrocyte development. SIRT2 inhibits SREBP-2 and the transcriptional factor Forkhead box protein O1 (Foxo1), mainly known to regulate gluconeogenesis and glycogenolysis, which in turn represses PPARγ (Zhu et al. 2022).

For what it concerns the IV class, as previously described in paragraph 4.1, mechanical strain on OLs can increase levels of HDAC11 and promote OLs differentiation (Koreman et al. 2018). HDAC11 has a main role in OLs development and remyelination, as it is involved in the transcription of two of the most important markers of myelination: MBP and PLP. HDAC11, by deacetylating lysine 9 (H3K9) and lysine 14 (H3K14) of histone H3, close to the regulatory elements of the *MBP* and *PLP* genes, pushes OLs maturation in vitro (Fig. 8.1) (Douvaras et al. 2016).

In conclusion, these studies have highlighted the fundamental role of HDACs in the regulation of OPC differentiation and myelination by modulating acetylation levels both at a protein and at a chromatin level. The various functions of the different HDAC isoforms highlight their stage-dependent influence on oligodendrocyte development. Class I HDACs, specifically HDAC1 and HDAC2, exhibit functional redundancy, underlining their fundamental role for OPC differentiation, while HDAC3 plays a significant role in fate determination and lipid metabolism. Meanwhile, HDAC6 and SIRT2, of classes II and III, respectively, contribute to key cellular processes like microtubule regulation and deacetylation of transcription factors, influencing both OPC differentiation and remyelination.

8.5 Histone Modifications in the Pathogenesis of Neurodegenerative Diseases

In the last decades, there has been increasing evidence regarding the role that histone modifications have in the pathogenesis of neurodegenerative diseases, such as PD, AD, Huntington's disease (HD), and amyotrophic lateral sclerosis (ALS) (Jablonska et al. 2022; Kabir et al. 2023; Z. Xu et al. 2012). Specifically, there is evidence that histone acetylation levels differ in various age-related neurodegenerative diseases such as PD and AD but also in mental disorders like schizophrenia. For instance, histone hyperacetylation or hypoacetylation has been shown to lead to an accumulation of pathophysiological proteins such as Tau, α-synuclein (aSyn), and Huntingtin (HTT) protein in AD, PD, and HD, respectively (Kabir et al. 2023).

Although the molecular bases of PD are still unknown, the loss of dopaminergic neurons is crucial in its pathogenesis. In the PD context, HDAC1–3 and 6 have been proven to be neurotoxic together with SIRT2, SIRT3, and SIRT6. This led to the development of several studies that use HDAC inhibitors for the treatment of PD, in particular, VPA and TSA have been extensively studied (Li et al. 2022a).

Regarding AD pathology, it has been observed that HDAC1 can trigger a DNA repair mechanism which eventually ameliorates cognitive deficits in AD mouse model (Pao et al. 2020). However, the most striking evidence of the involvement of HDACs in AD comes from the observation of HDAC2 activity, in which overexpression is associated with an increased level of Amyloid-β production and with hyperphosphorylation and aggregation of Tau protein (Guan et al. 2009; Ishizuka et al. 2014). Furthermore, HDAC2 is involved in neuronal apoptosis by impeding the transcription of the p21 promoter, whose expression protects neurons from oxidative stress (Peng et al. 2015).

For what it concerns schizophrenia, it has been observed that *HDAC2* mRNA levels are significantly lower in postmortem dorsolateral prefrontal cortex brain tissue from patients, compared with healthy controls. An altered HDAC expression has been confirmed also in vivo using positron emission tomography (PET) neuroimaging (Gilbert et al. 2019). Farrelly et al. (2022) applied an unbiased proteomics approach to evaluate combinatorial histone post-translational modifications in human induced pluripotent stem cell (hiPSC)-derived forebrain neurons from individuals with schizophrenia. They observed hyperacetylation of H2A.Z and H4, results that were then confirmed in the postmortem human brain, further proving aberrant histone acetylation in schizophrenia (Farrelly et al. 2022). In 2024, Martínez-Peula et al. also analysed postmortem dorsolateral prefrontal cortex samples of individuals with schizophrenia, compared to the controls matched for sex, age, and postmortem interval: the expression of permissive histone post-translational modifications as H3K9ac and H3K27ac has been found globally enhanced while HDAC activity and HDAC4 protein expression were downregulated (Martínez-Peula et al. 2024).

Because of the evidence listed above, histone acetylation has been suggested to be used as early disease biomarkers and/or as therapeutic targets for neurodegenerative diseases such as AD, PD, and HD, through the treatment with HDACi and other epigenetic drugs (Xu et al. 2012). Up to date, only a few clinical trials are assessing the employment of HDACi for the therapy of neurodegenerative diseases, mostly because of safety concerns related to HDACi toxicity (Park et al. 2022). The promising role of HDACi has been investigated both in vitro and in animal models, for their potential applications in the treatment of neurodegenerative diseases. For example, the treatment with Tubastatin A, an HDAC6 inhibitor, was shown to be effective in recovering cognitive deficits and reducing Tau levels in a mouse model of AD (Selenica et al. 2014). Similarly, VPA, an HDACi, can significantly reduce degeneration of dopaminergic neurons and increase aSyn expression in the substantia nigra in PD mouse models (Monti et al. 2012). Furthermore, Davis et al. (2024) demonstrated that the inhibition of HDAC3 through RGFP-966, a selective HDAC3 inhibitor, is associated with an increase in amyloid precursor protein (APP) and with a reduction of activated Iba-1-positive microglia and astrocytes in AD mouse model slices, suggesting a role of HDAC3 in the modulation of APP processing and the expression of neuroinflammatory genes (Davis et al. 2024). Another evidence of possible functional therapeutic intervention through HDACi comes from the study of Al-Griw et al. (2021) (Al-Griw et al. 2021). Their work showed that the pharmacological inhibition of HDAC2 and HDAC3, using the MI192 HDACi, restores NG2+ OPCs and MBP+ OLs from WM injury (Al-Griw et al. 2021). The usage of HDAC inhibitors in therapy is promising but must face two major challenges: the low capacity of HDACi to cross the blood-brain barrier (BBB) and the strong homology of the different HDAC isoforms, which represent an obstacle to selectivity and specificity of the inhibitors (Gräff and Tsai 2013). For these reasons, the design of new HDACi, more selective and specific, and with better pharmacological properties, is of primary importance to ameliorate their potential in the therapeutic intervention of neurodegenerative diseases.

Other than neurodegenerative diseases and mental disorders, changes in the levels of acetylation are correlated with NAA metabolism dysfunction in the early stages of MS and Canavan disease (CD). In paragraph 2, we highlighted the importance of NAA in post-transcriptional modification. NAA depletion is used as an early biomarker in nuclear magnetic resonance (NMR) for MS diagnosis, as well as its upregulation characterizes CD patients (Aboul-Enein et al. 2010; Grønbæk-Thygesen and Hartmann-Petersen 2024). Disrupted metabolic regulation leads to various conditions, such as hypoacetylaspartia, caused by N-acetyltransferase 8 like (NAT8L) malfunction or CD, an autosomal genetic leukodystrophy caused by mutations in the ASPA gene (Grønbæk-Thygesen and Hartmann-Petersen 2024; Lin et al. 2024; Matalon et al. 1988; Wiame et al. 2009). Further investigation of the role of HATs or HDACs in these diseases can pose the basis for a promising therapeutical approach.

Finally, the understanding of epigenetic changes in the pathogenesis of oligodendrocytes-related diseases, such as MS, neuro-myelitis optica (NMO), acute disseminated encephalomyelitis (ADE), and the progression of injuries, can be of primary importance to provide new molecular tools to improve the diagnosis and the treatment of these disorders.

For instance, an analysis of postmortem brains of MS patients showed higher levels of histone acetylation in oligodendrocytes in comparison to healthy individuals and early-stage MS patients. Simultaneously, this increase was associated with an upregulation of *HATs* transcript levels and high expression of inhibitory regulators such as *TCF7L2*, *Id2*, and *SOX2*. As the acetylation alters the differentiation and myelination of OLs, this result highlights the role of histone acetylation in decreasing the OLs' ability to remyelinate in primary progressive MS (PPMS) patients (Pedre et al. 2011).

It has been shown that an increased activity of HDAC1/2, together with the inhibition of β-catenin mediated Wingless/int-1 (Wnt) signalling, can promote OLs regeneration and myelin repair (Ye et al. 2009). Furthermore, the balance between SIRT1 and SIRT2 activity is essential for oligodendrogenesis during development, and these proteins could be potential therapeutic targets for WM injury repair (Jablonska et al. 2022).

Interestingly, histone acetylation plays a role also in OLs recovery after an ischemic stroke. After the injury, oligodendrogenesis starts, together with other endogenous repair processes. Levels of HDAC1 and HDAC2 increase in WM OPCs and the region surrounding the infarction (Felling and Song 2015; Kassis et al. 2014). Developing OLs retain the increase of HDAC2 until maturation while HDAC1 levels progressively return to pre-stroke condition, meaning that HDAC1 and 2 play different roles (Kassis et al. 2014).

In summary, these findings underscore the therapeutic potential of HDACi for demyelinating diseases and age-related disorders of the CNS but also further emphasize the complexity of their molecular mechanisms. Although pharmacological inhibition of HDACs offers valuable insights, studies using genetic depletion in oligodendrocytes may provide clearer interpretations of the effective role of HDACs in such diseases.

8.6 Conclusion

Epigenetic regulation of myelin plasticity through reversible acetylation has a great role in preserving cognitive functions and the brain's regenerative ability, particularly during ageing. HDACs play a fundamental role in the differentiation of OPCs into OLs, by acting through chromatin remodelling and protein deacetylation at multiple stages of OPC development. HDACs influence various important transcription factors, myelin gene expression, and metabolic processes necessary for correct myelination. Specifically, HDAC1 and HDAC2 are involved in the earlier stages of differentiation, while HDAC3 helps regulate the lineage switch between

OLs and astrocytes, other than lipid metabolism which is essential for myelin production. HDAC6, SIRT1, SIRT2, and HDAC11 contribute to myelin maintenance and OL maturation. Dysregulation in the balance between acetylation and deacetylation can lead to impaired remyelination, contributing to neurodegenerative pathologies such as MS, AD, and PD as well as psychiatric conditions like schizophrenia. These findings highlight the complex roles of HDACs in adult brain oligodendrogenesis and suggest that targeting specific HDACs can be an opportunity for novel therapeutic strategies for demyelinating diseases and cognitive decline. Indeed, HDAC inhibitors have been proposed for their therapeutical potential but their clinical potency in regenerative processes remains to be clarified. The promising positive effects observed in animal models are countered from two major limitations: the low capacity of HDACi to cross the BBB and the strong homology of the different HDAC isoforms, which represents an obstacle to their selectivity and specificity properties. For these reasons, the design of more selective and specific HDACi is crucial to improve their pharmacological properties and open the way to their therapeutic use in age-related neurodegenerative diseases. Further understanding of the epigenetic mechanisms at the basis of oligodendrocyte lineage progression will offer valuable insights into potential therapeutic strategies to fight age-related cognitive decline and neurodegeneration.

Competing Interests The authors have no conflicts of interest to declare that are relevant to the content of this chapter.

Financial Support This study was supported by FISM - Fondazione Italiana Sclerosi Multipla – cod. 2023/R-Multi/001 and financed or co-financed with the '5 per mille' public funding and partially supported by Università degli Studi di Roma Tor Vergata (Rome, Rome, IT) GRANT_NUMBER: E84I20000580005.

The author Alessandra Dominicis has received a research PhD grant from PhD program in Cellular and Molecular Biology, Department of Biology, University of Rome "Tor Vergata", Rome, Italy. The author Aland Ibrahim Ahmed Al Jaf has received a Fellowship "*Studio del meccanismo di azione di farmaci promielinanti e miglioramento delle tecniche di rilascio via nanoparticelle*" from the University of Rome "Tor Vergata", Rome, Italy. The author Tommaso Fabiano has received a Fellowship "Analisi di farmaci promielinanti in modelli cellulari di oligodendrociti" from University of Rome "Tor Vergata", Rome, Italy supported by project FISM 2023/R_Multi/001.

References

Aboul-Enein F, Krššák M, Höftberger R, Prayer D, Kristoferitsch W (2010) Reduced NAA-levels in the NAWM of patients with MS is a feature of progression. A study with quantitative magnetic resonance spectroscopy at 3 tesla. PLoS One 5(7):e11625. https://doi.org/10.1371/journal.pone.0011625

Al-Griw MA, Shmela ME, Elhensheri MM, Bennour EM (2021) HDAC2/3 inhibitor MI192 mitigates oligodendrocyte loss and reduces microglial activation upon injury: a potential role of epigenetics. Open Veterin J 11(3):447–457. https://doi.org/10.5455/OVJ.2021.v11.i3.18

Amaral AI, Hadera MG, Tavares JM, Kotter MRN, Sonnewald U (2016) Characterization of glucose-related metabolic pathways in differentiated rat oligodendrocyte lineage cells. Glia 64(1):21–34. https://doi.org/10.1002/glia.22900

Amaral I, Hadera MG, Kotter M, Sonnewald U (2017) Oligodendrocytes do not export NAA-derived aspartate in vitro. Neurochem Res 42(3):827–837. https://doi.org/10.1007/s11064-016-1985-y
Arede L, Pina C (2021) Buffering noise: KAT2A modular contributions to stabilization of transcription and cell identity in cancer and development. Exp Hematol 93:25–37. https://doi.org/10.1016/j.exphem.2020.10.003
Ariyannur PS, Moffett JR, Madhavarao CN, Arun P, Vishnu N, Jacobowitz DM, Hallows WC, Denu JM, Namboodiri AMA (2010) Nuclear-cytoplasmic localization of acetyl coenzyme A synthetase-1 in the rat brain. J Comp Neurol 518(15):2952–2977. https://doi.org/10.1002/cne.22373
Balboni N, Babini G, Poeta E, Protti M, Mercolini L, Magnifico MC, Barile SN, Massenzio F, Pignataro A, Giorgi FM, Lasorsa FM, Monti B (2024) Transcriptional and metabolic effects of aspartate-glutamate carrier isoform 1 (AGC1) downregulation in mouse oligodendrocyte precursor cells (OPCs). Cell Mol Biol Lett 29(1):44. https://doi.org/10.1186/s11658-024-00563-z
Bebo BF, Allegretta M, Landsman D, Zackowski KM, Brabazon F, Kostich WA, Coetzee T, Ng AV, Marrie RA, Monk KR, Bar-Or A, Whitacre CC (2022) Pathways to cures for multiple sclerosis: a research roadmap. Multiple Sclerosis (Houndmills, Basingstoke, England) 28(3):331–345. https://doi.org/10.1177/13524585221075990
Berndsen CE, Denu JM (2008) Catalysis and substrate selection by histone/protein lysine acetyltransferases. Curr Opin Struct Biol 18(6):682–689. https://doi.org/10.1016/j.sbi.2008.11.004
Berry K, Wang J, Lu QR (2020) Epigenetic regulation of oligodendrocyte myelination in developmental disorders and neurodegenerative diseases. F1000Research, *9*, F1000 Faculty Rev-105 9:105. https://doi.org/10.12688/f1000research.20904.1
Boshans LL, Sherafat A, Nishiyama A (2020) The effects of developmental and current niches on oligodendrocyte precursor dynamics and fate. Neurosci Lett 715:134593. https://doi.org/10.1016/j.neulet.2019.134593
Burrell JA, Stephens JM (2021) KAT8, lysine acetyltransferase 8, is required for adipocyte differentiation in vitro. Biochim Biophys Acta (BBA) Mol Basis Dis 1867(6):166103. https://doi.org/10.1016/j.bbadis.2021.166103
Castelo-Branco G, Lilja T, Wallenborg K, Falcão AM, Marques SC, Gracias A, Solum D, Paap R, Walfridsson J, Teixeira AI, Rosenfeld MG, Jepsen K, Hermanson O (2014) Neural stem cell differentiation is dictated by distinct actions of nuclear receptor corepressors and histone deacetylases. Stem Cell Reports 3(3):502–515. https://doi.org/10.1016/j.stemcr.2014.07.008
Chakraborty G, Mekala P, Yahya D, Wu G, Ledeen RW (2001) Intraneuronal N-acetylaspartate supplies acetyl groups for myelin lipid synthesis: evidence for myelin-associated aspartoacylase. J Neurochem 78(4):736–745. https://doi.org/10.1046/j.1471-4159.2001.00456.x
Chen J-F, Liu K, Hu B, Li R-R, Xin W, Chen H, Wang F, Chen L, Li R-X, Ren S-Y, Xiao L, Chan JR, Mei F (2021) Enhancing myelin renewal reverses cognitive dysfunction in a murine model of Alzheimer's disease. Neuron 109(14):2292–2307.e5. https://doi.org/10.1016/j.neuron.2021.05.012
Choi Y-H, Park S-H, Kim S-S, Lee M-W, Yu J-H, Shin K-S (2023) Functional characterization of the GNAT family histone acetyltransferase Elp3 and GcnE in aspergillus fumigatus. Int J Mol Sci 24(3):2179. https://doi.org/10.3390/ijms24032179
Ciccone L, Piragine E, Brogi S, Camodeca C, Fucci R, Calderone V, Nencetti S, Martelli A, Orlandini E (2022) Resveratrol-like Compounds as SIRT1 Activators. Int J Mol Sci 23(23):15105. https://doi.org/10.3390/ijms232315105. PMID: 36499460; PMCID: PMC9738298.
Coni S, Antonucci L, D'Amico D, Di Magno L, Infante P, De Smaele E, Giannini G, Di Marcotullio L, Screpanti I, Gulino A, Canettieri G (2013) Gli2 acetylation at lysine 757 regulates hedgehog-dependent transcriptional output by preventing its promoter occupancy. PLoS One 8(6):e65718. https://doi.org/10.1371/journal.pone.0065718
Conway GD, O'Bara MA, Vedia BH, Pol SU, Sim FJ (2012) Histone deacetylase activity is required for human oligodendrocyte progenitor differentiation. Glia 60(12):1944–1953. https://doi.org/10.1002/glia.22410

Cunliffe VT, Casaccia-Bonnefil P (2006) Histone deacetylase 1 is essential for oligodendrocyte specification in the zebrafish CNS. Mech Dev 123(1):24–30. https://doi.org/10.1016/j.mod.2005.10.005

Dai J, Bercury KK, Jin W, Macklin WB (2015) Olig1 acetylation and nuclear export mediate oligodendrocyte development. J Neurosci Off J Soc Neurosci 35(48):15875–15893. https://doi.org/10.1523/JNEUROSCI.0882-15.2015

Davis N, Taylor B, Abelleira-Hervas L, Karimian-Marnani N, Aleksynas R, Syed N, Di Giovanni S, Palmisano I, Sastre M (2024) Histone deacetylase-3 regulates the expression of the amyloid precursor protein and its inhibition promotes neuroregenerative pathways in Alzheimer's disease models. FASEB J 38(10):e23659. https://doi.org/10.1096/fj.202301762RR

Dean DC, Sojkova J, Hurley S, Kecskemeti S, Okonkwo O, Bendlin BB, Theisen F, Johnson SC, Alexander AL, Gallagher CL (2016) Alterations of myelin content in Parkinson's disease: a cross-sectional neuroimaging study. PLoS One 11(10):e0163774. https://doi.org/10.1371/journal.pone.0163774

Dekker FJ, Haisma HJ (2009) Histone acetyl transferases as emerging drug targets. Drug Discov Today 14(19–20):942–948. https://doi.org/10.1016/j.drudis.2009.06.008

Dimou L, Simons M (2017) Diversity of oligodendrocytes and their progenitors. Curr Opin Neurobiol 47:73–79. https://doi.org/10.1016/j.conb.2017.09.015

Ding L, Zhou J, Ye L, Sun Y, Jiang Z, Gan D, Xu L, Luo Q, Wang G (2020) PPAR-γ is critical for HDAC3-mediated control of oligodendrocyte progenitor cell proliferation and differentiation after focal demyelination. Mol Neurobiol 57(11):4810–4824. https://doi.org/10.1007/s12035-020-02060-8

Dominicis A, Del Giovane A, Torreggiani M, Recchia AD, Ciccarone F, Ciriolo MR, Ragnini-Wilson A (2023) N-Acetylaspartate drives Oligodendroglial differentiation via histone deacetylase activation. Cells 12(14):Articolo 14. https://doi.org/10.3390/cells12141861

Douet V, Kerever A, Arikawa-Hirasawa E, Mercier F (2013) Fractone-heparan sulphates mediate FGF-2 stimulation of cell proliferation in the adult subventricular zone. Cell Prolif 46(2):137–145. https://doi.org/10.1111/cpr.12023

Douvaras P, Rusielewicz T, Kim KH, Haines JD, Casaccia P, Fossati V (2016) Epigenetic modulation of human induced pluripotent stem cell differentiation to oligodendrocytes. Int J Mol Sci 17(4):614. https://doi.org/10.3390/ijms17040614

Duman M, Vaquié A, Nocera G, Heller M, Stumpe M, Siva Sankar D, Dengjel J, Meijer D, Yamaguchi T, Matthias P, Zeis T, Schaeren-Wiemers N, Hayoz A, Ruff S, Jacob C (2020) EEF1A1 deacetylation enables transcriptional activation of remyelination. Nat Commun 11(1):3420. https://doi.org/10.1038/s41467-020-17243-z

Egawa N, Shindo A, Hikawa R, Kinoshita H, Liang AC, Itoh K, Lok J, Maki T, Takahashi R, Lo EH, Arai K (2019) Differential roles of epigenetic regulators in the survival and differentiation of oligodendrocyte precursor cells. Glia 67(4):718–728. https://doi.org/10.1002/glia.23567

Emery B, Agalliu D, Cahoy JD, Watkins TA, Dugas JC, Mulinyawe SB, Ibrahim A, Ligon KL, Rowitch DH, Barres BA (2009) Identification of myelin-gene regulatory factor as a critical transcriptional regulator required for CNS myelination. Cell 138(1):172–185. https://doi.org/10.1016/j.cell.2009.04.031

Falk MJ, Li D, Gai X, McCormick E, Place E, Lasorsa FM, Otieno FG, Hou C, Kim CE, Abdel-Magid N, Vazquez L, Mentch FD, Chiavacci R, Liang J, Liu X, Jiang H, Giannuzzi G, Marsh ED, Yiran G et al (2014) AGC1 deficiency causes infantile epilepsy, abnormal myelination, and reduced N-acetylaspartate. JIMD Reports 14:77–85. https://doi.org/10.1007/8904_2013_287

Fang N, Cheng J, Zhang C, Chen K, Zhang C, Hu Z, Bi R, Furber KL, Thangaraj M, Nazarali AJ, Ji S (2019) Sirt2 epigenetically down-regulates PDGFRα expression and promotes CG4 cell differentiation. Cell Cycle (Georgetown, TX) 18(10):1095–1109. https://doi.org/10.1080/15384101.2019.1609818

Farrelly LA, Zheng S, Schrode N, Topol A, Bhanu NV, Bastle RM, Ramakrishnan A, Chan JC, Cetin B, Flaherty E, Shen L, Gleason K, Tamminga CA, Garcia BA, Li H, Brennand KJ, Maze I (2022) Chromatin profiling in human neurons reveals aberrant roles for histone acetylation and BET family proteins in schizophrenia. Nat Commun 13(1):2195. https://doi.org/10.1038/s41467-022-29922-0

Felling RJ, Song H (2015) Epigenetic mechanisms of neuroplasticity and the implications for stroke recovery. Exp Neurol 268:37–45. https://doi.org/10.1016/j.expneurol.2014.09.017

Fiorentino F, Sementilli S, Menna M, Turrisi F, Tomassi S, Pellegrini FR, Iuzzolino A, D'Acunzo F, Feoli A, Wapenaar H, Taraglio S, Fraschetti C, Del Bufalo D, Sbardella G, Dekker FJ, Paiardini A, Trisciuoglio D, Mai A, Rotili D (2023) First-in-class selective inhibitors of the lysine acetyltransferase KAT8. J Med Chem 66(10):6591–6616. https://doi.org/10.1021/acs.jmedchem.2c01937

Fischle W, Dequiedt F, Hendzel MJ, Guenther MG, Lazar MA, Voelter W, Verdin E (2002) Enzymatic activity associated with class II HDACs is dependent on a multiprotein complex containing HDAC3 and SMRT/N-CoR. Mol Cell 9(1):45–57. https://doi.org/10.1016/s1097-2765(01)00429-4

Foerster S, Hill MFE, Franklin RJM (2019) Diversity in the oligodendrocyte lineage: plasticity or heterogeneity? Glia 67(10):1797–1805. https://doi.org/10.1002/glia.23607

Fournier M, Orpinell M, Grauffel C, Scheer E, Garnier J-M, Ye T, Chavant V, Joint M, Esashi F, Dejaegere A, Gönczy P, Tora L (2016) KAT2A/KAT2B-targeted acetylome reveals a role for PLK4 acetylation in preventing centrosome amplification. Nat Commun 7(1):13227. https://doi.org/10.1038/ncomms13227

Francis JS, Wojtas I, Markov V, Gray SJ, McCown TJ, Samulski RJ, Bilaniuk LT, Wang D-J, De Vivo DC, Janson CG, Leone P (2016) N-acetylaspartate supports the energetic demands of developmental myelination via oligodendroglial aspartoacylase. Neurobiol Dis 96:323–334. https://doi.org/10.1016/j.nbd.2016.10.001

Galvez-Contreras AY, Zarate-Lopez D, Torres-Chavez AL, Gonzalez-Perez O (2020) Role of oligodendrocytes and myelin in the pathophysiology of autism spectrum disorder. Brain Sci 10(12):951. https://doi.org/10.3390/brainsci10120951

Gilbert TM, Zürcher NR, Wu CJ, Bhanot A, Hightower BG, Kim M, Albrecht DS, Wey H-Y, Schroeder FA, Rodriguez-Thompson A, Morin TM, Hart KL, Pellegrini AM, Riley MM, Wang C, Stufflebeam SM, Haggarty SJ, Holt DJ, Loggia ML et al (2019) PET neuroimaging reveals histone deacetylase dysregulation in schizophrenia. J Clin Invest 129(1):364–372. https://doi.org/10.1172/JCI123743

González-Martínez R, Márquez-Galera A, Del Blanco B, López-Atalaya JP, Barco A, Herrera E (2022) CBP and p300 jointly maintain neural progenitor viability but play unique roles in the differentiation of neural lineages. Cells 11(24):4118. https://doi.org/10.3390/cells11244118

Graciarena M, Seiffe A, Nait-Oumesmar B, Depino AM (2019) Hypomyelination and Oligodendroglial alterations in a mouse model of autism Spectrum disorder. Front Cell Neurosci 12:517. https://doi.org/10.3389/fncel.2018.00517

Gräff J, Tsai L-H (2013) The potential of HDAC inhibitors as cognitive enhancers. Annu Rev Pharmacol Toxicol 53:311–330. https://doi.org/10.1146/annurev-pharmtox-011112-140216

Gregath A, Lu QR (2018) Epigenetic modifications-insight into oligodendrocyte lineage progression, regeneration, and disease. FEBS Lett 592(7):1063–1078. https://doi.org/10.1002/1873-3468.12999

Grønbæk-Thygesen M, Hartmann-Petersen R (2024) Cellular and molecular mechanisms of aspartoacylase and its role in Canavan disease. Cell Biosci 14:45. https://doi.org/10.1186/s13578-024-01224-6

Guan J-S, Haggarty SJ, Giacometti E, Dannenberg J-H, Joseph N, Gao J, Nieland TJF, Zhou Y, Wang X, Mazitschek R, Bradner JE, DePinho RA, Jaenisch R, Tsai L-H (2009) HDAC2 negatively regulates memory formation and synaptic plasticity. Nature 459(7243):55–60. https://doi.org/10.1038/nature07925

Gupta A, Hossain MM, Miller N, Kerin M, Callagy G, Gupta S (2016) NCOA3 coactivator is a transcriptional target of XBP1 and regulates PERK–eIF2α–ATF4 signalling in breast cancer. Oncogene 35(45):5860–5871. https://doi.org/10.1038/onc.2016.121

He W, Zhang M-G, Wang X-J, Zhong S, Shao Y, Zhu Y, Shen Z-J (2013) KAT5 and KAT6B are in positive regulation on cell proliferation of prostate cancer through PI3K-AKT signaling. Int J Clin Exp Pathol 6(12):2864–2871

Hill RA, Li AM, Grutzendler J (2018) Lifelong cortical myelin plasticity and age-related degeneration in the live mammalian brain. Nat Neurosci 21(5):683–695. https://doi.org/10.1038/s41593-018-0120-6

Hisahara S, Iwahara N, Matsushita T, Suzuki S, Matsumura A, Fujikura M, Yokokawa K, Saito T, Manabe T, Kawamata J, Horio Y, Shimohama S (2021) SIRT1 decelerates morphological processing of oligodendrocyte cell lines and regulates the expression of cytoskeleton-related oligodendrocyte proteins. Biochem Biophys Res Commun 546:7–14. https://doi.org/10.1016/j.bbrc.2021.01.095

Hsieh J, Nakashima K, Kuwabara T, Mejia E, Gage FH (2004) Histone deacetylase inhibition-mediated neuronal differentiation of multipotent adult neural progenitor cells. Proc Natl Acad Sci USA 101(47):16659. https://doi.org/10.1073/pnas.0407643101

Huang B, Peng X, Zhai X, Hu J, Chen J, Yang S, Huang Q, Deng E, Li H, Barakat TS, Chen J, Pei D, Fan X, Chambers I, Zhang M (2024) Inhibition of HDAC activity directly reprograms murine embryonic stem cells to trophoblast stem cells. Dev Cell 59(16):2101–2117.e8. https://doi.org/10.1016/j.devcel.2024.05.009. Epub 2024 May 31. PMID: 38823394.

Hughes EG, Stockton ME (2021) Premyelinating oligodendrocytes: mechanisms underlying cell survival and integration. Front Cell Develop Biol 9:714169. https://doi.org/10.3389/fcell.2021.714169

Hughes EG, Orthmann-Murphy JL, Langseth AJ, Bergles DE (2018) Myelin remodeling through experience-dependent oligodendrogenesis in the adult somatosensory cortex. Nat Neurosci 21(5):696–706. https://doi.org/10.1038/s41593-018-0121-5

Ishizuka Y, Shimizu H, Takagi E, Kato M, Yamagata H, Mikuni M, Shirao T (2014) Histone deacetylase mediates the decrease in drebrin cluster density induced by amyloid beta oligomers. Neurochem Int 76:114–121. https://doi.org/10.1016/j.neuint.2014.07.005

Jablonska B, Adams KL, Kratimenos P, Li Z, Strickland E, Haydar TF, Kusch K, Nave K-A, Gallo V (2022) Sirt2 promotes white matter oligodendrogenesis during development and in models of neonatal hypoxia. Nat Commun 13(1):4771. https://doi.org/10.1038/s41467-022-32462-2

Jagielska A, Lowe AL, Makhija E, Wroblewska L, Guck J, Franklin RJM, Shivashankar GV, Van Vliet KJ (2017) Mechanical strain promotes oligodendrocyte differentiation by global changes of gene expression. Front Cell Neurosci 11:93. https://doi.org/10.3389/fncel.2017.00093

Jäkel S, Agirre E, Mendanha Falcão A, van Bruggen D, Lee KW, Knuesel I, Malhotra D, Ffrench-Constant C, Williams A, Castelo-Branco G (2019) Altered human oligodendrocyte heterogeneity in multiple sclerosis. Nature 566(7745):543–547. https://doi.org/10.1038/s41586-019-0903-2

Jaworska J, Zalewska T, Sypecka J, Ziemka-Nalecz M (2019) Effect of the HDAC inhibitor, sodium butyrate, on neurogenesis in a rat model of neonatal hypoxia-ischemia: potential mechanism of action. Mol Neurobiol 56(9):6341–6370. https://doi.org/10.1007/s12035-019-1518-1

Ji S, Doucette JR, Nazarali AJ (2011) Sirt2 is a novel in vivo downstream target of Nkx2.2 and enhances oligodendroglial cell differentiation. J Mol Cell Biol 3(6):351–359. https://doi.org/10.1093/jmcb/mjr009

Juarez AL, He D, Lu QR (2016) Oligodendrocyte progenitor programming and reprogramming: toward myelin regeneration. Brain Res 1638(Pt B):209–220. https://doi.org/10.1016/j.brainres.2015.10.051

Kabir F, Atkinson R, Cook AL, Phipps AJ, King AE (2023) The role of altered protein acetylation in neurodegenerative disease. Front Aging Neurosci 14. https://doi.org/10.3389/fnagi.2022.1025473

Kamei Y, Xu L, Heinzel T, Torchia J, Kurokawa R, Gloss B, Lin SC, Heyman RA, Rose DW, Glass CK, Rosenfeld MG (1996) A CBP integrator complex mediates transcriptional activation and AP-1 inhibition by nuclear receptors. Cell 85(3):403–414. https://doi.org/10.1016/s0092-8674(00)81118-6

Kanakasabai S, Pestereva E, Chearwae W, Gupta SK, Ansari S, Bright JJ (2012) PPARγ agonists promote oligodendrocyte differentiation of neural stem cells by modulating Stemness and differentiation genes. PLoS One 7(11):e50500. https://doi.org/10.1371/journal.pone.0050500

Kassis H, Chopp M, Liu XS, Shehadah A, Roberts C, Zhang ZG (2014) Histone deacetylase expression in white matter oligodendrocytes after stroke. Neurochem Int 77:17–23. https://doi.org/10.1016/j.neuint.2014.03.006

Keil JM, Qalieh A, Kwan KY (2018) Brain transcriptome databases: A user's guide. J Neurosci Off J Soc Neurosci 38(10):2399–2412. https://doi.org/10.1523/JNEUROSCI.1930-17.2018

Koreman E, Sun X, Lu QR (2018) Chromatin remodeling and epigenetic regulation of oligodendrocyte myelination and myelin repair. Mol Cell Neurosci 87:18–26. https://doi.org/10.1016/j.mcn.2017.11.010

Leyk J, Goldbaum O, Noack M, Richter-Landsberg C (2015) Inhibition of HDAC6 modifies tau inclusion body formation and impairs autophagic clearance. J Mol Neurosci MN 55(4):1031–1046. https://doi.org/10.1007/s12031-014-0460-y

Li W, Zhang B, Tang J, Cao Q, Wu Y, Wu C, Guo J, Ling E-A, Liang F (2007) Sirtuin 2, a mammalian homolog of yeast silent information regulator-2 longevity regulator, is an oligodendroglial protein that decelerates cell differentiation through deacetylating alpha-tubulin. J Neurosci Off J Soc Neurosci 27(10):2606–2616. https://doi.org/10.1523/JNEUROSCI.4181-06.2007

Li Y, Gu Z, Lin S, Chen L, Dzreyan V, Eid M, Demyanenko S, He B (2022a) Histone deacetylases as epigenetic targets for treating Parkinson's disease. Brain Sci 12(5):672. https://doi.org/10.3390/brainsci12050672

Li M, Xiao L, Chen X (2022b) Histone acetylation and methylation underlie Oligodendroglial and myelin susceptibility in schizophrenia. Front Cell Neurosci 16. https://doi.org/10.3389/fncel.2022.823708

Lin C, Jiang T, Wang L, Chen T, Wang X, Xiang Y (2024) LncRNA CRNDE promotes hepatoma cell proliferation by regulating the metabolic reprogramming of M2 macrophages via ERK pathway. Cancer Cell Int 24:193. https://doi.org/10.1186/s12935-024-03380-8

Liu J, Casaccia P (2010) Epigenetic regulation of oligodendrocyte identity. Trends Neurosci 33(4):193–201. https://doi.org/10.1016/j.tins.2010.01.007

Liu J, Magri L, Zhang F, Marsh NO, Albrecht S, Huynh JL, Kaur J, Kuhlmann T, Zhang W, Slesinger PA, Casaccia P (2015) Chromatin landscape defined by repressive histone methylation during oligodendrocyte differentiation. J Neurosci Off J Soc Neurosci 35(1):352–365. https://doi.org/10.1523/JNEUROSCI.2606-14.2015

Liu T, Wang T, Qi L, Liu Y, Shan M, Wang F, Fang Y, Liu S, Wen L, Chen S, Wu D, Xu Y (2025) CPSF6-RARγ interacts with histone deacetylase 3 to promote myeloid transformation in RARG-fusion acute myeloid leukemia. Nat Commun 16(1):616. https://doi.org/10.1038/s41467-024-54860-4. PMID: 39805830; PMCID: PMC11729889.

Llorente V, Velarde P, Desco M, Gómez-Gaviro MV (2022) Current understanding of the neural stem cell niches. Cells 11(19):3002. https://doi.org/10.3390/cells11193002

Long KLP, Breton JM, Barraza MK, Perloff OS, Kaufer D (2021) Hormonal regulation of Oligodendrogenesis I: effects across the lifespan. Biomol Ther 11(2):283. https://doi.org/10.3390/biom11020283

Lu C, Dong L, Zhou H, Li Q, Huang G, Bai S, Liao L (2018) G-protein-coupled receptor Gpr17 regulates oligodendrocyte differentiation in response to Lysolecithin-induced demyelination. Sci Rep 8(1):4502. https://doi.org/10.1038/s41598-018-22452-0

Luo Y, Yan Z, Chu X, Zhang Y, Qiu Y, Li H (2025) Binding mechanism and distant regulation of histone deacetylase 8 by PCI-34051. Commun Biol 8(1):221. https://doi.org/10.1038/s42003-025-07649-0. PMID: 39939814; PMCID: PMC11821889.

Lyssiotis CA, Walker J, Wu C, Kondo T, Schultz PG, Wu X (2007) Inhibition of histone deacetylase activity induces developmental plasticity in oligodendrocyte precursor cells. Proc Natl Acad Sci USA 104(38):14982–14987. https://doi.org/10.1073/pnas.0707044104

Ma X-R, Zhu X, Xiao Y, Gu H-M, Zheng S-S, Li L, Wang F, Dong Z-J, Wang D-X, Wu Y, Yang C, Jiang W, Yao K, Yin Y, Zhang Y, Peng C, Gao L, Meng Z, Hu Z et al (2022) Restoring nuclear entry of Sirtuin 2 in oligodendrocyte progenitor cells promotes remyelination during ageing. Nat Commun 13(1):1225. https://doi.org/10.1038/s41467-022-28844-1

Magistretti PJ, Allaman I (2018) Lactate in the brain: from metabolic end-product to signalling molecule. Nat Rev Neurosci 19(4):235–249. https://doi.org/10.1038/nrn.2018.19

Marin-Husstege M, Muggironi M, Liu A, Casaccia-Bonnefil P (2002) Histone deacetylase activity is necessary for oligodendrocyte lineage progression. J Neurosci Off J Soc Neurosci 22(23):10333–10345. https://doi.org/10.1523/JNEUROSCI.22-23-10333.2002

Marmorstein R, Trievel RC (2009) Histone modifying enzymes: structures, mechanisms, and specificities. Biochimica et Biophysica Acta (BBA) Gene Regulat Mechan 1789(1):58–68. https://doi.org/10.1016/j.bbagrm.2008.07.009

Marques S, Zeisel A, Codeluppi S, van Bruggen D, Mendanha Falcão A, Xiao L, Li H, Häring M, Hochgerner H, Romanov RA, Gyllborg D, Muñoz-Manchado AB, La Manno G, Lönnerberg P, Floriddia EM, Rezayee F, Ernfors P, Arenas E, Hjerling-Leffler J et al (2016) Oligodendrocyte heterogeneity in the mouse juvenile and adult central nervous system. Science 352(6291):1326–1329. https://doi.org/10.1126/science.aaf6463

Martínez-Peula O, Morentin B, Callado LF, Meana JJ, Rivero G, Ramos-Miguel A (2024) Permissive epigenetic regulatory mechanisms at the histone level are enhanced in postmortem dorsolateral prefrontal cortex of individuals with schizophrenia. J Psychiat Neurosci JPN 49(1):E35–E44. https://doi.org/10.1503/jpn.230054

Matalon R, Michals K, Sebesta D, Deanching M, Gashkoff P, Casanova J, Optiz JM, Reynolds JF (1988) Aspartoacylase deficiency and N-acetylaspartic aciduria in patients with canavan disease. Am J Med Genet 29(2):463–471. https://doi.org/10.1002/ajmg.1320290234

McKenzie AT, Moyon S, Wang M, Katsyv I, Song W-M, Zhou X, Dammer EB, Duong DM, Aaker J, Zhao Y, Beckmann N, Wang P, Zhu J, Lah JJ, Seyfried NT, Levey AI, Katsel P, Haroutunian V, Schadt EE et al (2017) Multiscale network modeling of oligodendrocytes reveals molecular components of myelin dysregulation in Alzheimer's disease. Mol Neurodegener 12(1):82. https://doi.org/10.1186/s13024-017-0219-3

Mi Y, Ji Y, Zhang L, Sun C, Wei B, Yang D, Wan H, Qi X, Wu S, Zhu L (2023) A first-in-class HBO1 inhibitor WM-3835 inhibits castration-resistant prostate cancer cell growth in vitro and in vivo. Cell Death Dis 14(1):67. https://doi.org/10.1038/s41419-023-05606-5

Miron VE, Kuhlmann T, Antel JP (2011) Cells of the oligodendroglial lineage, myelination, and remyelination. Biochim Biophys Acta (BBA) Mol Basis Dis 1812(2):184–193. https://doi.org/10.1016/j.bbadis.2010.09.010

Moffett JR, Ross B, Arun P, Madhavarao CN, Namboodiri AMA (2007) N-Acetylaspartate in the CNS: from neurodiagnostics to neurobiology. Prog Neurobiol 81(2):89–131. https://doi.org/10.1016/j.pneurobio.2006.12.003

Montani L (2021) Lipids in regulating oligodendrocyte structure and function. Semin Cell Dev Biol 112:114–122. https://doi.org/10.1016/j.semcdb.2020.07.016

Montgomery RL, Hsieh J, Barbosa AC, Richardson JA, Olson EN (2009) Histone deacetylases 1 and 2 control the progression of neural precursors to neurons during brain development. Proc Natl Acad Sci USA 106(19):7876–7881. https://doi.org/10.1073/pnas.0902750106

Monti B, Mercatelli D, Contestabile A (2012) Valproic acid neuroprotection in 6-OHDA lesioned rat, a model for parkinson's disease. HOAJ Biol 1:article 4. https://doi.org/10.7243/2050-0874-1-4

Moyon S, Holloman M, Salzer JL (2023) Neural stem cells and oligodendrocyte progenitor cells compete for remyelination in the corpus callosum. Front Cell Neurosci 17:1114781. https://doi.org/10.3389/fncel.2023.1114781

Mukohara T, Park YH, Sommerhalder D, Yonemori K, Hamilton E, Kim S-B, Kim JH, Iwata H, Yamashita T, Layman RM, Mita M, Clay T, Chae YS, Oakman C, Yan F, Kim GM, Im S-A, Lindeman GJ, Rugo HS et al (2024) Inhibition of lysine acetyltransferase KAT6 in ER+HER2− metastatic breast cancer: a phase 1 trial. Nat Med 30(8):2242–2250. https://doi.org/10.1038/s41591-024-03060-0

Namchaiw P, Wen H, Mayrhofer F, Chechneva O, Biswas S, Deng W (2019) Temporal and partial inhibition of GLI1 in neural stem cells (NSCs) results in the early maturation of NSC derived oligodendrocytes in vitro. Stem Cell Res Ther 10:272. https://doi.org/10.1186/s13287-019-1374-y

Nishiyama A, Shimizu T, Sherafat A, Richardson WD (2021) Life-long oligodendrocyte development and plasticity. Semin Cell Dev Biol 116:25–37. https://doi.org/10.1016/j.semcdb.2021.02.004

Noack M, Leyk J, Richter-Landsberg C (2014) HDAC6 inhibition results in tau acetylation and modulates tau phosphorylation and degradation in oligodendrocytes. Glia 62(4):535–547. https://doi.org/10.1002/glia.22624

Nocera S, Marchena MA, Fernández-Gómez B, Gómez-Martín P, Sánchez-Jiménez E, Macías-Castellano A, Laó Y, Cordano C, Gómez-Torres Ó, Luján R, de Castro F (2024) Activation of Shh/Smo is sufficient to maintain oligodendrocyte precursor cells in an undifferentiated state and is not necessary for myelin formation and (re)myelination. Glia. https://doi.org/10.1002/glia.24540

Pan S, Chan JR (2021) Clinical applications of myelin plasticity for remyelinating therapies in multiple sclerosis. Ann Neurol 90(4):558–567. https://doi.org/10.1002/ana.26196

Pao P-C, Patnaik D, Watson LA, Gao F, Pan L, Wang J, Adaikkan C, Penney J, Cam HP, Huang W-C, Pantano L, Lee A, Nott A, Phan TX, Gjoneska E, Elmsaouri S, Haggarty SJ, Tsai L-H (2020) HDAC1 modulates OGG1-initiated oxidative DNA damage repair in the aging brain and Alzheimer's disease. Nat Commun 11(1):2484. https://doi.org/10.1038/s41467-020-16361-y

Papuć E, Rejdak K (2018) The role of myelin damage in Alzheimer's disease pathology. Arch Med Sci AMS 16(2):345–351. https://doi.org/10.5114/aoms.2018.76863

Park J, Lee K, Kim K, Yi S-J (2022) The role of histone modifications: from neurodevelopment to neurodiseases. Signal Transduct Target Ther 7(1):1–23. https://doi.org/10.1038/s41392-022-01078-9

Pavlou S, Foskolou S, Patikas N, Field SF, Papachristou EK, Santos CD, Edwards AR, Kishore K, Ansari R, Rajan SS, Fernandes HJR, Metzakopian E (2023) CRISPR-Cas9 genetic screen leads to the discovery of L-Moses, a KAT2B inhibitor that attenuates Tunicamycin-mediated neuronal cell death. Sci Rep 13(1):3934. https://doi.org/10.1038/s41598-023-31141-6

Pedre X, Mastronardi F, Bruck W, López-Rodas G, Kuhlmann T, Casaccia P (2011) Changed histone acetylation patterns in normal-appearing white matter and early multiple sclerosis lesions. J Neurosci 31(9):3435–3445. https://doi.org/10.1523/JNEUROSCI.4507-10.2011

Peng S, Zhao S, Yan F, Cheng J, Huang L, Chen H, Liu Q, Ji X, Yuan Z (2015) HDAC2 selectively regulates FOXO3a-mediated gene transcription during oxidative stress-induced neuronal cell death. J Neurosci Off J Soc Neurosci 35(3):1250–1259. https://doi.org/10.1523/JNEUROSCI.2444-14.2015

Petralla S, Peña-Altamira LE, Poeta E, Massenzio F, Virgili M, Barile SN, Sbano L, Profilo E, Corricelli M, Danese A, Giorgi C, Ostan R, Capri M, Pinton P, Palmieri F, Lasorsa FM, Monti B (2019) Deficiency of mitochondrial aspartate-glutamate carrier 1 leads to oligodendrocyte precursor cell proliferation defects both in vitro and in vivo. Int J Mol Sci 20(18):4486. https://doi.org/10.3390/ijms20184486

Phipps AJ, Dwyer S, Collins JM, Kabir F, Atkinson RA, Chowdhury MA, Matthews L, Dixit D, Terry RS, Smith J, Gueven N, Bennett W, Cook AL, King AE, Perry S (2024) HDAC6 inhibition as a mechanism to prevent neurodegeneration in the mSOD1G93A mouse model of ALS. Heliyon 10(14):e34587. https://doi.org/10.1016/j.heliyon.2024.e34587

Poeta E, Petralla S, Babini G, Renzi B, Celauro L, Magnifico MC, Barile SN, Masotti M, De Chirico F, Massenzio F, Viggiano L, Palmieri L, Virgili M, Lasorsa FM, Monti B (2021) Histone acetylation defects in brain precursor cells: a potential pathogenic mechanism causing proliferation and differentiation dysfunctions in mitochondrial aspartate-glutamate carrier isoform 1 deficiency. Front Cell Neurosci 15:773709. https://doi.org/10.3389/fncel.2021.773709

Prozorovski T, Ingwersen J, Lukas D, Göttle P, Koop B, Graf J, Schneider R, Franke K, Schumacher S, Britsch S, Hartung HP, Küry P, Berndt C, Aktas O (2019) Regulation of sirtuin expression in autoimmune neuroinflammation: Induction of SIRT1 in oligodendrocyte progenitor cells. Neurosci Lett 704:116–125. https://doi.org/10.1016/j.neulet.2019.04.007. Epub 2019 Apr 4. PMID: 30953735)

Qiao Y, Yang X, Jing N (2016) Epigenetic regulation of early neural fate commitment. Cell Mol Life Sci 73(7):1399–1411. https://doi.org/10.1007/s00018-015-2125-6

Qin L, Wu Y-L, Toneff MJ, Li D, Liao L, Gao X, Bane FT, Tien JC-Y, Xu Y, Feng Z, Yang Z, Xu Y, Theissen SM, Li Y, Young L, Xu J (2014) NCOA1 directly targets *M-CSF1* expression to promote breast cancer metastasis. Cancer Res 74(13):3477–3488. https://doi.org/10.1158/0008-5472.CAN-13-2639

Qu Q, Chen Y, Wang Y, Wang W, Long S, Yang HY, Wu J, Li M, Tian X, Wei X, Liu YH, Xu S, Xiong J, Yang C, Wu Z, Huang X, Xie C, Wu Y, Xu Z, Zhang C, Zhang B, Feng JW, Chen J, Feng Y, Fang H, Lin L, Xie ZK, Sun B, Tian H, Yu Y, Piao HL, Xie XS, Deng X, Zhang CS, Lin SC (2024) Lithocholic acid binds TULP3 to activate sirtuins and AMPK to slow down ageing. Nature. https://doi.org/10.1038/s41586-024-08348-2. Epub ahead of print. PMID: 39695235.

Radecki DZ, Messling HM, Haggerty-Skeans JR, Bhamidipati SK, Clawson ED, Overman CA, Thatcher MM, Salzer JL, Samanta J (2020) Relative levels of Gli1 and Gli2 determine the response of ventral neural stem cells to demyelination. Stem Cell Reports 15(5):Articolo 5. https://doi.org/10.1016/j.stemcr.2020.10.003

Richter-Landsberg C (2016) Protein aggregate formation in oligodendrocytes: tau and the cytoskeleton at the intersection of neuroprotection and neurodegeneration. Biol Chem 397(3):185–194. https://doi.org/10.1515/hsz-2015-0157

Roth SY, Denu JM, Allis CD (2001) Histone acetyltransferases. Ann Rev Biochem 70:81–120. https://doi.org/10.1146/annurev.biochem.70.1.81

Sajad M, Zahoor I, Rashid F, Cerghet M, Rattan R, Giri S (2024) Pyruvate dehydrogenase-dependent metabolic programming affects the oligodendrocyte maturation and remyelination. Mol Neurobiol 61(1):397–410. https://doi.org/10.1007/s12035-023-03546-x

Samanta J, Grund EM, Silva HM, Lafaille JJ, Fishell G, Salzer JL (2015) Inhibition of Gli1 mobilizes endogenous neural stem cells for remyelination. Nature 526(7573):448–452. https://doi.org/10.1038/nature14957

Selenica M-L, Benner L, Housley SB, Manchec B, Lee DC, Nash KR, Kalin J, Bergman JA, Kozikowski A, Gordon MN, Morgan D (2014) Histone deacetylase 6 inhibition improves memory and reduces total tau levels in a mouse model of tau deposition. Alzheimers Res Ther 6(1):12. https://doi.org/10.1186/alzrt241

Shen S, Li J, Casaccia-Bonnefil P (2005) Histone modifications affect timing of oligodendrocyte progenitor differentiation in the developing rat brain. J Cell Biol 169(4):577–589. https://doi.org/10.1083/jcb.200412101

Shen S, Sandoval J, Swiss VA, Li J, Dupree J, Franklin RJM, Casaccia-Bonnefil P (2008) Age-dependent epigenetic control of differentiation inhibitors is critical for remyelination efficiency. Nat Neurosci 11(9):1024–1034. https://doi.org/10.1038/nn.2172

Sherafat A, Pfeiffer F, Nishiyama A (2021) Shaping of regional differences in oligodendrocyte dynamics by regional heterogeneity of the Pericellular microenvironment. Front Cell Neurosci 15. https://doi.org/10.3389/fncel.2021.721376

Singhal NK, Huang H, Li S, Clements R, Gadd J, Daniels A, Kooijman EE, Bannerman P, Burns T, Guo F, Pleasure D, Freeman E, Shriver L, McDonough J (2017) The neuronal metabolite NAA regulates histone H3 methylation in oligodendrocytes and myelin lipid composition. Exp Brain Res 235(1):279–292. https://doi.org/10.1007/s00221-016-4789-z

Steadman PE, Xia F, Ahmed M, Mocle AJ, Penning ARA, Geraghty AC, Steenland HW, Monje M, Josselyn SA, Frankland PW (2020) Disruption of oligodendrogenesis impairs memory consolidation in adult mice. Neuron 105(1):150–164.e6. https://doi.org/10.1016/j.neuron.2019.10.013

Stein LR, Imai S (2014) Specific ablation of Nampt in adult neural stem cells recapitulates their functional defects during aging. EMBO J 33(12):1321–1340. https://doi.org/10.1002/embj.201386917

Sterner DE, Berger SL (2000) Acetylation of histones and transcription-related factors. Microbiol Mol Biol Rev 64(2):435–459. https://doi.org/10.1128/MMBR.64.2.435-459.2000
Sun Z, Xu Y (2020) Nuclear receptor coactivators (NCOAs) and corepressors (NCORs) in the brain. Endocrinology 161(8):bqaa083. https://doi.org/10.1210/endocr/bqaa083
Swiss VA, Nguyen T, Dugas J, Ibrahim A, Barres B, Androulakis IP, Casaccia P (2011) Identification of a gene regulatory network necessary for the initiation of oligodendrocyte differentiation. PLoS One 6(4):e18088. https://doi.org/10.1371/journal.pone.0018088
Tang BL, Chua CEL (2008) SIRT2, tubulin deacetylation, and oligodendroglia differentiation. Cell Motil 65(3):179–182. https://doi.org/10.1002/cm.20253
Tapias A, Wang Z-Q (2017) Lysine acetylation and Deacetylation in brain development and neuropathies. Genomics Proteomics Bioinformatics 15(1):19–36. https://doi.org/10.1016/j.gpb.2016.09.002
Tegla CA, Azimzadeh P, Andrian-Albescu M, Martin A, Cudrici CD, Trippe R 3rd, Sugarman A, Chen H, Boodhoo D, Vlaicu SI, Royal W 3rd, Bever C, Rus V, Rus H (2014) SIRT1 is decreased during relapses in patients with multiple sclerosis. Exp Mol Pathol 96(2):139–48. https://doi.org/10.1016/j.yexmp.2013.12.010. Epub 2014 Jan 5. PMID: 24397908)
Tiane A, Schepers M, Rombaut B, Hupperts R, Prickaerts J, Hellings N, van den Hove D, Vanmierlo T (2019) From OPC to oligodendrocyte: an epigenetic Journey. Cells 8(10):1236. https://doi.org/10.3390/cells8101236
Ullah M, Pelletier N, Xiao L, Zhao SP, Wang K, Degerny C, Tahmasebi S, Cayrou C, Doyon Y, Goh S-L, Champagne N, Côté J, Yang X-J (2008) Molecular architecture of quartet MOZ/MORF histone acetyltransferase complexes. Mol Cell Biol 28(22):6828–6843. https://doi.org/10.1128/MCB.01297-08
Valdés-Tovar M, Rodríguez-Ramírez AM, Rodríguez-Cárdenas L, Sotelo-Ramírez CE, Camarena B, Sanabrais-Jiménez MA, Solís-Chagoyán H, Argueta J, López-Riquelme GO (2022) Insights into myelin dysfunction in schizophrenia and bipolar disorder. World J Psychiatr 12(2):264–285. https://doi.org/10.5498/wjp.v12.i2.264
Vaquero A, Sternglanz R, Reinberg D (2007) NAD+-dependent deacetylation of H4 lysine 16 by class III HDACs. Oncogene 26(37):5505–5520. https://doi.org/10.1038/sj.onc.1210617
Voss AK, Thomas T (2009) MYST family histone acetyltransferases take center stage in stem cells and development. BioEssays 31(10):1050–1061. https://doi.org/10.1002/bies.200900051
Wang J, Weaver ICG, Gauthier-Fisher A, Wang H, He L, Yeomans J, Wondisford F, Kaplan DR, Miller FD (2010) CBP histone acetyltransferase activity regulates embryonic neural differentiation in the normal and Rubinstein-Taybi syndrome brain. Dev Cell 18(1):114–125. https://doi.org/10.1016/j.devcel.2009.10.023
Wang Y, Lonard DM, Yu Y, Chow D-C, Palzkill TG, O'Malley BW (2011) Small molecule inhibition of the steroid receptor coactivators, SRC-3 and SRC-1. Mol Endocrinol 25(12):2041–2053. https://doi.org/10.1210/me.2011-1222
Wang Y, Lonard DM, Yu Y, Chow D-C, Palzkill TG, Wang J, Qi R, Matzuk AJ, Song X, Madoux F, Hodder P, Chase P, Griffin PR, Zhou S, Liao L, Xu J, O'Malley BW (2014) Bufalin is a potent small-molecule inhibitor of the steroid receptor coactivators SRC-3 and SRC-1. Cancer Res 74(5):1506–1517. https://doi.org/10.1158/0008-5472.CAN-13-2939
Wang F, Ren S-Y, Chen J-F, Liu K, Li R-X, Li Z-F, Hu B, Niu J-Q, Xiao L, Chan JR, Mei F (2020) Myelin degeneration and diminished myelin renewal contribute to age-related deficits in memory. Nat Neurosci 23(4):481–486. https://doi.org/10.1038/s41593-020-0588-8
Wapenaar H, Dekker FJ (2016) Histone acetyltransferases: challenges in targeting bi-substrate enzymes. Clin Epigenetics 8:59. https://doi.org/10.1186/s13148-016-0225-2
Wheeler NA, Lister JA, Fuss B (2015) The Autotaxin-lysophosphatidic acid Axis modulates histone acetylation and gene expression during oligodendrocyte differentiation. J Neurosci Off J Soc Neurosci 35(32):11399–11414. https://doi.org/10.1523/JNEUROSCI.0345-15.2015
Wiame E, Tyteca D, Pierrot N, Collard F, Amyere M, Noel G, Desmedt J, Nassogne M-C, Vikkula M, Octave J-N, Vincent M-F, Courtoy PJ, Boltshauser E, van Schaftingen E (2009) Molecular

identification of aspartate N-acetyltransferase and its mutation in hypoacetylaspartia. Biochem J 425(1):127–136. https://doi.org/10.1042/BJ20091024
Wu YJ, Zhang SS, Yin Q, Lei M, Wang QH, Chen WG, Luo TT, Zhou P, Ji CL (2023) α-Mangostin Inhibited M1 Polarization of Macrophages/Monocytes in Antigen-Induced Arthritis Mice by Up-Regulating Silent Information Regulator 1 and Peroxisome Proliferators-Activated Receptor γ Simultaneously. Drug Des Devel Ther 17:563–577. https://doi.org/10.2147/DDDT.S397914. PMID: 36860800; PMCID: PMC9969869.
Xiao L, Ohayon D, McKenzie IA, Sinclair-Wilson A, Wright JL, Fudge AD, Emery B, Li H, Richardson WD (2016) Rapid production of new oligodendrocytes is required in the earliest stages of motor-skill learning. Nat Neurosci 19(9):1210–1217. https://doi.org/10.1038/nn.4351
Xie S, Yang J, Huang S, Fan Y, Xu T, He J, Guo J, Ji X, Wang Z, Li P, Chen J, Zhang Y (2022) Disrupted myelination network in the cingulate cortex of Parkinson's disease. IET Syst Biol 16(3–4):98–119. https://doi.org/10.1049/syb2.12043
Xie Z, Shao J, Shen Z, Ye Z, Okada Y, Okuzaki D, Okada N, Tachibana M (2025) HDAC1-3 inhibition triggers NEDD4-mediated CCR2 downregulation and attenuates immunosuppression in myeloid-derived suppressor cells. Cancer Immunol Immunother 74(3):81. https://doi.org/10.1007/s00262-024-03931-y. PMID: 39891718; PMCID: PMC11787094.
Xing YL, Röth PT, Stratton JAS, Chuang BHA, Danne J, Ellis SL, Ng SW, Kilpatrick TJ, Merson TD (2014) Adult neural precursor cells from the subventricular zone contribute significantly to oligodendrocyte regeneration and remyelination. J Neurosci 34(42):14128–14146. https://doi.org/10.1523/JNEUROSCI.3491-13.2014
Xu Z, Li H, Jin P (2012) Epigenetics-based therapeutics for neurodegenerative disorders. Curr Transl Geriatrics Experim Gerontol Rep 1(4):229–236. https://doi.org/10.1007/s13670-012-0027-0
Xu W, Xie L, Yang Y, Xu J, Cai S, Tian Y (2022) KAT5 inhibitor NU9056 suppresses anaplastic thyroid carcinoma progression through c-Myc/miR-202 pathway. Int J Endocrinol 2022:1–14. https://doi.org/10.1155/2022/2014568
Yan H, Rivkees SA (2006) Hypoglycemia influences oligodendrocyte development and myelin formation. Neuroreport 17(1):55–59. https://doi.org/10.1097/01.wnr.0000192733.00535.b6
Yan MS, Turgeon PJ, Man H-SJ, Dubinsky MK, Ho JJD, El-Rass S, Wang Y-D, Wen X-Y, Marsden PA (2018) Histone acetyltransferase 7 (KAT7)-dependent intragenic histone acetylation regulates endothelial cell gene regulation. J Biol Chem 293(12):4381–4402. https://doi.org/10.1074/jbc.RA117.001383
Yang K, Wu Z, Long J, Li W, Wang X, Hu N, Zhao X, Sun T (2023) White matter changes in Parkinson's disease. Npj Parkinson's Dis 9(1):1–10. https://doi.org/10.1038/s41531-023-00592-z
Ye F, Chen Y, Hoang T, Montgomery RL, Zhao X, Bu H, Hu T, Taketo MM, van Es JH, Clevers H, Hsieh J, Bassel-Duby R, Olson EN, Lu QR (2009) HDAC1 and HDAC2 regulate oligodendrocyte differentiation by disrupting β-catenin-TCF interaction. Nat Neurosci 12(7):829–838. https://doi.org/10.1038/nn.2333
Zeng Q, Wang K, Zhao Y, Ma Q, Chen Z, Huang W (2023) Effects of the acetyltransferase p300 on tumour regulation from the novel perspective of posttranslational protein modification. Biomol Ther 13(3):417. https://doi.org/10.3390/biom13030417
Zhang Y, Chen K, Sloan SA, Bennett ML, Scholze AR, O'Keeffe S, Phatnani HP, Guarnieri P, Caneda C, Ruderisch N, Deng S, Liddelow SA, Zhang C, Daneman R, Maniatis T, Barres BA, Wu JQ (2014) An RNA-sequencing transcriptome and splicing database of glia, neurons, and vascular cells of the cerebral cortex. J Neurosci 34(36):11929–11947. https://doi.org/10.1523/JNEUROSCI.1860-14.2014
Zhang L, He X, Liu L, Jiang M, Zhao C, Wang H, He D, Zheng T, Zhou X, Hassan A, Ma Z, Xin M, Sun Z, Lazar MA, Goldman SA, Olson EN, Lu QR (2016) Hdac3 interaction with p300 histone acetyltransferase regulates the oligodendrocyte and astrocyte lineage fate switch. Dev Cell 36(3):316–330. https://doi.org/10.1016/j.devcel.2016.01.002

Zhang Y, Gao Y, Ding Y, Jiang Y, Chen H, Zhan Z, Liu X (2023) Targeting KAT2A inhibits inflammatory macrophage activation and rheumatoid arthritis through epigenetic and metabolic reprogramming. MedComm 4(3):e306. https://doi.org/10.1002/mco2.306

Zhao L, Cao J, Hu K, He X, Yun D, Tong T, Han L (2020) Sirtuins and their biological relevance in aging and age-related diseases. Aging Dis 11(4):927–945. https://doi.org/10.14336/AD.2019.0820

Zhou P, Li D, Luo F, Wan X (2022) NCOA2 coordinates with the transcriptional KAT2B-NF-κB partner to trigger inflammation response in acute kidney injury. Gene 832:146583. https://doi.org/10.1016/j.gene.2022.146583

Zhou P, Peng X, Zhang K, Cheng J, Tang M, Shen L, Zhou Q, Li D, Yang L (2025) HAT1/HDAC2 mediated ACSL4 acetylation confers radiosensitivity by inducing ferroptosis in nasopharyngeal carcinoma. Cell Death Dis 16(1):160. https://doi.org/10.1038/s41419-025-07477-4. PMID: 40050614; PMCID: PMC11885570.

Zhu C, Dong X, Wang X, Zheng Y, Qiu J, Peng Y, Xu J, Chai Z, Liu C (2022) Multiple roles of SIRT2 in regulating physiological and pathological signal transduction. Genet Res 2022:9282484. https://doi.org/10.1155/2022/9282484

Zuchero JB, Fu M, Sloan SA, Ibrahim A, Olson A, Zaremba A, Dugas JC, Wienbar S, Caprariello AV, Kantor C, Leonoudakis D, Lariosa-Willingham K, Kronenberg G, Gertz K, Soderling SH, Miller RH, Barres BA (2015) CNS myelin wrapping is driven by actin disassembly. Dev Cell 34(2):152–167. https://doi.org/10.1016/j.devcel.2015.06.011

Chapter 9
Acetylation in Cardiac Aging: Molecular Mechanism and Therapeutic Approaches

Aditi Dattatraya Hase and Sanjay K. Banerjee

Abstract This chapter highlights the hallmarks of cardiac aging, distinguishing characteristics between cardiac aging and cardiac senescence. An overview of the molecular mechanisms underlying cardiac aging, with a particular focus on the role of reversible protein acetylation, emphasizes the role of sirtuins in regulating heart function and structure. The chapter explores how alterations in energy metabolism contribute to heart dysfunction, with a focus on the impact of mitochondrial dysfunction and phenomena of protein acetylation, along with the role of acetylase and deacetylase in an aging heart. Additionally, the chapter discusses the regulation of cardiomyocyte proliferation and the potential for enhancing cardiac regeneration. Finally, therapeutic strategies, including caloric restriction and HDAC inhibitors, microRNAs, stem cells, and other pharmacological agents are examined as potential approaches to slow or reverse the effects of cardiac aging.

Abbreviations

ACC	Acetyl-CoA Carboxylase
AMP	Adenosine Mono-Phosphate
AMPK	Adenosine Monophosphate-Activated Protein Kinase
ATGL	Adipose Triglyceride Lipase
ATP	Adenosine Triphosphate
BAX	Bcl-2-Associated X Protein
CD38	Cluster of Differentiation 38
CMs	Cardiomyocytes
CoA	Coenzyme A

A. D. Hase · S. K. Banerjee (✉)
Department of Biotechnology, National Institute of Pharmaceutical Education and Research (NIPER), Guwahati, Assam, India
e-mail: sanjay@niperguwahati.in

M. Halasa, A. Wawruszak (eds.), *Histone and Non-Histone Reversible Acetylation in Development, Aging and Disease*, Results and Problems in Cell Differentiation 75,
https://doi.org/10.1007/978-3-031-91459-1_9

CR	Calorie Restriction
CVD	Cardiovascular Disease
CypD	Cyclophilin D
DHS	Dihydrosphingosine
DHS1P	Dihydrosphinganine-1-phosphate
DNA	Deoxyribose Nucleic Acid
Drp1	Dynamin-Related Protein 1
ECHA	Hydroxy Acyl-CoA Dehydrogenase
eNOS	Endothelial Nitric Oxide Synthase
ETC	Electron Transport Chain
FDA	Food and Drug Administration
FOXO3	Forkhead box class O3a
GABPβ1	GA Binding Protein Transcription Factor Subunit Beta 1
GATA4	GATA Binding Protein 4
GCN5L1	General Control of Amino Acid Synthesis 5-Like 1
GDF11	Growth Differentiation Factor 11
GSK3β	Glycogen Synthase Kinase-3 Beta
H3K27	Histone H3 Lysine 27
H3K4	Histone H3 Lysine 4
H3K9	Histone H3 Lysine 4
HAT	Histone Acetyltransferases
HDAC	Histone Deacetylases
HDACi	Histone Deacetylases Inhibitors
HF	Heart Failure
HSF1	Heat Shock Factor 1
ICDH2	Isocitrate Dehydrogenase 2
IGF	Insulin-Like Growth Factor
IGF-AKT	Insulin-Like Growth Factor-Ak Strain Transforming
IL-1	Interleukin-1
IR	Ischemia-Reperfusion
JNK	c-Jun N-Terminal Kinase
KEAP1	Kelch-Like ECH-Associated Protein 1
KLF15	Kruppel-Like Factor 15
KO	Knockout
LAM	Lymphangioleiomyomatosis
LKB1	Liver Kinase B1
MCD	Malonyl CoA Decarboxylase
MEF2	Myocyte Enhancer Factor 2
MI	Myocardial Infarction
miRNA	microRNA
MnSOD	Manganese Superoxide Dismutase
mTOR	Mechanistic Target of Rapamycin
MyoD	Myoblast Determination Protein 1
NAD+	Nicotinamide Adenine Dinucleotide
NADPH	Nicotinamide Adenine Dinucleotide Phosphate Hydrogen

Nampt	Nicotinamide Phosphoribosyl Transferase
NFATc2	Nuclear Factor of Activated T cells, Cytoplasmic, Calcineurin-Dependent 2
NF-Kb	Nuclear Factor-Kappa B
NO	Nitric Oxide
Nrf2	Nuclear Factor Erythroid 2-Related Factor 2
PARP-1	Poly (ADP-Ribose) Polymerase 1
PDGF	Platelet-Derived Growth Factor
PGC-1α	Peroxisome Proliferator-Activated Receptor-γ Coactivator
PTM	Post-Translational Modifications
RNA	Ribose Nucleic Acid
ROS	Reactive Oxygen Species
SAHA	Suberoylanilide Hydroxamic Acid
SAMP8	Senescence-Accelerated Mouse-Prone 8
SASP	Senescence-Associated Secretory Phenotype
SD	Sprague–Dawley
SDF-1	Stromal Cell-Derived Factor 1
SIR2	Silent Information Regulator 2
SIRT	Silent Information Regulator
Sirt1	Silent Information Regulator 1
Smad3	Mothers Against Decapentaplegic Homolog 3
SPHK2	Sphingosine Kinase 2
TAC	Transverse Aortic Constriction
TGF-β	Transforming Growth Factor-Beta
Tip60	Tat-Interactive Protein 60
TNF-α	Tumor Necrosis Factor-alpha
TSA	Trichostatin A
VDAC1	Voltage-Dependent Anion Channel 1
VEGF	Vascular Endothelial Growth Factor
VLCAD	Very Long-Chain Acyl-CoA Dehydrogenase
VPA	Valproic Acid
WNT	Wingless-Related Integration Site
β-OHB	Beta-hydroxybutyrate

9.1 Introduction

The heart's primary function is to promote blood circulation, ensuring an adequate supply of oxygen and essential nutrients to various organs and tissues, while concurrently expelling metabolic waste products to maintain metabolic homeostasis (Ventura-Clapier et al. 2004). As individuals age there is an increase in the risk of blocked and narrowed arteries and a weakened or thickened heart muscle. Age is an independent risk factor for several chronic diseases including cardiovascular disease for both men and women (Ren and Zhang 2018). Cardiovascular disease is the world's leading cause of death in developed and developing nations. Over the past few decades, there has been a gradual increase in the average lifespan of humans, which has led to the rise in the old population and the prevalence of aging-related disorders, particularly cardiovascular diseases (Lutz et al. 2008). Significant increases in the burden of age-related cardiac, arterial, and microvascular diseases are correlated with longer life expectancies (Alfaras et al. 2016). With age proteins get modified by different chemical groups. One of the important protein modifications that increase during aging is protein acetylation. Proteins undergo acetylation when an acetyl group from acetyl-CoA (coenzyme-A) is transferred to a lysine residue, neutralizing the positive charge of the lysine (Parodi-Rullán et al. 2018). More than one hundred proteins are implicated in regulating the process through mechanisms that include the addition, removal, or binding of acetyl groups. Acetylation of protein occurs in both histone and non-histone proteins. The changes in acetylation of histone proteins lead to functional dysregulation of genes involved in the process of inflammation, and oxidative stress. Acetylation of non-histone proteins is involved in the process of energy metabolism and various signaling pathways responsible for the accelerated aging process. Studies showed that both histone and non-histone protein modifications are responsible for the development of cardiac aging. The phenomenon of protein acetylation is of crucial importance in the context of age-related physiological alterations (Francois et al. 2021; Sun et al. 2016) and has been correlated with the onset of cardiac pathologies such as myocardial infarction and heart failure, primarily due to its effects on mitochondrial proteins (Parodi-Rullán et al. 2018). This book chapter is focused on elucidating the mechanisms involved in acetylation underlying cardiac aging and presents promising interventions to attenuate or potentially reverse these age-associated transformations.

9.2 Hallmark of Cardiac Aging

Aging is characterized by the loss of cell renewal and repair capacity, which leads to increased morbidity and mortality (Yan et al. 2021). Cardiac aging is associated with a range of gradual changes to the structure and function of the heart, including elevated levels of collagen, reduced heart rate, and diastolic filling rate, and compromised left ventricular function (Schulman et al. 1992). Aging promotes structural and functional changes in the pericardium, myocardium, valves, atria, ventricles, the cardiac conduction system, and the vasculature (Obas and Vasan 2018). Growing data indicates that during cardiac aging, the heart experiences a variety of complicated phenotypic alterations, such as myocardial remodeling, cardiac hypertrophy, microcirculatory dysfunction, arrhythmia, and heart failure (HF) (Yan et al. 2021). The cellular and molecular mechanisms of cardiac aging include mitochondrial oxidative stress, disruption of proteostasis, age-dependent decreases in autophagy and ubiquitin-proteasome degradation, extracellular matrix remodeling, increased apoptosis, impaired nitric oxide (NO) bioavailability, activation of poly(ADP-ribose) polymerase 1 (PARP-1), and cellular energetic dysfunction, telomere damage, alteration of noncoding RNAs (ribose nucleic acid) and the activation of the renin-angiotensin-aldosterone system (Johnson et al. 2013; Morimoto and Cuervo 2009). The aforementioned physiological modifications can act as a mediator for a deterioration in cardiac performance, rendering the heart more susceptible to stressors. Consequently, this leads to a substantial augmentation in the likelihood of cardiovascular disease (CVD). As individuals age, there is a marked exponential rise in the occurrence of CVD, including conditions such as coronary heart disease, myocardial infarction (MI), stroke, and atherosclerosis (Paneni et al. 2017). The impairment of calcium homeostasis and adrenergic signal transduction, which are related to the process of cell aging, can also lead to incompetence and diminished muscle strength of the left ventricle. Additionally, it has an impact on exercise endurance (Li et al. 2021a). The alterations in mitochondrial adaptor p66Shc due to age have a profound impact on the equilibrium of the cardiovascular system (Capell et al. 2007). The p66Shc has been identified as a longevity-modifying factor that is linked to the dysregulation of reactive oxygen species (ROS) within pathological contexts (Haslem et al. 2022). Mice deficient in p66Shc (p66Shc KO mice) demonstrate a notable decrease in the accumulation of triglycerides within adipocytes (Berniakovich et al. 2008), improved health status in advanced age (Tomilov et al. 2010), increased resistance to ischemic injury, and reduced susceptibility to atherosclerosis (Carpi et al. 2009; Graiani et al. 2005). Genomic instability poses a significant hazard for cardiac diseases. A prime example of this is Hutchinson-Gilford progeria syndrome, which is characterized by Lamin A/C mutation and extensive damage to nuclear DNA (deoxyribose nucleic acid) and is connected to premature occurrences of atherosclerosis and CVD. These conditions often lead to fatal MI or strokes by the age of 13, on average. Furthermore, the occurrence of CVD can also be influenced by abnormal alterations in epigenetic modifications (Xu et al. 2016). The hallmarks of cardiac aging are shown in Fig. 9.1.

9.3 Cardiac Cell Senescence

The term senescence is etymologically derived from the Latin word senescere, which literally translates to "to grow old." The concepts of aging and senescence are frequently employed interchangeably and synonymously; however, they exhibit distinctions in their evolutionary and biological foundations, as well as their underlying mechanisms. Aging delineates the biological, behavioral, and social variability that occurs throughout the lifespan, which does not inherently elevate mortality risk. In contrast, senescence is characterized as a biological phenomenon involving inexorable dysfunctional modifications that lead to a diminished probability of reproduction and an augmented vulnerability to death (Crews 2003; Siddiqi and Sussman 2013). In aggregate, age-associated modifications that coincide with exposure to detrimental risk factors that compromise vitality signify a markedly elevated risk of senescence and the advancement of degenerative diseases within the populace.

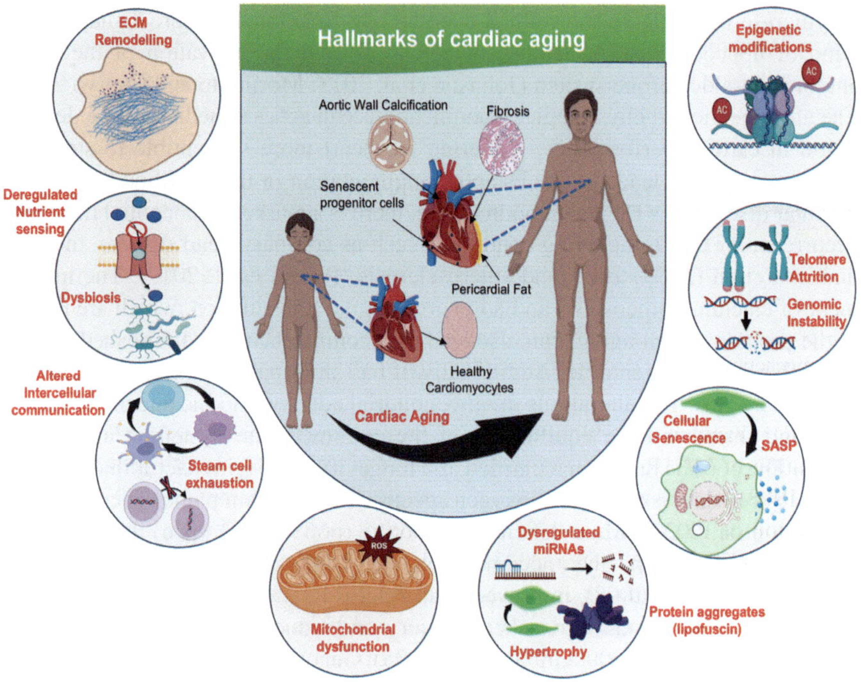

Fig. 9.1 Hallmarks of cardiac aging. (Created with BioRender.com)

Additionally, advancements in medical science have supplanted early mortality with a rise in chronic morbidity.

A fundamental distinction between senescent and quiescent cells lies in the fact that, under physiological conditions, quiescent cells possess the capacity to re-enter the cell cycle. In contrast, senescent cells necessitate stringent non-physiological interventions to facilitate cell cycle entry (e.g., inactivation of p53). A substantial proportion of senescent cells express markers indicative of cell cycle inhibition, such as p16^{Ink4a} and p53, and exhibit an increased lysosomal content, as evidenced by senescence-associated beta-galactosidase activity (Rayess et al. 2012; Serrano et al. 1997; Severino et al. 2000; Dimri and Campisi 1994). In contrast to quiescent cells, the permanent arrest of the cell cycle in senescent cells results in an augmentation of cell size and metabolic activity that exceeds two-fold compared to their healthy counterparts. Perhaps the most salient characteristic of a senescent cell is its incapacity to contribute to tissue repair, regeneration, and the maintenance of normative homeostasis. Moreover, senescent cells possess the distinctive capability to propagate their senescent phenotype through the proactive secretion of proteases, growth factors, and cytokines, ultimately culminating in the senescence-associated secretory phenotype (SASP) (Acosta et al. 2013; Aoshiba et al. 2013; Salminen et al. 2012; Coppé et al. 2010).

The phenomena of aging and cardiovascular disease exhibit a positive correlation, with cardiac ailments reaching epidemic proportions among the elderly population. As aging progresses, the myocardium experiences degenerative changes and processes that culminate in myocyte death. Pathological alterations in the expression of contractile protein isoforms and modified levels of calcium-handling proteins result in a gradual yet progressive deterioration of contractile function (Lakatta 1993; Lakatta 1990). Longitudinal echocardiographic evaluations reveal significant age-associated hypertrophy and a decline in diastolic function (Hees et al. 2002; Czuriga et al. 2012). Age-related histopathological changes in cardiac tissue manifest as interstitial and subendocardial fibrosis, hyaline cytoplasmic alterations, vacuolization, and collapsed sarcomeres, among others (Czuriga et al. 2012). Morphometric analyses of aging cardiac tissue demonstrate myocyte hypertrophy, an increase in myocyte apoptosis, and heightened fibrosis and amyloid deposition. The mechanistic underpinnings of the accumulation of age-related modifications and senescence in myocytes involve multiple pathways that are influenced at various levels within the cells (Anversa et al. 1990).

9.4 Protein Acetylation and Aging Heart

Recent research elucidates the pervasive occurrence of acetyl modifications in mitochondria, which can significantly impact intermediary metabolic processes. Such modifications imply that variations in acetylation could contribute to the pathogenesis of age-associated diseases, including heart failure and malignancies, in which mitochondrial carbon utilization is affected (Wagner and Payne 2011). Initial investigations utilizing *Saccharomyces cerevisiae* indicated that the presence of additional copies of the deacetylase gene, Sir2 (Silent Information Regulator 2) resulted in a 50% increase in lifespan, whereas the absence of this gene correspondingly diminished longevity (Kaeberlein et al. 1999). This pivotal finding has generated substantial academic interest in the exploration of mammalian sirtuins (SIRTs), owing to their prospective capacity to enhance health and offer protection against age-associated pathologies (de Oliveira et al. 2012).

Acetylation emerges as a reversible and highly dynamic post-translational modification that mitigates and diminishes the charge of lysine residues, consequently altering protein structures, which in turn affects DNA-binding affinity, enzymatic activity, protein stability, and subcellular localization as shown in Fig. 9.2 (Glozak et al. 2005). Lysine acetylation was initially identified on histone proteins during the latter part of the 1960s. In the subsequent decades of the 1970s and 1980s, numerous non-histone proteins were also implicated (Gershey et al. 1968; Li et al. 2021b). The acetylation of p53 and tubulin represents initial instances wherein protein acetylation is demonstrated to encompass a broader range of proteins beyond merely histone proteins (Piperno et al. 1987; Gu and Roeder 1997). This modification is critical in numerous cellular processes, including gene transcription, mRNA splicing, signal transduction, and cellular survival (Kouzarides 2000). Acetylation

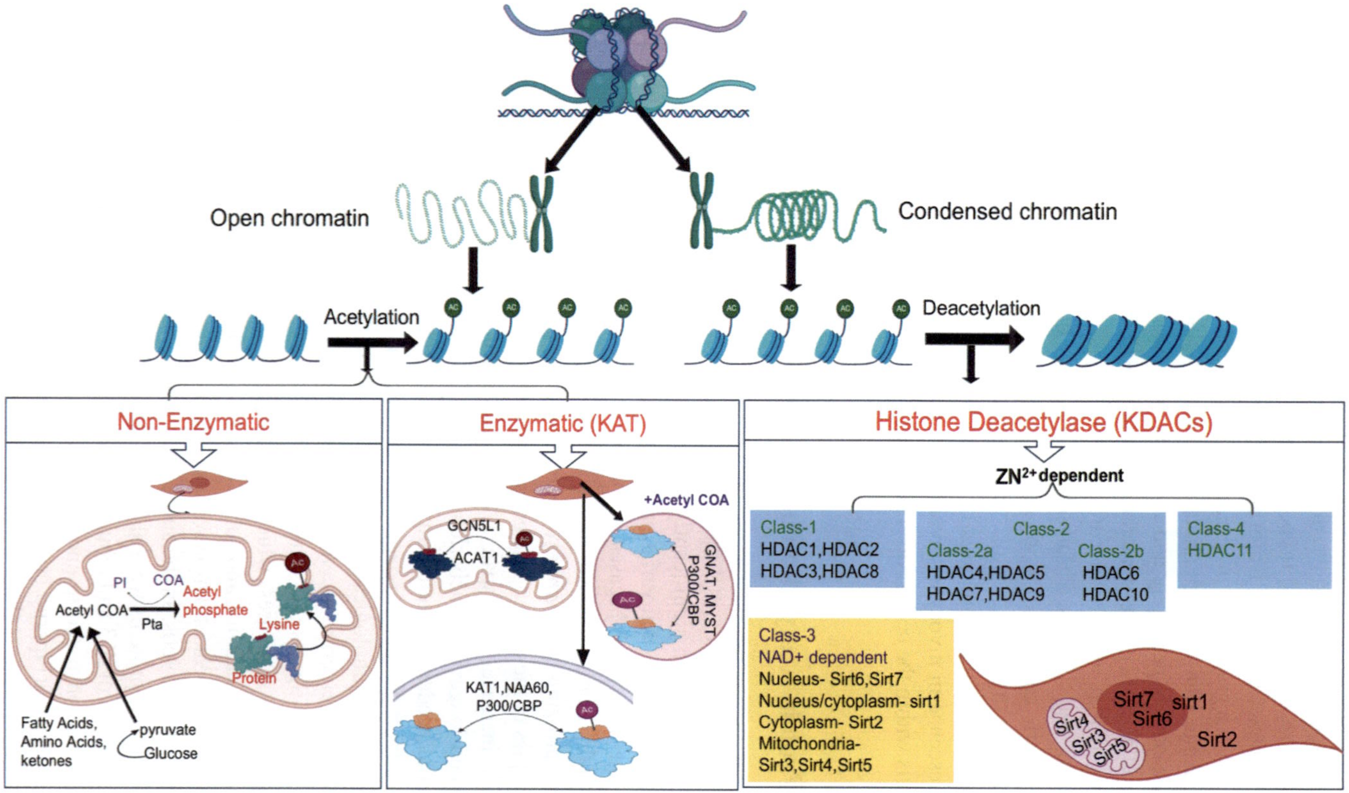

Fig. 9.2 The biochemical mechanisms of protein lysine acetylation (enzymatic-histone acetyltransferase; non-enzymatic) and deacetylation (histone deacetylase). (Created with BioRender.com)

represents a reversible biochemical modification characterized by the transfer of an acetyl moiety to the ε-amino group of lysine residues. This modification was initially delineated in the context of histone proteins and has been recognized for several decades as a crucial regulatory mechanism governing chromatin dynamics (Strahl and Allis 2000). The homeostasis of acetylation represents the fundamental epigenetic mechanism that underpins cardiac dysfunction (McKinsey and Olson 2005; Li et al. 2021b). In recent investigations, numerous extensive proteomic analyses have elucidated the prevalence of lysine acetylation in proteins, particularly those localized within mitochondria and metabolic enzymes broadly (Choudhary et al. 2009; Hebert et al. 2013; Kim et al. 2006; Hirschey et al. 2010). Functional analyses have further demonstrated that mitochondrial enzymes are subjected to regulatory control via acetylation (Zhao et al. 2010; Hirschey et al. 2010; Choudhary et al. 2009). Notably, this regulatory mechanism encompasses enzymes involved in the urea cycle, the tricarboxylic acid (TCA) cycle, fatty acid oxidation, and ketogenesis (Kouzarides 2000; Nakagawa et al. 2009; Shimazu et al. 2010; Yu et al. 2009). The physiological relevance of lysine acetylation is additionally emphasized by its remarkable evolutionary conservation across species, extending from prokaryotes to humans (Hirschey et al. 2010; Weinert et al. 2013; Zhang et al. 2009).

Recent investigations indicate that the majority of acetylation transpires in a nonenzymatic manner due to interaction with acetyl-CoA, especially within the alkaline conditions prevalent in the mitochondrial matrix (Wagner and Payne 2013). In this study, Weinert BT et al. demonstrated that the stoichiometry of acetylation of liver mitochondrial proteins is markedly elevated in comparison to that of cytoplasmic and nuclear proteins. These observations align with the hypothesis of a distinct mitochondrial acetyl-CoA reservoir and an alkaline environment that enhances nonenzymatic acetylation. (Weinert et al. 2015) Moreover, Weinert BT et al found that the fasting-induced enhancement of mitochondrial acetylation was mitigated by the depletion of ATGL (Adipose triglyceride lipase), implying that fatty acid oxidation facilitates mitochondrial protein acetylation. A separate investigation elucidated that the fasting-induced mitochondrial acetylation within hepatic tissue is contingent upon very long-chain acyl-CoA dehydrogenase (VLCAD), which serves as a pivotal enzyme in fatty acid oxidation (Weinert et al. 2014). Pougovkina et al. further elucidated that the stimulation of fatty acid oxidation in fibroblasts results in an elevation of protein acetylation and that fatty acids represent a direct substrate for mitochondrial protein acetylation. These findings corroborate the notion that acetyl-CoA, generated via fatty acid oxidation, influences mitochondrial acetylation levels through a nonenzymatic pathway. Additionally, the concept of nonenzymatic acetylation is congruent with the overarching alterations in mitochondrial acetylation instigated by fasting and the absence of ATGL and is in line with the observed low stoichiometry of acetylation (Pougovkina et al. 2014). Alterations in the acetylation of histone and non-histone proteins significantly influence protein conformation and functionality, which in turn may modify various other post-translational modifications (PTMs), such as protein phosphorylation. Acetylation represents a burgeoning and significant epigenetic modification that is modulated by histone deacetylases (HDACs) and histone acetyltransferases (HATs)(Li et al. 2021b) (Fig. 9.2).

9.5 Acetylation of Histone Proteins in Heart

Eukaryotic genomic DNA is organized into a highly structured form referred to as chromatin, which arises from the assembly of a fundamental unit known as the nucleosome, an octameric complex composed of eight distinct proteins termed histones. Approximately 146 base pairs of DNA are sequentially coiled around a singular nucleosome unit. The interactions between the DNA and the terminal tails of histones are pivotal in regulating the activation or repression of gene transcription. Various post-translational modifications can alter the state of histones, thereby influencing gene transcription. Specifically, the N-acetylation of lysine residues located within histones is modulated by two classes of enzymes: HATs and HDACs (Constantinou et al. 2023).

Abnormal acetylation of core histones, a phenomenon presumably correlated with environmental determinants, has been correlated with significant cardiovascular pathologies (Milan et al. 2018). Pan- and selective histone deacetylase inhibitors (HDACi) have been demonstrated to maintain cardiac functionality in pathological conditions by implementing an anti-inflammatory response and mitigating cardiac hypertrophy and fibrosis (Liao et al. 2010; McKinsey 2012). Givinostat (ITF2357) represents a potent pan-histone deacetylase inhibitor (pan-HDACi) that has attracted significant scholarly interest owing to its diverse range of applications, demonstrated efficacy, and favorable safety profile in human subjects (Leoni et al. 2005). Empirical evidence indicates that the compound effectively lowers the levels of tumor necrosis factor-alpha (TNF-α), interleukin-6 (IL-6), and interleukin-1 (IL-1), thereby achieving a remarkable attenuation of the inflammatory response while concurrently promoting pro-angiogenic effects (Consalvi et al. 2014). According to the research conducted by Marika Milan et al., Givinostat possesses the capacity to enhance cardiac performance by modulating the expression of transforming growth factor-beta (TGF-β) at the cardiac fibroblast level, executing a protective mechanism against apoptosis in myocytes, and fostering increased angiogenesis (Bertrand 2010; Milan et al. 2018).

9.6 Acetylation of Non-histone Proteins in Heart

Cardiac metabolic processes experience substantial transformations throughout the aging process, with the acetylation of metabolic enzymes serving as a pivotal factor in these modifications (Sack and Finkel 2012; Wagner and Payne 2011). The acetylation of non-histone proteins exerts several cellular functions like encompassing metabolic activities, altering energy metabolism, modulating signaling cascades, and adaptive responses to stress. In the context of senescent cardiac tissue, the acetylation modifications of transcription factors including NF-kB (Nuclear Factor-Kappa B) (Costantino et al. 2016), p53 (Gu and Roeder 1997), FOXO3 (Forkhead box class O3a), p65 (Chen et al. 2005; Kwon and Ott 2008) and HSF1 (Heat Shock Factor 1) significantly influences their regulatory capacity to express target genes associated with inflammatory responses, oxidative stress management, and cellular reparative processes. Notable enzymes, including acetyl-CoA carboxylase (ACC), isocitrate dehydrogenase, succinate dehydrogenase, and sirtuins (SIRTs) are influenced by acetylation, thereby affecting the metabolic activities within the cardiac tissue (Tomczyk et al. 2022; Ogura et al. 2018). The acetylation of pivotal proteins exerts a modulatory effect on signaling pathways that significantly impact cardiac functionality and the aging process. Prominent signaling entities that are influenced by acetylation encompass AMPK (adenosine monophosphate activated protein kinase, which governs energy homeostasis) and mTOR (mechanistic target of rapamycin, which is integral to cellular growth and metabolic processes), both of which are essential for maintaining cardiac health and promoting longevity (Kim et al. 2011; Harrison et al. 2009).

9.7 Deacetylation of Proteins, Sirtuins, and Cardiac Aging

Sirtuins, referred to as class III HDACs, represent homologous proteins of the mammalian silencing information regulator 2 (Sir2) that were initially identified in *Saccharomyces cerevisiae* (Chen et al. 2020a). In the mammalian system, the Sirtuin family comprises seven distinct members Sirt1–7 (Silent Information Regulator 1 to 7), each exhibiting specific subcellular localization and functional characteristics (Frye 1999). Sirt1, Sirt6, and Sirt7 (Fig. 9.2) localize within the nucleus and are implicated in the deacetylation of non-histone and histone proteins as well as the modulation of gene expression (Frye 2000). Moreover, Sirt1 is also located in the cytoplasm, where it plays a significant role in promoting apoptosis (Haigis and Guarente 2006). Sirt2 predominantly occupies the cytoplasmic compartment but possesses the capability to influence cell cycle dynamics within the nucleus. Sirt3, Sirt4, and Sirt5 are localized in the mitochondria (Fig. 9.2), where they are integral to the regulation of energy metabolism (Serrano et al. 2013).

The coenzyme NAD+(nicotinamide adenine dinucleotide) is critical for an array of cellular functions, inclusive of Sirtuin activity and mitochondrial operations (Santo-Domingo and Demaurex 2012). The concentration of NAD+ diminishes in aged and senescent cellular environments; however, the restoration of NAD+ has been shown to confer protective effects against the aging process (Mills et al. 2016). The enzyme CD38 (cluster of differentiation 38), which consumes NAD+ and catalyzes the cyclization of NAD, is notably upregulated in aging cardiac and neural tissues, and may contribute to the aging phenomenon by diminishing both NAD+ levels and sirtuin functionality (Wang et al. 2019). The pharmacological inhibition of CD38 utilizing the compound 78c (Table 9.2) has demonstrated the potential to prolong the lifespan, enhance cardiac performance, and decrease protein acylation, while simultaneously activating antioxidant signaling pathways involving Nrf2 (nuclear factor erythroid 2-related factor 2) and Foxo1(forkhead box class O-1) (Tarragó et al. 2018).

Among the seven sirtuins, Sirt1 and Sirt2 are mostly acknowledged for their role as cardioprotective deacetylases, linked with aging (Tang et al. 2017). Sirt1 and other nuclear sirtuins exert a profound influence on the aging of the cardiovascular system. Sirt1 facilitates the deacetylation of critical substrates such as p53, JNK(c-Jun N-terminal kinase), FOXO1 (forkhead box class O1), and eNOS (endothelial nitric oxide synthase) (Table 9.1). All of which are implicated in mechanisms associated with cardiac aging (Brodsky et al. 2004; Ota et al. 2007; Orimo et al. 2009). Sirt1 plays a pivotal role in modulating p53, thereby averting stress-induced apoptosis and cellular senescence through its deacetylation following DNA damage (Liu et al. 1999). This protective mechanism through Sirt1 exhibits a decline with advancing age (Sasaki et al. 2006). Moreover, Sirt1 enhances the process of autophagy, which is essential for cellular homeostasis during the aging process (Linton et al. 2014). In murine models, moderate overexpression of Sirt1 (up to 7.5-fold) demonstrates a protective effect against cardiac aging, characterized by a reduction in hypertrophy, fibrosis, and the maintenance of cardiac function, while excessive overexpression (more than 13-fold) may lead to cardiac dysfunction (Alcendor et al. 2007). Sirt1 is postulated to be instrumental in the preservation of telomeres throughout the aging process of stem cells, primarily through its role in gene-silencing mechanisms (Guarente 1999). Sirt1 is associated with telomeric regions and may enhance telomere stability by facilitating histone deacetylation and subsequent chromatin alterations. This particular mechanism may serve to mitigate the aging of stem cells and promote an extended lifespan (Cockell et al. 1995; Mantel and Broxmeyer 2008). As the cardiac tissue starts aging the cardioprotective properties of Sirt1 become attenuated, attributable to its diminished expression levels and altered downstream signaling cascades. In response to the decline in tissue Sirt1 levels observed in advanced heart failure, the expression of Nampt (nicotinamide phosphoribosyl transferase) is observed to increase (Lu et al. 2014). Nonetheless, the compromised functionality of the AMPK-Nampt-Sirt1 signaling pathway may contribute to the manifestation of these age-related alterations. Empirical research has demonstrated that the aging process adversely affects AMPK activation, resulting in diminished mitochondrial functionality and impaired lipid metabolic processes (Reznick et al. 2007; Ma et al. 2010). The administration of Sirt1 agonists presents a potentially efficacious strategy to mitigate these detrimental changes and confer protection to the heart against the deleterious effects of aging and heart failure (Lu et al. 2014).

Similar to Sirt1, Sirt6 is instrumental in regulating DNA repair, metabolic processes, inflammatory responses, and the maintenance of telomeres (Sundaresan et al. 2012). Within the cardiac context, Sirt6 confers protection against cardiac hypertrophy by suppressing c-Jun expression and facilitating the deacetylation and subsequent degradation of p300, which in turn diminishes the expression of NF-kB target genes associated with hypertrophy (Shen et al. 2016). Mice deficient in Sirt6 exhibit spontaneous hypertrophy, whereas overexpression of Sirt6 effectively prevents this condition. Additionally, Sirt6 activates NRF2 target genes through direct interaction and deacetylation of NRF2, while concurrently inhibiting KEAP1 (Kelch-like ECH-associated protein 1), and plays a crucial role to regulate mitochondrial sirtuins, particularly Sirt3(Stratton et al. 2022). At the cellular echelon, Sirt6 exerts an inhibitory effect on IGF (Insulin-like Growth Factor) signaling through its interaction with c-JUN and the subsequent deacetylation of H3K9 at the downstream targets of IGF (Kugel and Mostoslavsky 2014). Sirt6 is crucial for preserving glucose and lipid homeostasis, maintaining genomic integrity, and facilitating resistance against oxidative stress and DNA damage, all of which are associated with age-related pathologies. Empirical evidence has demonstrated its capacity to prolong lifespan in murine models (Etchegaray et al. 2023). Sirt6 emerges as a pivotal anti-aging molecule that modulates a variety of cellular mechanisms pertinent to the aging process and safeguards the cardiac tissue from age-related hypertrophic and fibrotic changes (Pillai et al. 2021; Liu et al. 2023). Sirt6 is integral to the repair of DNA double-strand breaks, a highly conserved biological mechanism that mitigates cellular senescence and the aging process, especially in species characterized by prolonged lifespans (Tian et al. 2019a). Despite the noted decline in Sirt6 activity with advancing age, its genetic overexpression in rodent models has been demonstrated to prolong lifespan through the modulation of IGF-AKT (insulin-like growth factor-Ak strain transforming) signaling pathways, thereby enhancing cardiac health and postponing the onset of heart failure (Sundaresan et al. 2012). In non-human primates, the ablation of Sirt6 function has been associated with delays in neuronal differentiation and a concomitant reduction in lifespan (Shen et al. 2016) (Weinert et al. 2015). Furthermore, Sirt6 plays a pivotal role in maintaining hematopoietic stem cell homeostasis in murine models by deacetylating H3K56ac, which subsequently influences WNT(wingless-related integration site) signaling and the overall functionality of stem cells (Wang et al. 2016). The Sirt6 activator MDL-800 has exhibited potential in enhancing genomic stability and stem cell pluripotency in aged mice, thereby underscoring Sirt6 as a significant target in the context of aging-related pathologies (Chen et al. 2020b; Ren et al. 2022).

Sirt7 operates within the nucleus and nucleolus, contributing to ribosomal RNA (rRNA) expression (Ford et al. 2006). Mice lacking Sirt7 display heightened acetylation of cardiac Akt, Ras, and p53, culminating in cardiac hypertrophy, inflammation, and cellular apoptosis (Vakhrusheva et al. 2008). Sirt7 deacetylates GABPβ1 (GA binding protein transcription factor subunit beta 1) (also known as NRF2) (Table 9.1), thereby facilitating the activation of mitochondrial genes, which bolster mitochondrial adaptations in response to stress and aging (Ryu et al. 2014). Furthermore, Sirt7 deacetylates the transcription factor GATA4 (GATA binding protein 4) (Table 9.1), inhibiting the expression of pro-hypertrophic genes and thereby reinforcing its cardioprotective function (Yamamura et al. 2020).

Sirt2, located in the cytoplasm, exhibits elevated expression levels in both cerebral and cardiac tissues; however, its expression diminishes in response to stress and the aging process. Several studies indicate a decline in Sirt2 expression with advancing age in both animal models and human peripheral blood cells, thereby implying its potential as a biomarker for aging and health monitoring (Yudoh et al. 2015). Mice lacking Sirt2 exhibit the development of age-associated cardiac fibrosis, hypertrophy, and dysfunction, whereas the overexpression of Sirt2 mitigates hypertrophic responses in cardiac myocytes. The cytoprotective mechanisms of Sirt2 are associated with the deacetylation of three pivotal substrates (Table 9.1): NFATc2 (nuclear factor of activated T cells, cytoplasmic, calcineurin-dependent 2), whose activity is inhibited in deacetylase form and prevents maladaptive cardiac remodeling (Sarikhani et al. 2018a); LKB1 (liver kinase B1), which serves to activate AMPK signaling pathways that counteract hypertrophy; and GSK3β (glycogen synthase kinase-3 beta), which contributes to the reduction of cardiomyocyte size and possesses anti-hypertrophic properties (Tang et al. 2017; Sarikhani et al. 2018b; Antos et al. 2002).

Sirt3 is recognized for its cardioprotective role by suppressing factors that facilitate hypertrophy, diminishing ROS, and enhancing the activity of antioxidant enzymes via the activation of FOXO3A (Porter et al. 2014). The susceptibility of the aging myocardium to ischemia-reperfusion (IR) injury is associated with a deterioration in mitochondrial functionality and a diminished expression of Sirt3. It modulates the electron transport chain through the deacetylation of subunits within complexes I and II, it confers protection against ischemia-reperfusion injury in aged murine models (Chouchani et al. 2013; Wojtovich and Brookes 2009; Ahn et al. 2008). Sirt3 exerts its effects by deacetylating cyclophilin D (CypD), thereby bolstering mitochondrial integrity and decreasing permeability, which in turn prevents the depletion of NAD+ and mitigates mitochondrial calcium and ROS accumulation (Hafner et al. 2010). Sirt3 confers protection against age-associated pathologies through the deacetylation of lysine residues (Sack 2011). Sirt3 may also bolster antioxidant defenses through the deacetylation and subsequent activation of manganese superoxide dismutase (MnSOD) in mitochondria. Moreover, the deacetylation and activation of mitochondrial isocitrate dehydrogenase 2 (ICDH2) (Table 9.1) amplify NADPH (nicotinamide adenine dinucleotide phosphate hydrogen)

Table 9.1 The classification of sirtuins, their location, role in cardiac diseases, and effect on protein acetylation

NAD+ dependent histone deacetylase	Location	Effect on the cardiac system	Animal model	Proteins deacetylated by sirtuins	References
SIRTUIN 1 (Sirt1)	Nucleus/ Cytoplasm	Protective against cardiac dysfunction, cardiac hypertrophy, Ischemic heart disease	Neonatal (1–2 days old) Sprague–Dawley rats, male C57BL/6 mice young (4–6 months) and old (22–24 months)	p53, JNK, FOXO1, and eNOS	Planavila et al. (2011) and Nadtochiy et al. (2010)
SIRTUIN 2 (Sirt2)	Cytoplasm	Protective against cardiac hypertrophy, fibrosis, and adverse remodeling	Male C57BL/6 J wild-type (WT) and *Sirt2* knockout (Sirt2-KO) mice 8–12 weeks old, 8 weeks old 129/Sv mice	α-Tubulin, NFATc2, GSK3β, LKB1	North et al. (2003), Tang et al. (2017), Sarikhani et al. (2018b), and Funamoto et al. (2023)
SIRTUIN 3 (Sirt3)	Mitochondria	Protective against ischemia-reperfusion (IR) injury, hypertrophy, and fibrosis	Sirt3 knock-out mice	Deacetylation of subunits within mitochondrial complexes I and II, isocitrate dehydrogenase 2 (ICDH2), Ku70, Cyclophilin D (CypD)	Ahn et al. (2008), Hafner et al. (2010), Qiu et al. (2010), Yu et al. (2012), Sundaresan et al. (2008), Sundaresan et al. (2016), and Samarakoon et al. (2013)

(continued)

Table 9.1 (continued)

NAD+ dependent histone deacetylase	Location	Effect on the cardiac system	Animal model	Proteins deacetylated by sirtuins	References
SIRTUIN 4 (Sirt4)	Mitochondria	Preventing apoptosis of cardiomyocytes by hypoxia, reducing cardiac pathologies	Male C57BL/6 Sirt4 knock-out mice	Malonyl CoA decarboxylase (MCD)	Luo et al. (2017), Liu et al. (2013a), Laurent et al. (2013), Han et al. (2019), and Luo et al. 2017)
SIRTUIN 5 (Sirt5)	Mitochondria	Protection against ischemia-reperfusion injury	Sirt5 knock-out mice	Carbamoyl phosphate synthetase 1	Boylston et al. (2015), Nakagawa et al. (2009), and Hershberger et al. (2018)
SIRTUIN 6 (Sirt6)	Nucleus	Protection against cardiac hypertrophy, reducing cellular senescence and aging	Sirt6 knock-out mice	NRF2, p300, TNF-alpha, removal of the acetyl moiety from the lysine residues of histone H3, particularly at the H3K9ac, H3K18ac, and H3K27ac sites	Shen et al. (2016), Kugel and Mostoslavsky (2014)), Tian et al. (2019a), Etchegaray et al. (2023), and Baran et al. (2023)
SIRTUIN 7 (Sirt7)	Nucleus/ cytoplasm	Inhibiting inflammatory cardiomyopathy, apoptosis, reducing stress-induced cardiac hypertrophy	Sirt7 knock-out mice	GABPβ1, GATA4, MyoD(myoblast determination protein 1), p53, NFκB	Vakhrusheva et al. (2008), Ryu et al. (2014) and (Yamamura et al. (2020)

reservoirs that enhance the antioxidant capacity (Qiu et al. 2010; Yu et al. 2012). Additionally, Sirt3 deacetylates Ku70 (DNA binding regulatory subunit) (Table 9.1), which facilitates its interaction with the BAX (Bcl-2-associated X protein) protein, thereby obstructing its translocation to the mitochondria (Sundaresan et al. 2008). As a critical regulator of mitochondrial homeostasis, Sirt3 a deacetylase enzyme, assumes a vital and irreplaceable role in the context of heart disease. The absence of sirt3 adversely affects the contractile capability of the heart (Wei et al. 2017) and subsequent investigations revealed that Sirt3 expression is down-regulated in instances of heart failure. Caloric restriction, a well-established promoter of healthy lifespan extension, induces the upregulation of Sirt3 and increases resilience against oxidative stress, implying that Sirt3 activators or caloric restriction mimetics could confer protection to the aging heart from IR injury through the promotion of deacetylation (Ahmet et al. 2011; Minor et al. 2010; Shinmura et al. 2008; Porter et al. 2014; Long et al. 2002). In Sirt3-KO murine models exhibited fibrosis across various organs, including hepatic, cardiac, pulmonary, and renal tissues, in contrast to their age-matched wild-type counterparts. The absence of Sirt3 resulted in an elevation of TGF-β1 expression alongside a hyperacetylation of glycogen synthase kinase 3β (GSK3β) specifically at the K15 residue. This alteration inhibited the functional activity of GSK3β, thereby diminishing the phosphorylation of β-catenin and Smad3 (mothers against decapentaplegic homolog 3), which consequently stabilized and activated these transcription factors, facilitating the upregulation of profibrotic gene expression (Sundaresan et al. 2016; Jin et al. 2009; Samarakoon et al. 2013).

Sirt4, a distinctive histone ADP-ribosyl transferase, in contrast to other sirtuins, lacks NAD + -dependent deacetylase activity (Mahlknecht and Voelter-Mahlknecht 2009) and serves to safeguard cardiac myocytes from apoptosis induced by hypoxia by inhibiting the translocation of Bax to the mitochondria. Nevertheless, a hypoxic environment results in reduced levels of Sirt4, which consequently leads to mitochondrial uncoupling (Liu et al. 2013a). In contradiction to these findings, in models simulating heart failure, sirt4 knockout mice displayed diminished cardiac injury and enhanced functional capacity, whereas the overexpression of Sirt4 exacerbated these pathological conditions. Sirt4 has been shown to inhibit the capacity of Sirt3 to activate the antioxidant enzyme manganese superoxide dismutase (MnSOD) (Luo et al. 2017).

Sirt5, which is prominently expressed in cardiac tissue, exhibits both cardioprotective and anti-aging attributes. Animals deficient in Sirt5 demonstrate increased susceptibility to heart failure and mortality under pressure overload conditions, and they develop spontaneous hypertrophy and cardiac dysfunction as they age (Hershberger et al. 2018). Proteomic analysis has revealed that in Sirt5 knockout models, excessive succinylation of 3-hydroxy acyl-CoA dehydrogenase, a protein integral to fatty acid oxidation, hampers this metabolic process, resulting in diminished ATP (adenosine triphosphate) levels within the heart (Sadhukhan et al. 2016).

9.8 Expression of Acetylases and Deacetylases Enzymes in Aging

Scientific literature indicates that both HATs and HDACs in cells were altered with aging. Studies showed that the expression of deacetylase enzymes is mostly altered in pathological conditions and responsible for enhanced protein acetylation. Sirt1 has been linked to longevity and aging in humans and other mammals. In particular, Sirt1 expression and activity in a number of organs and tissues, including the cardiovascular system, significantly decline with age (Braidy et al. 2011; Ministrini et al. 2021; Donato et al. 2011). During aging Sirt2 is downregulated in the vasculature and heart. Sirt2 expression is also altered in several diseases including cardiac hypertrophy and heart failure (Ministrini et al. 2021). Previous studies showed that angiotensin II (Ang II)- and aging-induced hypertrophy hearts showed decreased Sirt2 expression and compromised heart function (Tang et al. 2017). Serum Sirt3 levels are higher in younger populations relative to their older counterparts (Pradhan et al. 2017). Aging was correlated with a decrease in NAD+ concentrations, which subsequently resulted in diminished Sirt3 activity (Silaghi et al. 2021). Senescent cell accumulation is a hallmark of aging, and Sirt4 has been demonstrated to be increased in a variety of cell lines either during or after senescence (Betsinger and Cristea 2019; Anwar et al. 2016). The role of Sirt4 in the cardiovascular system required more research. It is worth noting that Sirt4 has been shown to ameliorate ischemic heart injury (Liu et al. 2013a), whereas another study reported that it promotes cardiac hypertrophy (Luo et al. 2017). In cardiomyocytes, oxidative stress downregulates Sirt5, which acts as a protective factor against ROS by reducing ROS-induced cell death (Liu et al. 2013b). Sirt expression is markedly downregulated in cardiomyopathy caused by isoproterenol (Galindo et al. 2009). Research findings indicate that Sirt6 is known as a longevity protein that helps in maintaining the stability of the genome and telomeres while delaying the aging process. Sirt6 overexpression resulted in increased longevity in male mice (Li et al. 2021c; Kanfi et al. 2012). Sirt6 may lessen cardiomyocyte aging and cardiac hypertrophy (Pillai et al. 2021). By controlling NAD+ metabolic balance, and glucose homeostasis Sirt6 slows down the aging process (Roichman et al. 2021; Li et al. 2022). Interestingly, progeria, an accelerated aging condition, was linked to Sirt6 deficiency (Liao and Kennedy 2012). Another study showed that Sirt-deficient monkeys show severe fetal developmental retardation and leading to death within hours after delivery (Zhang et al. 2018a). Mice with Sirt deletion have a shorter lifespan; they pass away at roughly one year of age, presumably as a result of inflammatory cardiomyopathy, and increased cardiomyocyte death. These findings highlight the significant influence of Sirt7 on cardiac functioning (Cencioni et al. 2015; Favero et al. 2015). Sirt7 has been shown to be crucial in halting the heart's progressive functional decline. Deletion of Sirt7 causes a number of pathological alterations in the heart, and these get worse as people age (Vakhrusheva et al. 2008).

Acetylase enzyme, p300 plays a crucial role in the aging process. Normalization of p300 activity may prevent or stop the progression of cardiac aging diseases and is evident from a number of investigations conducted by several researchers utilizing cellular and animal models (Ghosh 2020). In the case of Angiotensin II-induced hypertension, myocardial tissues have markedly higher levels of p300, which is linked to increased acetylation of H3K9, and responsible for cardiac hypertrophy, fibrosis, and aging (Rai et al. 2019). In contrast, in dilated and decompensated heart failure induced by thoracic aorta constriction (TAC), there was a considerable drop in the levels of the mitochondrial acetyltransferase protein GCN5L1 (general control of amino acid synthesis 5-like 1) (Zhang et al. 2023). Compared to younger women, postmenopausal women's hearts have lower levels of GCN5L1, which causes cardiac complications in older women (Manning et al. 2022). A study showed that enhanced GCN5L1 levels were associated with increased fatty acid acetylation and elevated acetylation of mitochondrial protein in older people's hearts. Decreased acetylation of mitochondrial proteins enhances fatty acid oxidation (FAO) and mitigates cardiac diastolic dysfunction in aging hearts in cardiac-specific GCN5L1 mutant mice (Stewart et al. 2024).

9.9 Protein Acetylation in Cardiac Aging: Molecular Mechanisms

9.9.1 Accelerated Cardiac Aging

As epigenetics plays a crucial role in the development of cardiovascular disease, epigenetic regulators are excellent therapeutic targets for several cardiac diseases. It is important to note that research into developing epidrugs to treat various malignancies is now underway (Franci et al. 2010). But there hasn't been much focus on creating epidrugs for cardiac conditions. MicroRNAs, methylases, demethylases, acetyltransferases, and deacetylases are the main epigenetic regulators and focus for drug discovery.

Epigenetics significantly impacts embryonic heart development and adult-onset heart disease. The activation of specific pathways due to stress-induced vascular damage contributes to the progression of cardiac aging, characterized by various manifestations such as cellular anomalies, pathological cardiac enlargement, diabetic-related heart muscle disease, restructuring of the cardiac matrix, compromised cardiac function, and ultimate heart failure (Ghosh 2020).

Acetyltransferase p300 (p300), a significant epigenetic modulator, serves as a critical factor in cardiac development in embryonic stages. The regulation of heart development is overseen by p300 through the process of histone acetylation-mediated chromatin remodeling. This mechanism involves the transcriptional regulation of genes that are essential for cardiac development. The regulation of GATA4 expression during embryonic mouse cardio genesis is associated with acetylation of H3K4 (histone H3 lysine 4), H3K9 (histone H3 lysine 9), and H3K27 (histone H3 lysine 27) within the GATA4 promoter region by p300 (Takaya et al. 2008). This highlights the crucial role of p300 in modulating transcription factors associated with cardiac development (Zhou et al. 2018). Deregulation of p300, triggered by prohypertrophic and profibrogenic stress signals, enhances p300 recruitment to various genes, including those encoding transcription factors. This leads to increased acetylation of specific lysine residues in histones and transcription factors, altering chromatin structure. Consequently, hypertrophic and fibrogenic gene expression is upregulated. Cardiac hypertrophy and myocardial fibrosis are key pathological features in various stress-induced conditions associated with accelerated cardiac aging (Ghosh 2020). Notably, p300 expression patterns differ between the left and right ventricles, correlating with the distinct expression of cardiac genes such as atrial natriuretic peptide (ANP) and brain natriuretic peptide (BNP), suggesting a region-specific role of p300 in the heart. These unique expression patterns and the differential regulation of genes may be influenced by the varying workload of the left and right ventricles under both physiological and pathophysiological conditions (Mathiyalagan et al. 2010). Kruppel-like factor 15 (KLF15) acts as a key regulator of cardiac lipid metabolism in vivo. Its expression is significantly reduced in dysfunctional human cardiac muscle and abnormal tissue formations, which concurrently show increased p300-mediated acetylation of p53 (Haldar et al. 2010).

Abnormal activity of p300, resulting from genetic mutations or elevated enzyme levels, is associated with diseases such as senescence and cardiovascular disorders. Serving as an epigenetic modulator and transcriptional coactivator, p300 exerts influence over a range of pathological pathways (Giles et al. 1998). Doxorubicin has been observed to drive cellular senescence in cardiomyocytes, fibroblasts, and endothelial cells, characterized by the upregulation of senescence-associated markers and an increase in p300 expression. In both young and aged SAMP8 (senescence-accelerated mouse-prone 8) mice, doxorubicin treatment was found to reduce Sirt1 deacetylase levels while simultaneously elevating p300 levels (Sin et al. 2015), confirming its role in promoting cardiac aging.

It can be concluded that the creation of harmless, non-hazardous, and small chemical compounds that act as inhibitors or epidrugs specifically targeting the p300 represents a highly favorable strategy for managing the increased incidence of cardiac-related mortality due to aging on a global scale. Data suggests that p300 inhibitors can be viewed as anti-aging agents. Sirt1 activated by resveratrol downregulates p300 by increasing its ubiquitination and proteasomal degradation, suggesting that increased p300 plays a crucial role in the pathophysiology of cardiomyopathy (Kuno et al. 2013).

Furthermore, in addition to its direct function in chromatin remodeling and transcriptional activation via particular acetylation of histones and transcription factors in target gene regulatory domains, p300 engages in multiple cellular signaling pathways that crosstalk to regulate various cellular and organ functions. Therefore, in order to prevent unintended side effects, it is crucial to look at how various dosages of small molecule inhibitors of p300 affect these pathways in a specific preclinical environment (Henry et al. 2019).

9.9.2 Cardiac Energy Metabolism

Obesity and diabetes may precipitate cardiac dysfunction through both indirect pathways, such as hypertension and atherosclerosis, as well as direct mechanisms that upset cardiomyocyte metabolism and function (Reaven et al. 1996; Beckman and Creager 2016). A primary instigator of this phenomenon is lipid overload, which impairs glucose metabolism and results in lipotoxicity (Wende and Abel 2010). While the myocardium predominantly relies on fatty acids for adenosine triphosphate (ATP) synthesis, an excessive influx of lipids can surpass the capacity for fatty acid oxidation, thereby disrupting cellular energetic homeostasis (Zhang et al. 2010).

Mitochondrial dysfunction represents another salient characteristic of metabolic syndrome, wherein an augmented substrate supply leads to the depletion of nicotinamide adenine dinucleotide (NAD+), the accumulation of intermediates such as acetyl-CoA, an elevation of ROS, and subsequent mitochondrial damage (Kraus et al. 2014). The excessive uptake of lipids within the cardiac tissue influences protein function such as dynamin-related protein 1 (Drp1), which initiates mitochondrial fission and dysfunction (Tsushima et al. 2018). Qingxun Hu et al. proved that upon activation through acetylation, Drp1 translocates to the mitochondria, facilitating the opening of the mitochondrial permeability transition pore (mPTP), triggering apoptosis, and compromising cardiomyocyte contractile functionality via the voltage-dependent anion channel 1 (VDAC1). These observations underscore Drp1 acetylation as a pivotal mechanism in the pathophysiology of heart dysfunction associated with metabolic stress (Hu et al. 2020).

Cardiac dysfunctions exhibit a marked increase of acetylation with advancing age. The age-related deterioration in both structural and functional integrity of the heart constitutes a significant risk factor for a variety of cardiovascular diseases (Perrino et al. 2017). In elucidating a previously unidentified contributor to cardiac aging, Gaurav Ahuja et al. present evidence of the age-dependent, cardiac-specific accumulation of the lysosphingolipid sphinganine (dihydrosphingosine, DHS) as an evolutionarily conserved characteristic of the senescent vertebrate heart. Mechanistically, the DHS derivative sphinganine-1-phosphate (dihydrosphinganine-1-phosphate, DHS1P) inhibits HDAC1 activity, resulting in an anomalous elevation of histone acetylation and transcriptional activity, thereby leading to DNA damage. Consequently, pharmacological interventions that (i) inhibit the accumulation of DHS1P through the application of SPHK2 (sphingosine kinase 2) inhibitors, (ii) mitigate the aberrant increase in histone acetylation through the use of histone acetyltransferase (HAT) inhibitors, and (iii) obstruct the DHS1P-dependent rise in transcription via an RNA polymerase II inhibitor, effectively prevent DHS-induced DNA damage in human cardiomyocytes. The age-associated deterioration of cardiac function may be at least partially attributable to the accumulation of DHS, which can be counteracted by curcumin (a HAT inhibitor) (Zhang et al. 2018b). These findings underscore the critical role of lysosphingolipid-mediated loss of genomic integrity in the cardiac aging phenomenon.

9.9.3 Cardiomyocyte Proliferation

Bing Li et al. unveiled a new post-translational modification pattern observed on the p21 in cardiac myocyte proliferation. Hyperacetylation of p21 in cardiac myocytes suggests a potential correlation between p21 acetylation and cardiac myocyte cell cycle arrest. This study demonstrates that increased p21 acetylation reduces cardiac myocyte proliferation and alters the levels of cell cycle proteins, particularly cyclins.

Conversely, decreased p21 acetylation significantly enhances cardiac myocyte proliferation. These findings suggest that p21 deacetylation may serve as a crucial regulatory mechanism for controlling cardiac myocyte proliferation (Li et al. 2019a).

Understanding the role of p21 acetylation and deacetylation in regulating its stability and functionality within cardiomyocytes (CMs) is crucial. Specifically, the acetylation of p21 by Tip60 (Tat-interactive protein 60) enhances its stability through the prevention of ubiquitination and subsequent proteasomal degradation, resulting in an elevated abundance of p21 (Lee et al. 2012), which contributes to the arrest of the cell cycle, thereby impeding cardiomyocyte proliferation. Conversely, p21 deacetylation promotes its ubiquitination and degradation, leading to decreased p21 levels and relieving proliferation arrest in cardiomyocytes. These findings suggest that modulating p21 acetylation could be a promising strategy for enhancing cardiac regeneration. Overall, Sirt1 significantly promotes cardiac myocyte proliferation by downregulating p21 through both transcriptional and post-translational mechanisms (Li et al. 2019a). Previous research has indicated that young postnatal hearts exhibit elevated levels of Sirt1 deacetylase activity compared to mature hearts, implying a role for Sirt1 deacetylase activity in modulating cardiac development (Mu et al. 2014). Acetylation and deacetylation of proteins have a significant impact on diverse cellular biological processes including cell proliferation, cell survival, and apoptosis (Spange et al. 2009).

9.10 Therapeutic Approaches to Reverse Cardiac Aging

9.10.1 Calorie Restriction

Caloric restriction of approximately 30–50% enhances health span, prolongs the initiation of senescence and problems related to aging, and enhances metabolic well-being in the majority of organisms (Solon-Biet et al. 2015). Calorie restriction (CR) entails the reduction of caloric consumption while maintaining adequate nutritional intake. CR is a replicable dietary strategy that has been significantly lowers the incidence of a variety of age-related disorders encompassing neurological conditions, and cardiovascular disorders (Yan et al. 2021). It is commonly believed that CR is primarily influenced by the direct reduction in energy consumption that affects cellular components such as NAD+ and AMP (adenosine mono-phosphate), subsequently leading to cascading effects on pathways involved in nutrient sensing, including sirtuin (Sirt1), AMP-activated protein kinase (AMPK), mechanistic target of rapamycin (mTOR), and insulin/IGF-1(Couteur et al. 2011). It is commonly accepted that the aforementioned consequence is indicative of a legitimate impact of caloric restriction on the process of aging because a majority of the observable characteristics and medical conditions associated with aging are postponed (Lee et al. 2008).

The occurrence of CVD is diminished by implementing CR to regulate the mechanisms that maintain cardiac functionality, such as proteasome-mediated turnover, autophagy, mitochondrial quality, and apoptosis, thereby exerting its effects during the aging process (Rattan 2014). The expression of PGC-1α (peroxisome proliferator-activated receptor-γ coactivator) and the promotion of mitochondrial biogenesis induced by Sirt1 and AMPK in muscle tissue, have been linked to the reduction of insulin resistance. The utilization of CR or CR mimickers, such as metformin or resveratrol, has been identified as a contributing factor in the prevention of disease onset and progression (Li et al. 2021a). The original theory behind Clive McCay's research on CR in the 1920s and 1930s was that it functioned by delaying development (Yan et al. 2021). The idea that CR functioned by lowering oxidative stress brought on by food consumption emerged with Harman's free radical hypothesis of aging (Sohal and Weindruch 1996). Further research verified that CR is linked to lower levels of free radical generation and oxidative stress indicators. The preliminary investigations into mitochondrial acetylation transitioned towards comprehensive proteomics assessments. In a proteomics survey carried out in 2006 by Kim et al., it was unveiled that more than 20% of proteins and enzymes within liver mitochondria undergo acetylation, and are linked with fasting state (Kim et al. 2006).

Literature strongly supports the fact that advancing senescence leads to alteration in cardiac functionality in mice, rats, and healthy humans; however, calorie restriction (CR) can prevent cardiac aging and associated complications. Taffet et al. investigated the consequences of chronic caloric restriction on both systolic and diastolic cardiac performance in ad libitum-fed female B6D2-F hybrid mice compared to calorically restricted mice aged between 32 and 35 months. Lifelong caloric restriction prevented the age-related impairments in late diastolic function but did not show any improvement in systolic or early diastolic cardiac function (Taffet et al. 1997). In male Dahl salt-sensitive rats, a high-salt diet causes decompensated heart failure, characterized by pressure-overload hypertrophy. Nevertheless, high-sodium rats subjected to a 15% caloric restriction exhibited notable improvements: diminished blood pressure, reduced cardiac mass, lowered plasma concentrations of IL-6 and TNF-α, decreased oxidative injury, and enhanced diastolic function (Seymour et al. 2006). Caloric restriction has been shown to diminish the steady-state levels of the byproducts resulting from oxidative damage to DNA, proteins, and lipids through the enhancement of antioxidative defenses and the reduction of hydrogen peroxide and superoxide generation within the mitochondria (Sohal et al. 1994a). In mice models, calorie restriction is effective in mitigating age-associated DNA damage within postmitotic tissues such as the cardiac muscles and skeletal muscles (Weindruch and Sohal 1997; Sohal et al. 1994b).

Caloric restriction (CR) has the potential to mitigate the age-associated elevation of atherosclerosis through the attenuation of proinflammatory processes in the geriatric population. Atherosclerosis, often referred to as "accelerated arterial aging," encompasses inflammatory reactions within vascular smooth muscle cells (VSMCs) (Head et al. 2017; McGill Jr et al. 2000). Young (6-month-old) and old (24-month-old) Fischer 344 rats were used in an experiment by Mingyi Wang et al. (2018). One group was fed ad libitum, while the other group was kept on a 40% CR starting at one month of age. The study showed that CR reduces arterial aging by lowering proinflammatory molecules in cells and tissues. It reduces elastin degradation, fibrosis, and arterial thickness by blocking the platelet-derived growth factor (PDGF) signaling cascade. CR maintains a younger aortic wall phenotype and delays biological aging (Wang et al. 2018).

Caloric restriction (CR) protects cardiac health against age-associated alterations, including shifts in myosin isoforms, fibrosis, and apoptosis of cardiomyocytes, while simultaneously maintaining or enhancing left ventricular diastolic function (Weiss and Fontana 2011). Both prolonged and short-term CR induce expression of several genes that mitigate age-related diminished contractility, fibrosis, myocardial remodeling, and disturbances in lipid metabolism (Dhahbi et al. 2006). Normal aging is linked with increased catecholamine biosynthesis enzymes in the adrenal medulla and enhanced levels of circulating catecholamines (Hoeldtke and Cilmi 1985). Caloric restriction is responsible for decreased expressions of catecholamine biosynthetic enzymes and lessens oxidative damage to the adrenal medulla (Erdös et al. 2007).

9.10.2 HDAC Inhibitors

Small molecule inhibitors targeting HDACs can be classified into four structurally heterogeneous categories: short-chain fatty acids (e.g., phenylbutyrate, valproic acid); hydroxamic acids (e.g., Trichostatin A [TSA], suberoylanilide hydroxamic acid [SAHA]), benzamides (e.g., MS-275), and cyclic peptides (e.g., depsipeptides) (Lee et al. 2007).

The use of Trichostatin A (TSA) mitigates angiotensin II-induced (Kee et al. 2006), isoproterenol-induced (Kook et al. 2003), and transverse aortic constriction (TAC)-induced ventricular hypertrophy. Administration of TSA was also effective in reversing established TAC-induced cardiac hypertrophy (Kee et al. 2006; Kong et al. 2006). Both TSA and scriptaid, an alternative hydroxamic acid-derived HDAC inhibitor, attenuated cardiac hypertrophy in a pressure-overload TAC model; a reduction in cardiomyocyte cross-sectional area was observed, alongside notable enhancements in systolic function (Kong et al. 2006).

Fibrosis constitutes a significant characteristic of the aging myocardium, and pan-HDAC inhibitors have demonstrated efficacy in attenuating the overaccumulation of extracellular matrix in models of cardiac pathology (Schuetze et al. 2014). The class I HDAC inhibitor MGCD0103 inhibits angiotensin II-induced cardiac fibrosis and has been evidenced to diminish fibrosis while enhancing ventricular performance in a rat model subjected to chronic coronary artery ligation (Williams et al. 2014; Nural-Guvener et al. 2014).

In cardiomyocytes, autophagy is pivotal to mitochondrial metabolism; however, it is inhibited during the reperfusion phase, which exacerbates injury following ischemic episodes (Ma et al. 2012). Autophagy, a critical lysosomal degradation mechanism, is indispensable for eliminating defective organelles and preserving cellular homeostasis, particularly under conditions of nutrient scarcity, wherein it facilitates the recycling of cytosolic components for energy production (Whitworth and Pallanck 2017; Bingol et al. 2014). Autophagy, encompassing the specific process of mitophagy that removes compromised mitochondria, is fundamental in the context of ischemic cardiovascular pathologies, as dysfunctional mitochondria precipitate oxidative stress, resulting in cellular deterioration. β-OHB (beta-hydroxy butyrate), an endogenous inhibitor of HDAC1 and HDAC2, significantly enhances mitochondrial quality in aged rabbit hearts through the activation of parkin translocation, albeit it concurrently hinders mitophagy in the setting of heart failure (Bingol et al. 2014; Shimazu et al. 2013; Thai et al. 2019; Chu et al. 2021). Additionally, it mitigates oxidative stress during ischemia-reperfusion events and influences mitochondrial functionality by diminishing the mitochondrial NAD couple and oxidizing coenzyme Q (Veech 2014; Xie et al. 2016). β-OHB also elevates the expression of FOXO3a, thereby promoting the synthesis of antioxidant enzymes such as SOD2 and catalase, which collectively reduce ROS levels. A brief engagement in a ketogenic dietary regimen has been shown to augment mitochondrial density within the cardiac tissue, thereby contributing to the cardioprotective properties associated with β-OHB (Nagao et al. 2016; Al-Zaid et al. 2007).

The identification of SAHA (Vorinostat), an FDA (Food and Drug Administration)-sanctioned pan-HDAC inhibitor, demonstrated a 40% reduction in infarct size alongside the preservation of cardiac function (Xie et al. 2019; Przyklenk et al. 2012). In an ex-vivo rat model examining cardiac ischemia-reperfusion injury, the class I HDAC inhibitor MS-275 was observed to maintain cardiac functionality and diminish infarct size, indicating that the activity of class I HDACs is implicated in ischemic-induced cardiac injury (Aune et al. 2014). The initial investigation was conducted utilizing a rat model of myocardial infarction, in which the coronary artery is subjected to a permanent surgical occlusion. The administration of HDAC inhibitors, specifically valproic acid (VPA) and tributyrin demonstrated a significant attenuation of cardiomyocyte hypertrophy and collagen accumulation within both the remote and border regions of the infarcted left ventricular tissue (Lee et al. 2007).

Class IIa HDACs, including HDAC4, HDAC5, and HDAC9, interact with MEF2 (myocyte enhancer factor 2) transcription factors, thereby inhibiting hypertrophic growth in the context of cardiac stress (Han et al. 2005; Han et al. 2003; McKinsey et al. 2002). The overexpression of these HDACs within cardiomyocytes results in a diminution of MEF2 activity, consequently averting hypertrophic responses (Backs et al. 2006; Bush et al. 2004; Vega et al. 2004; Zhang et al. 2002). In contrast, the silencing of HDAC9 or HDAC5 leads to an exacerbation of hypertrophy during stress, thus suggesting that class IIa HDACs generally confer cardioprotective effects (Zhang et al. 2002; Chang et al. 2004).

Class II HDACs, with particular emphasis on HDAC6, play a critical role in the modulation of cardiac redox equilibrium and inflammatory processes. In the absence of active HDAC6, there is an accumulation of acetylated Prx, which subsequently potentiates its enzymatic activity, confers resistance to superoxidative stress, and enhances its overall stability. A designated HDAC6 inhibitor may facilitate the increased accumulation of acetylated Prx while concurrently safeguarding these proteins from excessive overoxidation (Parmigiani et al. 2008; Li et al. 2021a). Furthermore, the inhibition of HDAC6 within macrophages has been shown to potentiate p38 kinase signaling, thereby increasing the expression of the anti-inflammatory cytokine IL-10 (Wang et al. 2014). Evidence suggests that the inhibition of HDAC6 confers protective benefits in a murine model of desmin-related cardiomyopathy (McLendon et al. 2014). In murine subjects exhibiting mutant αB-crystallin, which leads to the formation of toxic protein aggregates and subsequent heart failure, the inhibition of HDAC6 led to a reduction in aggresome formation and an enhancement in ventricular functionality. It is hypothesized that selective inhibitors targeting HDAC6 may offer therapeutic benefits for improving cardiac function in the aging population (De Meyer et al. 2010).

9.10.3 Other Pharmacological Agents and Cell Therapy

Cardiac aging is a complex process influenced by various internal and external factors and regulated by multiple aging-related signaling networks. With increasing morbidity and mortality linked to cardiac aging, there is a strong need for effective therapeutic strategies. A better understanding of the mechanisms behind cardiac aging can aid in developing treatments and reducing the risk of cardiovascular disease (CVD) (Yan et al. 2021).

We have previously explained regarding the role of HDACi in cardiac aging. Similar to HDACi, sirtuin activators like resveratrol, SRT1720, and honokiol are potential molecules that may be used for alleviating cardiac aging. Sirt1, the most researched target against aging, improves cardiac pathological states by activating AMPK activity. The AMPK activity was observed to decline in the aging myocardium, and the absence of AMPK exacerbates cardiac senescence in murine models (Hua et al. 2011; Turdi et al. 2010). Nevertheless, AMPK is activated in the myocardial tissue of mice undergoing caloric restriction (Zheng et al. 2015). Metformin, a pharmacological agent utilized for diabetes management, has been shown to activate AMPK and stimulate autophagy, although the precise mechanism of its activation remains a subject of debate. Historically, it has been posited that metformin inhibits the activity of respiratory chain complex 1, resulting in diminished ATP levels and subsequent AMPK activation (Zhang et al. 2016). A contemporary investigation proposes that metformin facilitates the translocation of LKB1 to the lysosomal compartment, thereby activating AMPK (Zhang et al. 2016). Irrespective of the underlying mechanism, metformin has been demonstrated to enhance both health span and lifespan in mice, while also mitigating age-associated dysfunction in cardiomyocytes (Martin-Montalvo et al. 2013; Turdi et al. 2010; Miyamoto 2019).

Rapamycin, a macrolide pharmaceutical agent sanctioned by the FDA, exerts a direct inhibitory effect on mTORC1 and was initially utilized as an antifungal (Sehgal et al. 1975) and immunosuppressive therapeutic, particularly in the context of renal transplantation. In 2015, the FDA granted approval for rapamycin in the treatment of lymphangioleiomyomatosis (LAM), a rare pathological condition (Wang et al. 2017). There is further scope to use rapamycin in other disease conditions.

Rapamycin has been recognized for its profound influence on lifespan extension, inhibiting cellular senescence and modulating conserved aging pathways across various organisms, including primates and rodents. Empirical studies indicate that it possesses the capacity to reverse age-associated diastolic dysfunction in rodent models, as well as myocardial stiffness, even after the cessation of treatment (Urfer et al. 2017). Additionally, short-term administration of rapamycin has been demonstrated to mitigate left ventricular hypertrophy in the context of cardiac aging (Dai et al. 2014).

Rapamycin facilitates autophagy within cardiac tissue by inhibiting mTORC1, activating AMPK signaling pathways, diminishing oxidative stress, and promoting mitochondrial biogenesis (Chiao et al. 2016). It can be used against cardiac aging as it improves organelle function and reduces oxidative stress and inflammation (Xia et al. 2017). Treatment with rapamycin in murine models results in a reduction of protein lipoxidation and lipofuscin accumulation, inhibition of cardiac hypertrophy, and mitochondrial ROS (Xia et al. 2017; Dai et al. 2014).

Other pharmacological agents also exhibit potential in addressing the challenges associated with cardiac aging as mentioned in Table 9.2. Spermidine, a naturally occurring polyamine responsible for inducing autophagy, diminishes inflammation and oxidative stress, and enhances diastolic function, thereby decelerating the progression of heart failure (Zhang et al. 2017). Agelastatin A, which functions as an osteopontin inhibitor, has been identified as effective in combating cardiac aging and inducing fibroblast senescence in murine subjects. A comprehensive understanding of these underlying mechanisms offers novel avenues for the preservation of cardiac function throughout the aging process (Sawaki et al. 2018).

Proteomic examination of plasma derived from both juvenile and senescent murine subjects has elucidated growth differentiation factor 11 (GDF11) as a circulating biomarker that diminishes with advancing age. Evidence suggests that GDF11 can ameliorate age-associated hypertrophy, as evidenced in heterochronic parabiosis experiments. Despite the ambiguity surrounding its specific mechanism of action, its implications for diastolic dysfunction, and its involvement in the aging of human cardiac tissue, the murine model indicates considerable therapeutic promise for GDF11 in the context of cardiac senescence (Loffredo et al. 2013; Obas and Vasan 2018).

Recent investigations indicate that microRNAs (miRNAs) exert a significant influence on the process of cardiac senescence. As organisms age, there is a notable decline in the miR-17-92 cluster, concomitant with an elevation in the levels of miR-22 and miR-34a within cardiac tissue. Boon et al. (2013) provided evidence that the in vivo silencing of miR-34a through the use of antisense oligonucleotides can partially ameliorate the effects of cardiac aging in murine models, thereby underscoring the potential for gene therapeutic applications. Nevertheless, the strategic targeting of miRNAs may elicit adverse effects attributable to their diverse

Table 9.2 Pharmacological interventions to reverse cardiac aging their mechanism action and role in cardioprotection

Pharmacological agents	Mechanism of actions	Role in cardioprotection	Animal models	References
Rapamycin	Inhibiting mTOR pathways (enhance mitophagy, decrease mitochondrial ROS)	Improving age-related diastolic function and reducing left ventricular hypertrophy	Old mice	Sohal and Weindruch (1996), Taffet et al. (1997, and Dai et al. 2014)
Metformin	AMPK activation	Reduce myocardial contractile dysfunction during cardiac aging	Male Wistar rats	Garg et al. (2017)
CD38-IN-78c	CD38 inhibitor	Enhancing the level of NAD^+, stimulating autophagy, and mitochondrial phagocytosis	One-year-old male C57BL/6 J mice, human	Mitchell et al. (2018), Tarragó et al. (2018), and Yan et al. (2021)
Spermidine	Inducing autophagy, inhibiting, histone acetyltransferase, oxidative stress, and inflammation	Reducing cardiac hypertrophy and remodeling, improving cardiac function	Young and old male Wistar rats	Zhang et al. (2017)
Agelastatin A	Inhibitor of osteopontin	Inducing selective modulation of fibroblast senescence in mice, reducing cardiac aging	C57BL/6 J male mice (11 months old)	Sawaki et al. (2018)
IGF-1	Increase IGF-1 signaling (growth hormone therapy)	Reducing mitochondrial oxidative stress in the heart, reducing age-dependent cardiovascular dysfunction, and aging-related cardioprotection	Healthy male Wistar rats, (17 weeks old), young and old C57BL/6 male mice	Puche et al. (2008) and Li et al. (2008)
Growth differentiation factor 11	Reversing of age-related hypertrophy in heterochronic parabiosis	Restoring age-dependent diastolic dysfunction	Young (2 months) and old (23 month) C57BL/6 mice	Loffredo et al. (2013)
Curcumin	Inhibits the acetylation occurred through histone acetyltransferase p300	Reduces hypertension-induced left ventricular hypertrophy, and perivascular fibrosis along with preserved ejection fraction	Male dahl salt-sensitive (DS) and salt-resistant (DR) rats (6 weeks old)	Sunagawa et al. (2021) and Dubois-Deruy et al. (2022)

(continued)

Table 9.2 (continued)

Pharmacological agents	Mechanism of actions	Role in cardioprotection	Animal models	References
Givinostat (pan-HDAC inhibitor)	Modulating the expression of transforming growth factor-beta (TGF-β)	Reducing levels of inflammatory cytokines, and inhibiting degeneration of cardiac tissues	Female C57BL/6 mice (10 weeks old)	Milan et al. (2018)
Extract of Ecklonia stolonifera Okamura	Inhibits the acetylation of histone H3K9, occurring through histone acetyltransferase p300	Repressing left ventricular systolic dysfunction induced by myocardial infarction	Male SD (Sprague–Dawley) rats (7–8 weeks old)	Katagiri et al. (2022)
Eicosapentaenoic acid and docosahexaenoic acid	Blocking histone acetyltransferase activity of p300	Suppressing myocardial infarction-induced heart failure and reducing the risk of coronary artery disease	Male SD rats neonatal and old (8 weeks)	Sunagawa et al. (2022)
Resveratrol	Activation of Sirt1, deacetylation of smad3	Mitigating adverse alterations associated with aging in the cardiac tissue	Female and male Wistar rats (20-month-old)	Chen et al. (2022) and Börzsei et al. (2022)
Tubastatin A	HDAC6 inhibitor	Inhibiting cardiac fibrosis and cardiac remodeling	Adult male Sprague–Dawley (SD) rats	Tao et al. (2016)
Trichostatin A	Class 1 and class 2 HDAC inhibitors.	Inhibiting hypertension-induced cardiac hypertrophy, preserving systolic function	Male C57BL6 mice (6–8 weeks)	Kong et al. (2006)
MGCD0103	Class I HDAC inhibitor	Inhibits angiotensin II-induced cardiac fibrosis	Male C57Bl/6 J mice (ten-week-old).	Williams et al. (2014)
β-Hydroxybutyrate	Endogenous class I HDAC inhibitor	Protective effect in sepsis-induced cardiomyopathy and reducing myocardial oxidative stress	Male C57BL/6 J mice (8 to 10 weeks)	Chu et al. (2021) and Ji et al. (2022)

(continued)

Table 9.2 (continued)

Pharmacological agents	Mechanism of actions	Role in cardioprotection	Animal models	References
Vorinostat	Class I and II HDAC enzymes	Inhibiting reperfusion injury, reducing myocardial infarct size, and inducing mitochondrial biogenesis	C57BL/6 wild-type mice and ATG7 KO mice (8 to 12-week-old).	Yang et al. (2019)
MS-275 (Entinostat)	Class I HDAC inhibitor	Protection in ischemia-reperfusion injury	Male Sprague–Dawley (SD) rats	Aune et al. (2014)
Valproic acid	Class I HDAC inhibitor	Protective against myocardial infarction	Male Sprague–Dawley (SD) rats	Cacabelos and Torrellas (2015) and Tian et al. (2019b)
MPT0E014	Novel HDAC inhibitor	Inhibiting adverse cardiac remodeling, preventing cardiac hypertrophy and fibrosis	Male Wistar rats	Kao et al. (2013) and Han et al. (2022)
Honokiol	Increasing Sirt3 activity	Preventing adverse cardiac remodeling	Adult male CD-1 mice (6–8 weeks)	Pillai et al. (2015)

array of molecular targets (Obas and Vasan 2018; Boon et al. 2013). Further research may find its application in cardiac aging.

CS1477 functions as a small-molecule inhibitor of p300/CBP bromodomain and demonstrates considerable potential in the therapeutic management of prostate cancer through the attenuation of cellular proliferation (Welti et al. 2021). Preliminary clinical investigations utilizing patient-derived xenografts indicate its potential as an epidrug. In light of convincing prospective evidence regarding its efficacy and safety profile, the repurposing of CS1477 for the treatment of cardiovascular disease may represent a strategically advantageous approach (Ghosh 2021). Anacardic acid, a plant-derived acetyltransferase inhibitor, efficiently decreases the activation of p300 and reduces H3K9 acetylation in the heart tissues of mice subjected to left ventricular pressure overload (Li et al. 2019b). Similarly, at low doses, the curcumin analog GO-Y030 reduces TAC-induced ventricular hypertrophy, and systolic dysfunction in the murine model by inhibiting p300 activity and reduce of H3K9 acetylation (Shimizu et al. 2020; Ghosh 2021).

Preclinical and clinical investigations indicate that cellular therapy constitutes a promising methodology for cardiac repair and regeneration, particularly in the context of addressing cardiac senescence. A variety of cellular types, encompassing cardiac progenitor cells, mesenchymal stem cells, and pluripotent stem cells, have exhibited potential advantages, such as the mitigation of scar tissue formation and enhancement of cardiac output (Broughton and Sussman 2016; Halme and Kessler 2006; Yoshida and Yamanaka 2017; Burridge et al. 2015; Wollert et al. 2004). Two principal strategies are (i) the direct administration of stem cells to the cardiac tissue and (ii) the utilization of agents such as IGF-1, PDGF, VEGF (vascular endothelial growth factor), and SDF-1 (stromal cell-derived factor 1) to augment the functionality of endogenous stem cells (Ballard and Edelberg 2007). The challenges associated with this approach include the enhancement of proliferation, engraftment, and survival rates of transplanted cellular entities (Obas and Vasan 2018). Nonetheless, the safety and efficacy of cellular therapy in the prevention and treatment of cardiac aging remains a subject of considerable debate (Burridge et al. 2015).

9.11 Conclusion

An increase in protein acetylation is observed with advancing age in cardiac muscle tissues. Mitochondrial protein hyperacetylation is identified as a principal contributor to the dysfunction of mitochondrial enzymes involved in the electron transport chain (ETC) and tricarboxylic acid (TCA) cycle, as well as disruption in redox homeostasis in cardiomyocytes. Key enzymes and transcription factors that are susceptible to hyperacetylation may significantly facilitate the processes associated with cardiac aging. A discernible pattern has been established, suggesting that calorie restriction, HDAC inhibition, and interventions aimed at activating sirtuins and AMPK, application of microRNAs, stem cells, and other pharmacological agents confer protective effects on cardiac health in the context of aging. It is anticipated that innovative therapeutic strategies targeting deacetylase activity will soon be introduced into clinical practice for cardiac conditions related to aging. Although the phenomenon of protein acetylation has been recognized for more than 50 years, accurate detection methodologies have only been developed in the past 20 years owing to previous technical constraints. The majority of research has predominantly utilized genetically altered or pharmacologically treated animal models to assess modulators of protein acetylation in cardiac disorders. Further human research is requisite prior to the potential consideration of these acetylation-targeting therapeutics for clinical application.

Acknowledgments The authors thank their collaborator Dr. Bhabatosh Das, Translational Health Science and Technology Institute, Faridabad, India for assisting us with the preparation of Fig. 9.1 (Agreement no. AJ27VL6TNJ, Citation - https://BioRender.com/z91s412) and Fig. 9.2 (Agreement No. YN27VL7MFW, Citation – https://BioRender.com/q68a466) for this book chapter. The authors are also thankful to the Department of Biotechnology (DBT, Govt. of India) for funding support (BT/NER/143/SP44413/2021) and the Department of Pharmaceuticals (DoP), Govt. of India for supporting Junior Research Fellow to Aditi Hase.

References

Acosta JC, Banito A, Wuestefeld T, Georgilis A, Janich P, Morton JP et al (2013) A complex secretory program orchestrated by the inflammasome controls paracrine senescence. Nat Cell Biol 15(8):978–990

Ahmet I, Tae HJ, de Cabo R, Lakatta EG, Talan MI (2011) Effects of calorie restriction on Cardioprotection and cardiovascular health. J Mol Cell Cardiol 51(2):263–271

Ahn BH, Kim HS, Song S, Lee IH, Liu J, Vassilopoulos A et al (2008) A role for the mitochondrial deacetylase Sirt3 in regulating energy homeostasis. Proc Natl Acad Sci USA 105(38):14447–14452

Alcendor RR, Gao S, Zhai P, Zablocki D, Holle E, Yu X et al (2007) Sirt1 regulates aging and resistance to oxidative stress in the heart. Circ Res 100(10):1512–1521

Alfaras I, Di Germanio C, Bernier M, Csiszar A, Ungvari Z, Lakatta EG et al (2016) Pharmacological strategies to retard cardiovascular aging. Circ Res 118(10):1626–1642

Al-Zaid NS, Dashti HM, Mathew TC, Juggi JS (2007) Low carbohydrate ketogenic diet enhances cardiac tolerance to global ischaemia. Acta Cardiol 62(4):381–389

Antos CL, McKinsey TA, Frey N, Kutschke W, McAnally J, Shelton JM et al (2002) Activated glycogen synthase-3β suppresses cardiac hypertrophy in vivo. Proc Natl Acad Sci USA 99(2):907–912

Anversa P, Palackal T, Sonnenblick EH, Olivetti G, Meggs LG, Capasso JM (1990) Myocyte cell loss and myocyte cellular hyperplasia in the hypertrophied aging rat heart. Circ Res 67(4):871–885

Anwar T, Khosla S, Ramakrishna G (2016) Increased expression of SIRT2 is a novel marker of cellular senescence and is dependent on wild type p53 status. Cell Cycle 15(14):1883–1897

Aoshiba K, Tsuji T, Kameyama S, Itoh M, Semba S, Yamaguchi K et al (2013) Senescence-associated secretory phenotype in a mouse model of bleomycin-induced lung injury. Exp Toxicol Pathol 65(7):1053–1062

Aune SE, Herr DJ, Mani SK, Menick DR (2014) Selective inhibition of class I but not class IIb histone deacetylases exerts cardiac protection from ischemia reperfusion. J Mol Cell Cardiol 72:138–145

Backs J, Song K, Bezprozvannaya S, Chang S, Olson EN (2006) CaM kinase II selectively signals to histone deacetylase 4 during cardiomyocyte hypertrophy. J Clin Invest 116(7):1853–1864

Ballard VLT, Edelberg JM (2007) Stem cells and the regeneration of the aging cardiovascular system. Circ Res 100(8):1116–1127

Baran M, Miziak P, Stepulak A, Cybulski M (2023) The role of Sirtuin 6 in the deacetylation of histone proteins as a factor in the progression of neoplastic disease. Int J Mol Sci 25(1):497

Beckman JA, Creager MA (2016) Vascular complications of diabetes. Circ Res 118(11):1771–1785

Berniakovich I, Trinei M, Stendardo M, Migliaccio E, Minucci S, Bernardi P et al (2008) p66Shc-generated oxidative signal promotes fat accumulation. J Biol Chem 283(49):34283

Bertrand P (2010) Inside HDAC with HDAC inhibitors. Eur J Med Chem 45(6):2095–2116

Betsinger CN, Cristea IM (2019) Mitochondrial function, metabolic regulation, and human disease viewed through the prism of Sirtuin 4 (SIRT4) functions. J Proteome Res 18(5):1929

Bingol B, Tea JS, Phu L, Reichelt M, Bakalarski CE, Song Q et al (2014) The mitochondrial deubiquitinase USP30 opposes parkin-mediated mitophagy. Nature 510(7505):370–375

Boon RA, Iekushi K, Lechner S, Seeger T, Fischer A, Heydt S et al (2013) MicroRNA-34a regulates cardiac ageing and function. Nature 495(7439):107–110

Börzsei D, Sebestyén J, Szabó R, Lesi ZN, Pálszabó A, Pálszabó P et al (2022) Resveratrol as a promising polyphenol in age-associated cardiac alterations. Oxidative Med Cell Longev 2022:7911222

Boylston JA, Sun J, Chen Y, Gucek M, Sack MN, Murphy E (2015) Characterization of the cardiac succinylome and its role in ischemia-reperfusion injury. J Mol Cell Cardiol 88:73

Braidy N, Guillemin GJ, Mansour H, Chan-Ling T, Poljak A, Grant R (2011) Age related changes in NAD+ metabolism oxidative stress and Sirt1 activity in wistar rats. PLoS One 6(4):e19194

Brodsky SV, Gealekman O, Chen J, Zhang F, Togashi N, Crabtree M et al (2004) Prevention and reversal of premature endothelial cell senescence and vasculopathy in obesity-induced diabetes by Ebselen. Circ Res 94(3):377–384

Broughton K, Sussman MA (2016) Empowering adult stem cells for myocardial regeneration V2.0: success in small steps. Circ Res 118(5):867–880

Burridge PW, Sharma A, Wu JC (2015) Genetic and epigenetic regulation of human cardiac reprogramming and differentiation in regenerative medicine. Annu Rev Genet 49:461

Bush E, Fielitz J, Melvin L, Martinez-Arnold M, McKinsey TA, Plichta R et al (2004) A small molecular activator of cardiac hypertrophy uncovered in a chemical screen for modifiers of the calcineurin signaling pathway. Proc Natl Acad Sci 101(9):2870–2875

Cacabelos R, Torrellas C (2015) Epigenetics of aging and Alzheimer's disease: implications for pharmacogenomics and drug response. Int J Mol Sci 16(12):30483

Capell BC, Collins FS, Nabel EG (2007) Mechanisms of cardiovascular disease in accelerated aging syndromes. Circ Res 101(1):13–26

Carpi A, Menabò R, Kaludercic N, Pelicci P, Di Lisa F, Giorgio M (2009) The cardioprotective effects elicited by p66Shc ablation demonstrate the crucial role of mitochondrial ROS formation in ischemia/reperfusion injury. Biochimica et Biophysica Acta (BBA) Bioenergetics 1787(7):774–780

Cencioni C, Spallotta F, Mai A, Martelli F, Farsetti A, Zeiher AM et al (2015) Sirtuin function in aging heart and vessels. J Mol Cell Cardiol 83:55–61

Chang S, McKinsey TA, Zhang CL, Richardson JA, Hill JA, Olson EN (2004) Histone deacetylases 5 and 9 govern responsiveness of the heart to a subset of stress signals and play redundant roles in heart development. Mol Cell Biol 24(19):8467–8476

Chen WY, Wang DH, Yen RC, Luo J, Gu W, Baylin SB (2005) Tumor suppressor HIC1 directly regulates SIRT1 to modulate p53-dependent DNA-damage responses. Cell 123(3):437–448

Chen C, Zhou M, Ge Y, Wang X (2020a) SIRT1 and aging related signaling pathways. Mech Ageing Dev 1(187):111215

Chen Y, Chen J, Sun X, Yu J, Qian Z, Wu L et al (2020b) The SIRT6 activator MDL-800 improves genomic stability and pluripotency of old murine-derived iPS cells. Aging Cell 19(8):e13185

Chen Q, Zeng Y, Yang X, Wu Y, Zhang S, Huang S et al (2022) Resveratrol ameliorates myocardial fibrosis by regulating Sirt1/Smad3 deacetylation pathway in rat model with dilated cardiomyopathy. BMC Cardiovasc Disord 22:17

Chiao YA, Kolwicz SC, Basisty N, Gagnidze A, Zhang J, Gu H et al (2016) Rapamycin transiently induces mitochondrial remodeling to reprogram energy metabolism in old hearts. Aging (Albany NY) 8(2):314

Chouchani ET, Methner C, Nadtochiy SM, Logan A, Pell VR, Ding S et al (2013) Cardioprotection by S-nitrosation of a cysteine switch on mitochondrial complex I. Nat Med 19(6):753

Choudhary C, Kumar C, Gnad F, Nielsen ML, Rehman M, Walther TC et al (2009) Lysine acetylation targets protein complexes and co-regulates major cellular functions. Science 325(5942):834–840

Chu Y, Zhang C, Xie M (2021) Beta-Hydroxybutyrate, friend or foe for stressed hearts. Frontiers in Aging 2:681513

Cockell M, Palladino F, Laroche T, Kyrion G, Liu C, Lustig AJ et al (1995) The carboxy termini of Sir4 and Rap1 affect Sir3 localization: evidence for a multicomponent complex required for yeast telomeric silencing. J Cell Biol 129(4):909–924

Consalvi S, Saccone V, Mozzetta C (2014) Histone deacetylase inhibitors: a potential epigenetic treatment for Duchenne muscular dystrophy. Epigenomics. 6(5):547–560

Constantinou M, Klavaris A, Koufaris C, Kirmizis A (2023) Cellular effects of NAT-mediated histone N-terminal acetylation. J Cell Sci 136(7):260801

Coppé JP, Desprez PY, Krtolica A, Campisi J (2010) The senescence-associated secretory phenotype: the dark side of tumor suppression. Annu Rev Pathol 5:99–118

Costantino S, Paneni F, Cosentino F (2016) Ageing, metabolism and cardiovascular disease. J Physiol 594(8):2061–2073

Couteur DGL, McLachlan AJ, Quinn RJ, Simpson SJ, de Cabo R (2011) Aging biology and novel targets for drug discovery. J Gerontol Ser A Biol Med Sci 67A(2):168

Crews D (2003) Human senescence: evolutionary and biocultural. Perspectives 11

Czuriga D, Paulus W, Czuriga I, Edes I, Papp Z, Borbély A (2012) Cellular mechanisms for diastolic dysfunction in the human heart. Curr Pharm Biotechnol 20:13

Dai DF, Karunadharma PP, Chiao YA, Basisty N, Crispin D, Hsieh EJ et al (2014) Altered proteome turnover and remodeling by short-term caloric restriction or rapamycin rejuvenate the aging heart. Aging Cell 13(3):529–539

De Meyer GRY, De Keulenaer GW, Martinet W (2010) Role of autophagy in heart failure associated with aging. Heart Fail Rev 15(5):423–430

de Oliveira RM, Sarkander J, Kazantsev AG, Outeiro TF (2012) SIRT2 as a therapeutic target for age-related disorders. Front Pharmacol 3(3):82

Dhahbi JM, Tsuchiya T, Kim HJ, Mote PL, Spindler SR (2006) Gene expression and physiologic responses of the heart to the initiation and withdrawal of caloric restriction. J Gerontol Ser A Biol Med Sci 61(3):218–231

Dimri GP, Campisi J (1994) Molecular and cell biology of replicative senescence. Cold Spring Harb Symp Quant Biol 59(0):67–73

Donato AJ, Magerko KA, Lawson BR, Durrant JR, Lesniewski LA, Seals DR (2011) SIRT-1 and vascular endothelial dysfunction with ageing in mice and humans. J Physiol 589(18):4545–4554

Dubois-Deruy E, El Masri Y, Turkieh A, Amouyel P, Pinet F, Annicotte JS (2022) Cardiac Acetylation in Metabolic Diseases Biomedicines 10(8):1834

Erdös B, Broxson CS, Landa T, Scarpace PJ, Leeuwenburgh C, Zhang Y et al (2007) Effects of life-long caloric restriction and voluntary exercise on age-related changes in levels of catecholamine biosynthetic enzymes and angiotensin II receptors in the rat adrenal medulla and hypothalamus. Exp Gerontol 42(8):745–752

Etchegaray JP, Zhong L, Mostoslavsky R (2023) The histone deacetylase SIRT6: at the crossroads between epigenetics, metabolism and disease. Curr Top Med Chem 13(23):2991–3000

Favero G, Franceschetti L, Rodella LF, Rezzani R (2015) Sirtuins, aging, and cardiovascular risks. Age (Dordr) 37(4):65

Ford E, Voit R, Liszt G, Magin C, Grummt I, Guarente L (2006) Mammalian Sir2 homolog SIRT7 is an activator of RNA polymerase I transcription. Genes Dev 20(9):1075–1080

Franci G, Miceli M, Altucci L (2010) Targeting epigenetic networks with Polypharmacology: a new avenue to tackle cancer. Epigenomics 2(6):731–742

Francois A, Canella A, Marcho L, Stratton MS (2021) Protein acetylation in cardiac aging. J Mol Cell Cardiol 157:90–97

Frye RA (1999) Characterization of five human cDNAs with homology to the yeast SIR2 gene: Sir2-like proteins (Sirtuins) metabolize NAD and may have protein ADP-Ribosyltransferase activity. Biochem Biophys Res Commun 260(1):273–279

Frye RA (2000) Phylogenetic classification of prokaryotic and eukaryotic Sir2-like proteins. Biochem Biophys Res Commun 273(2):793–798

Funamoto M, Imanishi M, Tsuchiya K, Ikeda Y (2023) Roles of histone acetylation sites in cardiac hypertrophy and heart failure. Front Cardiovas Med 10:1133611

Galindo CL, Skinner MA, Errami M, Olson LD, Watson DA, Li J et al (2009) Transcriptional profile of isoproterenol-induced cardiomyopathy and comparison to exercise-induced cardiac hypertrophy and human cardiac failure. BMC Physiol 9:1–22

Garg G, Singh S, Singh AK, Rizvi SI (2017) Antiaging effect of metformin on brain in naturally aged and accelerated senescence model of rat. Rejuvenation Res 20(3):173–182

Gershey EL, Vidali G, Allfrey VG (1968) Chemical studies of histone acetylation: the occurrence of ε-N-Acetyllysine in the f2a1 histone. J Biol Chem 243(19):5018–5022

Ghosh AK (2020) p300 in cardiac development and accelerated cardiac aging. Aging Dis 11(4):916–926

Ghosh AK (2021) Acetyltransferase p300 is a putative Epidrug target for amelioration of cellular aging-related cardiovascular disease. Cells 10(11):2839

Giles RH, Peters DJ, Breuning MH (1998) Conjunction dysfunction: CBP/p300 in human disease. Trends Genet 14(5):178–183

Glozak MA, Sengupta N, Zhang X, Seto E (2005) Acetylation and deacetylation of non-histone proteins. Gene 19(363):15–23

Graiani G, Lagrasta C, Migliaccio E, Spillmann F, Meloni M, Madeddu P et al (2005) Genetic deletion of the p66Shc adaptor protein protects from angiotensin II–induced myocardial damage. Hypertension 46(2):433–440

Gu W, Roeder RG (1997) Activation of p53 sequence-specific DNA binding by acetylation of the p53 C-terminal domain. Cell 90(4):595–606

Guarente L (1999) Diverse and dynamic functions of the sir silencing complex. Nat Genet 23(3):281–285

Hafner AV, Dai J, Gomes AP, Xiao CY, Palmeira CM, Rosenzweig A et al (2010) Regulation of the mPTP by SIRT3-mediated deacetylation of CypD at lysine 166 suppresses age-related cardiac hypertrophy. Aging (Albany NY) 2(12):914–923

Haigis MC, Guarente LP (2006) Mammalian sirtuins—emerging roles in physiology, aging, and calorie restriction. Genes Dev 20(21):2913–2921

Haldar SM, Lu Y, Jeyaraj D, Kawanami D, Cui Y, Eapen SJ et al (2010) Klf15 deficiency is a molecular link between heart failure and aortic aneurysm formation. Sci Transl Med 2(26):26ra26

Halme DG, Kessler DA (2006) FDA regulation of stem-cell–based therapies. N Engl J Med 355(16):1730–1735

Han A, Pan F, Stroud JC, Youn HD, Liu JO, Chen L (2003) Sequence-specific recruitment of transcriptional co-repressor Cabin1 by myocyte enhancer factor-2. Nature 422(6933):730–734

Han A, He J, Wu Y, Liu JO, Chen L (2005) Mechanism of recruitment of class II histone deacetylases by myocyte enhancer Factor-2. J Mol Biol 345(1):91–102

Han Y, Zhou S, Coetzee S, Chen A (2019) SIRT4 and its roles in energy and redox metabolism in health, disease and during exercise. Front Physiol 10:1006

Han Y, Nie J, Wang DW, Ni L (2022) Mechanism of histone deacetylases in cardiac hypertrophy and its therapeutic inhibitors. Front Cardiovasc Med 9:931475

Harrison DE, Strong R, Sharp ZD, Nelson JF, Astle CM, Flurkey K et al (2009) Rapamycin fed late in life extends lifespan in genetically heterogeneous mice. Nature 460(7253):392–395

Haslem L, Hays JM, Hays FA (2022) p66Shc in cardiovascular pathology. Cells 11(11):1855

Head T, Daunert S, Goldschmidt-Clermont PJ (2017) The aging risk and atherosclerosis: a fresh look at arterial homeostasis. Front Genet 8:216

Hebert AS, Dittenhafer-Reed KE, Yu W, Bailey DJ, Selen ES, Boersma MD et al (2013) Calorie restriction and SIRT3 trigger global reprogramming of the mitochondrial protein acetylome. Mol Cell 49(1):186–199

Hees PS, Fleg JL, Lakatta EG, Shapiro EP (2002) Left ventricular remodeling with age in normal men versus women: novel insights using three-dimensional magnetic resonance imaging. Am J Cardiol 90(11):1231–1236

Henry RA, Kuo YM, Siegel ZS, Yen TJ, Rhodes J, Taylor EA et al (2019) Discordant effects of putative lysine acetyltransferase inhibitors in biochemical and living systems. Cells 8(9):1022

Hershberger KA, Abraham DM, Liu J, Locasale JW, Grimsrud PA, Hirschey MD (2018) Ablation of Sirtuin5 in the postnatal mouse heart results in protein succinylation and normal survival in response to chronic pressure overload. J Biol Chem 293(27):10630–10645

Hirschey MD, Shimazu T, Goetzman E, Jing E, Schwer B, Lombard DB et al (2010) SIRT3 regulates fatty acid oxidation via reversible enzyme deacetylation. Nature 464(7285):121–125

Hoeldtke RD, Cilmi KM (1985) Effects of aging on catecholamine metabolism. J Clin Endocrinol Metabol 60(3):479–484

Hu Q, Zhang H, Gutiérrez Cortés N, Wu D, Wang P, Zhang J et al (2020) Increased Drp1 acetylation by lipid overload induces cardiomyocyte death and heart dysfunction. Circ Res 126(4):456–470

Hua Y, Zhang Y, Ceylan-Isik AF, Wold LE, Nunn JM, Ren J (2011) Chronic akt activation accentuates aging-induced cardiac hypertrophy and myocardial contractile dysfunction: role of autophagy. Basic Res Cardiol 106(6):1173–1191

Piperno G, LeDizet M, Chang XJ (1987) Microtubules containing acetylated alpha-tubulin in mammalian cells in culture. J Cell Biol 104(2):289–302

Ji L, He Q, Liu Y, Deng Y, Xie M, Luo K et al (2022) Ketone body β-Hydroxybutyrate prevents myocardial oxidative stress in septic cardiomyopathy. Oxidative Med Cell Longev 2022(1):2513837

Jin J, Wang GL, Timchenko L (2009) GSK3β and aging liver. Aging (Albany NY) 1(6):582–585

Johnson SC, Rabinovitch PS, Kaeberlein M (2013) mTOR is a key modulator of ageing and age-related disease. Nature 493(7432):338–345

Kaeberlein M, McVey M, Guarente L (1999) The SIR2/3/4 complex and SIR2 alone promote longevity in Saccharomyces cerevisiae by two different mechanisms. Genes Dev 13(19):2570

Kanfi Y, Naiman S, Amir G, Peshti V, Zinman G, Nahum L et al (2012) The sirtuin SIRT6 regulates lifespan in male mice. Nature 483(7388):218–221

Kao YH, Liou JP, Chung CC, Lien GS, Kuo CC, Chen SA et al (2013) Histone deacetylase inhibition improved cardiac functions with direct antifibrotic activity in heart failure. Int J Cardiol 168(4):4178–4183

Katagiri T, Sunagawa Y, Maekawa T, Funamoto M, Shimizu S, Shimizu K et al (2022) Ecklonia stolonifera Okamura extract suppresses myocardial infarction-induced left ventricular systolic dysfunction by inhibiting p300-HAT activity. Nutrients 14(3):580

Kee HJ, Sohn IS, Nam KI, Park JE, Qian YR, Yin Z et al (2006) Inhibition of histone deacetylation blocks cardiac hypertrophy induced by angiotensin II infusion and aortic banding. Circulation 113(1):51–59

Kim SC, Sprung R, Chen Y, Xu Y, Ball H, Pei J et al (2006) Substrate and functional diversity of lysine acetylation revealed by a proteomics survey. Mol Cell 23(4):607–618

Kim J, Kundu M, Viollet B, Guan KL (2011) AMPK and mTOR regulate autophagy through direct phosphorylation of Ulk1. Nat Cell Biol 13(2):132–141

Kong Y, Tannous P, Lu G, Berenji K, Rothermel BA, Olson EN et al (2006) Suppression of class I and II histone deacetylases blunts pressure-overload cardiac hypertrophy. Circulation 113(22):2579–2588

Kook H, Lepore JJ, Gitler AD, Lu MM, Wing-Man Yung W, Mackay J et al (2003) Cardiac hypertrophy and histone deacetylase–dependent transcriptional repression mediated by the atypical homeodomain protein Hop. J Clin Invest 112(6):863–871

Kouzarides T (2000) Acetylation: a regulatory modification to rival phosphorylation? EMBO J 19(6):1176–1179

Kraus D, Yang Q, Kong D, Banks AS, Zhang L, Rodgers JT et al (2014) Nicotinamide N-methyltransferase knockdown protects against diet-induced obesity. Nature 508(7495):258–262

Kugel S, Mostoslavsky R (2014 Feb) Chromatin and beyond: the multitasking roles for SIRT6. Trends Biochem Sci 39(2):72–81

Kuno A, Hori YS, Hosoda R, Tanno M, Miura T, Shimamoto K et al (2013) Resveratrol improves cardiomyopathy in dystrophin-deficient mice through SIRT1 protein-mediated modulation of p300 protein. J Biol Chem 288(8):5963–5972

Kwon HS, Ott M (2008) The ups and downs of SIRT1. Trends Biochem Sci 33(11):517–525

Lakatta EG (1990) Changes in cardiovascular function with aging. Eur Heart J 11(suppl_C):22–29

Lakatta EG (1993) Cardiovascular regulatory mechanisms in advanced age. Physiol Rev 73(2):413–467

Laurent G, German NJ, Saha AK, de Boer VCJ, Davies M, Koves TR et al (2013) SIRT4 coordinates the balance between lipid synthesis and catabolism by repressing Malonyl CoA decarboxylase. Mol Cell 50(5):686–698

Lee TM, Lin MS, Chang NC (2007 Aug) Inhibition of histone deacetylase on ventricular remodeling in infarcted rats. Am J Phys Heart Circ Phys 293(2):H968–H977

Lee KP, Simpson SJ, Clissold FJ, Brooks R, Ballard JWO, Taylor PW et al (2008) Lifespan and reproduction in drosophila: new insights from nutritional geometry. Proc Natl Acad Sci USA 105(7):2498
Lee MS, Seo J, Choi DY, Lee EW, Ko A, Ha NC et al (2012) Stabilization of p21 (Cip1/WAF1) following Tip60-dependent acetylation is required for p21-mediated DNA damage response. Cell Death Differ 20(4):620
Leoni F, Fossati G, Lewis EC, Lee JK, Porro G, Pagani P et al (2005) The histone deacetylase inhibitor ITF2357 reduces production of pro-inflammatory cytokines in vitro and systemic inflammation in vivo. Mol Med 11(1–12):1–15
Li Q, Ceylan-Isik AF, Li J, Ren J (2008) Deficiency of insulin-like growth factor 1 reduces sensitivity to aging-associated cardiomyocyte dysfunction. Rejuvenation Res 11(4):725–733
Li B, Li M, Li X, Li H, Lai Y, Huang S et al (2019a) Sirt1-inducible deacetylation of p21 promotes cardiomyocyte proliferation. Aging (Albany NY) 11(24):12546–12567
Li S, Peng B, Luo X, Sun H, Peng C (2019b) Anacardic acid attenuates pressure-overload cardiac hypertrophy through inhibiting histone acetylases. J Cell Mol Med 23(4):2744–2752
Li Z, Zhang Z, Ren Y, Wang Y, Fang J, Yue H et al (2021a) Aging and age-related diseases: from mechanisms to therapeutic strategies. Biogerontology 22(2):165–187
Li D, Yang Y, Wang S, He X, Liu M, Bai B et al (2021b) Role of acetylation in doxorubicin-induced cardiotoxicity. Redox Biol 46:102089
Li X, Liu L, Li T, Liu M, Wang Y, Ma H et al (2021c) SIRT6 in senescence and aging-related cardiovascular diseases. Front Cell Dev Biol 9:641315
Li Y, Jin J, Wang Y (2022) SIRT6 widely regulates aging, immunity, and cancer. Front Oncol 12:861334
Liao CY, Kennedy BK (2012) Will the real aging Sirtuin please stand up? Cell Res 22(8):1215–1217
Liao C, Akazawa H, Tamagawa M, Ito K, Yasuda N, Kudo Y et al (2010) Cardiac mast cells cause atrial fibrillation through PDGF-A–mediated fibrosis in pressure-overloaded mouse hearts. J Clin Invest 120(1):242–253
Linton PJ, Gurney M, Sengstock D, Mentzer RM, J, Gottlieb RA. (2014) This old heart: cardiac aging and autophagy. J Mol Cell Cardiol 24(83):44
Liu L, Scolnick DM, Trievel RC, Zhang HB, Marmorstein R, Halazonetis TD et al (1999) p53 sites acetylated in vitro by PCAF and p300 are acetylated in vivo in response to DNA damage. Mol Cell Biol 19(2):1202–1209
Liu B, Che W, Xue J, Zheng C, Tang K, Zhang J et al (2013a) SIRT4 prevents hypoxia-induced apoptosis in H9c2 Cardiomyoblast cells. Cell Physiol Biochem 32(3):655–662
Liu B, Che W, Zheng C, Liu W, Wen J, Fu H et al (2013b) SIRT5: a safeguard against oxidative stress-induced apoptosis in cardiomyocytes. Cell Physiol Biochem 32(4):1050–1059
Liu YP, Wen R, Liu CF, Zhang TN, Yang N (2023) Cellular and molecular biology of sirtuins in cardiovascular disease. Biomed Pharmacother 164:114931
Loffredo FS, Steinhauser ML, Jay SM, Gannon J, Pancoast JR, Yalamanchi P et al (2013) Growth differentiation factor 11 is a circulating factor that reverses age-related cardiac hypertrophy. Cell 153(4):828–839
Long P, Nguyen Q, Thurow C, Broderick TL (2002) Caloric restriction restores the cardioprotective effect of preconditioning in the rat heart. Mech Ageing Dev 123(10):1411–1413
Lu TM, Tsai JY, Chen YC, Huang CY, Hsu HL, Weng CF et al (2014) Downregulation of Sirt1 as aging change in advanced heart failure. J Biomed Sci 21(1):57
Luo YX, Tang X, An XZ, Xie XM, Chen XF, Zhao X et al (2017) SIRT4 accelerates Ang II-induced pathological cardiac hypertrophy by inhibiting manganese superoxide dismutase activity. Eur Heart J 38(18):1389–1398
Lutz W, Sanderson W, Scherbov S (2008) The coming acceleration of global population ageing. Nature 451(7179):716–719
Ma H, Wang J, Thomas DP, Tong C, Leng L, Wang W et al (2010) Impaired macrophage migration inhibitory factor (MIF)-AMPK activation and ischemic recovery in the senescent heart. Circulation 122(3):282–292

Ma X, Liu H, Foyil SR, Godar RJ, Weinheimer CJ, Diwan A (2012) Autophagy is impaired in cardiac ischemia-reperfusion injury. Autophagy 8(9):1394–1396

Mahlknecht U, Voelter-Mahlknecht S (2009) Fluorescence in situ hybridization and chromosomal organization of the sirtuin 4 gene (Sirt4) in the mouse. Biochem Biophys Res Commun 382(4):685–690

Manning JR, Thapa D, Zhang M, Stoner MW, Sembrat JC, Rojas M et al (2022) GPER-dependent estrogen signaling increases cardiac GCN5L1 expression. Am J Physiol Heart Circ Physiol 322(5):H762–H768

Mantel C, Broxmeyer HE (2008) SIRT1, stem cells, aging, and stem cell aging. Curr Opin Hematol 15(4):326–331

Martin-Montalvo A, Mercken EM, Mitchell SJ, Palacios HH, Mote PL, Scheibye-Knudsen M et al (2013) Metformin improves healthspan and lifespan in mice. Nat Commun 4:2192

Mathiyalagan P, Chang L, Du XJ, El-Osta A (2010) Cardiac ventricular chambers are epigenetically distinguishable. Cell Cycle 9(3):612–617

McGill HC Jr, McMahan CA, Herderick EE, Malcom GT, Tracy RE, Strong JP et al (2000) Origin of atherosclerosis in childhood and adolescence. Am J Clin Nutr 72(5):1307s–1315s

McKinsey TA (2012) Therapeutic potential for HDAC inhibitors in the heart. Annu Rev Pharmacol Toxicol 52(1):303–319

McKinsey TA, Olson EN (2005) Toward transcriptional therapies for the failing heart: chemical screens to modulate genes. J Clin Invest 115(3):538–546

McKinsey TA, Zhang CL, Olson EN (2002) MEF2: a calcium-dependent regulator of cell division, differentiation and death. Trends Biochem Sci 27(1):40–47

McLendon PM, Ferguson BS, Osinska H, Bhuiyan MS, James J, McKinsey TA et al (2014) Tubulin hyperacetylation is adaptive in cardiac proteotoxicity by promoting autophagy. Proc Natl Acad Sci USA 111(48):E5178–E5186

Milan M, Pace V, Maiullari F, Chirivì M, Baci D, Maiullari S et al (2018) Givinostat reduces adverse cardiac remodeling through regulating fibroblasts activation. Cell Death Dis 9(2):108

Mills KF, Yoshida S, Stein LR, Grozio A, Kubota S, Sasaki Y et al (2016) Long-term administration of nicotinamide mononucleotide mitigates age-associated physiological decline in mice. Cell Metab 24(6):795–806

Ministrini S, Puspitasari YM, Beer G, Liberale L, Montecucco F, Camici GG (2021) Sirtuin 1 in endothelial dysfunction and cardiovascular aging. Front Physiol 12:733696

Minor RK, Allard JS, Younts CM, Ward TM, de Cabo R (2010) Dietary interventions to extend life span and health span based on calorie restriction. J Gerontol A Biol Sci Med Sci 65A(7):695–703

Mitchell SJ, Bernier M, Aon MA, Cortassa S, Kim EY, Fang EF et al (2018) Nicotinamide improves aspects of healthspan but not lifespan in mice. Cell Metab 27(3):667–676.e4

Miyamoto S (2019) Autophagy and cardiac aging. Cell Death Differ 26(4):653–664

Morimoto RI, Cuervo AM (2009 Feb) Protein homeostasis and aging: taking care of proteins from the cradle to the grave. J Gerontol Ser A Biol Med Sci 64A(2):167

Mu W, Zhang Q, Tang X, Fu W, Zheng W, Lu Y et al (2014) Overexpression of a dominant-negative mutant of SIRT1 in mouse heart causes cardiomyocyte apoptosis and early-onset heart failure. Sci China Life Sci 57(9):915–924

Nadtochiy SM, Redman E, Rahman I, Brookes PS (2010) Lysine deacetylation in ischaemic preconditioning: the role of SIRT1. Cardiovasc Res 89(3):643

Nagao M, Toh R, Irino Y, Mori T, Nakajima H, Hara T et al (2016) Β-Hydroxybutyrate elevation as a compensatory response against oxidative stress in cardiomyocytes. Biochem Biophys Res Commun 475(4):322–328

Nakagawa T, Lomb DJ, Haigis MC, Guarente L (2009) SIRT5 deacetylates carbamoyl phosphate Synthetase 1 and regulates the urea cycle. Cell 137(3):560–570

North BJ, Marshall BL, Borra MT, Denu JM, Verdin E (2003) The human Sir2 Ortholog, SIRT2, is an NAD+-dependent tubulin deacetylase. Mol Cell 11(2):437–444

Nural-Guvener HF, Zakharova L, Nimlos J, Popovic S, Mastroeni D, Gaballa MA (2014) HDAC class I inhibitor, Mocetinostat, reverses cardiac fibrosis in heart failure and diminishes CD90+ cardiac myofibroblast activation. Fibrogenesis Tissue Repair 7:10

Obas V, Vasan RS (2018) The aging heart. Clin Sci 132(13):1367–1382

Ogura Y, Kitada M, Monno I, Kanasaki K, Watanabe A, Koya D (2018) Renal mitochondrial oxidative stress is enhanced by the reduction of Sirt3 activity. Zucker diabetic fatty rats Redox Rep 23(1):153–159

Orimo M, Minamino T, Miyauchi H, Tateno K, Okada S, Moriya J et al (2009) Protective role of SIRT1 in diabetic vascular dysfunction. Arterioscler Thromb Vasc Biol 29(6):889–894

Ota H, Akishita M, Eto M, Iijima K, Kaneki M, Ouchi Y (2007) Sirt1 modulates premature senescence-like phenotype in human endothelial cells. J Mol Cell Cardiol 43(5):571–579

Paneni F, Diaz Cañestro C, Libby P, Lüscher TF, Camici GG (2017) The aging cardiovascular system: understanding it at the cellular and clinical levels. J Am Coll Cardiol 69(15):1952–1967

Parmigiani RB, Xu WS, Venta-Perez G, Erdjument-Bromage H, Yaneva M, Tempst P et al (2008) HDAC6 is a specific deacetylase of peroxiredoxins and is involved in redox regulation. Proc Natl Acad Sci USA 105(28):9633

Parodi-Rullán RM, Chapa-Dubocq XR, Javadov S (2018) Acetylation of mitochondrial proteins in the heart: the role of SIRT3. Front Physiol 7(9):1094

Perrino C, Barabási AL, Condorelli G, Davidson SM, De Windt L, Dimmeler S et al (2017) Epigenomic and transcriptomic approaches in the post-genomic era: path to novel targets for diagnosis and therapy of the ischaemic heart? Position paper of the European Society of Cardiology Working Group on cellular biology of the heart. Cardiovasc Res 113(7):725–736

Pillai VB, Samant S, Sundaresan NR, Raghuraman H, Kim G, Bonner MY et al (2015) Honokiol blocks and reverses cardiac hypertrophy in mice by activating mitochondrial SIRT3. Nat Commun 6:6656

Pillai VB, Samant S, Hund S, Gupta M, Gupta MP (2021) The nuclear sirtuin SIRT6 protects the heart from developing aging-associated myocyte senescence and cardiac hypertrophy. Aging (Albany NY) 13(9):12334

Planavila A, Iglesias R, Giralt M, Villarroya F (2011) Sirt1 acts in association with PPARα to protect the heart from hypertrophy, metabolic dysregulation, and inflammation. Cardiovasc Res 90(2):276–284

Porter GA, Urciuoli WR, Brookes PS, Nadtochiy SM. Cardiovascular and Cerebrovascular Aging–New Mechanisms and Insights: SIRT3 deficiency exacerbates ischemia-reperfusion injury: implication for aged hearts. Am J Physiol Heart Circ Physiol 2014 306(12):H1602

Pougovkina O, te Brinke H, Ofman R, van Cruchten AG, Kulik W, Wanders RJA et al (2014) Mitochondrial protein acetylation is driven by acetyl-CoA from fatty acid oxidation. Hum Mol Genet 23(13):3513–3522

Pradhan R, Kumar R, Shekhar S, Rai N, Ambashtha A, Banerjee J et al (2017) Longevity and healthy ageing genes FOXO3A and SIRT3: serum protein marker and new road map to burst oxidative stress by Withania somnifera. Exp Gerontol 95:9–15

Przyklenk K, Dong Y, Undyala VV, Whittaker P (2012) Autophagy as a therapeutic target for ischaemia/reperfusion injury? Concepts, controversies, and challenges. Cardiovasc Res 94(2):197–205

Puche JE, García-Fernández M, Muntané J, Rioja J, González-Barón S, Castilla CI (2008) Low doses of insulin-like growth factor-I induce mitochondrial protection in aging rats. Endocrinology 149(5):2620–2627

Qiu X, Brown K, Hirschey MD, Verdin E, Chen D (2010) Calorie restriction reduces oxidative stress by SIRT3-mediated SOD2 activation. Cell Metab 12(6):662–667

Rai R, Sun T, Ramirez V, Lux E, Eren M, Vaughan DE et al (2019) Acetyltransferase p300 inhibitor reverses hypertension-induced cardiac fibrosis. J Cell Mol Med 23(4):3026–3031

Rayess H, Wang MB, Srivatsan ES (2012) Cellular senescence and tumor suppressor gene p16. Int J Cancer 130(8):1715–1725

Reaven GM, Lithell H, Landsberg L (1996) Hypertension and associated metabolic abnormalities – the role of insulin resistance and the Sympathoadrenal system (Epstein FH, editor). N Engl J Med 334(6):374–382
Ren J, Zhang Y (2018) Targeting autophagy in aging and aging-related cardiovascular diseases. Trends Pharmacol Sci 39(12):1064–1076
Ren SC, Chen X, Gong H, Wang H, Wu C, Li PH et al (2022) SIRT6 in vascular diseases, from bench to bedside. Aging Dis 13(4):1015
Reznick RM, Zong H, Li J, Morino K, Moore IK, Yu HJ et al (2007) Aging-associated reductions in AMP-activated protein kinase activity and mitochondrial biogenesis. Cell Metab 5(2):151–156
Roichman A, Elhanati S, Aon M, Abramovich I, Di Francesco A, Shahar Y et al (2021) Restoration of energy homeostasis by SIRT6 extends healthy lifespan. Nat Commun 12(1):3208
Ryu D, Jo YS, Lo Sasso G, Stein S, Zhang H, Perino A et al (2014) A SIRT7-dependent acetylation switch of GABPβ1 controls mitochondrial function. Cell Metab 20(5):856–869
Sack MN (2011) Emerging characterization of the role of SIRT3-mediated mitochondrial protein deacetylation in the heart. Am J Phys Heart Circ Phys 301(6):H2191–H2197
Sack MN, Finkel T (2012) Mitochondrial metabolism, Sirtuins, and aging. Cold Spring Harb Perspect Biol 4(12):a013102
Sadhukhan S, Liu X, Ryu D, Nelson OD, Stupinski JA, Li Z et al (2016) Metabolomics-assisted proteomics identifies succinylation and SIRT5 as important regulators of cardiac function. Proc Natl Acad Sci USA 113(16):4320–4325
Salminen A, Kauppinen A, Kaarniranta K (2012) Emerging role of NF-κB signaling in the induction of senescence-associated secretory phenotype (SASP). Cell Signal 24(4):835–845
Samarakoon R, Overstreet JM, Higgins PJ (2013) TGF-β signaling in tissue fibrosis: redox controls, target genes and therapeutic opportunities. Cell Signal 25(1):264–268
Santo-Domingo J, Demaurex N (2012) The renaissance of mitochondrial pH. J Gen Physiol 139(6):415
Sarikhani M, Maity S, Mishra S, Jain A, Tamta AK, Ravi V et al (2018a) SIRT2 deacetylase represses NFAT transcription factor to maintain cardiac homeostasis. J Biol Chem 293(14):5281–5294
Sarikhani M, Mishra S, Maity S, Kotyada C, Wolfgeher D, Gupta MP et al (2018b) SIRT2 deacetylase regulates the activity of GSK3 isoforms independent of inhibitory phosphorylation. eLife 5(7):e32952
Sasaki T, Maier B, Bartke A, Scrable H (2006) Progressive loss of SIRT1 with cell cycle withdrawal. Aging Cell 5(5):413–422
Sawaki D, Czibik G, Pini M, Ternacle J, Suffee N, Mercedes R et al (2018) Visceral adipose tissue drives cardiac aging through modulation of fibroblast senescence by Osteopontin production. Circulation 138(8):809–822
Schuetze KB, McKinsey TA, Long CS (2014) Targeting cardiac fibroblasts to treat fibrosis of the heart: focus on HDACs. J Mol Cell Cardiol 70:100–107
Schulman SP, Lakatta EG, Fleg JL, Lakatta L, Becker LC, Gerstenblith G (1992) Age-related decline in left ventricular filling at rest and exercise. Am J Phys Heart Circ Phys 263(6):H1932–H1938
Sehgal SN, Baker H, Vézina C (1975) Rapamycin (AY-22,989), a new antifungal antibiotic. II. Fermentation, isolation and characterization. J Antibiot 28(10):727–732
Serrano M, Lin AW, McCurrach ME, Beach D, Lowe SW (1997) Oncogenic *ras* provokes premature cell senescence associated with accumulation of p53 and p16INK4a. Cell 88(5):593–602
Serrano L, Martínez-Redondo P, Marazuela-Duque A, Vazquez BN, Dooley SJ, Voigt P et al (2013) The tumor suppressor SirT2 regulates cell cycle progression and genome stability by modulating the mitotic deposition of H4K20 methylation. Genes Dev (27):639–653
Severino J, Allen RG, Balin S, Balin A, Cristofalo VJ (2000) Is β-galactosidase staining a marker of senescence in vitro and in vivo? Exp Cell Res 257(1):162–171
Seymour EM, Parikh RV, Singer AAM, Bolling SF (2006) Moderate calorie restriction improves cardiac remodeling and diastolic dysfunction in the Dahl-SS rat. J Mol Cell Cardiol 41(4):661–668

Shen P, Feng X, Zhang X, Huang X, Liu S, Lu X et al (2016) SIRT6 suppresses phenylephrine-induced cardiomyocyte hypertrophy though inhibiting p300. J Pharmacol Sci 132(1):31–40

Shimazu T, Hirschey MD, Hua L, Dittenhafer-Reed KE, Schwer B, Lombard DB et al (2010) SIRT3 deacetylates mitochondrial 3-Hydroxy-3-Methylglutaryl CoA synthase 2 and regulates ketone body production. Cell Metab 12(6):654–661

Shimazu T, Hirschey MD, Newman J, He W, Shirakawa K, Le Moan N et al (2013) Suppression of oxidative stress by β-Hydroxybutyrate, an endogenous histone deacetylase inhibitor. Science 339(6116):211–214

Shimizu K, Sunagawa Y, Funamoto M, Wakabayashi H, Genpei M, Miyazaki Y et al (2020) The synthetic curcumin analogue GO-Y030 effectively suppresses the development of pressure overload-induced heart failure in mice. Sci Rep 10(1):7172

Shinmura K, Tamaki K, Bolli R (2008) Impact of 6-mo caloric restriction on myocardial ischemic tolerance: possible involvement of nitric oxide-dependent increase in nuclear Sirt1. Am J Physiol Heart Circ Physiol 295(6):H2348–H2355

Siddiqi S, Sussman MA (2013) Cardiac hegemony of senescence. Curr Transl Geriatr Exp Gerontol Rep 2(4). https://doi.org/10.1007/s13670-013-0064-3

Silaghi CN, Farcaş M, Crăciun AM (2021) Sirtuin 3 (SIRT3) pathways in age-related cardiovascular and neurodegenerative diseases. Biomedicines 9(11):1574

Sin TK, Tam BT, Yung BY, Yip SP, Chan LW, Wong CS et al (2015) Resveratrol protects against doxorubicin-induced cardiotoxicity in aged hearts through the SIRT1-USP7 axis. J Physiol 593(8):1887–1899

Sohal RS, Weindruch R (1996) Oxidative stress, caloric restriction, and aging. Science 273(5271):59–63

Sohal RS, Ku HH, Agarwal S, Forster MJ, Lal H (1994a) Oxidative damage, mitochondrial oxidant generation and antioxidant defenses during aging and in response to food restriction in the mouse. Mech Ageing Dev 74(1–2):121–133

Sohal RS, Agarwal S, Candas M, Forster MJ, Lal H (1994b) Effect of age and caloric restriction on DNA oxidative damage in different tissues of C57BL/6 mice. Mech Ageing Dev 76(2–3):215–224

Solon-Biet SM, Mitchell SJ, Coogan SCP, Cogger VC, Gokarn R, McMahon AC et al (2015) Dietary protein to carbohydrate ratio and caloric restriction: comparing metabolic outcomes in mice. Cell Rep 11(10):1529–1534

Spange S, Wagner T, Heinzel T, Krämer OH (2009) Acetylation of non-histone proteins modulates cellular signalling at multiple levels. Int J Biochem Cell Biol 41(1):185–198

Stewart JE, Crawford JM, Mullen WE, Jacques A, Stoner MW, Scott I et al (2024) Cardiomyocyte-specific deletion of GCN5L1 reduces lysine acetylation and attenuates diastolic dysfunction in aged mice by improving cardiac fatty acid oxidation. Biochem J 481(6):423–436

Strahl BD, Allis CD (2000) The language of covalent histone modifications. Nature 403(6765):41–45

Stratton M, López-Domínguez JA, Canella A, Ramsey J, Cortopassi G. Long-term ketogenic diet increases expression of SOX2-dependent oligodendrocyte- and myelination-associated genes in the aged mouse brain. 2022

Sun N, Youle RJ, Finkel T (2016) The mitochondrial basis of aging. Mol Cell 61(5):654–666

Sunagawa Y, Funamoto M, Shimizu K, Shimizu S, Sari N, Katanasaka Y et al (2021) Curcumin, an inhibitor of p300-HAT activity, suppresses the development of hypertension-induced left ventricular hypertrophy with preserved ejection fraction in dahl rats. Nutrients 13(8):2608

Sunagawa Y, Katayama A, Funamoto M, Shimizu K, Shimizu S, Sari N et al (2022) The polyunsaturated fatty acids, EPA and DHA, ameliorate myocardial infarction-induced heart failure by inhibiting p300-HAT activity in rats. J Nutr Biochem 106:109031

Sundaresan NR, Samant SA, Pillai VB, Rajamohan SB, Gupta MP (2008) SIRT3 is a stress-responsive deacetylase in cardiomyocytes that protects cells from stress-mediated cell death by deacetylation of Ku70. Mol Cell Biol 28(20):6384–6401

Sundaresan NR, Vasudevan P, Zhong L, Kim G, Samant S, Parekh V et al (2012) The sirtuin SIRT6 blocks IGF-Akt signaling and development of cardiac hypertrophy by targeting c-Jun. Nat Med 18(11):1643

Sundaresan NR, Bindu S, Pillai VB, Samant S, Pan Y, Huang JY et al (2016) SIRT3 blocks aging-associated tissue fibrosis in mice by Deacetylating and activating glycogen synthase kinase 3β. Mol Cell Biol 36(5):678–692

Taffet GE, Pham TT, Hartley CJ (1997) The age-associated alterations in late diastolic function in mice are improved by caloric restriction. J Gerontol Ser A Biol Med Sci 52A(6):B285–B290

Takaya T, Kawamura T, Morimoto T, Ono K, Kita T, Shimatsu A et al (2008) Identification of p300-targeted acetylated residues in GATA4 during hypertrophic responses in cardiac myocytes. J Biol Chem 283(15):9828–9835

Tang X, Chen XF, Wang NY, Wang XM, Liang ST, Zheng W et al (2017) SIRT2 acts as a CARDIOPROTECTIVE deacetylase in pathological cardiac hypertrophy. Circulation 136(21):2051

Tao H, Yang JJ, Shi KH, Li J (2016) Epigenetic factors MeCP2 and HDAC6 control α-tubulin acetylation in cardiac fibroblast proliferation and fibrosis. Inflamm Res 65(5):415–426

Tarragó MG, Chini CCS, Kanamori KS, Warner GM, Caride A, de Oliveira GC et al (2018) A potent and specific CD38 inhibitor ameliorates age-related metabolic dysfunction by reversing tissue NAD+ decline. Cell Metab 27(5):1081–1095.e10

Thai PN, Seidlmayer LK, Miller C, Ferrero M, Dorn GW, Schaefer S et al (2019) Mitochondrial quality control in aging and heart failure: influence of ketone bodies and Mitofusin-stabilizing peptides. Front Physiol 10:382

Tian X, Firsanov D, Zhang Z, Cheng Y, Luo L, Tombline G et al (2019a) SIRT6 is responsible for more efficient DNA double-strand break repair in long-lived species. Cell 177(3):622

Tian S, Lei I, Gao W, Liu L, Guo Y, Creech J et al (2019b) HDAC inhibitor valproic acid protects heart function through Foxm1 pathway after acute myocardial infarction. EBioMedicine 39:83–94

Tomczyk MM, Cheung KG, Xiang B, Tamanna N, Teixeira ALF, Agarwal P et al (2022) Mitochondrial SIRT3 prevents doxorubicin-induced dilated cardiomyopathy by modulating protein acetylation and oxidative stress. Circ Heart Fail 15(5):e008547

Tomilov AA, Ramsey JJ, Hagopian K, Giorgio M, Kim KM, Lam A et al (2010) The Shc locus regulates insulin signaling and adiposity in mammals. Aging Cell 10(1):55

Tsushima K, Bugger H, Wende AR, Soto J, Jenson GA, Tor AR et al (2018) Mitochondrial reactive oxygen species in Lipotoxic hearts induces post-translational modifications of AKAP121, DRP1 and OPA1 that promote mitochondrial fission. Circ Res 122(1):58–73

Turdi S, Fan X, Li J, Zhao J, Huff AF, Du M et al (2010) AMP-activated protein kinase deficiency exacerbates aging-induced myocardial contractile dysfunction. Aging Cell 9(4):592–606

Urfer SR, Kaeberlein TL, Mailheau S, Bergman PJ, Creevy KE, Promislow DEL et al (2017) A randomized controlled trial to establish effects of short-term rapamycin treatment in 24 middle-aged companion dogs. GeroScience 39(2):117–127

Vakhrusheva O, Smolka C, Gajawada P, Kostin S, Boettger T, Kubin T et al (2008) Sirt7 increases stress resistance of cardiomyocytes and prevents apoptosis and inflammatory cardiomyopathy in mice. Circ Res 102(6):703–710

Veech RL (2014) Ketone ester effects on metabolism and transcription. J Lipid Res 55(10):2004–2006

Vega RB, Harrison BC, Meadows E, Roberts CR, Papst PJ, Olson EN et al (2004) Protein kinases C and D mediate agonist-dependent cardiac hypertrophy through nuclear export of histone deacetylase 5. Mol Cell Biol 24(19):8374–8385

Ventura-Clapier R, Garnier A, Veksler V (2004) Energy metabolism in heart failure. J Physiol 555(Pt 1):1–13

Wagner GR, Payne RM (2011) Mitochondrial acetylation and diseases of aging. J Aging Res 7(2011):234875

Wagner GR, Payne RM (2013) Widespread and enzyme-independent Nϵ-acetylation and Nϵ-Succinylation of proteins in the chemical conditions of the mitochondrial matrix. J Biol Chem 288(40):29036–29045

Wang B, Rao YH, Inoue M, Hao R, Lai CH, Chen D et al (2014) Microtubule acetylation amplifies p38 kinase signalling and anti-inflammatory IL-10 production. Nat Commun 5(1):3479

Wang H, Diao D, Shi Z, Zhu X, Gao Y, Gao S et al (2016) SIRT6 controls hematopoietic stem cell homeostasis through epigenetic regulation of Wnt signaling. Cell Stem Cell 18(4):495–507

Wang R, Yu Z, Sunchu B, Shoaf J, Dang I, Zhao S et al (2017) Rapamycin inhibits the secretory phenotype of senescent cells by a Nrf2-independent mechanism. Aging Cell 16(3):564

Wang M, Zhang L, Zhu W, Zhang J, Kim SH, Wang Y et al (2018) Calorie restriction curbs Proinflammation that accompanies arterial aging, preserving a youthful phenotype. J Am Heart Assoc 7(18):e009112

Wang LF, Cao Q, Wen K, Xiao YF, Chen TT, Guan XH et al (2019) CD38 deficiency alleviates D-galactose-induced myocardial cell senescence through NAD+/Sirt1 signaling pathway. Front Physiol 3(10):1125

Wei T, Huang G, Gao J, Huang C, Sun M, Wu J et al (2017) Sirtuin 3 deficiency accelerates hypertensive cardiac remodeling by impairing angiogenesis. J Am Heart Assoc 6(8):e006114

Weindruch R, Sohal RS (1997) Caloric intake and aging. N Engl J Med 337(14):986

Weinert BT, Iesmantavicius V, Wagner SA, Schölz C, Gummesson B, Beli P et al (2013) Acetyl-phosphate is a critical determinant of lysine acetylation in *E. coli*. Mol Cell 51(2):265–272

Weinert BT, Iesmantavicius V, Moustafa T, Schölz C, Wagner SA, Magnes C et al (2014) Acetylation dynamics and stoichiometry in Saccharomyces cerevisiae. Mol Syst Biol 10(1):716

Weinert BT, Moustafa T, Iesmantavicius V, Zechner R, Choudhary C (2015) Analysis of acetylation stoichiometry suggests that SIRT3 repairs nonenzymatic acetylation lesions. EMBO J 34(21):2620–2632

Weiss EP, Fontana L (2011) Caloric restriction: powerful protection for the aging heart and vasculature. Am J Physiol Heart Circ Physiol 301(4):H1205

Welti J, Sharp A, Brooks N, Yuan W, McNair C, Chand SN et al (2021) Targeting p300/CBP in lethal prostate cancer. Cancer Discov 11(5):1118

Wende AR, Abel ED (2010) Lipotoxicity in the heart. Biochim Biophys Acta 1801(3):311–319

Whitworth AJ, Pallanck LJ (2017) PINK1/Parkin mitophagy and neurodegeneration—what do we really know *in vivo*? Curr Opin Genet Dev 44:47–53

Williams SM, Golden-Mason L, Ferguson BS, Douglas KB, Cavasin MA, Demos-Davies K et al (2014 Feb) Class I HDACs regulate angiotensin II-dependent cardiac fibrosis via fibroblasts and circulating fibrocytes. J Mol Cell Cardiol 67:112

Wojtovich AP, Brookes PS (2009) The complex II inhibitor atpenin A5 protects against cardiac ischemia-reperfusion injury via activation of mitochondrial KATP channels. Basic Res Cardiol 104(2):121–129

Wollert KC, Meyer GP, Lotz J, Ringes Lichtenberg S, Lippolt P, Breidenbach C et al (2004) Intracoronary autologous bone-marrow cell transfer after myocardial infarction: the BOOST randomised controlled clinical trial. Lancet 364(9429):141–148

Xia Y, Sun M, Xie Y, Shu R (2017) mTOR inhibition rejuvenates the aging gingival fibroblasts through alleviating oxidative stress. Oxidative Med Cell Longev 2017:6292630

Xie Z, Zhang D, Chung D, Tang Z, Huang H, Dai L et al (2016) Metabolic regulation of gene expression by histone lysine β-hydroxybutyrylation. Mol Cell 62(2):194–206

Xie M, Tang Y, Hill JA (2019) HDAC inhibition as a therapeutic strategy in myocardial ischemia/reperfusion injury. J Mol Cell Cardiol 129:188–192

Xu S, Yin M, Koroleva M, Mastrangelo MA, Zhang W, Bai P et al (2016) SIRT6 protects against endothelial dysfunction and atherosclerosis in mice. Aging (Albany NY) 8(5):1064–1078

Yamamura S, Izumiya Y, Araki S, Nakamura T, Kimura Y, Hanatani S et al (2020) Cardiomyocyte Sirt (Sirtuin) 7 ameliorates stress-induced cardiac hypertrophy by interacting with and Deacetylating GATA4. Hypertension 75(1):98–108

Yan M, Sun S, Xu K, Huang X, Dou L, Pang J et al (2021) Cardiac aging: from basic research to therapeutics. Oxidative Med Cell Longev 9(2021):9570325

Yang J, He J, Ismail M, Tweeten S, Zeng F, Gao L et al (2019) HDAC inhibition induces autophagy and mitochondrial biogenesis to maintain mitochondrial homeostasis during cardiac ischemia/reperfusion injury. J Mol Cell Cardiol 130:36

Yoshida Y, Yamanaka S (2017) Induced pluripotent stem cells 10 years later. Circ Res 120(12):1958–1968

Yu W, Lin Y, Yao J, Huang W, Lei Q, Xiong Y et al (2009) Lysine 88 acetylation negatively regulates ornithine Carbamoyltransferase activity in response to nutrient signals. J Biol Chem 284(20):13669–13675

Yu W, Dittenhafer-Reed KE, Denu JM (2012) SIRT3 protein deacetylates Isocitrate dehydrogenase 2 (IDH2) and regulates mitochondrial redox status. J Biol Chem 287(17):14078–14086

Yudoh K, Karasawa R, Ishikawa J (2015) Age-related decrease of sirtuin 2 protein in human peripheral blood mononuclear cells. Curr Aging Sci 8(3):256–258

Zhang CL, McKinsey TA, Chang S, Antos CL, Hill JA, Olson EN (2002) Class II histone deacetylases act as signal-responsive repressors of cardiac hypertrophy. Cell 110(4):479–488

Zhang J, Sprung R, Pei J, Tan X, Kim S, Zhu H et al (2009 Feb) Lysine acetylation is a highly abundant and evolutionarily conserved modification in *Escherichia Coli*. Mol Cell Proteomics 8(2):215–225

Zhang L, Keung W, Samokhvalov V, Wang W, Lopaschuk GD (2010) Role of fatty acid uptake and fatty acid β-oxidation in mediating insulin resistance in heart and skeletal muscle. Biochimica et Biophysica Acta (BBA) Mol Cell Biol Lipids 1801(1):1–22

Zhang CS, Li M, Ma T, Zong Y, Cui J, Feng JW et al (2016) Metformin activates AMPK through the lysosomal pathway. Cell Metab 24(4):521–522

Zhang H, Wang J, Li L, Chai N, Chen Y, Wu F et al (2017) Spermine and spermidine reversed age-related cardiac deterioration in rats. Oncotarget 8(39):64793–64808

Zhang W, Wan H, Feng G, Qu J, Wang J, Jing Y et al (2018a) SIRT6 deficiency results in developmental retardation in cynomolgus monkeys. Nature 560(7720):661–665

Zhang W, Song M, Qu J, Liu GH (2018b) Epigenetic modifications in cardiovascular aging and diseases. Circ Res 123(7):773–786

Zhang M, Feng N, Peng Z, Thapa D, Stoner MW, Manning JR et al (2023) Reduced acetylation of TFAM promotes bioenergetic dysfunction in the failing heart. iScience 26(6):106942

Zhao S, Xu W, Jiang W, Yu W, Lin Y, Zhang T et al (2010) Regulation of cellular metabolism by protein lysine acetylation. Science 327(5968):1000–1004

Zheng Q, Zhao K, Han X, Huff AF, Cui Q, Babcock SA et al (2015) Inhibition of AMPK accentuates prolonged caloric restriction-induced change in cardiac contractile function through disruption of compensatory autophagy. Biochim Biophys Acta (BBA) Mol Basis Dis 1852(2):332–342

Zhou W, Jiang D, Tian J, Liu L, Lu T, Huang X et al (2018) Acetylation of H3K4, H3K9, and H3K27 mediated by p300 regulates the expression of GATA4 in cardiocytes. Genes Dis 6(3):318–325

Chapter 10
Regulatory Roles for SIRT1 in Aging and Immunosenescence

Devonique Brissett-Lasalle, Zoya Syed, Yashitha Yarubandi, Jeff Leips, and Erin Green

Abstract Silent information regulator 1 (SIRT1), a conserved lysine deacetylase, is an important contributor to the function of macrophages, which are the scavengers of the innate immune system. Macrophages are part of the first line of defense against infection and key players in immunity due to their ability to survey tissue for infections or damage, release inflammatory cytokines, and clear pathogens. Macrophage function deteriorates with age and is a common indicator of immunosenescence. SIRT1 is known to influence multiple aspects of macrophage physiology, particularly proliferation, self-renewal, migration, the regulation of macrophage polarization, and the ability of macrophages to clear pathogens via phagocytosis and inflammasome signaling. Furthermore, mammalian SIRT1 and orthologous Sir2 in other organisms have well-defined roles in aging. Therefore, in this chapter, we discuss evidence that links SIRT1 to macrophage behavior and function, explore its role in inflammatory pathways linked to aging, and highlight key research questions in immunosenescence and the implications for epigenetic and non-epigenetic roles of SIRT1.

D. Brissett-Lasalle (✉) · Z. Syed · Y. Yarubandi · J. Leips
Department of Biological Sciences, University of Maryland Baltimore County, Baltimore, MD, USA
e-mail: dbrisse1@umbc.edu

E. Green (✉)
Department of Biological Sciences, University of Maryland Baltimore County, Baltimore, MD, USA

Marlene and Stewart Greenebaum Comprehensive Cancer Center, University of Maryland School of Medicine, Baltimore, MD, USA
e-mail: egreen@umbc.edu

M. Halasa, A. Wawruszak (eds.), *Histone and Non-Histone Reversible Acetylation in Development, Aging and Disease*, Results and Problems in Cell Differentiation 75,
https://doi.org/10.1007/978-3-031-91459-1_10

10.1 Introduction

Epigenetic dysfunction is a distinctive hallmark of the aging process. One epigenetic mechanism is the addition or removal of post-translational modifications (PTMs) to residues within histone proteins, a dynamic process mediated by enzymes that control the abundance and distribution of histone PTMs during aging. One group of histone-modifying enzymes are the sirtuins, which belong to a family of nicotinamide adenine dinucleotide (NAD^+)-dependent deacetylases with both histone and non-histone substrates. Since their discovery, sirtuins have been implicated in the aging process. Mammals have seven sirtuins (SIRT1-SIRT7), which have varying biological functions ranging from gene expression regulation to insulin secretion (Haigis and Sinclair 2010). Mostly found in the nucleus, SIRT1 deacetylates N-terminal tails of histones H1, H2A, H3, and H4 (Chen et al. 2006; Ryall et al. 2015; Vaquero et al. 2004). SIRT1 can also deacetylate non-histone proteins, for example, forkhead box transcription factors, tumor protein 53 (p53), and nuclear factor κB (NF-κB) (Motta et al. 2004; Vaziri et al. 2001; Yeung et al. 2004). The abundance of two histone targets of SIRT1, histone 3 lysine 9 acetylation (H3K9ac) and histone 4 lysine 16 acetylation (H4K16ac), varies during aging. The age-associated changes in H4K16ac are well-documented and associated with several hallmarks of aging (Dang et al. 2009; Delaney et al. 2011; Dube et al. 2022, 2024; Gong et al. 2015; Nativio et al. 2018). While not as well studied in the context of aging, the relevance of H3K9ac has been linked to metabolic pathways (Balan et al. 2008; Eisenberg et al. 2014; Kawakami et al. 2009; Peleg et al. 2016; Zhang et al. 2022). In many cases, sirtuins are studied in the context of age-related diseases like cancer, neurodegenerative disorders, or heart disease (Carafa et al. 2012). However, it is also critical that we assess how sirtuins modulate lifespan during normal aging and in a non-disease state to understand their causative role in organismal aging and lifespan determination.

Immunosenescence is the decline in the function of the immune system with age. Differences in rates of immunosenescence can determine the rates of aging, and thus lifespan. Immune cell dysfunction is hypothesized to contribute to impairment of both the innate and adaptive immune systems (Rodrigues et al. 2021). Immunosenescence can arise from direct changes in immune cell activity or indirect effects of tissue homeostasis. Features associated with cellular immunosenescence include the reduction in the ability to recognize and clear infections or cellular damage, as well as an increase in the release of pro-inflammatory cytokines, interleukin (IL) 1β, IL-6, Tumor Necrosis Factor α (TNF- α), driving a process known as "inflammaging" (Li et al. 2023). Macrophages were once thought to be the sole contributors to inflammaging (Franceschi et al. 2000). However, other hallmarks of aging are also known to be key in the development of inflammaging including cellular senescence (Coppé et al. 2008), mitochondrial dysfunction (Zampino et al. 2020), and dysbiosis (Bodogai et al. 2018).

SIRT1 plays various roles in the immune response and has been shown to modify expression of immune response genes in macrophages (Schug et al. 2010). The biological function of macrophages includes their recruitment to sites of infection (immune surveillance), differentiating into multiple functional states (polarization), and inflammation resolution, all of which are controlled in part by SIRT1 (Fig. 10.1). However, whether SIRT1 could mediate the inflammatory response to delay aging or maintain homeostasis within the aging innate immune system is unexplored. To begin to fill this gap, this chapter reviews the role of SIRT1 in macrophage biology, discusses how aging could alter this phenomenon, and highlights areas of future research that could uncover the anti-inflammatory potential of SIRT1.

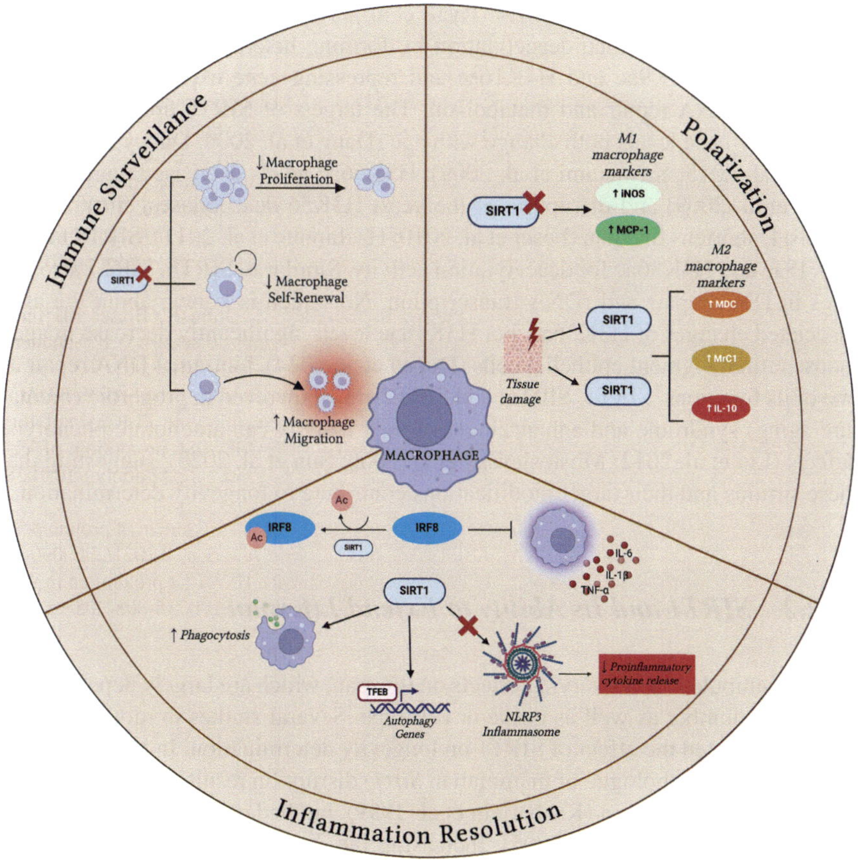

Fig. 10.1 Role for SIRT1 in macrophage function. *Immune surveillance*—SIRT1 inhibition decreases efficiency of macrophage proliferation and self-renewal but increases macrophage migration. *Polarization*—SIRT1 inhibition increases M1 polarization and in turn, an increase in proinflammatory cytokines. Upon tissue damage, SIRT1 can either be activated or inhibited where in either case there is an increase in M2 polarization to repair damage. *Inflammation resolution*—SIRT1 is required for IRF8 inflammatory signaling. SIRT1 inhibition blocks NLRP3 inflammasome activity and reduces pro-inflammatory cytokine release. Additionally, SIRT1 expression promotes phagocytic clearance and autophagy-related gene expression. (Created with BioRender.com)

10.2 Contribution of Sirtuins to Lifespan Regulation

10.2.1 Histone-Specific Roles of Nuclear SIRT1, SIRT6, and SIRT7 During Aging

The seven mammalian sirtuins are evolutionarily conserved and can be categorized based on their subcellular localization. SIRT1, SIRT6, and SIRT7 are predominantly localized in the nucleus, while SIRT3, SIRT4, and SIRT5 are found in the mitochondria, and SIRT2 is primarily cytosolic. SIRT1 deacetylates several lysine residues specifically on H3 and H4 (Ryall et al. 2015; Yang et al. 2023), and has a role in age-related histone deacetylation by forming heterochromatin regions via modulation of H3K9ac and H4K16ac and repressing gene expression. SIRT6 is involved in DNA repair and metabolism. The targets of SIRT6 are H3K9ac and H3K56ac, whose levels both change with age (Dang et al. 2009; Dubey et al. 2024; Gong et al. 2015; Kawakami et al. 2009). H3K56ac levels decrease in aging yeast (Dang et al. 2009) and disruption of the yeast H3K56 deacetylases, sirtuins Hst3 and Hst4, shortens lifespan (Feser et al. 2010; Hachinohe et al. 2011). SIRT7 targets H3K18ac and H3K36ac for deacetylation activity. Similar to SIRT6, SIRT7 also has roles in DNA repair and rDNA transcription. Not much is known about the age-associated changes of H3K36ac, but H3K18ac levels significantly decrease in aged mouse retinal pigment epithelial cells (Dubey et al. 2024), hinting at DNA repair as one of its functions. SIRT1, SIRT6, and SIRT7 are all involved in progeroid (premature aging) syndrome and enhancement of their activity can ameliorate phenotype defects (Liu et al. 2012; Mostoslavsky et al. 2006; Sun et al. 2020), suggesting that these sirtuins and their target modifications contribute to longevity determination.

10.2.2 SIRT1 and Its Ability to Extend Lifespan

SIRT1 manipulation has varying effects on lifespan, which are largely dependent on gene copy number as well as tissue or cell type. Several studies in model systems have investigated the effect of SIRT1 on longevity determination. In budding yeast, *SIR2* (the yeast orthologue of mammalian *Sirt1*) disruption resulted in a decrease in mean replicative lifespan (Kaeberlein et al. 1999), but had no effect on chronological lifespan (Fabrizio et al. 2005). Increasing the dosage of *sir-2.1* in *C. elegans* resulted in a lifespan extension of 11–50% among three transgenic lines (Tissenbaum and Guarente 2001). Similarly, studies in *Drosophila* showed that increasing *Sir2* expression in neurons, fat body, and whole bodies extends lifespan significantly (Banerjee et al. 2012; Bauer et al. 2008; Rogina and Helfand 2004; Whitaker et al. 2013). While there have been contradictory observations regarding the role of *Sir2* in lifespan extension, some of this has been determined to be due to gene dosage, transposon insertion site, and tissue specificity (Burnett et al. 2011; Hoffmann et al. 2013; Whitaker et al. 2013). Interestingly, whole-body *Sirt1* overexpression in mice did not affect lifespan, but was crucial for maintenance of healthspan (Herranz et al. 2010; Roichman et al. 2021), whereas brain-specific overexpression of *Sirt1* led to a significant extension in lifespan (Satoh et al. 2013). These data imply that loss of SIRT1 is detrimental to lifespan, while increasing SIRT1 expression can promote longevity, however the benefit of SIRT1 for longevity is both dose-dependent and tissue-specific. The endogenous levels of SIRT1 during aging complement its longevity benefit. Analysis of SIRT1 during natural aging reveals a consistent decrease in *Sirt1* expression (Liu et al. 2024; Ma et al. 2024; Piragine et al. 2024; Zhou et al. 2023). Although one of these studies was conducted in an immunodeficient mouse model, the fact that studies in healthy animals reflect similar age-related patterns aligns with the idea that preserving SIRT1 levels during aging promotes longevity.

SIRT1 deacetylates H4K16ac and H3K9ac, histone modifications which exhibit age-related changes, but whether these alterations directly control lifespan is still under investigation. H4K16ac, an active mark for transcription, is highly involved in several age-associated pathways shown to delay aging and promote longevity. In aged yeast, H4K16ac levels increased within subtelomeric regions (Dang et al. 2009). A later study linked the increase in H4K16ac to rDNA silencing as a contributor to longevity regulation (Delaney et al. 2011). In mammals, the patterns of H4K16ac during aging are tissue-dependent. In aged mice retina, there was a decrease in H4K16ac levels, which could be attributed to parallel downregulation of nuclear histone transcription factors (Dubey et al. 2024). Alongside this, an induction of senescence-associated secretory protein (SASP) markers was also observed, suggesting that the loss of H4K16ac may contribute to lifespan and aging through senescence-mediated pathways. Similarly, a decline in H4K16ac was reported in aged mouse basal epidermal cells, likely driven by significant downregulation of histone acetyltransferases, lysine acetyltransferase 5 and 8 (KAT5 and KAT8), during aging (Dube et al. 2022). In contrast, young 3-month and old 22-month mice brains showed no observable changes in H4K16ac with age, although levels of the mark decreased in young, diet-restricted mice versus old mice (Gong et al. 2015). Studies of brain tissue samples from postmortem humans demonstrated an increase in the genome-wide abundance of H4K16ac peaks, alongside enrichment for binding sites of transcription factor hypermethylated-in-cancer 1 (HIC1), a p53-mediated DNA damage response factor, at genes demonstrating age-dependent abundance of H4K16ac (Nativio et al. 2018). Given these observations, it is possible that the regulation of H4K16ac with age lies at the intersection of senescence, nutrient sensing, and telomere maintenance, where the modification functions to balance expression of SASP-related genes, downstream regulators of nutrient sensing, and/or prevent genomic instability.

Another histone target of SIRT1, H3K9ac, is also known to change with age. However, the pathways through which it acts to influence longevity are not fully understood. Lysine acetylation is highly dependent on levels of metabolites, NAD^+, and intracellular acetyl-CoA. One report in *Drosophila* noted a moderate increase in H3K9ac from young to midlife flies (Peleg et al. 2016), consistent with other observations that alterations in metabolic activity can prolong lifespan (Balan et al. 2008; Eisenberg et al. 2014). More recently, H3K9ac was correlated with expression changes in metabolic genes during aging and leaf senescence in rice (Zhang et al. 2022). Experiments in mouse models have yielded mixed results, with no change in H3K9ac in aged mouse retinal epithelium (Dubey et al. 2024) and a decrease in H3K9ac with age in rat liver (Kawakami et al. 2009). Further studies are required in mammalian systems to clarify the role of H3K9ac abundance in lifespan and aging.

10.3 Regulatory Functions of SIRT1 on Aging of the Innate Immune System

10.3.1 Immunosenescence and Inflammaging

A signature of aging is a loss of system robustness, which is particularly apparent in the immune system. "Immunosenescence" was theorized in 1964 and proposes that as somatic mutations arise during aging, the cells of the immune system are especially susceptible to dysfunction due to the diversity of cell types and their physiological roles (Walford 1964). A more recent interpretation of the concept of immunosenescence posits that it arises from induction of senescence within individual immune cells and at the tissue level (Callender et al. 2018; Hall et al. 2016). As an example, senescent macrophages are characterized by inhibition of inflammatory cytokine release (Murakami et al. 2012) and elevated levels of pathogens which could, in turn, lead to widespread tissue aging. These processes are interconnected as they not only drive aging, but also increase the risk of age-related diseases. Senescent immune cells have been shown to secrete SASP factors, which include interleukins, chemokines, growth factors, proteases, non-protein components (nitric oxide and reactive oxygen species), and extracellular matrix components. While SASPs promote resolution of inflammation through recruitment of immune cells, in aging systems they may have a negative effect as their accumulation could induce senescence of other cells in a paracrine manner, leading to their degeneration (Schafer et al. 2020).

A characteristic state of inflammation impinges upon organisms undergoing immunosenescence. SASP accumulation enhances chronic inflammation by causing cells to release pro-inflammatory cytokines, such as IL-1β, IL-6, and TNF-α.. This state of low-grade inflammation leads to a phenomenon known as "inflammaging". Circulating pro-inflammatory cytokines are known to increase during aging (Álvarez-Rodríguez et al. 2012). Additionally, genetic and pharmacologic inhibition of IL-11 in mice extended lifespan and healthspan of these animals (Widjaja et al. 2024). These and numerous other findings support the classification of chronic inflammation as a hallmark of aging (López-Otín et al. 2023).

The innate immune system is the first line of defense against infection. Neutrophils, the most abundant innate immune cells, are the first to migrate to sites of inflammation, and monocytes, which differentiate into macrophages and dendritic cells, are second in line to sites of damage. Macrophages are long-lived and reside in various tissues, enabling them to rapidly eliminate damage or invading pathogens, and are a critical component of the innate immune system. To bridge the innate and adaptive immune response, macrophages also function as professional antigen-presenting cells (APCs) by capturing exogenous antigens and presenting them to helper T cells through the major histocompatibility complex II (MHC-II), ultimately signaling other immune cells such as B cells to clear inflammation through several mechanisms (Gate et al. 2020). Macrophages are enriched in senescent environments induced by obesity (Matacchione et al. 2022), which promotes an inflammaging landscape. Furthermore, age-related alterations in macrophage function make them an ideal population for dissecting the role of innate immune cells in immunosenescence.

10.3.2 Interplay Between SIRT1 Activity and Age-Related Macrophage Function

Here, we consider a singular branch of the immune system to gain an understanding of how SIRT1 contributes to immunosenescence, while acknowledging that macrophages are not the sole effectors whose dysfunction leads to immunosenescence. Moreover, the mechanisms through which macrophages contribute to a decline in immune response with age are not fully understood. SIRT1 does appear to play a role in preventing age-related macrophage deterioration by maintaining macrophage homeostasis. In particular, SIRT1 functions in the induction of cytokine production, phagocytosis, tissue surveillance, and inflammation resolution (Fig. 10.1). Below, we review the role for SIRT1 in each of these processes.

Immune Surveillance For macrophages to optimally survey infected tissues, they require an orchestrated response among cytoskeleton components that influence their motility (Paterson and Lämmermann 2022), and there is evidence of the regulatory role of SIRT1 in macrophage migration, including that macrophage migration was enhanced upon SIRT1 deficiency (Hui et al. 2017; Ka et al. 2015). Furthermore, inactivating monocyte chemoattractant protein-1 (MCP-1) receptors suppressed the enhanced migration elicited by SIRT1-deficient adipocytes (Hui et al. 2017). Altogether, these data indicate that inactivation of SIRT1 positively regulates the migratory capability of macrophages. Another feature of macrophage biology that is essential for resolution of local and systemic infections is their capacity to proliferate, which also drives their self-renewal. Inhibition of SIRT1 led to a reduction in colony-forming ability and proliferation of peritoneal macrophages (Imperatore et al. 2017), and genetic and pharmacological inhibition of *Sirt1* downregulates cell cycle regulator genes and checkpoint genes, resulting in fewer cells in S/G2 phases and more cells in the G1 phase (Imperatore et al. 2017; Li et al. 2020). The reduced proliferation and self-renewal capacity upon SIRT1 depletion in macrophages may be a consequence of these cells entering a senescent state, while still allowing cells to migrate.

The proliferative and migratory capacity of macrophages changes with age. For example, alveolar macrophages show decreased proliferation during aging, which is also concomitant with an increase in cellular musculoaponeurotic fibrosarcoma (*c-Maf*), a repressor of macrophage self-renewal capacity (Wong et al. 2017). In macrophages derived from healthy donors, a decline in the rate of migration was observed from young to old age (Moss et al. 2024). This observation is likely due to the downregulation of two transcription factors, MYC and Upstream stimulatory factor 1 (USF1), which regulate the expression of cell migration genes. SIRT1 is a component of a feedback loop with MYC in which they both augment cell growth pathways (Yuan et al. 2009), suggesting a role for SIRT1 in macrophage proliferation and self-renewal via control of cell migration gene expression. Additional studies are required to investigate the age-related role of SIRT1 in maintaining macrophage migration, proliferation, and self-renewal to better understand its contribution to macrophage function during tissue surveillance, as well as to immunosenescence more broadly.

Macrophage Polarization Macrophages can adopt one of two phenotypes depending on the activation signal: M1 macrophages (or classically activated macrophages) are polarized by lipopolysaccharide and interferon-gamma (IFNγ) and produce pro-inflammatory cytokines such as IL-1β, IL-6, IL-12, IL-23, and TNF-α, while M2 macrophages (or alternatively activated macrophages) are polarized by IL-4 and IL-13 and subsequently produce anti-inflammatory cytokines such as IL-10, Transforming growth factor beta (TGF-β), and IL-1ra. Multiple instances of SIRT1 influencing macrophage polarization, and in turn their inflammatory function, have been reported. Depletion of SIRT1 causes an increase in abundance of inducible nitric oxide synthase (iNOS) and MCP-1, M1 macrophage markers (Hui et al. 2017). In rheumatoid arthritis patients with high systemic inflammation, activation of SIRT1 decreased expression of pro-inflammatory M1 macrophage markers, but increased expression of anti-inflammatory M2 macrophage markers, IL-10, macrophage-derived chemokine (MDC), and C-type mannose receptor 1 (MrC1) (Park et al. 2017). Likewise, in the context of brain injury, knockdown of SIRT1 elevated the amount of M2-type macrophages (Xia et al. 2021), potentially to ameliorate inflammation and repair damage. Altogether, these findings indicate the potential for SIRT1 activation to suppress inflammation through regulation of macrophage polarization.

SIRT1 has also been shown to limit pro-inflammatory mechanisms through interferon regulatory factor 8 (IRF8) and NF-κB signaling. In a mouse myeloid cell-specific SIRT1 knockout, NF-κB is hyperacetylated, enhancing its binding to the pro-inflammatory target gene TNF-α, and increasing its expression (Schug et al. 2010), indicating an anti-inflammatory function for SIRT1. Subsequent reports suggested that SIRT1 suppresses pro-inflammatory cytokine production by deacetylating the transcription factor IRF8 (Jia et al. 2017, 2022). Consistent with these observations, SIRT1 suppresses pro-inflammatory cytokines in an arthritis mouse model by blocking M1 polarization (Zhao et al. 2024). These observations not only suggest that SIRT1 is critical for maintaining the levels of M1 macrophage populations, but also that it has both pro- and anti-inflammatory properties.

Plasticity of macrophages to differentiate into M1/M2 macrophages is altered during aging, which in turn disrupts cytokine production (Mahbub et al. 2012). Moreover, during aging, macrophages shift from an M2 anti-inflammatory state to an M1 pro-inflammatory state (Becker et al. 2018), further supporting the enhanced inflammatory state during aging, or inflammaging. Given that aging triggers a shift in macrophage polarization, and in turn, cytokine production, future studies on the age-related effect of SIRT1 in cytokine production would reveal (1) the mechanisms by which this is controlled and (2) how SIRT1 can contribute to an anti-inflammatory phenotype to protect against aging and disease.

Inflammation Resolution Macrophages contain several pathogen recognition receptors that bind danger signals upon infection to eliminate pathogens by triggering proinflammatory factors and antimicrobial activity (Mukhopadhyay et al. 2009). A key activity during this process is phagocytosis, which is mediated in part by SIRT1 expression and activity. Prostaglandins are lipid signaling molecules that are known to inhibit macrophage-mediated phagocytosis and polarization (Cui et al. 2021). SIRT1 expression improves phagocytic activity of peritoneal macrophages through modulation of activator protein 1 (AP-1), a transcriptional regulator of prostaglandin production and bacterial clearance (Zhang et al. 2010). More recently, SIRT1 activation was associated with a reduction in macrophage phagocytosis when cultured under hyperosmotic stress (Meng et al. 2021) Furthermore, upon induction of an inflammatory response, extracellular vesicles derived from adipose mesenchymal stem cells promoted phagocytosis through a SIRT1-dependent mechanism (Wu et al. 2023), providing additional evidence linking SIRT1 to regulation of phagocytosis.

SIRT1 has also been implicated in autophagy, another line of defense against pathogens. For macrophages to destroy a pathogen, it must first be internalized, forming a phagosome, then fusion of the phagosome and lysosome leads to degradation of the pathogen by lysosomal enzymes. The ability of macrophages to clear pathogens via autophagy is enhanced upon SIRT1 activation. The expression of lysosomal and autophagy genes is dependent on SIRT1 due to its role in regulating the nuclear translocation of the transcription factor EB (TFEB), a major transcriptional regulator of autophagy (Zheng et al. 2021). Interestingly, it has also been shown that autophagy inhibition increases SIRT1 levels in macrophages (Wu et al. 2020), whereas SIRT1 inhibition decreases starvation-induced autophagy in undifferentiated macrophages (Takeda-Watanabe et al. 2012), suggesting that there is an interconnected SIRT1/autophagy axis within macrophages.

SIRT1 activity has been linked to the NLR family pyrin domain containing protein 3 (NLRP3) inflammasome, which is active in macrophages, as well as other cells of the innate immune system. This activation can be triggered by danger signals such as toll-like receptor agonists or mitochondrial reactive oxygen species. This leads to the activation of caspase-1, which in turn converts IL-1β and IL-18 precursor proteins to their mature forms, eliciting an inflammatory response (He et al. 2016). IL-1β secretion, upon treatment with compound K (an autophagy activator), is dependent on SIRT1 in macrophages (Wu et al. 2020), supporting a role for SIRT1 in NLRP3 inflammasome activity and highlighting an important connection between SIRT1 and NLRP3 in protecting against inflammation.

Collectively, the interplay between phagocytosis, SIRT1, autophagy, and inflammasome activity is crucial for maintaining tissue homeostasis and resolving inflammation. Both phagocytic ability and inflammasome activity are altered during the aging process. In *Drosophila* hemocytes, phagocytic ability is reduced with age (Horn et al. 2014). Evidence of hyperactivity of the NLRP3 inflammasome during aging (Andina et al. 2023) and longevity benefit upon NLRP3 suppression (Marín-Aguilar et al. 2020; Youm et al. 2013) imply that the NLRP3 inflammasome could be a target for maintaining tissue homeostasis. Future research into the age-dependent role of SIRT1 in phagocytosis and inflammasome signaling during aging could provide insight into mechanisms of delaying the aging process and improving health span

10.3.3 Connections Between Chronic Inflammatory Diseases and Macrophage Regulation by SIRT1

Older individuals are at an increased risk for chronic inflammatory illnesses. Chronic inflammatory diseases are among one of the major causes of death worldwide, with 60% of people dying due to conditions such as diabetes, arthritis and joint diseases, cardiovascular diseases, neurodegenerative diseases, chronic respiratory diseases, cancer, and obesity (Roth et al. 2018). Chronic inflammatory diseases have a distinctive pathophysiology, arising from a combination of factors, including persistent acute inflammation, infiltration of inflammatory immune cells, and progressive tissue damage. In the case of acute inflammation, the inflammatory response is rapid and short-lived within an injured tissue. In a disease state, inflammation is persistent and lacks suppressive mechanisms to dampen the inflammatory response.

SIRT1 is highly involved in arthritis conditions where it mediates inflammatory signaling. Rheumatoid arthritis (RA) is an autoimmune disease that not only damages joints, but also causes heightened inflammation and damage to other organ systems such as the eyes, heart, lungs, or skin. Differentiated macrophages and their pro-inflammatory cytokines are abundant in the synovial joints of RA patients (Kinne et al. 2007). Through inhibiting monocyte-to-macrophage differentiation, SIRT1 suppresses the inflammatory response in RA patient synovial joints (Park et al. 2016). One SIRT1 target, miR-223, was shown to be significantly upregulated in macrophages derived from RA patients (Ogando et al. 2016). Furthermore, in a mouse collagen-induced arthritis model, phenotypes of RA, SIRT1 appears to regulate early changes in macrophages that may drive pathogenesis (Li et al. 2022). Taken together, these findings suggest that SIRT1 modulation may be a target for treating and reducing RA symptoms by reprogramming the inflammatory state in RA patients.

Chronic inflammation is also characteristic of neurodegenerative disorders including Parkinson's disease, Alzheimer's disease (AD), amyotrophic lateral sclerosis, and multiple sclerosis. Microglia, resident macrophages of the central nervous system, are believed to exert a protective role in neuroinflammation via clearing cellular debris and repairing tissue injury (Tremblay et al. 2011). Amyloid-beta (Aβ) plaque aggregation drives development of AD and microglia migration to sites of Aβ deposition where they secrete pro-inflammatory cytokines, contributing to an inflammaging phenotype. In the presence of Aβ, cultured microglial cells show increased senescence and dysfunction, which was accompanied by downregulation of SIRT1. Reintroducing SIRT1 into these cells rescued pathological phenotypes in the presence of Aβ (An et al. 2022). Several neurodegenerative diseases are also characterized by aggregated and phosphorylated Tau, a microtubule-associated protein that destabilizes neurons when aggregated. *Sirt1* expression was reduced in microglia during aging, and its loss contributes to memory deficits and elevated IL-1R, which is enhanced even further in mouse models expressing mutant tau that is linked to neurodegenerative disorders (Cho et al. 2015). Collectively, these observations link SIRT1 activity to microglial function, suggesting it could play a role in modulating neuroinflammation in the context of neurodegeneration.

10.4 Concluding Remarks and Future Perspectives

The aging epigenome is characterized by dynamic changes in chromatin, which contribute to disease pathogenesis and regulation of healthspan and lifespan. Immunosenescence is highly associated with the development of several age-related diseases and lifespan. As discussed here, the lysine deacetylase SIRT1, is a key regulator of pathways linked to immunosenescence, including age-dependent transcriptional control, signaling, and metabolism.

Key epigenetic targets of SIRT1 are H3K9 and H4K16 acetylation, which change during aging and are linked to the regulation of longevity, although their causal role in controlling lifespan needs further investigation. More specifically, additional investigation is necessary to define the age-associated effect of SIRT1, H4K16ac, and H3K9ac abundance on the declining immune response with age. The evidence discussed here demonstrates a role for SIRT1 in macrophage function and maintaining a balanced inflammatory state. While the mechanisms through which SIRT1 functions are understood in certain cases, it is not known in all contexts whether SIRT1 is functioning via its activity on histone proteins or on non-histone targets, nor is it fully tested which roles are dependent on potential non-catalytic functions for the enzyme.

To understand the potential of targeting SIRT1 as a modulator of inflammation, it is imperative to clarify the mechanisms through which it functions in multiple aspects of macrophage biology, including its role in immune surveillance, polarization, and inflammation resolution via phagocytosis and autophagy. For example, the findings on non-histone roles of SIRT1 highlight the potential of restoring macrophage function using SIRT1 activators. Further investigation into the anti-inflammatory potential of SIRT1 would provide additional insights into the mechanism by which inflammaging can be ameliorated. Therefore, understanding how SIRT1 contributes to immune dysfunction with age is critical for identifying therapeutic targets to prevent or delay age-related diseases and increase healthy lifespan.

Acknowledgments We thank Shonda Campbell for her careful review and comments on the manuscript and the members of the Green lab for helpful discussions. The figure was created in BioRender (https://BioRender.com/r32v122).

References

Adjei-Mosi J, Sun Q, Smithson SB, Shealy GL, Amerineni KD, Liang Z, Chen H, Wang M, Ping Q, Han J, Morita M, Kamat A, Musi N, Zang M (2023) Age-dependent loss of hepatic SIRT1 enhances NLRP3 inflammasome signaling and impairs capacity for liver fibrosis resolution. Aging Cell 22(5):e13811. https://doi.org/10.1111/acel.13811

Álvarez-Rodríguez L, López-Hoyos M, Muñoz-Cacho P, Martínez-Taboada VM (2012) Aging is associated with circulating cytokine dysregulation. Cell Immunol 273(2):124–132. https://doi.org/10.1016/j.cellimm.2012.01.001

An Y, Li Y, Hou Y, Huang S, Pei G (2022) Alzheimer's Amyloid-β Accelerates Cell Senescence and Suppresses the SIRT1/NRF2 Pathway in Human Microglial Cells. Oxidative Med Cell Longev 2022:1–17. https://doi.org/10.1155/2022/3086010

Andina N, De Meuron L, Schnegg-Kaufmann AS, Sarangdhar MA, Ansermet C, Bombaci G, Batta K, Keller N, Porret NA, Angelillo-Scherrer A, Bonadies N, Allam R (2023) Increased inflammasome activation is associated with aging and chronic myelomonocytic leukemia disease severity. J Immunol 210(5):580–589. https://doi.org/10.4049/jimmunol.2200412

Balan V, Miller GS, Kaplun L, Balan K, Chong Z-Z, Li F, Kaplun A, VanBerkum MFA, Arking R, Freeman DC, Maiese K, Tzivion G (2008) Life span extension and neuronal cell protection by drosophila nicotinamidase. J Biol Chem 283(41):27810–27819. https://doi.org/10.1074/jbc.M804681200

Banerjee KK, Ayyub C, Ali SZ, Mandot V, Prasad NG, Kolthur-Seetharam U (2012) dSir2 in the adult fat body, but not in muscles, regulates life span in a diet-dependent manner. Cell Rep 2(6):1485–1491. https://doi.org/10.1016/j.celrep.2012.11.013

Bauer JH, Morris SNS, Chang C, Flatt T, Wood JG, Helfand SL (2008) dSir2 and Dmp53 interact to mediate aspects of CR-dependent life span extension in *D. melanogaster*. Aging 1(1):38–48. https://doi.org/10.18632/aging.100001

Becker L, Nguyen L, Gill J, Kulkarni S, Pasricha PJ, Habtezion A (2018) Age-dependent shift in macrophage polarisation causes inflammation-mediated degeneration of enteric nervous system. Gut 67(5):827–836. https://doi.org/10.1136/gutjnl-2016-312940

Bodogai M, O'Connell J, Kim K, Kim Y, Moritoh K, Chen C, Gusev F, Vaughan K, Shulzhenko N, Mattison JA, Lee-Chang C, Chen W, Carlson O, Becker KG, Gurung M, Morgun A, White J, Meade T, Perdue K et al (2018) Commensal bacteria contribute to insulin resistance in aging by activating innate B1a cells. Sci Transl Med 10(467):eaat4271. https://doi.org/10.1126/scitranslmed.aat4271

Burnett C, Valentini S, Cabreiro F, Goss M, Somogyvári M, Piper MD, Hoddinott M, Sutphin GL, Leko V, McElwee JJ, Vazquez-Manrique RP, Orfila A-M, Ackerman D, Au C, Vinti G,

Riesen M, Howard K, Neri C, Bedalov A et al (2011) Absence of effects of Sir2 overexpression on lifespan in *C. elegans* and Drosophila. Nature 477(7365):482–485. https://doi.org/10.1038/nature10296

Callender LA, Carroll EC, Beal RWJ, Chambers ES, Nourshargh S, Akbar AN, Henson SM (2018) Human CD 8^+ EMRA T cells display a senescence-associated secretory phenotype regulated by p38 MAPK. Aging Cell 17(1):e12675. https://doi.org/10.1111/acel.12675

Carafa V, Nebbioso A, Altucci L (2012) Sirtuins and disease: the road ahead. Front Pharmacol 3. https://doi.org/10.3389/fphar.2012.00004

Chen I-Y, Lypowy J, Pain J, Sayed D, Grinberg S, Alcendor RR, Sadoshima J, Abdellatif M (2006) Histone H2A.z Is essential for cardiac myocyte hypertrophy but opposed by silent information regulator 2α. J Biol Chem 281(28):19369–19377. https://doi.org/10.1074/jbc.M601443200

Cho S-H, Chen JA, Sayed F, Ward ME, Gao F, Nguyen TA, Krabbe G, Sohn PD, Lo I, Minami S, Devidze N, Zhou Y, Coppola G, Gan L (2015) SIRT1 deficiency in microglia contributes to cognitive decline in aging and neurodegeneration via epigenetic regulation of IL-1β. J Neurosci 35(2):807–818. https://doi.org/10.1523/JNEUROSCI.2939-14.2015

Coppé J-P, Patil CK, Rodier F, Sun Y, Muñoz DP, Goldstein J, Nelson PS, Desprez P-Y, Campisi J (2008) Senescence-associated secretory phenotypes reveal cell-nonautonomous functions of oncogenic RAS and the p53 tumor suppressor. PLoS Biol 6(12):e301. https://doi.org/10.1371/journal.pbio.0060301

Cui J, Shan K, Yang Q, Qi Y, Qu H, Li J, Wang R, Jia L, Chen W, Feng N, Chen YQ (2021) Prostaglandin E_3 attenuates macrophage-associated inflammation and prostate tumour growth by modulating polarization. J Cell Mol Med 25(12):5586–5601. https://doi.org/10.1111/jcmm.16570

Dang W, Steffen KK, Perry R, Dorsey JA, Johnson FB, Shilatifard A, Kaeberlein M, Kennedy BK, Berger SL (2009) Histone H4 lysine 16 acetylation regulates cellular lifespan. Nature 459(7248):802–807. https://doi.org/10.1038/nature08085

Delaney JR, Sutphin GL, Dulken B, Sim S, Kim JR, Robison B, Schleit J, Murakami CJ, Carr D, An EH, Choi E, Chou A, Fletcher M, Jelic M, Liu B, Lockshon D, Moller RM, Pak DN, Peng Q et al (2011) Sir2 deletion prevents lifespan extension in 32 long-lived mutants. Aging Cell 10(6):1089–1091. https://doi.org/10.1111/j.1474-9726.2011.00742.x

Dube CT, Jahan FRS, Lim CY (2022) Key changes in chromatin mark mammalian epidermal differentiation and ageing. Epigenetics 17(4):444–459. https://doi.org/10.1080/15592294.2021.1917812

Dubey SK, Dubey R, Prajapati SC, Jung K, Mohan K, Liu X, Roney J, Tian W, Abney J, Giarmarco MM, Hernandez AG, Liu J, Kleinman ME (2024) Histone deficiency and hypoacetylation in the aging retinal pigment epithelium. Aging Cell e14108:e14108. https://doi.org/10.1111/acel.14108

Eisenberg T, Schroeder S, Andryushkova A, Pendl T, Küttner V, Bhukel A, Mariño G, Pietrocola F, Harger A, Zimmermann A, Moustafa T, Sprenger A, Jany E, Büttner S, Carmona-Gutierrez D, Ruckenstuhl C, Ring J, Reichelt W, Schimmel K et al (2014) Nucleocytosolic depletion of the energy metabolite acetyl-coenzyme A stimulates autophagy and prolongs lifespan. Cell Metab 19(3):431–444. https://doi.org/10.1016/j.cmet.2014.02.010

Fabrizio P, Gattazzo C, Battistella L, Wei M, Cheng C, McGrew K, Longo VD (2005) Sir2 blocks extreme life-span extension. Cell 123(4):655–667. https://doi.org/10.1016/j.cell.2005.08.042

Feser J, Truong D, Das C, Carson JJ, Kieft J, Harkness T, Tyler JK (2010) Elevated histone expression promotes life span extension. Mol Cell 39(5):724–735. https://doi.org/10.1016/j.molcel.2010.08.015

Franceschi C, Bonafè M, Valensin S, Olivieri F, De Luca M, Ottaviani E, De Benedictis G (2000) Inflamm-aging: an Evolutionary Perspective on Immunosenescence. Ann N Y Acad Sci 908(1):244–254. https://doi.org/10.1111/j.1749-6632.2000.tb06651.x

Gate D, Saligrama N, Leventhal O, Yang AC, Unger MS, Middeldorp J, Chen K, Lehallier B, Channappa D, De Los Santos MB, McBride A, Pluvinage J, Elahi F, Tam GK-Y, Kim Y, Greicius M, Wagner AD, Aigner L, Galasko DR et al (2020) Clonally expanded CD8 T cells patrol the cerebrospinal fluid in Alzheimer's disease. Nature 577(7790):399–404. https://doi.org/10.1038/s41586-019-1895-7

Gong H, Qian H, Ertl R, Astle CM, Wang GG, Harrison DE, Xu X (2015) Histone modifications change with age, dietary restriction and rapamycin treatment in mouse brain. Oncotarget 6(18):15882–15890. https://doi.org/10.18632/oncotarget.4137

Hachinohe M, Hanaoka F, Masumoto H (2011) Hst3 and Hst4 histone deacetylases regulate replicative lifespan by preventing genome instability in Saccharomyces cerevisiae: Hst3 and Hst4 histone deacetylases regulate replicative lifespan. Genes Cells 16(4):467–477. https://doi.org/10.1111/j.1365-2443.2011.01493.x

Haigis MC, Sinclair DA (2010) Mammalian sirtuins: biological insights and disease relevance. Ann Rev Pathol Mechan Dis 5(1):253–295. https://doi.org/10.1146/annurev.pathol.4.110807.092250

Hall BM, Balan V, Gleiberman AS, Strom E, Krasnov P, Virtuoso LP, Rydkina E, Vujcic S, Balan K, Gitlin I, Leonova K, Polinsky A, Chernova OB, Gudkov AV (2016) Aging of mice is associated with p16(Ink4a)- and β-galactosidase-positive macrophage accumulation that can be induced in young mice by senescent cells. Aging 8(7):1294–1315. https://doi.org/10.18632/aging.100991

He Y, Zeng MY, Yang D, Motro B, Núñez G (2016) NEK7 is an essential mediator of NLRP3 activation downstream of potassium efflux. Nature 530(7590):354–357. https://doi.org/10.1038/nature16959

Herranz D, Muñoz-Martin M, Cañamero M, Mulero F, Martinez-Pastor B, Fernandez-Capetillo O, Serrano M (2010) Sirt1 improves healthy ageing and protects from metabolic syndrome-associated cancer. Nature Commun 1(1). https://doi.org/10.1038/ncomms1001

Hoffmann J, Romey R, Fink C, Yong L, Roeder T (2013) Overexpression of Sir2 in the adult fat body is sufficient to extend lifespan of male and female Drosophila. Aging 5(4):315–327. https://doi.org/10.18632/aging.100553

Horn L, Leips J, Starz-Gaiano M (2014) Phagocytic ability declines with age in adult D rosophila hemocytes. Aging Cell 13(4):719–728. https://doi.org/10.1111/acel.12227

Hui X, Zhang M, Gu P, Li K, Gao Y, Wu D, Wang Y, Xu A (2017) Adipocyte SIRT 1 controls systemic insulin sensitivity by modulating macrophages in adipose tissue. EMBO Rep 18(4):645–657. https://doi.org/10.15252/embr.201643184

Imperatore F, Maurizio J, Vargas Aguilar S, Busch CJ, Favret J, Kowenz-Leutz E, Cathou W, Gentek R, Perrin P, Leutz A, Berruyer C, Sieweke MH (2017) SIRT1 regulates macrophage self-renewal. EMBO J 36(16):2353–2372. https://doi.org/10.15252/embj.201695737

Jia Y, Han S, Li J, Wang H, Liu J, Li N, Yang X, Shi J, Han J, Li Y, Bai X, Su L, Hu D (2017) IRF8 is the target of SIRT1 for the inflammation response in macrophages. Innate Immun 23(2):188–195. https://doi.org/10.1177/1753425916683751

Jia Y, Shen K, Yang Y, Cai W, Wang K, Zhang Y, Zhao M, Luo L, Su L, Guan H, Hu D (2022) Deacetylation of IRF8 inhibits iNOS expression and inflammation via SIRT1 in macrophages. Immunobiology 227(6):152300. https://doi.org/10.1016/j.imbio.2022.152300

Ka S-O, Song M-Y, Bae EJ, Park B-H (2015) Myeloid SIRT1 regulates macrophage infiltration and insulin sensitivity in mice fed a high-fat diet. J Endocrinol 224(2):109–118. https://doi.org/10.1530/JOE-14-0527

Kaeberlein M, McVey M, Guarente L (1999) The SIR2/3/4 complex and SIR2 alone promote longevity in Saccharomyces cerevisiae by two different mechanisms. Genes Dev 13(19):2570–2580. https://doi.org/10.1101/gad.13.19.2570

Kawakami K, Nakamura A, Ishigami A, Goto S, Takahashi R (2009) Age-related difference of site-specific histone modifications in rat liver. Biogerontology 10(4):415–421. https://doi.org/10.1007/s10522-008-9176-0

Kinne RW, Stuhlmüller B, Burmester G-R (2007) Cells of the synovium in rheumatoid arthritis. Macroph Arthritis Res Therapy 9(6):224. https://doi.org/10.1186/ar2333

Li T, Garcia-Gomez A, Morante-Palacios O, Ciudad L, Özkaramehmet S, Van Dijck E, Rodríguez-Ubreva J, Vaquero A, Ballestar E (2020) SIRT1/2 orchestrate acquisition of DNA methylation and loss of histone H3 activating marks to prevent premature activation of inflammatory genes in macrophages. Nucleic Acids Res 48(2):665–681. https://doi.org/10.1093/nar/gkz1127

Li L, Freitag J, Asbrand C, Munteanu B, Wang B-T, Zezina E, Didier M, Thill G, Rocher C, Herrmann M, Biesemann N (2022) Multi-omics profiling of collagen-induced arthritis mouse model reveals early metabolic dysregulation via SIRT1 axis. Sci Rep 12(1):11830. https://doi.org/10.1038/s41598-022-16005-9

Li X, Li C, Zhang W, Wang Y, Qian P, Huang H (2023) Inflammation and aging: signaling pathways and intervention therapies. Signal Transduct Target Ther 8(1):239. https://doi.org/10.1038/s41392-023-01502-8

Liu B, Ghosh S, Yang X, Zheng H, Liu X, Wang Z, Jin G, Zheng B, Kennedy BK, Suh Y, Kaeberlein M, Tryggvason K, Zhou Z (2012) Resveratrol rescues SIRT1-dependent adult stem cell decline and alleviates progeroid features in laminopathy-based progeria. Cell Metab 16(6):738–750. https://doi.org/10.1016/j.cmet.2012.11.007

Liu L, Zhu L, Liang Q, Yu L, Hu L, Yu Y, Cheng X, Bao H (2024) Tissue-resident C1q + macrophages exert anti-aging potential through the Sirt1 pathway. Inflamm Res 73(7):1069–1080. https://doi.org/10.1007/s00011-024-01883-8

López-Otín C, Blasco MA, Partridge L, Serrano M, Kroemer G (2023) Hallmarks of aging: an expanding universe. Cell 186(2):243–278. https://doi.org/10.1016/j.cell.2022.11.001

Ma Z, Qiu L, Sun J, Wu Z, Liang S, Zhao Y, Yang J, Yue S, Hu M, Li Y (2024) A novel pulmonary fibrosis NOD/SCID murine model with natural aging. BMC Pulm Med 24(1):457. https://doi.org/10.1186/s12890-024-03268-3

Mahbub S, Deburghgraeve CR, Kovacs EJ (2012) Advanced age impairs macrophage polarization. J Interf Cytokine Res 32(1):18–26. https://doi.org/10.1089/jir.2011.0058

Marín-Aguilar F, Lechuga-Vieco AV, Alcocer-Gómez E, Castejón-Vega B, Lucas J, Garrido C, Peralta-Garcia A, Pérez-Pulido AJ, Varela-López A, Quiles JL, Ryffel B, Flores I, Bullón P, Ruiz-Cabello J, Cordero MD (2020) NLRP3 inflammasome suppression improves longevity and prevents cardiac aging in male mice. Aging Cell 19(1):e13050. https://doi.org/10.1111/acel.13050

Matacchione G, Perugini J, Di Mercurio E, Sabbatinelli J, Prattichizzo F, Senzacqua M, Storci G, Dani C, Lezoche G, Guerrieri M, Giordano A, Bonafè M, Olivieri F (2022) Senescent macrophages in the human adipose tissue as a source of inflammaging. GeroScience 44(4):1941–1960. https://doi.org/10.1007/s11357-022-00536-0

Meng Y-F, Pu Q, Dai S-Y, Ma Q, Li X, Zhu W (2021) Nicotinamide mononucleotide alleviates hyperosmolarity-induced IL-17a secretion and macrophage activation in corneal epithelial cells/macrophage co-culture system. J Inflamm Res 14:479–493. https://doi.org/10.2147/JIR.S292764

Moss CE, Johnston SA, Kimble JV, Clements M, Codd V, Hamby S, Goodall AH, Deshmukh S, Sudbery I, Coca D, Wilson HL, Kiss-Toth E (2024) Aging-related defects in macrophage function are driven by MYC and USF1 transcriptional programs. Cell Rep 43(4):114073. https://doi.org/10.1016/j.celrep.2024.114073

Mostoslavsky R, Chua KF, Lombard DB, Pang WW, Fischer MR, Gellon L, Liu P, Mostoslavsky G, Franco S, Murphy MM, Mills KD, Patel P, Hsu JT, Hong AL, Ford E, Cheng H-L, Kennedy C, Nunez N, Bronson R et al (2006) Genomic Instability and aging-like phenotype in the absence of mammalian SIRT6. Cell 124(2):315–329. https://doi.org/10.1016/j.cell.2005.11.044

Motta MC, Divecha N, Lemieux M, Kamel C, Chen D, Gu W, Bultsma Y, McBurney M, Guarente L (2004) Mammalian SIRT1 represses forkhead transcription factors. Cell 116(4):551–563. https://doi.org/10.1016/S0092-8674(04)00126-6

Murakami Y, Mizoguchi F, Saito T, Miyasaka N, Kohsaka H (2012) p16INK4a exerts an anti-inflammatory effect through accelerated irak1 degradation in macrophages. J Immunol 189(10):5066–5072. https://doi.org/10.4049/jimmunol.1103156

Nativio R, Donahue G, Berson A, Lan Y, Amlie-Wolf A, Tuzer F, Toledo JB, Gosai SJ, Gregory BD, Torres C, Trojanowski JQ, Wang L-S, Johnson FB, Bonini NM, Berger SL (2018)

Dysregulation of the epigenetic landscape of normal aging in Alzheimer's disease. Nat Neurosci 21(4):497–505. https://doi.org/10.1038/s41593-018-0101-9
Ogando J, Tardáguila M, Díaz-Alderete A, Usategui A, Miranda-Ramos V, Martínez-Herrera DJ, De La Fuente L, García-León MJ, Moreno MC, Escudero S, Cañete JD, Toribio ML, Cases I, Pascual-Montano A, Pablos JL, Mañes S (2016) Notch-regulated miR-223 targets the aryl hydrocarbon receptor pathway and increases cytokine production in macrophages from rheumatoid arthritis patients. Sci Rep 6(1):20223. https://doi.org/10.1038/srep20223
Park SY, Lee SW, Kim HY, Lee SY, Lee WS, Hong KW, Kim CD (2016) SIRT1 inhibits differentiation of monocytes to macrophages: amelioration of synovial inflammation in rheumatoid arthritis. J Mol Med 94(8):921–931. https://doi.org/10.1007/s00109-016-1402-7
Park SY, Lee SW, Lee SY, Hong KW, Bae SS, Kim K, Kim CD (2017) SIRT1/adenosine monophosphate-activated protein kinase α signaling enhances macrophage polarization to an anti-inflammatory phenotype in rheumatoid arthritis. Front Immunol 8:1135. https://doi.org/10.3389/fimmu.2017.01135
Paterson N, Lämmermann T (2022) Macrophage network dynamics depend on haptokinesis for optimal local surveillance. eLife 11:e75354. https://doi.org/10.7554/eLife.75354
Peleg S, Feller C, Ladurner AG, Imhof A (2016) The metabolic impact on histone acetylation and transcription in ageing. Trends Biochem Sci 41(8):700–711. https://doi.org/10.1016/j.tibs.2016.05.008
Piragine E, De Felice M, Germelli L, Brinkmann V, Flori L, Martini C, Calderone V, Ventura N, Da Pozzo E, Testai L (2024) The citrus flavanone naringenin prolongs the lifespan in *C. elegans* and slows signs of brain aging in mice. Exp Gerontol 194:112495. https://doi.org/10.1016/j.exger.2024.112495
Rodrigues LP, Teixeira VR, Alencar-Silva T, Simonassi-Paiva B, Pereira RW, Pogue R, Carvalho JL (2021) Hallmarks of aging and immunosenescence: connecting the dots. Cytokine Growth Factor Rev 59:9–21. https://doi.org/10.1016/j.cytogfr.2021.01.006
Rogina B, Helfand SL (2004) Sir2 mediates longevity in the fly through a pathway related to calorie restriction. Proc Natl Acad Sci 101(45):15998–16003. https://doi.org/10.1073/pnas.0404184101
Roichman A, Elhanati S, Aon MA, Abramovich I, Di Francesco A, Shahar Y, Avivi MY, Shurgi M, Rubinstein A, Wiesner Y, Shuchami A, Petrover Z, Lebenthal-Loinger I, Yaron O, Lyashkov A, Ubaida-Mohien C, Kanfi Y, Lerrer B, Fernández-Marcos PJ et al (2021) Restoration of energy homeostasis by SIRT6 extends healthy lifespan. Nat Commun 12(1):3208. https://doi.org/10.1038/s41467-021-23545-7
Roth GA, Abate D, Abate KH, Abay SM, Abbafati C, Abbasi N, Abbastabar H, Abd-Allah F, Abdela J, Abdelalim A, Abdollahpour I, Abdulkader RS, Abebe HT, Abebe M, Abebe Z, Abejie AN, Abera SF, Abil OZ, Abraha HN et al (2018) Global, regional, and national age-sex-specific mortality for 282 causes of death in 195 countries and territories, 1980–2017: a systematic analysis for the Global Burden of Disease Study 2017. Lancet 392(10159):1736–1788. https://doi.org/10.1016/S0140-6736(18)32203-7
Ryall JG, Dell'Orso S, Derfoul A, Juan A, Zare H, Feng X, Clermont D, Koulnis M, Gutierrez-Cruz G, Fulco M, Sartorelli V (2015) The NAD+-dependent SIRT1 deacetylase translates a metabolic switch into regulatory epigenetics in skeletal muscle stem cells. Cell Stem Cell 16(2):171–183. https://doi.org/10.1016/j.stem.2014.12.004
Satoh A, Brace CS, Rensing N, Cliften P, Wozniak DF, Herzog ED, Yamada KA, Imai S (2013) Sirt1 extends life span and delays aging in mice through the regulation of Nk2 homeobox 1 in the DMH and LH. Cell Metab 18(3):416–430. https://doi.org/10.1016/j.cmet.2013.07.013
Schafer MJ, Zhang X, Kumar A, Atkinson EJ, Zhu Y, Jachim S, Mazula DL, Brown AK, Berning M, Aversa Z, Kotajarvi B, Bruce CJ, Greason KL, Suri RM, Tracy RP, Cummings SR, White TA, LeBrasseur NK (2020) The senescence-associated secretome as an indicator of age and medical risk. JCI Insight 5(12):e133668. https://doi.org/10.1172/jci.insight.133668

Schug TT, Xu Q, Gao H, Peres-da-Silva A, Draper DW, Fessler MB, Purushotham A, Li X (2010) Myeloid deletion of SIRT1 induces inflammatory signaling in response to environmental stress. Mol Cell Biol 30(19):4712–4721. https://doi.org/10.1128/MCB.00657-10

Sun S, Qin W, Tang X, Meng Y, Hu W, Zhang S, Qian M, Liu Z, Cao X, Pang Q, Zhao B, Wang Z, Zhou Z, Liu B (2020) Vascular endothelium–targeted *Sirt7* gene therapy rejuvenates blood vessels and extends life span in a Hutchinson-Gilford progeria model. Sci Adv 6(8):eaay5556. https://doi.org/10.1126/sciadv.aay5556

Takeda-Watanabe A, Kitada M, Kanasaki K, Koya D (2012) SIRT1 inactivation induces inflammation through the dysregulation of autophagy in human THP-1 cells. Biochem Biophys Res Commun 427(1):191–196. https://doi.org/10.1016/j.bbrc.2012.09.042

Tissenbaum HA, Guarente L (2001) Increased dosage of a sir-2 gene extends lifespan in Caenorhabditis elegans. Nature 410(6825):227–230. https://doi.org/10.1038/35065638

Tremblay M-È, Stevens B, Sierra A, Wake H, Bessis A, Nimmerjahn A (2011) The role of microglia in the healthy brain: figure 1. J Neurosci 31(45):16064–16069. https://doi.org/10.1523/JNEUROSCI.4158-11.2011

Vaquero A, Scher M, Lee D, Erdjument-Bromage H, Tempst P, Reinberg D (2004) Human SirT1 Interacts with Histone H1 and Promotes Formation of Facultative Heterochromatin. Mol Cell 16(1):93–105. https://doi.org/10.1016/j.molcel.2004.08.031

Vaziri H, Dessain SK, Eaton EN, Imai S-I, Frye RA, Pandita TK, Guarente L, Weinberg RA (2001) hSIR2SIRT1 functions as an NAD-dependent p53 Deacetylase. Cell 107(2):149–159. https://doi.org/10.1016/S0092-8674(01)00527-X

Walford RL (1964) The immunologic theory of aging. The Gerontologist 4(4):195–197. https://doi.org/10.1093/geront/4.4.195

Whitaker R, Faulkner S, Miyokawa R, Burhenn L, Henriksen M, Wood JG, Helfand SL (2013) Increased expression of Drosophila Sir 2 extends life span in a dose-dependent manner. Aging 5(9):682–691. https://doi.org/10.18632/aging.100599

Widjaja AA, Lim W-W, Viswanathan S, Chothani S, Corden B, Dasan CM, Goh JWT, Lim R, Singh BK, Tan J, Pua CJ, Lim SY, Adami E, Schafer S, George BL, Sweeney M, Xie C, Tripathi M, Sims NA et al (2024) Inhibition of IL-11 signalling extends mammalian healthspan and lifespan. Nature 632(8023):157–165. https://doi.org/10.1038/s41586-024-07701-9

Wong CK, Smith CA, Sakamoto K, Kaminski N, Koff JL, Goldstein DR (2017) Aging impairs alveolar macrophage phagocytosis and increases influenza-induced mortality in mice. J Immunol 199(3):1060–1068. https://doi.org/10.4049/jimmunol.1700397

Wu C-Y, Hua K-F, Hsu W-H, Suzuki Y, Chu LJ, Lee Y-C, Takahata A, Lee S-L, Wu C-C, Nikolic-Paterson DJ, Ka S-M, Chen A (2020) IgA nephropathy benefits from compound K treatment by inhibiting NF-κB/NLRP3 inflammasome and enhancing autophagy and SIRT1. J Immunol 205(1):202–212. https://doi.org/10.4049/jimmunol.1900284

Wu G, Su Q, Li J, Xue C, Zhu J, Cai Q, Huang J, Ji S, Cheng B, Ge H (2023) NAMPT encapsulated by extracellular vesicles from young adipose-derived mesenchymal stem cells treated tendinopathy in a "One-Stone-Two-Birds" manner. J Nanobiotechnol 21(1):7. https://doi.org/10.1186/s12951-022-01763-5

Xia D-Y, Yuan J-L, Jiang X-C, Qi M, Lai N-S, Wu L-Y, Zhang X-S (2021) SIRT1 promotes M2 microglia polarization via reducing ROS-mediated NLRP3 inflammasome signaling after subarachnoid hemorrhage. Front Immunol 12:770744. https://doi.org/10.3389/fimmu.2021.770744

Yang Y, Peng W, Su X, Yue B, Shu S, Wang J, Fu C, Zhong J, Wang H (2023) Epigenomics analysis of the suppression role of SIRT1 via H3K9 deacetylation in preadipocyte differentiation. Int J Mol Sci 24(14):11281. https://doi.org/10.3390/ijms241411281

Yeung F, Hoberg JE, Ramsey CS, Keller MD, Jones DR, Frye RA, Mayo MW (2004) Modulation of NF-κB-dependent transcription and cell survival by the SIRT1 deacetylase. EMBO J 23(12):2369–2380. https://doi.org/10.1038/sj.emboj.7600244

Youm Y-H, Grant RW, McCabe LR, Albarado DC, Nguyen KY, Ravussin A, Pistell P, Newman S, Carter R, Laque A, Münzberg H, Rosen CJ, Ingram DK, Salbaum JM, Dixit VD (2013)

Canonical Nlrp3 inflammasome links systemic Low-grade inflammation to functional decline in aging. Cell Metab 18(4):519–532. https://doi.org/10.1016/j.cmet.2013.09.010
Yuan J, Minter-Dykhouse K, Lou Z (2009) A c-Myc–SIRT1 feedback loop regulates cell growth and transformation. J Cell Biol 185(2):203–211. https://doi.org/10.1083/jcb.200809167
Zampino M, Brennan NA, Kuo P-L, Spencer RG, Fishbein KW, Simonsick EM, Ferrucci L (2020) Poor mitochondrial health and systemic inflammation? Test of a classic hypothesis in the Baltimore Longitudinal Study of Aging. GeroScience 42(4):1175–1182. https://doi.org/10.1007/s11357-020-00208-x
Zhang R, Chen H-Z, Liu J-J, Jia Y-Y, Zhang Z-Q, Yang R-F, Zhang Y, Xu J, Wei Y-S, Liu D-P, Liang C-C (2010) SIRT1 suppresses activator protein-1 transcriptional activity and cyclooxygenase-2 expression in macrophages. J Biol Chem 285(10):7097–7110. https://doi.org/10.1074/jbc.M109.038604
Zhang Y, Li Y, Zhang Y, Zhang Z, Zhang D, Wang X, Lai B, Huang D, Gu L, Xie Y, Miao Y (2022) Genome-wide H3K9 acetylation level increases with age-dependent senescence of flag leaf in rice. J Exp Bot 73(14):4696–4715. https://doi.org/10.1093/jxb/erac155
Zhao X, Li M, Lu Y, Wang M, Xiao J, Xie Q, He X, Shuai S (2024) Sirt1 inhibits macrophage polarization and inflammation in gouty arthritis by inhibiting the MAPK/NF-κB/AP-1 pathway and activating the Nrf2/HO-1 pathway. Inflamm Res 73(7):1173–1184. https://doi.org/10.1007/s00011-024-01890-9
Zheng Y, Kou J, Wang P, Ye T, Wang Z, Gao Z, Cong L, Li M, Dong B, Yang W, Li Q, Li H, Wang R, Yang L (2021) Berberine-induced TFEB deacetylation by SIRT1 promotes autophagy in peritoneal macrophages. Aging 13(5):7096–7119. https://doi.org/10.18632/aging.202566
Zhou S, Zhao X, Wu L, Yan R, Sun L, Zhang Q, Gong C, Liu Y, Xiang L, Li S, Wang P, Yang Y, Ren W, Jiang J, Yang Y (2023) Parishin treatment alleviates cardiac aging in naturally aged mice. Heliyon 9(12):e22970. https://doi.org/10.1016/j.heliyon.2023.e22970

Chapter 11
Repurposing Histone Deacetylase Inhibitors for Management of Solid Organ Transplant Rejection

Marta Halasa, Anna Wawruszak, and Malgorzata Kloc

Abstract Histone deacetylase inhibitors (HDACi) have gained attention for their potential therapeutic effects in various immune-related diseases and solid organ transplantation. This review focuses on the role of HDACs' expression in immune responses, especially of T and B cells, in organ transplant rejection. We examined the expression levels of HDACs in T and B cells across different immune states, providing insight into their regulatory functions during immune activation and tolerance. Through an analysis of in vitro, in vivo, and preclinical studies, we explored the effects of HDACi on immune modulation, highlighting their impact on T- and B-cell function in diseases other than cancer. Finally, we discussed the outcomes of HDACi treatment in various organ transplantation models, assessing their potential to mitigate rejection, promote tolerance, and improve graft survival. The review synthesizes current evidence, identifies knowledge gaps, and offers a comprehensive outlook on HDACi for clinical applications in organ transplantation.

M. Halasa (✉)
Transplant Immunology, Houston Methodist Research Institute, Houston, TX, USA

Department of Surgery, Houston Methodist Hospital, University of Texas, Houston, TX, USA
e-mail: mhalasa@houstonmethodist.org

A. Wawruszak
Department of Biochemistry and Molecular Biology, Medical University of Lublin, Lublin, Poland

M. Kloc
Transplant Immunology, Houston Methodist Research Institute, Houston, TX, USA

Department of Surgery, Houston Methodist Hospital, University of Texas, Houston, TX, USA

Department of Genetics, University of Texas, MD Anderson Cancer Center, Houston, TX, USA

M. Halasa, A. Wawruszak (eds.), *Histone and Non-Histone Reversible Acetylation in Development, Aging and Disease*, Results and Problems in Cell Differentiation 75,
https://doi.org/10.1007/978-3-031-91459-1_11

11.1 Introduction

Solid organ transplantation (SOT) is a life-saving procedure for patients with end-stage diseases and/or fatal organ failure. After organ transplantation, patients must take immunosuppressive drugs for the rest of their lives to prevent organ rejection. The ideal immunosuppression approach assumes the induction of donor-specific tolerance without reducing host defenses, which makes patients less susceptible to infections and other immune-related diseases (Vaillant and Mohseni 2024). However, so far, the available post-transplant therapies do not meet expectations; thus, chronic rejection remains an unsolved challenge in modern transplantation. The main goals of the commonly used immunosuppressive regimens after SOT are to prevent T-cell proliferation, decrease their cytotoxic effect on allografts, and prevent B-cell-mediated production of antibodies. Post-transplant treatment typically involves a combination of medications such as calcineurin inhibitors, corticosteroids, antimetabolites, and mTOR inhibitors. The specific regimen depends on the type of organ and the individual needs of the patient (Nelson et al. 2022). While prevention of acute rejection (days/weeks after transplantation) has reached very high survival rates (Vaillant et al. 2024), the long-term (chronic) rejection significantly affects the post-transplant patient's outcomes. Chronic rejection accelerates arteriosclerosis and luminal narrowing of graft vessels and usually causes progressive parenchymal fibrosis (Kloc and Ghobrial 2014).

Allograft rejection is largely mediated by T lymphocytes or IgG alloantibodies, leading to T-cell-mediated rejection (TCMR) or antibody-mediated rejection (ABMR), respectively, where both types of rejection could occur separately or simultaneously. TCMR can develop weeks to years after transplantation, with high response rates to immunosuppression (Lusco et al. 2016). Molecularly, CD8+ T lymphocytes are activated in secondary lymphoid organs by antigen-presenting cells presenting donor antigens *via* major histocompatibility complex class I (MHC I). After activation, effector CD8+ T cells migrate into the graft tissue where they recognize donor antigens and execute cytotoxic functions. Activated CD8+ T cells start proliferating, differentiating, activating their effector functions, and releasing pro-inflammatory factors, which will activate other immune cells and, concurrently, cause target cell apoptosis (Short et al. 2023; Yap et al. 2015). For example, interferon γ (INF-γ) secreted by CD8+ T cells potentiates macrophage activation and their polarization into pro-inflammatory phenotype, contributing to organ rejection (Cantisán et al. 2013; Günthner and Anders 2013). Simultaneously, CD8+ T cells produce interleukin 2 (IL-2), which creates a positive feedback loop as they require IL-2 for extensive proliferation (Ross and Cantrell 2018).

While CD8+ T cells bind to MHC I, the T helper cells expressing CD4 antigen (CD4+ T cells) recognize antigens presented by MHC class II (MHC II) molecules, typically expressed by antigen-presenting cells such as macrophages, dendric cells, and B cells (Künzli and Masopust 2023). Upon activation, CD4+ T cells differentiate into cytokine-dependent specific subpopulations with different phenotypes and functions. Naïve CD4+ T cells differentiate into Th1 (pro-inflammatory) or Th2 (anti-inflammatory) cells. Th1 cells secrete IL-2, which enhances the proliferation of CD8+ T cells, activates B cells to produce alloreactive antibodies, and directly damages allografts through Fas/FasL-mediated signaling (Borst et al. 2018; Luckheeram et al. 2012; Akiyama et al. 2012). Th2 cells secrete a bunch of cytokines, including interleukin 4 (IL-4), interleukin 5 (IL-5), interleukin 9 (IL-9), and interleukin 13 (IL-13). IL-4 activates B cells and, together with IL-13, potentiates macrophage differentiation into pro-fibrotic phenotype (Walker and McKenzie 2017; Junttila 2018). Immune responses driven by Th1 and Th2 cells can regulate each other's activity, although both pathways may coexist during allograft rejection. Another subpopulation comprises Th17 cells, which secrete interleukin 17 (IL-17). They differentiate from naïve CD4+ T cells under distinct cytokine conditions, following a separate differentiation pathway from Th1 and Th2 cells. Biopsies of early post-lung transplants contain high levels of IL-17 (Snell et al. 2007). The suppression of T effector cells (including both CD4+ and CD8+ T cells) and, thus, prevention of autoimmunity is driven by regulatory T cells (Treg) expressing transcription factor forkhead box P3 (Foxp3). The Treg population is heterogeneous and can suppress recipient immune response to allograft via direct interaction with T cells and APCs cells and/or secreting immunosuppressive cytokines (IL-10, IL-35, and transforming growth factor beta (TGF-β)) (Wood et al. 2012; Cassano et al. 2023). Treg cells could compete with CD8+ T cells for IL-2. In this competitive scenario, Tregs often have a higher expression of CD25 (part of the IL-2 receptor), giving them an advantage in IL-2 uptake. By effectively utilizing IL-2, Tregs can promote their survival and function while potentially limiting the availability of this cytokine for CD8+ T cells, thereby influencing the balance between regulatory and effector T-cell responses (Shevach 2012; McNally et al. 2011). For simplicity, we only distinguished three central subpopulations of T helper cells and described the general role of T regulatory cells. However, the CD4+ T cell lineage is more complex, as it was nicely revised here (Short et al. 2023).

The primary drivers of ABMR are B cells. After differentiation into antibody-secreting plasma cells (PCs), they start producing antibodies against donor antigens, particularly human leukocyte antigens (HLAs), as well as non-donor-specific antigens (Schmitz et al. 2020). The presence of anti-HLA donor-specific antibodies (DSAs) can lead to acute and chronic rejection of transplanted organs, including hearts, lungs, kidneys, and livers (Sutanto et al. 2024; Schinstock et al. 2017; Wang et al. 2023; Zhang et al. 2018). Molecularly, DSAs secreted by B cells bind to the allograft endothelium (they interact with cell surface antigens), resulting in direct injury or recruiting of other immune cells such as natural killer (NK) cells and monocytes (Drachenberg and Papadimitriou 2013; Wei et al. 2020; Calabrese et al. 2023).

Solid organ rejection is complex and involves various immune and non-immune cells. In this chapter, we focus specifically on the impact of histone deacetylases (HDACs) on T and B cells, as they play crucial roles in TCMR and ABMR. Figure 11.1 illustrates T and B cells' molecular responses to organ transplantation.

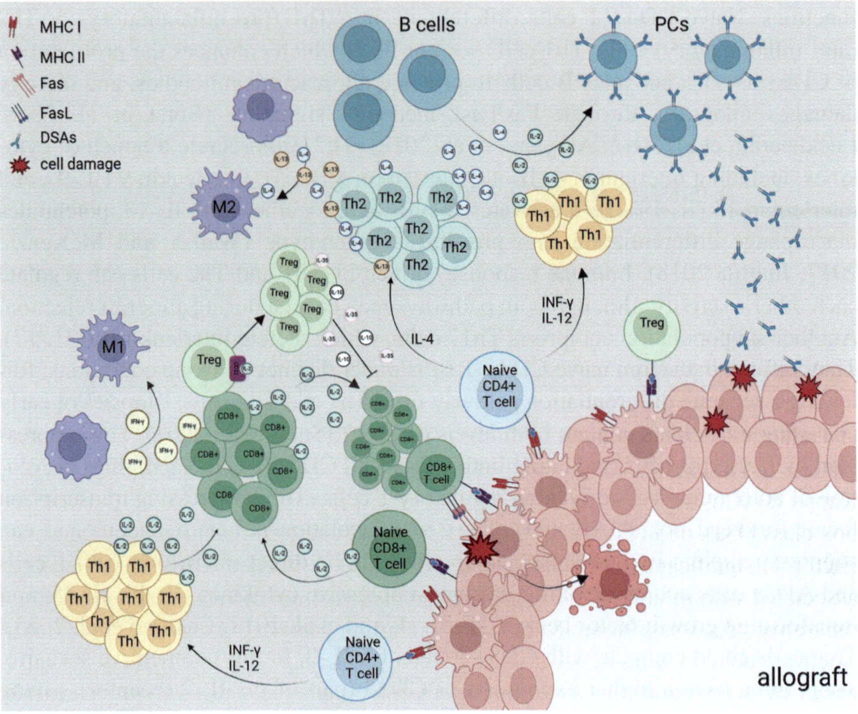

Fig. 11.1 T and B cell molecular responses to organ transplantation. Naïve $CD8^+$ and $CD4^+$ T cells recognize MHC class I and MHC class II molecules, respectively, presented on the surface of donor antigen-presenting cells (APCs). Upon activation, $CD8^+$ T cells undergo clonal expansion and secrete IL-2, which further enhances their proliferation, as well as IFN-γ, which promotes pro-inflammatory (M1-like) macrophage activation. Activated $CD8^+$ T cells can induce apoptosis of donor APCs via the FasL/Fas signaling pathway. Activated $CD4^+$ T cells differentiate into Th1 or Th2 subsets. Th1 cells secrete IL-2 and IFN-γ, supporting $CD8^+$ T cell responses, whereas Th2 cells produce IL-4 and IL-13, promoting polarization of macrophages toward a pro-fibrotic (M2-like) phenotype. B cells, stimulated by Th1- and Th2- derived cytokines, differentiate into plasma cells that produce donor-specific antibodies (DSAs), which contribute to graft injury by targeting the allograft endothelium. Regulatory T cells (Tregs) suppress effector T cell responses through cell–cell contact and inhibitory cytokines such as IL-10 and IL-35. Tregs also express high levels of CD25, allowing them to compete for IL-2 and limit $CD8^+$ T cell proliferation. In addition, Tregs may interact with APCs via MHC II binding, reducing their capacity to activate effector T cells. Created with BioRender.com

11.2 Histone Deacetylase Inhibitors in Clinics and Transplantation-Oriented Clinical Trials

Reversible acetylation of histones and non-histone proteins mediated by histone acetyltransferases (HATs) and histone deacetylases (HDACs) is critical in maintaining cellular homeostasis. The abnormal activity of HDACs is often associated with diseases and is the most intensively studied in cancer (Hałasa et al. 2019; Wang et al. 2020a; He et al. 2020; Tan et al. 2020). HDACs are divided into four classes: class I consists of HDAC1,2,3 and 8; class II consists of HDAC4,5,7,9 (subclass IIa) and HDAC6,10 (subclass IIb), class III consists of Sirtuins (SIRT1-SIRT7), and class IV has only one member: HDAC 11 (Wawruszak et al. 2019). According to The Human Protein Atlas (https://www.proteinatlas.org), class I and II of HDAC and Sirtuins show low tissue specificity, except *HDAC4*, with enhanced expression in skeletal muscle, and *SIRT2,* with enhanced expression in the brain, skeletal muscle, and tongue (Uhlén et al. 2015). At the cellular level, classes I and IV of HDACs are primarily localized in the nucleus, with HDAC1 and HDAC3 capable of shuttling between the nucleus and cytoplasm. Class IIa HDACs can also move between these compartments, while class IIb HDACs are predominantly found in the cytoplasm. In contrast, sirtuins (SIRT1, SIRT2, SIRT6, and SIRT7) are mainly localized in the nucleus, although SIRT1 and SIRT2 can also translocate to the cytoplasm. Meanwhile, SIRT3, SIRT4, and SIRT5 are specifically located in the mitochondria (He et al. 2023).

Due to HDACs' extensive function within the cells, the regulation of their activity has attracted the attention of many researchers, and histone deacetylase inhibitors (HDACi) have become the subject of many pre-clinical and clinical studies. For now, only 4 HDACi have been approved by the FDA for cutaneous T-cell lymphoma (Vorinostat and Romidepsin), T-cell lymphoma (Belinostat), and refractory multiple myeloma (Panobinostat) (Squarzoni et al. 2022). All FDA-approved HDACi are unselective (pan-inhibitors), and their mechanism of action results in the accumulation of acetylated histones (subsequently inducing alteration of gene expression) and non-histone proteins, which eventually leads to cell cycle arrest and cell apoptosis (Eckschlager et al. 2017; Kim and Bae 2010). While HDACi are effective in hematological cancers, some are also tested in clinical trials for solid tumors in combination with other chemotherapeutics (NCT02890069, NCT00719875). HDACi have also been approved for non-cancer diseases. Valproic acid (VPA) is broadly used in neurological and psychiatric disorders. This pan-inhibitor not only induces changes in histone acetylation levels but also inhibits voltage-gated sodium channels (blocking abnormal electrical impulses generated by neurons) and enhances the synthesis of gamma-aminobutyric (GABA), the primary inhibitory neurotransmitter in the brain (Chateauvieux et al. 2024). Just recently, the FDA has approved Giovinostat, another pan-inhibitor, for Duchenne muscular dystrophy, which seems to regenerate muscle fibers (Mercuri et al. 2024). HDAC pan-inhibitors, however, have severe dosage limitations due to their high toxicity; thus, HDAC-selective inhibitors are under deep pre-clinical and clinical investigation (Zamperla et al. 2024; Steimbach et al. 2022; Yang et al. 2019; Bondarev et al. 2021).

While no HDACi are currently being studied clinically to prevent allograft rejection after SOT, some have been evaluated in the context of autologous and allogeneic hematopoietic cell transplantation. Vorinostat, for instance, has been tested in patients with high-risk lymphoma undergoing high-dose chemotherapy followed by autologous hematopoietic stem cell transplantation. However, this trial failed due to the drug's high toxicity (Hofmeister et al. 2014). Conversely, both Vorinostat and Panobinostat have shown promise when combined with immunosuppressive agents like sirolimus and tacrolimus for treating acute graft-versus-host disease (aGVHD). Notably, the upregulation of acetylated histone H3 was consistently observed with these inhibitors, indicating their mechanism of action (Choi et al. 2017; Perez et al. 2021). Panobinostat, co-administered with corticosteroids, has also demonstrated potential benefits for aGVHD patients (Perez et al. 2013).

As the clinical evidence supporting the efficacy of HDACi in preventing organ rejection is minimal, in this chapter, we will focus on molecular and preclinical data to address whether HDACi could be repurposed and successfully translated into a clinic as a cure for posttransplant patients.

11.3 Impact of HDACi on the Immune Cells

This section will explore the impact of HDACi on immune cells, which play a crucial role in the initial response to solid organ rejection. We have previously discussed the role of HDACs in macrophage polarization and their involvement in mediating chronic rejection (Halasa and Wawruszak 2024). This discussion will shift to T and B cells, highlighting the molecular pathways triggered by HDACi. We will primarily focus on the FDA-approved pan inhibitors. We will also briefly discuss the selective inhibitors for HDACs in the context of allograft rejection.

11.3.1 HDACs' Expression in T Cells

The role of HDACs in T-cell proliferation, differentiation, and activation has been extensively studied and nicely reviewed here (Pieniawska and Iżykowska 2022). Many papers show the important role of HDAC1 and HDAC2 in normal T-cell development (Dovey et al. 2013) and in T-cell-mediated immune diseases. For example, T cell-specific deletion of *Hdac1* in mice caused resistance to encephalomyelitis (EAE) despite the ability of $CD4^+$ T cells to differentiate into Th17 cells (Göschl et al. 2018). HDAC1 and HDAC2 also regulate CD4+ cytotoxic T lymphocyte (CD4+ CTL) differentiation, and their deletion promotes a cytotoxic program, enhancing cytotoxic activity in both murine and human cells (Preglej et al. 2020). T-cell research primarily focuses on class I HDACs, which seems to be a logical approach given that this class is the most upregulated in T cells, regardless of their activation and differentiation states as revealed by sc-RNA (single-cell RNA) analysis of peripheral blood mononuclear cells (PBMCs) (Karlsson et al. 2021; Monaco et al. 2019). While class I HDACs expression in T cells could be classified as high (101–1000 nTPM), the SIRTs expression is relatively medium (11–100 nTPM) or low (1–10 nTPM), as presented in Fig. 11.2.

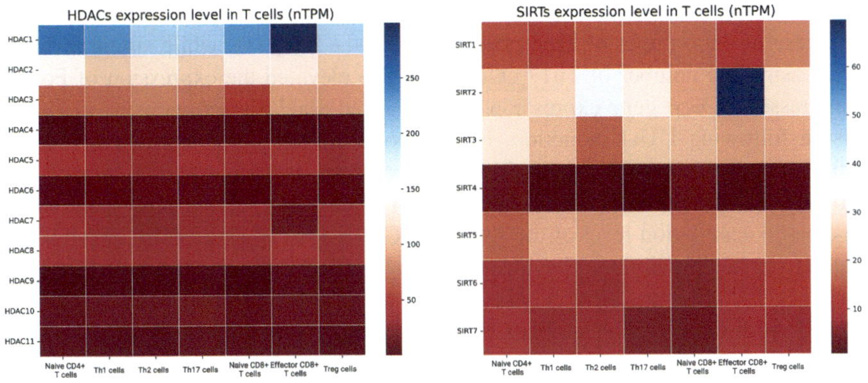

Fig. 11.2 The *HDACs* and *SIRTs* expression analysis in T cells. The gene expression analysis of all members of the HDAC family was performed based on the data provided by The Human Protein Atlas database (https://www.proteinatlas.org). Expression levels are categorized as low (1–10 nTPM), medium (11–100 nTPM), and high (101–1000 nTPM). nTPM is normalized Transcript Per Million

11.3.2 HDACi and T Cells in Immune-Related Diseases

Upregulation of HDACs class I and II has been observed in CD4+ T cells circulating in patients with autoimmune uveitis (AU), particularly in differentiated and effector T-cell subsets. In a mouse model of AU, treatment with Belinostat effectively alleviated AU symptoms and reduced retinal inflammation. HDAC inhibition was found to block Th17 differentiation while simultaneously restoring the population of Treg. Mechanistically, HDACs influence the cyclin-dependent kinase 6 (CDK6)/inhibitor of DNA binding 2 (ID2) and cell cycle pathways, which play a role in Th17 differentiation. Inhibition of HDACs blocked the CDK6/ID2 axis, decreased acetylation of H3, and reduced expression of p21 and p53, contributing to the modulation of T-cell responses (Zhang et al. 2024). These findings are supported by another research showing that Giovinostat suppressed Th17 differentiation and increased Treg population in vitro and in mice with developed experimental colitis. The suppression of Th17 differentiation was mediated by the downregulation of transcription factor: retinoic acid-related orphan receptor (RORγt) in naïve CD4+ cells (Glauben et al. 2014). RORγt plays a crucial role in Th17 differentiation by directly activating the transcription of IL-17 (Ivanov et al. 2006). Giovinostat also inhibited the secretion of IFNγ and IL-6 in a dose-dependent manner, while not affecting IL-4 levels in CD4+ T cells isolated from mice, suggesting a shift away from Th1 responses and a relative preservation of Th2-like cytokine expression. Mechanistically, HDACi suppressed interleukin 6 receptors (IL-6R) through increased chromatin acetylation at the IL6R gene and its promoter, as well as their downstream signaling, as evidenced by the downregulation of STAT3 phosphorylation (Glauben et al. 2014). Similar effects on T cells were observed when trichostatin A (TSA), a pan-inhibitor structurally related to Vorinostat, was administered to mice with induced unilateral ureteral obstruction (UUO). TSA treatment led to an increase in the number of Treg population (CD4+ Foxp3+), while simultaneously decreasing the number of Th17 cells. TSA also elevated the expression of Foxp3, suppressed RORγt gene expression, and inhibited Stat3 phosphorylation in kidney tissue following UUO induction. Furthermore, collagen deposition was reduced after TSA treatment (Wu et al. 2017). In contrast, another research showed that combining TSA with EX-527, an inhibitor of SIRT1, SIRT2, and SIRT3, caused an increased acetylation level of the RORγt in CD4+ T cells sorted from PBMCs isolated from healthy donors. Elevated acetylation of RORγt increased its stability and enhanced *IL-17* transcription, thus potentiating Th17 differentiation (Wu et al. 2015). Additionally, TSA has been found to downregulate Foxp3 mRNA expression in splenic T cells and significantly decrease the naturally occurring Treg (CD4+ CD25+ Foxp3+) population in mice, suggesting an anti-Treg differentiation role (Liu et al. 2010). Finally, the use of TS, in phytohemagglutinin (PHA)-stimulated T cells resulted in hyperacetylation and subsequent reduction of Th1 interleukins and upregulation of Th2 markers (INF-γ, CXCL10), suggesting that pan-HDAC inhibition favors Th2-like phenotype (Su et al. 2008).

The dysregulation of HDACs has been observed among patients suffering from pulmonary arterial hypertension (HAP) (Chen et al. 2023). HAP is frequently associated with varying levels of perivascular inflammatory infiltration, including T and B cells (Zhao et al. 2024). The PBMCs isolated from HAP patients showed upregulation of HDAC class II expression (HDAC5,6,9), downregulation of α-tubulin acetylation, and decrease in Treg cells population. Vorinostat treatment induced ex vivo differentiation of CD4+ CD25- T cells into CD4+ CD25+ Foxp3+ T cells (Treg cells). In in vivo models of pulmonary hypertension (PH), Vorinostat reduced PAH phenotypes and restored the diminished population of circulating Tregs. This effect was accompanied by increased acetylation of H3/H4 levels in the *FOXP3* promoter region, resulting in elevated *FOXP3* mRNA expression (Chen et al. 2023). Of note, conditioned medium from bovine PH fibroblasts treated with Vorinostat (CM-V) were shown to significantly suppress CD4+ T cell differentiation. This was evidenced by decreased IFNγ levels and increased IL4, IL10, and TGF-β levels, suggesting a rise in the Th2 (IL-4) and Treg population. Additionally, CM-V altered T-cell metabolism and reduced proliferation (Plecitá-Hlavatá et al. 2023). These findings indicate that Vorinostat may change the PH fibroblast response and stimuli-releasing profiles, leading to altered T-cell behaviors, although some effects may result from direct interactions between Vorinostat and T cells.

11.3.3 HDACs' Expression in B Cells

The role of HDACs in B-cell functions has been intensively studied over the last decade and was nicely reviewed here (Moran et al. 2023). Naïve B cells upon activation could differentiate into switched memory B cells or into plasmablasts (eventually differentiate into plasma cells) when encountering an antigen and receiving signals from T helper cells. Switched memory B cells are the main drivers of AMR. They produce antibodies that bind to the endothelium of the transplanted organ, leading to complement activation, inflammation, and tissue damage. Plasmablasts are short-lived, early-stage plasma cells that rapidly proliferate, whereas plasma cells are the long-lived mediators of lasting humoral immunity (Blume et al. 2012; Rodriguez-Ramirez et al. 2022). HDACs were found to be critical in B-cell differentiation and maturation (Stengel et al. 2019; Kienzler et al. 2013). scRNA-seq analysis of PBMCs revealed that class I HDACs exhibit the highest overall expression across B cells regardless of their activation or differentiation, except for *HDAC2*, which shows the highest expression specifically in plasmablasts and much lower expression in naïve and switched memory B cells. The expression of *SIRTs* remains at medium or low level, with the highest expression of *SIRT2* regardless of activation and differentiation state of B cells (Fig. 11.3) (Karlsson et al. 2021; Monaco et al. 2019).

11.3.4 HDACi and B Cells in Immune-Related Diseases

HDACi have diverse and significant effects on B-cell differentiation, proliferation, and function, influencing both harmful and regulatory B-cell subsets in autoimmune conditions and other immune responses. VPA, even at low concentrations, delayed and reduced the differentiation of naïve B cells into plasmablasts (CD27^{high} CD28^{high}) as well as inhibited the activation of switched memory B cells at early time points.

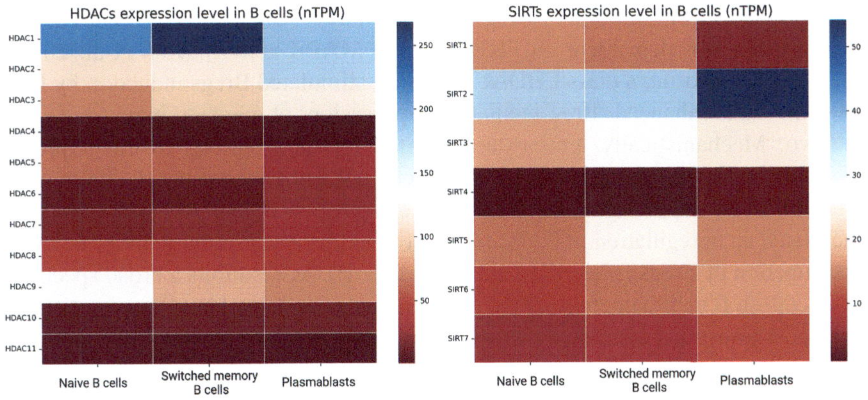

Fig. 11.3 The *HDACs* and *SIRTs* expression analysis in B cells. The gene expression analysis of all members of the HDAC family was performed based on the data provided by The Human Protein Atlas database (https://www.proteinatlas.org). Expression levels are categorized as low (1–10 nTPM), medium (11–100 nTPM), and high (101–1000 nTPM). nTPM is normalized Transcript Per Million

Consistently, early activation events in response to CD40L and IL-21 in the presence of VPA demonstrated impaired blast formation and reduced upregulation of CD69, a key lectin involved in the retention of lymphocytes within lymphoid organs. While VPA did not induce changes in the viability of naïve B cells, it significantly reduced proliferation in response to CD40L and IL-21 stimulation, and upon Toll-like receptor 9 (TLR9) activation (Kienzler et al. 2013). In contrast, Vorinostat and Panobinostat reduced B-cell viability in a dose-dependent manner after 24 hours, while Romidepsin (a class I HDAC inhibitor) exhibited similar effects after 48 hours. Consistently, in mice with B-cell-mediated autoimmune conditions (systemic lupus erythematosus), Panobinostat caused a significant reduction in naïve B cells and plasma cells from peripheral lymphoid organs and a reduction of renal pathology, typical for this disease. Interestingly, unlike B cells, T-cell levels remain unchanged after Panobinostat treatment, suggesting that B cells could be more sensitive than T cells to HDACi (Waibel et al. 2015).

B regulatory cells (Bregs), due to their immunomodulatory properties, could suppress organ rejection (Lee et al. 2022). They differentiate from naïve B cells (primary, yet not exclusive source) under specific conditions; cytokines (IL-10, TGF-β, IL-21), and signals from Treg cells, dendric cells, or macrophages (Bao et al. 2023). The secretion of IL-10 by Bregs inhibits several key immune responses including suppressing the activation, proliferation, and cytokine production by effector T cells and inhibiting the function of APCs (Elias et al. 2024). Bregs can be modulated by short-chain fatty acids (SCFAs), some of which are known to function as HDACis. For instance, sodium butyrate (NaBu) treatment has been shown to inhibit B-cell proliferation while simultaneously promoting their differentiation into Bregs. Consistent with this, NaBu treatment increased the number of Bregs in B cells isolated from the PBMCs of healthy mice. In a mouse model of induced colitis, NaBu treatment elevated the frequency of Bregs in B cells from peritoneal cavity fluids. The promotion of Bregs generation by NaBu, as well as other HDACi (Vorinostat), was dependent on ERK and p38 MAPK activity (Zou et al. 2021). Similarly, Entinostat, a class I HDAC inhibitor, stimulated Bregs induction by LPS in vitro and enhanced Breg-mediated suppression of contact hypersensitivity in vivo. Mechanistically, Entinostat inhibited HDAC1 binding to the proximal region of the IL-10 promoter in splenic B cells, thereby increasing the binding of Nuclear factor kappa-light-chain-enhancer of activated B cells (NF-κB) p65. This facilitated an upregulation in IL-10 expression, ultimately promoting the generation and function of Bregs (Min et al. 2021). Overall, HDACi exhibits a broad spectrum of effects on B cells, suppressing the proliferation of unfavorable subsets associated with post-transplant complications while promoting the expansion of Bregs. These dual actions underscore their promising therapeutic potential in managing autoimmune diseases and modulating the immune system.

11.4 HDACi in Transplantation Pre-clinical and Human Study

In this part of the chapter, we focus on HDACi in pre-clinical transplant models and post-transplant human studies.

11.4.1 HDACi in Cardiac Transplantation

Vorinostat, when used with the immunosuppressant FK506, prolonged cardiac allograft survival in a murine model by regulating the Th17/Treg balance. It significantly suppressed Th17 differentiation, as evidenced by reduced RORγt and STAT3 expression and upregulation of Foxp3 expression, while selectively inducing apoptosis in T effector cells. Notably, Vorinostat synergized with FK506 in suppressing IL-17 mRNA expression and simultaneously promoting Foxp3 and CTLA4 mRNA expression, suggesting its clinical potential in preventing allograft rejection (Zhang et al. 2012). These findings are confirmed by another research, where TSA treatment resulted in upregulation of FOXP3 expression in naïve T cells, promoting Treg differentiation and their activation in vitro (Moon et al. 2009). Another pan-inhibitor, FR276457, in combination with tacrolimus, prevented allograft rejection in a rat cardiac transplant model marked by intact myocardium and lack of immune cell infiltration in the isograft. Additionally, FR276457 reduced dose-dependent T-cell proliferation rates and showed immunosuppressive effects in vitro (Kinugasa et al. 2008). Pan inhibition of HDACs by TSA resulted in a higher number of Bregs producing IL-10 and TGF-β and significant prolongation of the survival rate in mice after allogeneic heart transplantation (Zhou et al. 2021).

The addition of VPA to the standard histidine-tryptophan-ketoglutarate (HTK) solution preserves histone acetylation during human donor heart preservation. In a mouse model, VPA enhanced heart contractility and relaxation, reduced markers of myocardial damage and leukocyte infiltration, and activated cardioprotective pathways. It achieved this by increasing *IRG1* expression and itaconate availability, which boosted antioxidant processes, inhibited inflammation, and promoted cellular survival, improving cardiac function after preservation (Lei et al. 2023).

The use of pan-HDACi is limited due to their high toxicity and broad range of targets. An alternative to pan-HDACi could be selective inhibitors that target particular HDACs. In the heart transplant model in mice, the Foxp3+ Treg-selective genetic deletion of *HDAC11* prolonged allograft survival and inhibited transplant arteriosclerosis. In line, selective HDAC11 inhibitors (Elevenostat or JB3-22) significantly increased Treg suppressive function in vitro and drastically enhanced post-transplant survival rates (Huang et al. 2017).

11.4.2 HDACi in Lung Transplantation

While the effect of HDACi on pulmonary fibrosis, a hallmark of chronic rejection, is well studied—often using the bleomycin model—there is less research on HDACi after lung transplantation (Ota et al. 2015; Kim et al. 2019; Conforti et al. 2017). However, some data showed that selective inhibition of particular HDACs could improve post-transplantation outcomes. For example, the selective inhibition of HDAC6 by Tubastatin A reduced Th17 cell activity during acute lung allograft rejection, leading to prolonged lung graft survival and lessening acute rejection, achieving this effect by limiting the accumulation of Th17 cells (Zhou et al. 2020). Other findings suggest that not only HDACi but also HDAC activators could be helpful for post-transplant patients. The HDAC2 expression was decreased in CD8+ T cells and natural killer T (NKT)-like cells in patients after lung transplantation compared to healthy controls. Loss of HDAC2 was associated with an increased percentage of CD8+ T and NKT-like cells expressing immune effectors and cytotoxic molecules (Hodge et al. 2017). Interestingly, a decrease in HDAC2 expression was found in proinflammatory lymphocytes in the small airways in patients suffering from bronchiolitis obliterans syndrome (BOS) (Hodge et al. 2018). BOS is a well-established phenotype of chronic lung allograft dysfunction after lung transplantation (Todd and Palmer 2011), indicating the potential for HDAC activation.

11.4.3 HDACi in Liver Transplantation

The ischemia/reperfusion injury (IRI) is a significant cause of organ loss after liver transplantation (Dar et al. 2019). Global inhibition of acetylation by TSA, as well as selective inhibition of HDAC6 by Tubacin A, significantly reduced hepatocellular injury after liver IRI in mice, suggesting that pan- and selective- HDACi could be potentially used in post-transplant cure (Dar et al. 2019). Another pan-inhibitor, Vorinostat, reduced inflammation and promoted autophagy in liver grafts in rats, alleviating transplantation-induced IRI. While this research focuses on Kupffer cells (resident liver macrophages), the overall effect of Vorinostat on IRI inhibition is promising and should be further explored in terms of other immune cells (Wang et al. 2020b).

11.4.4 HDACi in Kidney Transplantation

IRI remains the leading cause of renal dysfunction and failure after transplantation (Lasorsa et al. 2024). Recent evidence suggests that interventions like HDAC inhibition may effectively mitigate IRI-related damage and improve transplant outcomes. The use of TSA enhanced renal functional benefit by the reduction of blood urea nitrogen (BUN), a clinically used marker for assessment of kidney function in the warm IRI model. TSA-treated mice demonstrated lower tubular injury and inflammation 30 days after IRI induction and showed less fibrosis. In line, mice treated with TSA before IRI induction developed less fibrosis than untreated controls. In the cold ischemia renal transplant model, TSA-treated mice showed better survival and less fibrosis (Levine et al. 2015). Another pan-HDAC inhibitor FR276457, as a monotherapy and in combination with tacrolimus, significantly prolonged the survival of canine renal allografts. The combined treatment reduced tacrolimus dosages, effectively mitigating its associated toxicity. Histopathological analysis revealed that the combination therapy suppressed key indicators of acute rejection, including mononuclear cell infiltration and vasculitis (Kinugasa et al. 2009).

Additionally, HDAC inhibition shows potential in preserving kidneys from deceased donors during cold storage and rewarming before transplantation. The selective HDAC3 inhibitor RGFP966 significantly reduced acute tubular injury and cell death after prolonged cold storage in a mouse model. Furthermore, RGFP966 enhanced renal function, facilitated kidney repair, and preserved tubular integrity in transplanted kidneys. In vitro, cold storage followed by rewarming proximal tubular cells induced substantial cell death, effectively mitigated by RGFP966 treatment (Xiang et al. 2022).

11.5 Discussion and Future Perspectives

HDACi are currently FDA-approved for blood cancer treatment. For solid tumors, they are intensively studied in pre-clinical and clinical models, yet the outcomes remain limited. HDACi are highly synergistic with various immunotherapeutic approaches, exerting anti-tumor effects via modulation of immune cells, increasing immune surveillance, and forcing tumor cells to express PD-L1 (Shi et al. 2024). In clinics, HDACi are often used in combined therapy due to their limited efficacy when used alone, and frequently develop resistance to HDACi monotherapy, reducing their therapeutic impact (Bondarev et al. 2021). Compared with the widespread use of HDACi in cancer models, much less is known about HDACi's effects on post-transplant models. The current data indicates that the use of HDACi generally prolonged survival after allograft transplantation, suggesting that they could be considered as "extra help" in immunosuppression therapy to prevent organ rejection. The therapeutic potential of HDACi is highlighted by their ability to reduce disease symptoms and inflammation in vitro and in animal models, suggesting their promise as potential treatments for immune-related diseases, including post-transplant treatment. However, the effects of HDAC inhibition can vary depending on the specific context, disease model, and the used inhibitor, requiring further investigation to fully understand the therapeutic potential and limitations of these treatments.

While it is unlikely that HDACi in monotherapy could benefit posttransplant patients, the combined therapy with currently used anti-rejection drugs seems to be a promising approach. For example, Panobinostat, unselective HDACi, induced calcineurin degradation in multiple myeloma (Imai et al. 2016). Considering that calcineurin inhibitors are commonly used in post-transplantation to prevent T-cell activation and suppression of cytokine production (Shah and Thomas Ginat 2023), the combined treatment could enhance this effect. Panobinostat also synergistically interacted with PI3K/mTOR inhibitor (BEZ235) against non-Hodgkin lymphoma (Rahmani et al. 2014). Since mTOR inhibitors (Sirolimus, Everolimus) are used after organ transplantation (Granata et al. 2023), the synergistic effect could be translated into the transplantation field. Because most investigations focus on cancer, the knowledge about HDACi in the post-transplant application is still limited and requires further, deeper understanding.

While only pan-inhibitors are currently in clinical use, they exert severe side effects as they hit unintended enzymes within the cell, which remains a serious limitation. Thus, the concept of isoform-selective HDACi might broaden the clinical application of HDACi beyond their current use. However, the selective design could be challenging due to the high degree of similarity in the catalytic cores across the HDAC family. Most HDACs share a conserved zinc-dependent catalytic domain, which makes it difficult to achieve specificity for one HDAC isoform without affecting others (Park and Kim 2020). Improving HDAC isoform specificity is arguably one of the most critical challenges to address for reducing off-target toxicity. The cytosolic HDAC6 has gained scientific interest as a negative regulator of regulatory T-cell suppressive activity (Tsuji et al. 2014). HDAC6-selective inhibitors have been evaluated in clinical trials, demonstrating safety and high tolerability (Vogl et al. 2017; Bexon et al. 2024).

One of the key issues is to reveal to what extent HDACi regulate B- and T-cell function and the underlying mechanisms. Many questions remain, such as whether HDACs influence immune cells solely through chromatin-mediated effects (histone targeting), non-histone substrates, or a combination. Additionally, HDACi are known to affect tumor immune microenvironments (Yang et al. 2024). Do they similarly impact allograft microenvironments, and if so, what are the consequences? Addressing these questions is crucial for understanding the broader immunomodulatory effects of HDACi. While this chapter primarily focuses on T and B cells, other immune cells also play critical roles in orchestrating chronic rejection (Diebold et al. 2024). Therefore, the anti-rejection effects mediated by HDACi should also be explored in other cell types, such as macrophages and natural killer (NK) cells. Advances in omics technologies, including genomics, transcriptomics, proteomics, and metabolomics allow us to explore the effects of HDACi in the context of chronic rejection. For example, single-cell transcriptomic profiling can reveal how HDACi modulate gene expression patterns in immune cells, while proteomics can uncover alterations in protein acetylation. Additionally, metabolomics offers the ability to assess changes in metabolic pathways that may influence immune cell function and graft tolerance. By integrating these omics datasets, researchers can construct a detailed map of the immunomodulatory effects of HDACi, identify biomarkers of response, and pinpoint potential therapeutic targets. As omics technologies continue to evolve, they hold immense potential to deepen our understanding of HDACi mechanisms and guide the development of more effective and personalized anti-rejection therapies.

References

Akiyama K et al. (2012) Fas-FasL-mediated CD4+ T-cell apoptosis following stem cell Transplantation1 | cancer research | American Association for Cancer Research. Cell Stem Cell. 10(5): 544–555. Accessed 23 Jan 2025. [Online]. Available: https://aacrjournals.org/cancerres/article/59/13/3107/505237/Fas-FasL-mediated-CD4-T-Cell-Apoptosis-following

Bao J, Betzler AC, Hess J, Brunner C (2023) Exploring the dual role of B cells in solid tumors: implications for head and neck squamous cell carcinoma. Front Immunol 14. https://doi.org/10.3389/FIMMU.2023.1233085

Bexon M et al (2024) Phase 1 clinical trial of TN-301, a highly selective HDAC6 inhibitor with potential in HFPEF, shows target engagement. J Card Fail 30(1):295. https://doi.org/10.1016/J.CARDFAIL.2023.10.426

Blume OR, Yost SE, Kaplan B (2012) Antibody-mediated rejection: pathogenesis, prevention, treatment, and outcomes. J Transp Secur 2012:201754. https://doi.org/10.1155/2012/201754

Bondarev AD, Attwood MM, Jonsson J, Chubarev VN, Tarasov VV, Schiöth HB (2021) Recent developments of HDAC inhibitors: emerging indications and novel molecules. Br J Clin Pharmacol 87(12):4577–4597. https://doi.org/10.1111/BCP.14889

Borst J, Ahrends T, Bąbała N, Melief CJM, Kastenmüller W (2018) CD4+ T cell help in cancer immunology and immunotherapy. Nature Rev Immunol 18(10):635–647. https://doi.org/10.1038/s41577-018-0044-0

Calabrese DR et al (2023) CD16+ natural killer cells in bronchoalveolar lavage are associated with antibody-mediated rejection and chronic lung allograft dysfunction. Am J Transplant 23(1):37–44. https://doi.org/10.1016/J.AJT.2022.10.006

Cantisán S et al (2013) Pretransplant interferon-γ secretion by CMV-specific CD8+ T cells informs the risk of CMV replication after transplantation. Am J Transplant 13(3):738–745. https://doi.org/10.1111/AJT.12049

Cassano A, Chong AS, Alegre M-L (2023) Tregs in transplantation tolerance: role and therapeutic potential. Front Transplant 2:1217065. https://doi.org/10.3389/FRTRA.2023.1217065/BIBTEX

Chateauvieux S, Morceau F, Diederich M (2024) Valproic acid. Encyclop Toxicol Fourth Ed Vol 1–9 9:V9-705-V9-713. https://doi.org/10.1016/B978-0-12-824315-2.00578-9

Chen C-N et al (2023) Restoration of Foxp3+ regulatory T cells by HDAC-dependent epigenetic modulation plays a pivotal role in resolving pulmonary arterial hypertension pathology. Am J Respir Crit Care Med 208(8):879. https://doi.org/10.1164/RCCM.202301-0181OC

Choi SW et al (2017) Vorinostat plus tacrolimus/methotrexate to prevent GVHD after myeloablative conditioning, unrelated donor HCT. Blood 130(15):1760. https://doi.org/10.1182/BLOOD-2017-06-790469

Conforti F et al (2017) The histone deacetylase inhibitor, romidepsin, as a potential treatment for pulmonary fibrosis. Oncotarget 8(30):48737–48754. https://doi.org/10.18632/ONCOTARGET.17114

Dar WA, Sullivan E, Bynon JS, Eltzschig H, Ju C (2019) Ischemia reperfusion injury in liver transplantation: cellular and molecular mechanisms. Liver Int 39(5):788. https://doi.org/10.1111/LIV.14091

Diebold M, Mayer KA, Hidalgo L, Kozakowski N, Budde K, Böhmig GA (2024) Chronic rejection after kidney transplantation. Transplantation 109:610. https://doi.org/10.1097/TP.0000000000005187

Dovey OM et al (2013) Histone deacetylase 1 and 2 are essential for normal T-cell development and genomic stability in mice. Blood 121(8):1335–1344. https://doi.org/10.1182/BLOOD-2012-07-441949

Drachenberg CB, Papadimitriou JC (2013) Endothelial injury in renal antibody-mediated allograft rejection: a schematic view based on pathogenesis. Transplantation 95(9):1073–1083. https://doi.org/10.1097/TP.0B013E31827E6B45

Eckschlager T, Plch J, Stiborova M, Hrabeta J (2017) Histone deacetylase inhibitors as anticancer drugs. Int J Mol Sci 18(7):1414. https://doi.org/10.3390/IJMS18071414

Elias C, Chen C, Cherukuri A (2024) Regulatory B cells in solid organ transplantation: from immune monitoring to immunotherapy. Transplantation 108(5):1080–1089. https://doi.org/10.1097/TP.0000000000004798

Glauben R, Sonnenberg E, Wetzel M, Mascagni P, Siegmund B (2014) Histone deacetylase inhibitors modulate interleukin 6-dependent CD4+ T cell polarization in vitro and in vivo. J Biol Chem 289(9):6142. https://doi.org/10.1074/JBC.M113.517599

Göschl L et al (2018) A T cell-specific deletion of HDAC1 protects against experimental autoimmune encephalomyelitis. J Autoimmun 86:51–61. https://doi.org/10.1016/J.JAUT.2017.09.008

Granata S, Mercuri S, Troise D, Gesualdo L, Stallone G, Zaza G (2023) mTOR-inhibitors and post-transplant diabetes mellitus: a link still debated in kidney transplantation. Front Med (Lausanne) 10:1168967. https://doi.org/10.3389/FMED.2023.1168967/BIBTEX

Günthner R, Anders HJ (2013) Interferon-regulatory factors determine macrophage phenotype polarization. Mediat Inflamm 2013:731023. https://doi.org/10.1155/2013/731023

Halasa M, Wawruszak A (2024) Unlocking macrophage secrets: histone deacetylases in chronic transplant rejection. Results Probl Cell Differ 74:297–313. https://doi.org/10.1007/978-3-031-65944-7_12

Hałasa M et al (2019) H3K18Ac as a marker of cancer progression and potential target of anticancer therapy. Cells 8(5):485. https://doi.org/10.3390/CELLS8050485

He Q et al. (2020) Associations of sirtuins with clinicopathological variables and prognosis in human ovarian cancer. Oncol Lett 19(4):3278–3288. Accessed Jan 21 2025. [Online]. Available: https://www.spandidos-publications.com. https://doi.org/10.3892/ol.2020.11432?text=fulltext

He R, Liu B, Geng B, Li N, Geng Q (2023) The role of HDAC3 and its inhibitors in regulation of oxidative stress and chronic diseases. Cell Death Disc 9(1):1–11. https://doi.org/10.1038/s41420-023-01399-w

Hodge G, Hodge S, Holmes-Liew CL, Reynolds PN, Holmes M (2017) Histone deacetylase 2 is decreased in peripheral blood pro-inflammatory CD8+ T and NKT-like lymphocytes following lung transplant. Respirology 22(2):394–400. https://doi.org/10.1111/RESP.12933

Hodge G et al (2018) BOS is associated with decreased HDAC2 from steroid resistant lymphocytes in the small airways. Clin Exp Immunol 195(2):277. https://doi.org/10.1111/CEI.13221

Hofmeister CC et al (2014) A phase 1 study of Vorinostat maintenance after autologous transplant in high risk lymphoma. Leuk Lymphoma 56(4):1043. https://doi.org/10.3109/10428194.2014.963073

Huang J et al (2017) Histone/protein deacetylase 11 targeting promotes Foxp3+ Treg function. Sci Rep 7(1):8626. https://doi.org/10.1038/S41598-017-09211-3

Imai Y et al (2016) Histone deacetylase inhibitor panobinostat induces calcineurin degradation in multiple myeloma. JCI Insight 1(5):e85061. https://doi.org/10.1172/JCI.INSIGHT.85061

Ivanov II et al (2006) The orphan nuclear receptor RORγt directs the differentiation program of Proinflammatory IL-17+ T helper cells. Cell 126(6):1121–1133. https://doi.org/10.1016/J.CELL.2006.07.035/ATTACHMENT/AA3D23B0-636F-4D92-974B-809BD96BFE55/MMC1.PDF

Junttila IS (2018) Tuning the cytokine responses: an update on interleukin (IL)-4 and IL-13 receptor complexes. Front Immunol 9:338745. https://doi.org/10.3389/FIMMU.2018.00888/BIBTEX

Karlsson M et al (2021) A single–cell type transcriptomics map of human tissues. Sci Adv 7(31). https://doi.org/10.1126/SCIADV.ABH2169/SUPPL_FILE/SCIADV.ABH2169_SM.PDF

Kienzler AK, Rizzi M, Reith M, Nutt SL, Eibel H (2013) Inhibition of human B-cell development into plasmablasts by histone deacetylase inhibitor valproic acid. J Allergy Clin Immunol 131(6):1695–1699.e9. https://doi.org/10.1016/j.jaci.2013.01.018

Kim HJ, Bae SC (2010) Histone deacetylase inhibitors: molecular mechanisms of action and clinical trials as anti-cancer drugs. Am J Transl Res 3(2):166. Accessed Jan 21 2025. [Online]. Available: https://pmc.ncbi.nlm.nih.gov/articles/PMC3056563/

Kim YS, Cha H, Kim HJ, Cho JM, Kim HR (2019) The anti-fibrotic effects of CG-745, an HDAC inhibitor, in Bleomycin and PHMG-induced mouse models. Molecules 24(15). https://doi.org/10.3390/MOLECULES24152792

Kinugasa F et al (2008) Effect of a new immunosuppressant histon deacetylase (HDAC) inhibitor FR276457 in a rat cardiac transplant model. Biol Pharm Bull 31(9):1723–1726. https://doi.org/10.1248/BPB.31.1723

Kinugasa F et al (2009) Effect of the immunosuppressant histone deacetylase inhibitor FR276457 in a canine renal transplant model. Transpl Immunol 21(4):198–202. https://doi.org/10.1016/J.TRIM.2009.04.006

Kloc M, Ghobrial RM (2014) Chronic allograft rejection: a significant hurdle to transplant success. Burns Trauma 2(1):3. https://doi.org/10.4103/2321-3868.121646

Künzli M, Masopust D (2023) CD4+ T cell memory. Nature Immunol 24(6):903–914. https://doi.org/10.1038/s41590-023-01510-4

Lasorsa F et al (2024) Ischemia-reperfusion injury in kidney transplantation: mechanisms and potential therapeutic targets. Int J Mol Sci 25(8). https://doi.org/10.3390/IJMS25084332

Lee KM et al (2022) Suppression of allograft rejection by regulatory B cells induced via TLR signaling. JCI Insight 7(17). https://doi.org/10.1172/JCI.INSIGHT.152213

Lei I et al (2023) Metabolic reprogramming by immune-responsive gene 1 up-regulation improves donor heart preservation and function. Sci Transl Med 15(682):eade3782. https://doi.org/10.1126/SCITRANSLMED.ADE3782

Levine MH et al (2015) Class-specific histone/protein deacetylase inhibition protects against renal ischemia reperfusion injury and fibrosis formation. Am J Transplant 15(4):965. https://doi.org/10.1111/AJT.13106

Liu Z, Zhang C, Sun J (2010) Deacetylase inhibitor trichostatin A down-regulates Foxp3 expression and reduces CD4+CD25+ regulatory T cells. Biochem Biophys Res Commun 400(3):409–412. https://doi.org/10.1016/J.BBRC.2010.08.090

Luckheeram RV, Zhou R, Verma AD, Xia B (2012) CD4+T cells: differentiation and functions. Clin Dev Immunol 2012:925135. https://doi.org/10.1155/2012/925135

Lusco MA, Fogo AB, Najafian B, Alpers CE (2016) AJKD atlas of renal pathology: acute T-cell-mediated rejection. Am J Kidney Dis 67(5):e29–e30. https://doi.org/10.1053/J.AJKD.2016.03.004

McNally A, Hill GR, Sparwasser T, Thomas R, Steptoe RJ (2011) CD4+CD25+ regulatory T cells control CD8+ T-cell effector differentiation by modulating IL-2 homeostasis. Proc Natl Acad Sci USA 108(18):7529–7534. https://doi.org/10.1073/PNAS.1103782108/-/DCSUPPLEMENTAL

Mercuri E et al (2024) Safety and efficacy of givinostat in boys with Duchenne muscular dystrophy (EPIDYS): a multicentre, randomised, double-blind, placebo-controlled, phase 3 trial. Lancet Neurol 23(4):393–403. https://doi.org/10.1016/S1474-4422(24)00036-X

Min KY et al (2021) Entinostat, a histone deacetylase inhibitor, increases the population of IL-10+ regulatory B cells to suppress contact hypersensitivity. BMB Rep 54(10):534. https://doi.org/10.5483/BMBREP.2021.54.10.092

Monaco G et al (2019) RNA-Seq signatures normalized by mRNA abundance allow absolute deconvolution of human immune cell types. Cell Rep 26(6):1627–1640.e7. https://doi.org/10.1016/J.CELREP.2019.01.041

Moon C et al (2009) Use of epigenetic modification to induce FOXP3 expression in naïve T cells. Transplant Proc 41(5):1848–1854. https://doi.org/10.1016/J.TRANSPROCEED.2009.02.101

Moran B, Davern M, Reynolds JV, Donlon NE, Lysaght J (2023) The impact of histone deacetylase inhibitors on immune cells and implications for cancer therapy. Cancer Lett 559:216121. https://doi.org/10.1016/J.CANLET.2023.216121

Nelson J et al (2022) Consensus recommendations for use of maintenance immunosuppression in solid organ transplantation: Endorsed by the American College of Clinical Pharmacy, American Society of Transplantation, and the International Society for Heart and Lung Transplantation. Pharmacother J Hum Pharmacol Drug Therapy 42(8):599–633. https://doi.org/10.1002/PHAR.2716

Ota C et al (2015) Histone deacetylase inhibitor restores surfactant protein-C expression in alveolar-epithelial type II cells and attenuates bleomycin-induced pulmonary fibrosis in vivo. Exp Lung Res 41(8):422–434. https://doi.org/10.3109/01902148.2015.1060275

Park SY, Kim JS (2020) A short guide to histone deacetylases including recent progress on class II enzymes. Experim Mol Med 52(2):204–212. https://doi.org/10.1038/s12276-020-0382-4

Perez LE et al (2013) A phase I/II trial evaluating the use of a histone deacetylase inhibitor LBH589 in addition to corticosteroids in patients with acute graft versus host disease. Biol Blood Marrow Transplant 19(2):S337–S338. https://doi.org/10.1016/j.bbmt.2012.11.515

Perez L et al (2021) A phase 2 trial of the histone deacetylase inhibitor panobinostat for graft-versus-host disease prevention. Blood Adv 5(13):2740–2750. https://doi.org/10.1182/BLOODADVANCES.2021004225

Pieniawska M, Iżykowska K (2022) Role of histone deacetylases in T-cell development and function. Int J Mol Sci 23(14):7828. https://doi.org/10.3390/IJMS23147828

Plecitá-Hlavatá L et al (2023) Microenvironmental regulation of T-cells in pulmonary hypertension. Front Immunol 14. https://doi.org/10.3389/FIMMU.2023.1223122

Preglej T et al (2020) Histone deacetylases 1 and 2 restrain CD4+ cytotoxic T lymphocyte differentiation. JCI Insight 5(4). https://doi.org/10.1172/JCI.INSIGHT.133393

Rahmani M, Aust MM, Benson EC, Wallace L, Friedberg J, Grant S (2014) PI3K/mTOR inhibition markedly potentiates HDAC inhibitor activity in NHL cells through BIM- and MCL-1-dependent mechanisms in vitro and in vivo. Clin Cancer Res 20(18):4849–4860. https://doi.org/10.1158/1078-0432.CCR-14-0034/176353/AM/PI3K-MTOR-INHIBITION-MARKEDLY-POTENTIATES-HDAC

Rodriguez-Ramirez S, Al Jurdi A, Konvalinka A, Riella LV (2022) Antibody-mediated rejection: prevention, monitoring and treatment dilemmas. Curr Opin Organ Transplant 27(5):405. https://doi.org/10.1097/MOT.0000000000001011

Ross SH, Cantrell DA (2018) Signaling and function of interleukin-2 in T lymphocytes. Annu Rev Immunol 36:411. https://doi.org/10.1146/ANNUREV-IMMUNOL-042617-053352

Schinstock CA et al (2017) Kidney transplant with low levels of DSA or low positive B-flow Crossmatch: an underappreciated option for highly sensitized transplant candidates. Transplantation 101(10):2429–2439. https://doi.org/10.1097/TP.0000000000001619

Schmitz R et al (2020) B cells in transplant tolerance and rejection: friends or foes? Transpl Int 33(1):30. https://doi.org/10.1111/TRI.13549

Shah S, Thomas Ginat D (2023) Calcineurin Inhibitors. Neuroimag Pharmacop Second Ed:173–183. https://doi.org/10.1007/978-3-031-08774-5_22

Shevach EM (2012) Application of IL-2 therapy to target T regulatory cell function. Trends Immunol 33(12):626. https://doi.org/10.1016/J.IT.2012.07.007

Shi MQ et al (2024) Advances in targeting histone deacetylase for treatment of solid tumors. J Hematol Oncol 17(1):1–20. https://doi.org/10.1186/S13045-024-01551-8

Short S, Lewik G, Issa F (2023) An immune atlas of T cells in transplant rejection: pathways and therapeutic opportunities. Transplantation 107(11):2341–2352. https://doi.org/10.1097/TP.0000000000004572

Snell GI et al (2007) Interleukin-17 and airway inflammation: a longitudinal airway biopsy study after lung transplantation. J Heart Lung Transplant 26(7):669–674. https://doi.org/10.1016/J.HEALUN.2007.05.004

Squarzoni A, Scuteri A, Cavaletti G (2022) HDACi: the Columbus' egg in improving cancer treatment and reducing neurotoxicity? *Cancers* 14(21):5251. https://doi.org/10.3390/CANCERS14215251

Steimbach RR et al (2022) Aza-SAHA derivatives are selective histone deacetylase 10 chemical probes that inhibit polyamine Deacetylation and Phenocopy HDAC10 knockout. J Am Chem Soc 144(41):18861–18875. https://doi.org/10.1021/JACS.2C05030/SUPPL_FILE/JA2C05030_SI_006.PDB

Stengel KR et al (2019) Histone deacetylase 3 controls a transcriptional network required for B cell maturation. Nucleic Acids Res 47(20):10612–10627. https://doi.org/10.1093/NAR/GKZ816

Su RC, Becker AB, Kozyrskyj AL, HayGlass KT (2008) Epigenetic regulation of established human type 1 versus type 2 cytokine responses. J Allergy Clin Immunol 121(1):57-63.e3. https://doi.org/10.1016/j.jaci.2007.09.004

Sutanto H, Maimunah U, Fetarayani D (2024) Donor-specific antibodies and their impact on antibody-mediated rejection post-liver transplantation: a comprehensive review. J Liver Transplant 14:100214. https://doi.org/10.1016/J.LIVER.2024.100214

Tan Y, Li B, Peng F, Gong G, Li N (2020) Integrative analysis of Sirtuins and their prognostic significance in clear cell renal cell carcinoma. Front Oncol 10:503043. https://doi.org/10.3389/FONC.2020.00218/BIBTEX

Todd JL, Palmer SM (2011) Bronchiolitis obliterans syndrome: the final frontier for lung transplantation. Chest 140(2):502–508. https://doi.org/10.1378/CHEST.10-2838

Tsuji G, Okiyama N, Villarroel VA, Katz SI (2014) Histone deacetylase 6 inhibition impairs effector CD8 T-cell functions during skin inflammation. J Allergy Clin Immunol 135(5):1228. https://doi.org/10.1016/J.JACI.2014.10.002

Uhlén M et al (2015) Tissue-based map of the human proteome. Science (1979) 347(6220). https://doi.org/10.1126/SCIENCE.1260419/SUPPL_FILE/1260419_UHLEN.SM.PDF

Vaillant AAJ, Mohseni M (2024) Chronic transplantation rejection. StatPearls. Accessed: Jan 21 2025. [Online]. Available: https://www.ncbi.nlm.nih.gov/books/NBK535435/

Vaillant AAJ, Misra S, Fitzgerald BM (2024) Acute transplantation rejection. StatPearls. Accessed Jan 21 2025. [Online]. Available: https://www.ncbi.nlm.nih.gov/books/NBK535410/

Vogl DT et al (2017) Ricolinostat, the first selective histone deacetylase 6 inhibitor, in combination with Bortezomib and dexamethasone for relapsed or refractory multiple myeloma. Clin Cancer Res 23(13):3307–3315. https://doi.org/10.1158/1078-0432.CCR-16-2526

Waibel M et al (2015) Manipulation of B-cell responses with histone deacetylase inhibitors. Nat Commun 6(1):1–12. https://doi.org/10.1038/ncomms7838

Walker JA, McKenzie ANJ (2017) TH2 cell development and function. Nat Rev Immunol 18(2):121–133. https://doi.org/10.1038/nri.2017.118

Wang P, Wang Z, Liu J (2020a) Role of HDACs in normal and malignant hematopoiesis. Mol Cancer 19(1):1–21. https://doi.org/10.1186/S12943-019-1127-7

Wang J et al. (2020b) Suberoylanilide hydroxamic acid alleviates orthotopic liver transplantation-induced hepatic ischemia-reperfusion injury by regulating the AKT/GSK3β/NF-κB and AKT/mTOR pathways in rat Kupffer cells. Int J Mol Med 45(6):1875-1887 Accessed: 22 Jan 2025. [Online]. Available: https://www.spandidos-publications.com/ijmm/45/6/1875

Wang M et al (2023) Safety of crossing donor-specific antibodies in lung transplantation. Transplantation 107(5):1172–1179. https://doi.org/10.1097/TP.0000000000004413

Wawruszak A et al (2019) Histone deacetylase inhibitors and phenotypical transformation of cancer cells. Cancers 11(2):148. https://doi.org/10.3390/CANCERS11020148

Wei X et al (2020) Antibody-induced vascular inflammation skews infiltrating macrophages to a novel remodeling phenotype in a model of transplant rejection. Am J Transplant 20(10):2686–2702. https://doi.org/10.1111/AJT.15934

Wood KJ, Bushell A, Hester J (2012) Regulatory immune cells in transplantation. Nat Rev Immunol 12(6):417–430. https://doi.org/10.1038/NRI3227

Wu Q et al (2015) Reciprocal regulation of RORγt acetylation and function by p300 and HDAC1. Sci Rep 5:16355. https://doi.org/10.1038/SREP16355

Wu WP, Tsai YG, Lin TY, Wu MJ, Lin CY (2017) The attenuation of renal fibrosis by histone deacetylase inhibitors is associated with the plasticity of FOXP3+IL-17+ T cells. BMC Nephrol 18(1):1–12. https://doi.org/10.1186/S12882-017-0630-6/FIGURES/10

Xiang X, Dong G, Zhu J, Zhang G, Dong Z (2022) Inhibition of HDAC3 protects against kidney cold storage/transplantation injury and allograft dysfunction. Clin Sci 136(1):45–60. https://doi.org/10.1042/CS20210823

Yang F, Zhao N, Ge D, Chen Y (2019) Next-generation of selective histone deacetylase inhibitors. RSC Adv 9(34):19571–19583. https://doi.org/10.1039/C9RA02985K

Yang X et al (2024) HDAC inhibitor regulates the tumor immune microenvironment via pyroptosis in triple negative breast cancer. Mol Carcinog 63(9):1800–1813. https://doi.org/10.1002/MC.23773

Yap M, Brouard S, Pecqueur C, Degauque N (2015) Targeting CD8 T-cell metabolism in transplantation. Front Immunol 6, no. OCT:163610. https://doi.org/10.3389/FIMMU.2015.00547/BIBTEX

Zamperla MG et al (2024) HDAC6 inhibition disrupts HDAC6-P300 interaction reshaping the cancer chromatin landscape. Clin Epigenetics 16(1):1–20. https://doi.org/10.1186/S13148-024-01725-8/FIGURES/7

Zhang X et al (2012) SAHA, an HDAC inhibitor, synergizes with tacrolimus to prevent murine cardiac allograft rejection. Cell Mol Immunol 9(5):390. https://doi.org/10.1038/CMI.2012.28

Zhang Q et al (2018) Understanding the correlation between DSA, complement activation and antibody mediated rejection in heart transplant recipients. Transplantation 102(10):e431. https://doi.org/10.1097/TP.0000000000002333

Zhang C et al (2024) Histone deacetylases facilitate Th17-cell differentiation and pathogenicity in autoimmune uveitis via CDK6/ID2 axis. J Adv Res. https://doi.org/10.1016/J.JARE.2024.07.029

Zhao H et al (2024) The role of immune cells and inflammation in pulmonary hypertension: mechanisms and implications. Front Immunol 15:1374506. https://doi.org/10.3389/FIMMU.2024.1374506/BIBTEX

Zhou W et al (2020) HDAC6-specific inhibitor suppresses Th17 cell function via the HIF-1α pathway in acute lung allograft rejection in mice. Theranostics 10(15):6790. https://doi.org/10.7150/THNO.44961

Zhou B et al. (2021) Protective effect of trichostatin A on $CD19^{+}CD5^{+}CD1d^{high}$ regulatory B cells in heart transplantation. Mol Med Rep 23(5):339. Accessed: 22 Jan 2025. [Online]. Available: https://www.spandidos-publications.com/mmr/23/5/339

Zou F et al (2021) Effects of short-chain fatty acids in inhibiting HDAC and activating p38 MAPK are critical for promoting B10 cell generation and function. Cell Death Dis 12(6):582. https://doi.org/10.1038/S41419-021-03880-9

Chapter 12
Acetylation in Viral Infection and Disease

Matloob Husain

Abstract Viruses are acellular organisms and part of our ecosystem but exist at the interface of living and non-living. Furthermore, viruses are obligate intracellular parasites hence require the machinery of other organisms to multiply. Consequently, most viral infections result into a viral disease. Broadly, viruses cause two types of infection—acute and persistent (latent and chronic), in humans and other mammals that could lead to various lethal and non-lethal viral diseases. Acetylation is now known to be a ubiquitous protein (and nucleic acid) modification and is critical for cellular metabolism. An imbalance in acetylation has been associated with various cancers and diseases in humans. Likewise, the association of acetylation with viral infection and disease was observed soon after its discovery in twentieth century. Now, the literature accumulated in this space shows that acetylation promotes the infection of many viruses causing both acute and persistent infections. Furthermore, reduction in the acetylation level reduces viral clearance from the host and promotes viral persistency. The latter can be interrupted by increasing the acetylation level by using deacetylase inhibitors. Indeed, this approach has become a therapeutic tool to treat and clear the persistent viral infections as well as boost the oncolytic virus-mediated cancer therapy.

12.1 Introduction

12.1.1 Viruses

Viruses are microscopic organisms that exist at the interface of living and non-living (Willey et al. 2022). Viruses are characterized as living organisms because they possess own genetic material and can reproduce and encode enzymes to catalyze the

M. Husain (✉)
Department of Microbiology and Immunology, University of Otago, Dunedin, New Zealand
e-mail: matloob.husain@otago.ac.nz

M. Halasa, A. Wawruszak (eds.), *Histone and Non-Histone Reversible Acetylation in Development, Aging and Disease*, Results and Problems in Cell Differentiation 75,
https://doi.org/10.1007/978-3-031-91459-1_12

biochemical reactions. In contrast, viruses are characterized as non-living organisms because they have an acellular structure, do not possess own metabolism, and depend on other organisms for reproduction. Therefore, viruses are obligate intracellular parasites. Most viruses exist as a single particle, which has a simple structure where nucleic acid is surrounded by a protein coat called capsid. Depending on virus, the nucleic acid could be either RNA or DNA, and with any one of the four possible configurations—single-stranded or double-stranded and linear or circular. Furthermore, the protein molecules in capsid could be arranged in one of the three symmetries—helical, icosahedral, and complex, giving different shapes to the virus particle. In addition, in some viruses, the capsid is surrounded by a lipid bilayer, which is derived from a host cell membrane and called as envelope. Therefore, broadly, viruses are of two types—non-enveloped and enveloped (Willey et al. 2022).

Viruses infect most known living organisms (called host) and can be found wherever a living entity is present, hence viruses are the most abundant biological entity on earth (Breitbart and Rohwer 2005). Viruses are part of the ecosystem and, through their parasitic predisposition toward living organisms, play a critical role in host genome evolution, immune responses and diseases, and earth's biogeochemical cycles (Santiago-Rodriguez and Hollister 2022). Paradoxically, viruses have also played a critical role in the development of diagnostic and research tools and therapeutic and biocontrol strategies (Varanda et al. 2021). Only a fraction of known viruses have been characterized and only a couple of hundred of viruses are known to cause a disease in humans (de la Higuera and Lazaro 2022; Woolhouse et al. 2012; Wolfe et al. 2007). Nevertheless, a disease-causing virus can have the devastating consequences for human health, global economy, and world order. The earliest example of this is the influenza A virus H1N1 subtype, which caused the first recorded viral pandemic in history, in 1918 that killed estimated 50 million people (Martini et al. 2019). The latest example of this is the severe acute respiratory syndrome coronavirus-2 (SARS-CoV-2), which caused the coronavirus disease 2019 (COVID-19) pandemic in 2020. The COVID-19 pandemic lasted for almost 4 years, resulted in over 770 million infections, and killed over seven million people (WHO 2024). In between these two major events, several viruses, e.g., variola (smallpox) virus, human immunodeficiency virus (HIV), poliovirus, hepatitis B virus (HBV), hepatitis C virus (HCV), human papillomavirus (HPV), and Ebola virus have caused significant outbreaks in human population that, combinedly, killed hundreds of thousands of people (Willey et al. 2022). Furthermore, the respiratory viruses, e.g., influenza A and B virus, rhinovirus, parainfluenza virus, respiratory syncytial virus (RSV), mumps virus, measles virus, rubella virus, and varicella-zoster virus, cause recurring outbreaks of diseases with different symptoms in human population (Willey et al. 2022). Similarly, viruses like rotavirus, norovirus, and astrovirus cause regular outbreaks of gastrointestinal symptoms in human population (Willey et al. 2022). In addition, mosquito-borne viruses, like Dengue, Zika, and West Nile cause the outbreaks of fever, encephalitis, and other brain disorders in human population, particularly residing in tropical regions (de Souza and Weaver 2024). Therefore, viruses continue to pose a significant burden to global public health.

12.1.2 Virus Life Cycle in a Host Cell

As obligate intracellular parasites, viruses must be inside the host cells to complete their life cycle and produce progeny. The life cycle of viruses in a mammalian host cell has five common steps: attachment, entry, synthesis, assembly, and release (Willey et al. 2022). Viruses attach to mammalian host cells through interaction between a protein or carbohydrate on the surface of the host cell, called receptor, and a protein on the surface of the virus particle, called receptor-binding protein. Viruses then enter the host cell either by receptor-mediated endocytosis (common with both enveloped and non-enveloped viruses), fusing with the plasma membrane (common with enveloped viruses), or directly injecting the genome inside the host cell cytoplasm (common with bacteriophages and some non-enveloped mammalian viruses). Nevertheless, the goal of the virus entry is to deliver viral genome inside the host cell to synthesize viral components and produce viral progeny. Depending on the virus, viral genome replicates and viral progeny assembles either in the cytoplasm (coronavirus, poxvirus) or in the nucleus (adenovirus, polyomavirus). Further, some viruses replicate their genome in the nucleus and assemble on the plasma membrane (influenza virus, herpesvirus, HIV). Finally, viral progeny are released from the host cell either by budding (influenza virus, HIV), cell lysis (adenovirus), or exocytosis (coronavirus, herpesvirus) (Flint et al. 2015; Willey et al. 2022). Throughout their life cycle, viruses exploit and antagonize various cellular factors and transport and signaling pathways for successful multiplication (Flint et al. 2015).

12.1.3 Viral Infection and Disease

Although viruses as organisms were discovered in the early twentieth century, the viral diseases like polio, foot and mouth disease, and smallpox have been known since ancient times (Duckworth 1976; Scholthof 2004; Woolhouse et al. 2012; Wolfe et al. 2007). In humans, viruses cause two types of infection—acute and persistent; the latter is further divided into latent and chronic infection. These infections could lead to various short-term or life-long fatal or non-fatal viral diseases. Acute viral infection has a short incubation period, but is characterized by robust virus replication and progeny production and appearance of disease symptoms. Most acute symptoms are resolved within days and the virus is cleared from the host. However, some viruses like herpesvirus, HBV, HCV, and HIV, can persist in the host even after the resolution of acute symptoms. This can lead to either latent infection (herpesvirus), where mostly viral genome persists in the host or chronic infection (HBV, HCV, and HIV), where virus continues to replicate and make progeny albeit at a slower pace. In addition, some viruses like HPV and HBV, transform the infected cells into malignant cells which leads to the development of cancer. The host's innate immune response and its antagonism by the virus significantly influence the progression and severity of viral infection and disease (Flint et al. 2015; Willey et al. 2022).

12.1.4 Viral Infection and Cancer

Cancer is developed when the normal growth control mechanisms of a cell are disrupted, and the cell proliferates uncontrollably and stops performing its tissue-specific functions. Cancer is caused by multiple agents and mechanisms, and about 10% of all cancers are caused by viral infection (Plummer et al. 2016). Seven viruses—HPV, HBV, HCV, Epstein–Barr virus, Kaposi's sarcoma-associated herpesvirus, Human T-lymphotropic virus 1, and Merkel cell polyomavirus—are known to cause the majority of the viral infection-associated cancers in humans (Krump and You 2018; Kellogg et al. 2021; Mullen and Christofk 2022). In addition, HIV infection increases the risk of cancer development (Kellogg et al. 2021). These seven viruses, also known as oncogenic viruses, carry viral oncogenes which encode viral oncoproteins (Krump and You 2018; Mullen and Christofk 2022). The viral oncoproteins interfere with multiple cellular pathways, particularly associated with cell cycle and metabolism, to increase the host cell survival and anabolic processes for viral proliferation (Krump and You 2018; Mullen and Christofk 2022). Almost all oncogenic viruses encode viral oncoproteins that dysregulate the function of cell cycle-regulating proteins, p53 and retinoblastoma protein (pRB) (Krump and You 2018). Furthermore, viral oncoproteins interfere with the metabolic pathways which results, e.g., in increased glycolysis, oxidative stress, and amino acid, nucleotide and lipid synthesis (Mullen and Christofk 2022). In addition, oncogenic viruses also exploit and manipulate the host cell DNA damage response and innate immune response, respectively (Krump and You 2018).

12.1.5 Host Proteome Complexity and Viral Infection

The proteome (a complete set of expressed proteins) of a cell is much more complex and dynamic than the genome (a complete set of genetic instructions), because it constantly changes in response to various stimuli, e.g., infection and disease. More than one protein, called isoforms or proteoforms can be encoded by an mRNA through alternative initiation and splicing and ribosomal frameshifting. Furthermore, a protein isoform can exist in multiple species through protein modifications. Over 650 protein modifications, e.g., glycosylation, phosphorylation, methylation, acetylation, and ubiquitination are now known to occur on over 250,000 sites in human proteome (Doll and Burlingame 2015; Zhong et al. 2023). In addition, those protein isoforms and species, depending on the stimulus, can form multiple complexes and localize to and traffic through different cellular organelles and compartments (Harper and Bennett 2016; Carbonara et al. 2021). Since mammalian viruses utilize one or more cellular organelle and compartment to complete their life cycle, they must navigate such proteome complexity as well as evade the intricate cascades of innate immune signaling (Lum and Cristea 2016; Jean Beltran et al. 2017; Zhou et al. 2017; Locatelli and Faure-Dupuy 2023).

The protein modifications are one of the main contributors to complexity of proteome thereby the complexity of an organism (Zhong et al. 2023). Any imbalance and abnormality in the modifications of a protein changes the properties and functions of that protein, which results in the development of various diseases and disorders (Q. Zhong et al. 2023). Acetylation was initially discovered as an epigenetic modification of histones (Phillips 1963) and, consequently, an important regulator of gene expression (Allfrey et al. 1964; Kimura et al. 2005; MacDonald and Howe 2009). However, acetylation is now known to be ubiquitous and rivals the phosphorylation in prevalence and significance. Acetylation occurs on both histone and non-histone proteins and is known to play a critical role in diverse cellular processes (Narita et al. 2019; Norris et al. 2009; Verdin and Ott 2015).

12.1.6 Acetylation

Acetylation is the transfer of an acetyl group to an amino, thiol, or hydroxyl group of a protein (Christensen et al. 2019; Drazic et al. 2016). It is known to occur at lysine, serine, threonine, or histidine side chains or at the N-terminus of a polypeptide (Christensen et al. 2019; Ree et al. 2018). Hence, it is known as lysine acetylation, O-acetylation, or N-terminal acetylation (Diallo et al. 2019). The lysine and N-terminal acetylation are common in humans and occur post-translationally and co-translationally, respectively (Verdin and Ott 2015; Ree et al. 2018) though N-terminal acetylation can also occur post-translationally (Aksnes et al. 2016). Lysine acetylation and N-terminal acetylation are catalyzed by the histone acetyltransferases (HATs) (also known as lysine acetyltransferases or KATs) and N-terminal acetyltransferases (NATs), respectively. Further, lysine acetylation is reversible and can be removed by histone deacetylases (HDACs) (also known as lysine deacetylases or KDACs and class III HDACs are also known as sirtuins or SIRT). So far, N-terminal acetylation is believed to be irreversible, and a "N-terminal deacetylase" is yet to be identified (Ree et al. 2018).

Acetylation influences protein stability, protein folding, protein-protein interactions, protein trafficking, and protein localization; therefore, acetylation is critical for cellular metabolism and cellular response to stress (Aksnes et al. 2016; Ree et al. 2018; Verdin and Ott 2015; Xue et al. 2022). An imbalance in acetylation as a result of aberrant activity or expression of HATs, NATs, or HDACs contributes to the development of cancers and cardiovascular, inflammatory and infectious diseases (including viral diseases) and developmental, neurodegenerative and genetic disorders in humans (Seto and Yoshida 2014; Park and Kim 2020). Consequently, acetylation is a therapeutic target for treating such diseases and disorders (Verdin and Ott 2015; Ashton et al. 2022).

As early as 1976, a role of acetylation was observed in virus infection (Schaffhausen and Benjamin 1976). Since then, the role and significance of acetylation have been extensively studied in the life cycle of viruses which cause both acute and persistent infection (Murray et al. 2019; Knipe and Cliffe 2008; Jeng et al. 2015; Tsukuda and Watashi 2020; Xue et al. 2022; Husain 2024).

12.2 Acetylation and Adenovirus

Adenoviruses infect and cause a range of acute symptoms in humans. These include common cold, sore throat, bronchitis, conjunctivitis (red eye), and gastroenteritis. Adenoviruses can also cause persistent infection in immunocompromised patients (Lion 2014) and have been found to be associated with malignancy in humans (Lion 2014). Acetylation promotes adenovirus replication and some HATs and HDACs have been identified as proviral and antiviral factors, respectively (Kamel et al. 2024; Wang et al. 2015; Xie et al. 2020; Taura et al. 2010; Chiocca et al. 2002). Further, HDAC-mediated repression of adenovirus replication also promotes persistent infection (Snaider et al. 2022). To facilitate their replication, adenoviruses induce the acetylation of microtubules (Warren et al. 2006) and viral core proteins (Fedor and Daniell 1980; Avgousti et al. 2016). However, adenoviruses also selectively inhibit p300/CBP-mediated histone acetylation to repress the expression of host genes which interfere with viral replication (E. Hsu et al. 2018; Lynch et al. 2019; Savelyeva and Dobbelstein 2011).

12.3 Acetylation and Coronavirus

Coronaviruses have become of significant interest since the emergence of severe acute respiratory syndrome coronavirus 1 (SARS-CoV-1) in 2003, Middle East Respiratory Syndrome Coronavirus (MERS-CoV) in 2012, and SARS-CoV-2 in 2019 (Fehr and Perlman 2015; Zhu et al. 2020b). In addition to humans, coronaviruses infect other mammals as well as birds and cause acute respiratory (mild and severe) and gastrointestinal symptoms (Fehr and Perlman 2015; Zhu et al. 2020b). Latest findings indicate that acetylation plays a proviral role during coronavirus infection as different coronaviruses antagonize multiple HDACs, which exhibit antiviral properties during infection (Xu et al. 2021; Li et al. 2022, Li et al. 2024a, b; Song et al. 2023; Li et al. 2023d). In agreement, HDAC inhibitors augment the replication of SARS-CoV-2 and porcine epidemic diarrhea virus (PEDV) (Ravindran et al. 2022; Trionfetti et al. 2023; Yang et al. 2023). Furthermore, MERS-CoV polyprotein and SARS-CoV-2 nucleocapsid (N) and envelope (E) proteins are acetylated on multiple conserved sites (L. Zhu et al. 2020a; Hatakeyama et al. 2021; Wang et al. 2023; Vann et al. 2022; Pinto et al. 2023). Importantly, the increased acetylation of a host protein—transforming growth factor–beta (TGF-β)–induced protein (TGFBIp)—has been linked with the increased COVID-19 severity in patients, and acetylated TGFBIp has been suggested as a diagnostic biomarker of COVID-19 severity (Park et al. 2020).

12.4 Acetylation and Hepatitis Virus

Hepatitis viruses cause viral hepatitis, an inflammation of liver in humans. Five types, but biologically unrelated hepatitis viruses—A, B, C, D, and E, are known to cause viral hepatitis. The infection with all hepatitis viruses results in similar acute symptoms—malaise, fatigue, nausea, abdominal pain, and jaundice. While hepatitis A virus (HAV) and hepatitis E virus (HEV) are cleared from almost all patients, the hepatitis B virus (HBV) and hepatitis C virus (HCV) persist in a subset of patients, which develop chronic viral hepatitis. Further, some chronic HBV and HCV hepatitis patients could develop liver cancer or hepatocellular carcinoma (HCC) (Dienstag and Delemos 2015). Acetylation has been discovered to be involved in or promote the replication of most hepatitis viruses though the majority of the data in this space are available with HBV infection.

HBV is the only hepatitis virus that contains a DNA genome as a relaxed circular, partially double-stranded DNA, which is converted into covalently closed circular DNA (cccDNA) in the host cell nucleus. The cccDNA is a transcriptionally active replicative intermediate and exists as a stable episome organized into minichromosome by histones and non-histone proteins. Various HATs (p300/CBP/GCN5) and HDACs (HDAC1, SIRT1) are recruited to this minichromosome, and the kinetics of their recruitment is co-related with HBV replication and viremia in patients (Levrero et al. 2009). An increased recruitment of HATs or increased acetylation of histones results in the increased HBV replication, whereas an increased recruitment of HDACs or decreased acetylation of histones results in the decreased HBV replication (Levrero et al. 2009; Koumbi et al. 2016; Lucifora et al. 2011; Riviere et al. 2015; Cougot et al. 2012; Qin et al. 2024). Furthermore, KAT2A (Qin et al. 2021) and HDAC3 (Zhao et al. 2022a), HDAC4 (Zhang et al. 2011), HDAC11 (Yuan et al. 2019), and SIRT3 (Ren et al. 2018) have been individually demonstrated to promote and inhibit the HBV transcription and replication, respectively. In addition, many proteins in HBV-induced HCC liver tissues are hyperacetylated (Xue et al. 2020; Chai et al. 2021), indicating that acetylation promotes HCC. Nevertheless, some reports indicate a contradictory role of HATs and HDACs in HBV transcription and replication (Ren et al. 2014; Zhang et al. 2019; Wang et al. 2021).

Consistent with the role of acetylation in HBV infection, acetylation is positively co-related with the replication of hepatitis D virus (HDV), which is a satellite virus and only replicates in HBV-infected cells (Mu et al. 2004; Liao et al. 2009). Furthermore, HDAC6 has been shown to inhibit HCV replication (H. M. Liu et al. 2016) and microtubules are hyperacetylated in HEV-infected cells (Kannan et al. 2009).

12.5 Acetylation and Herpesvirus

Herpesviruses are some of the most common and prevalent viruses, and infect humans, other mammals, birds, and reptiles. Over 100 herpesviruses are known, and eight of them—herpes simplex virus (HSV)-1, HSV-2, human cytomegalovirus (HCMV), varicella-zoster virus (VZV), Epstein–Barr virus (EBV), Kaposi's sarcoma-associated herpesvirus (KSHV), human herpesvirus (HHV)-6 and HHV-7—are known to infect humans. Herpesviruses exhibit cellular tropism and cause different diseases (cold sores, genital herpes, warts, lymphoma, skin rash, malignancy) in humans. A hallmark of herpesviruses is that following the acute (also referred as "lytic") infection phase, most of them establish a life-long latent infection in their host. However, this latent infection can be reactivated into acute infection in response to a stimulus such as stress and hormonal imbalance (van Lint and Knipe 2014).

Multiple herpesviruses have been shown to disrupt the function and activity of HDACs, indicating an antiviral role of HDACs and a proviral role of acetylation during herpesvirus infection (Guise et al. 2013). Indeed, several HDACs have been demonstrated to play a direct antiviral role during the infection of different herpesviruses (Sheng and Cristea 2021; Koyuncu et al. 2014; Murphy et al. 2002; Tang and Maul 2003; Tang et al. 2021; Nakashima et al. 2015; Soday et al. 2019). Furthermore, a reduced histone acetylation or recruitment of HDACs on lytic gene promoters has been associated with herpesvirus latency and, conversely, an increased histone acetylation or the inhibition of HDACs is linked with herpesvirus reactivation (Jenkins et al. 2000; Murphy et al. 2002; Lu et al. 2006; Chang and Liu 2000; Gruffat et al. 2002; Yang et al. 2009; Kubat et al. 2004b; Poon et al. 2006; Shin et al. 2014). Likewise, several HATs have been linked with the transcription of herpesvirus lytic genes and the reactivation of herpesviruses from latency (Cotter 2nd and Robertson 2000; Wang et al. 2000; Subramanian et al. 2002; Oussaief et al. 2009; Adamson and Kenney 1999; Kalamvoki and Roizman 2010; Yan et al. 2023). Paradoxically, acetylation also promotes herpesvirus latency by increasing the transcription of latency-associated genes (Gerle et al. 2007; Kubat et al. 2004a). Also, increased acetylation is associated with the herpesvirus-mediated malignancy (Huang et al. 2015; Ding et al. 2019). In addition, several viral and host proteins are acetylated in response to herpesvirus infection and the acetylated host proteins facilitate herpesvirus infection (Murray et al. 2018; Elliott and O'Hare 1998; Zhong et al. 2014; Naranatt et al. 2005). Nevertheless, in contrast to the above, some acetylated host proteins and SIRT2 have recently been shown to inhibit and promote herpesvirus infection, respectively (Betsinger et al. 2023; Murray et al. 2018; Hsu et al. 2004; Li et al. 2023c).

12.6 Acetylation and Influenza Virus

Influenza virus causes an acute febrile respiratory disease called "influenza" in humans. Influenza mostly occurs in winter season and is commonly known as "flu", but flu name has become synonymous with any acute respiratory or gastrointestinal symptoms even though they may have been caused by a different virus or microbe. Humans, worldwide, have been suffering with the epidemics of flu for centuries throughout the year, because winter season alternates between the Northern and Southern hemispheres. In addition, influenza virus causes occasional but unpredictable pandemics and zoonotic outbreaks in humans. Influenza virus exists in four types—A, B, C, and D. Influenza A virus (IAV) is the most significant among four types as it causes all influenza events—winter epidemics, occasional pandemics, and zoonotic outbreaks, in humans. Furthermore, IAV has a broad range of hosts that inhabit all three media of environment, the land, water, and air (Husain 2024). Influenza virus possesses the error-prone RNA genome but in segmented configuration. These characteristics, combined with broad host range, enable IAV to constantly evolve into new variants and continuously circulate in nature, making it difficult to eradicate.

Acetylation promotes IAV infection (Husain 2024). The first indication of this phenomenon came to light just over a decade and a half ago when it was discovered that tubulin is hyperacetylated and the activity of tubulin deacetylase, HDAC6 is downregulated in IAV-infected epithelial cells (Husain and Harrod 2009, 2011). In addition, an HDAC inhibitor enhanced the infection of IAV in epithelial cells (Husain and Harrod 2011), indicating a proviral role of HATs/NATs and an antiviral role of HDACs during IAV infection. Indeed, several HATs/NATs (Jin et al. 2014; Zhao et al. 2015; Oishi et al. 2018; Ahmed and Husain 2021) and all class I HDACs (Yang et al. 2022; Xia et al. 2018; Nagesh et al. 2017; Nagesh and Husain 2016), three class II HDACs (Galvin and Husain 2019; Husain and Cheung 2014; Wang et al. 2015; Daly et al. 2022; Choi et al. 2016; Zanin et al. 2020) (Henzeler et al., unpublished), all class III HDACs (Koyuncu et al. 2014; Kim et al. 2016), and sole class IV HDAC (Nutsford et al. 2019) have been identified as pro-IAV and anti-IAV factors, respectively. Furthermore, IAV has been discovered to antagonize the HDACs by downregulating their expression (Hussain et al. 2023; Nutsford et al. 2019; Galvin and Husain 2019; Nagesh et al. 2017; Nagesh and Husain 2016; Xia et al. 2018; De Angelis et al. 2021) as well as activity (Nagesh and Husain 2016; Husain and Cheung 2014; Husain and Harrod 2011) in infected cells. To facilitate this, IAV employs both viral (PA, PA-X) and host (micro-RNA, caspases, proteasome) components (Galvin and Husain 2019; Hussain et al. 2023; Xia et al. 2018; Nagesh et al. 2017; Nagesh and Husain 2016).

So far, three mechanisms have been identified, through which acetylation promotes IAV infection. Acetylation, (a) reduces the sensing of IAV infection by host cells, (b) dampens the host innate anti-IAV response, and (c) increases the stability and function of IAV proteins. Acetylation of RIG-I, one of the host cell pattern recognition receptors (PRRs) that sense IAV infection, prevents the RIG-I activation and, consequently, the sensing of IAV infection and subsequent induction of host innate anti-IAV response cascade (Choi et al. 2016; Liu et al. 2016). Similarly, acetylation downregulates the expression of interferon alpha and beta and other cytokines (CXCL10, interleukin 6 and 15) and interferon-stimulated genes (*IFITM3*, *ISG15*, *ISG20*, and *RSAD2*) in IAV-infected cells (Jin et al. 2014; Ahmed and Husain 2021; Choi et al. 2016; Zanin et al. 2020; Yang et al. 2022; Nutsford et al. 2019; Galvin and Husain 2019; Nagesh et al. 2017; Nagesh and Husain 2016; Zhang et al. 2020). Furthermore, many IAV proteins (M1, M2, NP, NS1, NS2, PA, PA-X, PB1) have been identified to be acetylated (Ahmed et al. 2021; Desuzinges Mandon et al. 2017; Hatakeyama et al. 2018b; Giese et al. 2017; Ma et al. 2020; Hatakeyama et al. 2022; Chen et al. 2019; Oishi et al. 2018). Acetylation promotes the stability and function of IAV proteins and, consequently, the IAV replication (Chen et al. 2019; Hatakeyama et al. 2022; Giese et al. 2017; Ma et al. 2020; Oishi et al. 2018; Hatakeyama et al. 2018b; Daly et al. 2024). In addition, several host proteins, which facilitate different stages of IAV life cycle are also hyperacetylated in IAV-infected cells, e.g., nucleophosmin, SF3B2, and NHP2L1 (viral RNA synthesis and processing); mTOR and eIF4GI (viral protein synthesis); importin, exportin, and tubulin (viral protein trafficking through nucleus and cytoplasm); and actin (viral budding) (Ahmed et al. 2021; Nagesh and Husain 2016; Husain and Harrod 2011; Husain and Cheung 2014).

12.7 Acetylation and Papillomavirus

Papillomaviruses infect the basal epithelial cells of anogenital tract and oral cavity and cause anogenital and oropharyngeal cancers. One of the well-known papillomaviruses is human papillomavirus (HPV), which is associated with cervical cancer in humans. The life cycle of papillomaviruses is tightly regulated and dependent upon the differentiation of infected cells. Early in infection, papillomaviruses maintain a low copy number of their circular DNA genome as episome in basal epithelial cells. However, this number proliferates as the basal epithelial cells differentiate, and viral protein E2, which is essential for viral transcription and replication, drives this proliferation. Subsequently, the viral genome integrates into the host chromosome. This leads to the loss of E2 function but a high level expression of late viral oncoproteins E6 and E7, resulting in viral persistency and oncogenesis (Galloway and Laimins 2015).

Acetylation promotes E2 function but has been reported to play a differential role in oncogenesis. E2 is acetylated and forms complex with various HATs, and the acetylation and interaction of E2 with HATs promote early viral transcription and replication (Lee et al. 2000; Lee et al. 2002; Muller et al. 2002; Kruppel et al. 2008; Quinlan et al. 2013; Thomas and Androphy 2018, 2019). Conversely, SIRT1 deacetylates E2 to inhibit early viral transcription and replication (Das et al. 2017; Prabhakar et al. 2024). Consistent with this, acetylation has been reported to promote cervical cancer progression (Long et al. 2023). In contrast, viral oncoproteins

E6 and E7 have been reported to inhibit the function of HATs and augment the function of SIRT1 to promote viral persistency and transformation of infected cells (Thomas and Chiang 2005; Hebner et al. 2007; Allison et al. 2009; Guo et al. 2013; Langsfeld et al. 2015).

12.8 Acetylation and Retrovirus

Retroviruses are another group of viruses that cause persistent infection in humans albeit by a different mechanism. Using a unique enzyme called reverse transcriptase, retroviruses reverse transcribe their single-stranded RNA genome into double-stranded DNA, which is then integrated into host genome. The integrated viral DNA is then transcribed and translated into viral transcripts and proteins, respectively, for viral replication. Consequently, retroviruses are two types: exogenous and endogenous. The exogenous retroviruses are complete and infectious viruses which are transmitted horizontally or vertically from host to host. Whereas, the endogenous retroviruses are the remnant integrated genetic sequences of the ancient exogenous retroviral infections that are transmitted vertically as inherited sequences (Cloyd 1996). Two prominent species of exogenous retroviruses which infect humans are HIV (HIV-1 and -2) and human T-lymphotropic or T-cell leukemia virus (HTLV-1, −2, −3, and − 4). The HIV infection in humans has three phases: asymptomatic (or latent) phase, acute phase with symptoms like fever, sweats, sore throat, lymphadenopathy, nausea, headaches and rash, and chronic phase which develops into acquired immune deficiency syndrome (AIDS). The chronic phase is also accompanied by a wide range of opportunistic infections and neoplasia (Cloyd 1996). The HTLV causes an asymptomatic infection in humans but in a subset of individuals, it can lead to adult T-cell leukemia/lymphoma (ATL) and a neurologic disease called tropical spastic paraparesis (Bryan and Tadi 2024).

From genome integration to activation of transcription to viral progeny production, acetylation plays a proviral role in HIV infection (Jeng et al. 2015; Tsai et al. 2020). Furthermore, acetylation is also critical for HIV latency. The integration of reverse-transcribed viral DNA is catalyzed by a viral protein, integrase. Acetylation of integrase by HATs—p300 and GCN5—enhances its integration activity (Cereseto et al. 2005; Topper et al. 2007; Terreni et al. 2010; Winans and Goff 2020) and, conversely, deacetylation of integrase inhibits its integration activity (Allouch et al. 2011; H. Kim et al. 2018). Next step in HIV life cycle, the activation of transcription facilitated by essential viral protein transactivator of transcription (Tat), is also promoted by acetylation. Tat itself undergoes the acetylation by various HATs as well as recruits them to viral promoter to activate the viral transcription (Marzio et al. 1998; Kiernan et al. 1999; Dorr et al. 2002; Mujtaba et al. 2002; Dormeyer et al. 2003; Lusic et al. 2003; Col et al. 2001; Boehm et al. 2013). Further, acetylation of Tat is critical for its viral transcription activation function (Kiernan et al. 1999; Ott et al. 1999; Deng et al. 2000; Kaehlcke et al. 2003; Huo et al. 2011b). Furthermore, acetylated Tat regulates the HIV mRNA splicing (Berro et al. 2006) and promotes the apoptosis of T cells during AIDS (Huo et al. 2011a). In addition, HIV infection promotes the acetylation of several host proteins which, in turn, promote HIV replication (Gadad et al. 2011; Sabo et al. 2013). Consistent with this, several HDACs inhibit HIV infection and, in turn, HIV antagonizes them to promote infection (Valenzuela-Fernandez et al. 2005; Marrero-Hernandez et al. 2019; Ran et al. 2019; Kwon et al. 2008; Huo et al. 2011b). Finally, like herpesviruses, several HDACs promote HIV latency and the HDAC inhibitors reactivate HIV from latency (Jiang et al. 2007; Kuai et al. 2018; Peterson et al. 2023; Armani-Tourret et al. 2024; Boehm et al. 2013).

Like HIV Tat protein, the HTLV Tax protein is acetylated by HATs and the acetylated Tax recruits HATs to HTLV promoter to activate viral transcription (Nyborg et al. 2010; Lodewick et al. 2009). Furthermore, acetylation is associated with T cell transformation during HTLV infection toward the development of adult T cell leukemia (Wurm et al. 2012; Lodewick et al. 2013; Romeo et al. 2015). Similarly, the foamy virus Tas protein is acetylated by p300/PCAF and the acetylated Tas facilitates viral transcription (Bodem et al. 2007; Chang et al. 2011).

12.9 Acetylation and Other Viruses

12.9.1 African Swine Fever Virus (ASFV)

ASFV, a large DNA virus, causes lethal hemorrhagic fever in pigs and is transmitted by ticks. ASFV infection induces the acetylation of microtubules in infected cells that aid the intracellular transport of virions (Jouvenet et al. 2004). Further, an ASFV protein essential for such transport is acetylated (Alfonso et al. 2007).

12.9.2 Borna Disease Virus (BDV)

BDV is a neurotropic virus and infects various mammalian hosts, including humans. It causes persistent infection which leads to cognitive impairment. BDV infections lead to reduction in global acetylation level in infected cells due to downregulation and upregulation of HATs and HDACs expression, respectively (Liu et al. 2014; Bonnaud et al. 2015), and such reduction in acetylation level is associated with viral persistency and cognitive impairment (Jie et al. 2018).

12.9.3 Classic Swine Fever Virus (CSFV)

CSFV is an RNA virus and, like ASFV, causes hemorrhagic fever in pigs. Acetylation is pro-CSFV as CSFV infection induces the acetylation of microtubules in infected cells that promote virus replication (Lou et al. 2023). Consistent with this, HDAC1 inhibits CSFV replication but is antagonized by CSFV (Chen et al. 2023). Furthermore, CSFV induces the acetylation of antiviral protein, High Mobility Group Box 1 (HMGB1) to antagonize its antiviral function (Han et al. 2024).

12.9.4 Ebola Virus

Ebola virus causes viral hemorrhagic fever in humans and primates and is one of the deadliest pathogens known. Acetylation plays a proviral role in Ebola virus infection and facilitates the release of Ebola virus progeny from infected cells (Zhang et al. 2021). Furthermore, Ebola viral proteins, NP and VP40 can be acetylated *in vitro* (Hatakeyama et al. 2018a).

12.9.5 Enterovirus

Enteroviruses (coxsackievirus B, enterovirus 71) cause mild to severe diseases, like herpangina, hand, foot and mouth disease, viral myocarditis, and various neurological complications in humans. Acetylation promotes enterovirus infection by increasing the stability of viral RNA and proteins and replication sites in infected cells (Zhou et al. 2015; Han et al. 2016; Zhao et al. 2022b; Yang et al. 2024; Hao et al. 2022). Furthermore, acetylation exacerbates the coxsackievirus B3-induced myocarditis (Zhou et al. 2015).

12.9.6 Flavivirus

Flaviviruses are vector-borne viruses which pose significant burden to human health. The flaviviruses, Zika, West Nile, Dengue, and Yellow Fever are transmitted by mosquitoes and cause severe complications like microcephaly, encephalitis, hemorrhagic fever, and liver damage, respectively, in humans. Flavivirus NS3 protein, which is essential and performs multiple functions like protease and RNA helicase during virus replication, is acetylated on its RNA-binding domain by KAT5 (Serman et al. 2023). Acetylation of NS3 is essential for its RNA helicase activity and, consequently, flavivirus replication (Serman et al. 2023). Furthermore, Zika virus NS5 protein induces the acetylation of microtubules in infected cells (Pérez-Yanes et al. 2024).

12.9.7 Paramyxovirus

Paramyxoviruses (human parainfluenza virus, Sendai virus) cause mild respiratory disease in humans and rodents. The HDAC3, HDAC6, and SIRT5 have been identified to promote innate antiviral response in Sendai virus-infected cells (Zhu et al. 2011; He et al. 2021; Tang et al. 2021). Further, Sendai virus M protein is acetylated at the N-terminus (Blumberg et al. 1984). Furthermore, acetylation of microtubules stabilizes the cytoplasmic replication sites (inclusion bodies) of parainfluenza virus type 3 and promotes viral RNA synthesis and replication (Liu et al. 2022; Zhang et al. 2017). This indicates that acetylation promotes paramyxovirus infection.

12.9.8 Polyomavirus

Polyomaviruses are widespread in nature and cause mostly latent infection in a range of hosts, including humans, other mammals, birds, reptiles, and fish. Polyomaviruses are particularly known for their ability to transform infected cells, and the simian virus 40 (SV40) is the most characterized and studied polyomavirus. All polyomaviruses encode a multifunctional viral oncoprotein called tumor (T) antigen, which is essential for virus replication and sufficient for the transformation and immortalization of mammalian cells. T antigen interacts with CBP/p300 and promotes their autoacetylation and protein levels in transformed cells (Saenz Robles et al. 2013). Further, VP1 capsid protein of mouse polyomavirus induces the acetylation of microtubules in infected cells (Hornikova et al. 2020). Also, T antigen is acetylated by CBP and deacetylated by HDAC1, HDAC3, and SIRT1 in infected cells; however, acetylation destabilizes T antigen (Poulin et al. 2004; Shimazu et al. 2006).

12.9.9 Porcine Reproductive and Respiratory Syndrome Virus (PRRSV)

PRRSV infection, as the name suggests, causes reproductive and respiratory syndromes in pigs resulting in reproductive failure and severe respiratory disease, respectively. Acetylation promotes PRRSV infection as transgenic pigs overexpressing the HDAC6 (Lu et al. 2017) or pig cells overexpressing the HDAC2 (Zhang et al. 2024a) resist PRRSV infection. Further, PRRSV antagonizes endogenous

HDAC2 to evade the host innate immunity (Zhang et al. 2024a). However, Li et al. reported that the N-terminal acetylation of PRRSV structural protein, GP5 promotes its degradation and reduces viral assembly (Li et al. 2023b).

12.9.10 Poxvirus

Poxviruses cause a disease in humans that is characterized by skin lesions, which could lead to permanent skin marks and disfigurement. Poxviruses have a special place in the history of modern medicine. During the late eighteenth century, an experiment with a poxvirus (cowpox virus) by Edward Jenner led to the discovery of vaccination. Then, a poxvirus (vaccinia virus) was used as first vaccine to eradicate the deadly smallpox disease caused by another poxvirus, variola virus. Now, another poxvirus, mpox (formerly known as monkeypox) has recently re-emerged in humans and is spreading rapidly into many countries and, in August 2024, the World Health Organization (WHO) declared the mpox a global health emergency (Editorial 2024). Recent studies have shown that HDAC4 and HDAC5 restrict vaccinia virus infection and, in turn, vaccinia virus antagonizes them through viral protein C6 (Y. Lu et al. 2019; Soday et al. 2019). Furthermore, mouse poxvirus, ectromelia induces the acetylation of microtubules to facilitate intracellular transport of virions (Szulc-Dabrowska et al. 2019). These findings indicate that acetylation plays a supportive role during poxvirus infection.

12.9.11 Vesicular Stomatitis Virus (VSV)

VSV causes an acute febrile vesicular disease in cattle, horses, and pigs. However, VSV is best known for being used as a model to understand the biology of RNA viruses due to its broad cellular tropism. Furthermore, VSV G protein has been extensively used as a model protein to understand the intracellular protein trafficking. The HDAC3, HDAC8, and SIRT5 have been identified to promote host innate antiviral immunity in response to VSV infection, indicating a pro-VSV role for acetylation (He et al. 2021; Tang et al. 2021; Zhang et al. 2024b).

12.9.12 Viruses of Plants, Fungi, and Insects

Viruses also cause a significant burden on plant, fungal, and insect hosts, and acetylation has been identified to play an important role in the infection of viruses of plants, fungi, as well as insects. The viral proteins of plant viruses have been discovered to be acetylated (McGarry et al. 2003; Cilia et al. 2014) and acetylation of many host proteins is upregulated in response to the infection of plant viruses

(B. Yuan et al. 2021; Song et al. 2022; Gong et al. 2023). Similarly, several host proteins are also hyperacetylated in response to the infection of fungal (Li et al. 2023a) and insect (Hu et al. 2018; Gao et al. 2022; M. Hu et al. 2022; Zhang et al. 2024d) viruses. Also, several proteins of insect viruses are acetylated (Gao et al. 2021; Ngowo et al. 2022; Wu et al. 2023). However, the acetylation of insect host and virus proteins has been identified to play both proviral and antiviral functions. While the acetylation of host fructose-bisphosphate aldolase (ALDO) in response to Bombyx mori nucleopolyhedrovirus (BmNPV) infection promotes BmNPV infection (Gao et al. 2022), the acetylation of host citrate synthase (BmCS) and heat shock protein 90 inhibits BmNPV infection (Hu et al. 2022; Zhang et al. 2024d). Similarly, the acetylation of viral late expression factor (LEF) 4 and 11 promotes BmNPV infection (Ngowo et al. 2022; Wu et al. 2023), but the acetylation of viral LEF3 inhibits BmNPV infection (Gao et al. 2021).

Finally, acetylation promotes the infection of parvoviruses, the single-stranded DNA viruses which cause respiratory and gastrointestinal diseases in dogs (Mantyla et al. 2016; Zhang et al. 2024c), foot-and-mouth disease virus (FMDV) which causes foot-and-mouth disease in cloven-hoofed animals (Hou et al. 2022), and mosquito-borne Sindbis virus which causes Sindbis fever in humans (Dang et al. 2024). In addition, two Sindbis virus proteins are acetylated (Bell and Strauss 1981).

12.10 Acetylation and Treatment of Viral Infection and Disease

It is clear from this review that acetylation promotes the infection of the majority of viruses. Furthermore, a decrease in acetylation level hinders clearance of viruses like EBV and HIV from the host and promotes persistent viral infection, which is difficult to cure. However, such persistent viral infections can be re-activated into acute viral infections by increasing the acetylation level through the use of HDAC inhibitors (Jenkins et al. 2000; Van Lint et al. 1996). This observation led to the "Shock and Kill" approach (Deeks 2012), where HDAC inhibitors are used to reactivate a latent viral infection into an acute (or lytic) infection, which then can be cleared with regular antiviral therapies (Lehrman et al. 2005; Archin et al. 2012; Ghosh et al. 2012). In addition, this approach also has the potential to treat viral infection-associated cancers (Hui et al. 2016). Several studies and trials using a range of HDAC inhibitors have adopted this approach to cure patients with EBV infection (Hui et al. 2016; Haverkos et al. 2023) and HIV infection (Rasmussen et al. 2014; Kuai et al. 2018; Li et al. 2020; Luling Wu et al. 2024). Paradoxically, HDAC inhibitors are also being utilized to boost the infection of oncolytic viruses for cancer therapy (Kawamura et al. 2022; Nakatake et al. 2024; Nakashima et al. 2015).

12.11 Summary

As obligate intracellular parasites, viruses exploit and antagonize the host machinery to infect, multiply and cause disease. Therefore, a ubiquitous protein modification such as acetylation, which contributes to the complexity of host proteome, is bound to play an integral role in viral infection and disease. Barely a decade after the discovery of acetylation, a role of acetylation in virus infection was observed (Schaffhausen and Benjamin 1976). However, for the next several decades, such a role was mostly studied for the viruses that cause persistent (latent and chronic) infection. Lately, the significance of acetylation in the infection of viruses that cause acute infection has been investigated. Overall, by reviewing the literature generated in this space so far, it can be concluded that acetylation plays a proviral role in the infection of most viruses and is one of the determinants of viral pathogenesis in the host. For this, acetylation influences both host (Table 12.1) and viral (Table 12.2) components. Therefore, acetylation can be targeted to treat various viral infections and diseases.

Table 12.1 Host factors or pathways influenced by the acetylation

Virus	Host factor or pathway
Adenovirus	Microtubules
Coronavirus	Transforming growth factor beta-induced protein (SARS-CoV-2)
Hepatitis virus	Histones (HBV); microtubules (HEV)
Herpesvirus	Histones, microtubules (HSV-1, KSHV)
Influenza virus	Microtubules, RIG-I, innate immune response (IAV)
Retrovirus	Microtubules, nucleophosmin (HIV-1)
Other viruses	Microtubules (ASFV, CSFV, Zika, Parainfluenza, Polyoma, Ectromelia); high mobility group box 1 (CSFV); innate immune response (Sendai, PRRSV, VSV)

Table 12.2 Viral proteins influenced by acetylation

Virus	Viral protein
Adenovirus	Core proteins V and pVII
Coronavirus	Polyprotein (MERS-CoV); Nucleocapsid and envelope (SARS-CoV-2)
Herpesvirus	pUL26 (CMV)
Influenza virus	M1, M2, NP, NS1, NS2, PA, PA-X, and PB1 (IAV)
Papillomavirus	E2 (HPV)
Retrovirus	Integrase, Tat (HIV); Tax (HTLV); Tas (foamy virus)
Other viruses	pE120R (ASFV); Non-structural proteins (enterovirus 71); NP and VP40 (Ebola); NS3 (Flavivirus); M (Sendai); T antigen (Polyoma); GP5 (PRRSV)

Acknowledgments The author acknowledges recent funding support from the School of Biomedical Sciences (University of Otago), J C and H S Anderson Charitable Trust (New Zealand), Maurice & Paykel Charitable Trust (New Zealand), and Maurice Wilkins Centre (New Zealand).

References

Adamson AL, Kenney S (1999) The Epstein-Barr virus BZLF1 protein interacts physically and functionally with the histone acetylase CREB-binding protein. J Virol 73(8):6551–6558. https://doi.org/10.1128/JVI.73.8.6551-6558.1999

Ahmed F, Husain M (2021) Human N-alpha-acetyltransferase 60 promotes influenza A virus infection by dampening the interferon alpha signaling (original research). Front Immunol 12(5833):771792. https://doi.org/10.3389/fimmu.2021.771792

Ahmed F, Kleffmann T, Husain M (2021) Acetylation, methylation and Allysine modification profile of viral and host proteins during influenza A virus infection. Viruses 13(7). https://doi.org/10.3390/v13071415

Aksnes H, Drazic A, Marie M, Arnesen T (2016) First things first: vital protein marks by N-terminal acetyltransferases. Trends Biochem Sci 41(9):746–760. https://doi.org/10.1016/j.tibs.2016.07.005

Alfonso P, Quetglas JI, Escribano JM, Alonso C (2007) Protein pE120R of African swine fever virus is post-translationally acetylated as revealed by post-source decay MALDI mass spectrometry. Virus Genes 35(1):81–85. https://doi.org/10.1007/s11262-006-0015-6

Allfrey VG, Faulkner R, Mirsky AE (1964) Acetylation and methylation of histones and their possible role in the regulation of Rna synthesis. Proc Natl Acad Sci USA 51(5):786–794. https://doi.org/10.1073/pnas.51.5.786

Allison SJ, Jiang M, Milner J (2009) Oncogenic viral protein HPV E7 up-regulates the SIRT1 longevity protein in human cervical cancer cells. Aging (Albany NY) 1(3):316–327. https://doi.org/10.18632/aging.100028

Allouch A, Di Primio C, Alpi E, Lusic M, Arosio D, Giacca M et al (2011) The TRIM family protein KAP1 inhibits HIV-1 integration. Cell Host Microbe 9(6):484–495. https://doi.org/10.1016/j.chom.2011.05.004

Archin NM, Liberty AL, Kashuba AD, Choudhary SK, Kuruc JD, Crooks AM et al (2012) Administration of vorinostat disrupts HIV-1 latency in patients on antiretroviral therapy. Nature 487(7408):482–485. https://doi.org/10.1038/nature11286

Armani-Tourret M, Gao C, Hartana CA, Sun W, Carrere L, Vela L et al (2024) Selection of epigenetically privileged HIV-1 proviruses during treatment with panobinostat and interferon-alpha2a. Cell 187(5):1238–1254 e14. https://doi.org/10.1016/j.cell.2024.01.037

Ashton AW, Dhanjal HK, Rossner B, Mahmood H, Patel VI, Nadim M et al (2022) Acetylation of nuclear receptors in health and disease: an update. FEBS J. https://doi.org/10.1111/febs.16695

Avgousti DC, Herrmann C, Kulej K, Pancholi NJ, Sekulic N, Petrescu J et al (2016) A core viral protein binds host nucleosomes to sequester immune danger signals. Nature 535(7610):173–177. https://doi.org/10.1038/nature18317

Bell JR, Strauss JH (1981) In vivo NH2-terminal acetylation of Sindbis virus proteins. J Biol Chem 256(15):8006–8011. https://doi.org/10.1016/S0021-9258(18)43379-0

Berro R, Kehn K, de la Fuente C, Pumfery A, Adair R, Wade J et al (2006) Acetylated tat regulates human immunodeficiency virus type 1 splicing through its interaction with the splicing regulator p32. J Virol 80(7):3189–3204. https://doi.org/10.1128/JVI.80.7.3189-3204.2006

Betsinger CN, Justice JL, Tyl MD, Edgar JE, Budayeva HG, Abu YF et al (2023) Sirtuin 2 promotes human cytomegalovirus replication by regulating cell cycle progression. mSystems 8(6):e0051023. https://doi.org/10.1128/msystems.00510-23

Blumberg BM, Rose K, Simona MG, Roux L, Giorgi C, Kolakofsky D (1984) Analysis of the Sendai virus M gene and protein. J Virol 52(2):656–663. https://doi.org/10.1128/JVI.52.2.656-663.1984

Bodem J, Krausslich HG, Rethwilm A (2007) Acetylation of the foamy virus transactivator Tas by PCAF augments promoter-binding affinity and virus transcription. J Gen Virol 88(Pt 1):259–263. https://doi.org/10.1099/vir.0.82169-0

Boehm D, Conrad RJ, Ott M (2013) Bromodomain proteins in HIV infection. Viruses 5(6):1571–1586. https://doi.org/10.3390/v5061571

Bonnaud EM, Szelechowski M, Betourne A, Foret C, Thouard A, Gonzalez-Dunia D et al (2015) Borna disease virus phosphoprotein modulates epigenetic signaling in neurons to control viral replication. J Virol 89(11):5996–6008. https://doi.org/10.1128/JVI.00454-15

Breitbart M, Rohwer F (2005) Here a virus, there a virus, everywhere the same virus? Trends Microbiol 13(6):278–284. https://doi.org/10.1016/j.tim.2005.04.003

Bryan ES, Tadi P (2024) Human T-cell lymphotropic virus. *StatPearls* Treasure Island (FL) ineligible companies Disclosure: Prasanna Tadi declares no relevant financial relationships with ineligible companies.: StatPearls Publishing. Copyright © 2024, StatPearls Publishing LLC., https://www.ncbi.nlm.nih.gov/pubmed/32809660

Carbonara K, Andonovski M, Coorssen JR (2021) Proteomes are of Proteoforms: embracing the complexity. Proteomes 9(3). https://doi.org/10.3390/proteomes9030038

Cereseto A, Manganaro L, Gutierrez MI, Terreni M, Fittipaldi A, Lusic M et al (2005) Acetylation of HIV-1 integrase by p300 regulates viral integration. EMBO J 24(17):3070–3081. https://doi.org/10.1038/sj.emboj.7600770

Chai X, Guo J, Dong R, Yang X, Deng C, Wei C et al (2021) Quantitative acetylome analysis reveals histone modifications that may predict prognosis in hepatitis B-related hepatocellular carcinoma. Clin Transl Med 11(3):e313. https://doi.org/10.1002/ctm2.313

Chang LK, Liu ST (2000) Activation of the BRLF1 promoter and lytic cycle of Epstein-Barr virus by histone acetylation. Nucleic Acids Res 28(20):3918–3925. https://doi.org/10.1093/nar/28.20.3918

Chang R, Tan J, Xu F, Han H, Geng Y, Li Y et al (2011) Lysine acetylation sites in bovine foamy virus transactivator BTas are important for its DNA binding activity. Virology 418(1):21–26. https://doi.org/10.1016/j.virol.2011.07.003

Chen H, Qian Y, Chen X, Ruan Z, Ye Y, Chen H et al (2019) HDAC6 restricts influenza A virus by Deacetylation of the RNA polymerase PA subunit. J Virol 93(4). https://doi.org/10.1128/JVI.01896-18

Chen R, Han X, Xu H, Xu J, Cao T, Shan Y et al (2023) N-terminal domain of classical swine fever virus N(pro) induces proteasomal degradation of specificity protein 1 with reduced HDAC1 expression to evade from innate immune responses. J Virol 97(10):e0111523. https://doi.org/10.1128/jvi.01115-23

Chiocca S, Kurtev V, Colombo R, Boggio R, Sciurpi MT, Brosch G et al (2002) Histone deacetylase 1 inactivation by an adenovirus early gene product. Curr Biol 12(7):594–598. https://doi.org/10.1016/s0960-9822(02)00720-0

Choi SJ, Lee HC, Kim JH, Park SY, Kim TH, Lee WK et al (2016) HDAC6 regulates cellular viral RNA sensing by deacetylation of RIG-I. EMBO J 35(4):429–442. https://doi.org/10.15252/embj.201592586

Christensen DG, Xie X, Basisty N, Byrnes J, McSweeney S, Schilling B et al (2019) Post-translational protein acetylation: an elegant mechanism for bacteria to dynamically regulate metabolic functions (review). Front Microbiol 10:1604. https://doi.org/10.3389/fmicb.2019.01604

Cilia M, Johnson R, Sweeney M, DeBlasio SL, Bruce JE, MacCoss MJ et al (2014) Evidence for lysine acetylation in the coat protein of a polerovirus. J Gen Virol 95(Pt 10):2321–2327. https://doi.org/10.1099/vir.0.066514-0

Cloyd MW (1996) Human retroviruses. In: Baron S (ed) Medical microbiology. University of Texas Medical Branch at Galveston, Galveston. Copyright © 1996, The University of Texas Medical Branch at Galveston. https://www.ncbi.nlm.nih.gov/pubmed/21413279

Col E, Caron C, Seigneurin-Berny D, Gracia J, Favier A, Khochbin S (2001) The histone acetyltransferase, hGCN5, interacts with and acetylates the HIV transactivator, Tat. J Biol Chem 276(30):28179–28184. https://doi.org/10.1074/jbc.M101385200

Cotter MA 2nd, Robertson ES (2000) Modulation of histone acetyltransferase activity through interaction of epstein-barr nuclear antigen 3C with prothymosin alpha. Mol Cell Biol 20(15):5722–5735. https://doi.org/10.1128/MCB.20.15.5722-5735.2000

Cougot D, Allemand E, Riviere L, Benhenda S, Duroure K, Levillayer F et al (2012) Inhibition of PP1 phosphatase activity by HBx: a mechanism for the activation of hepatitis B virus transcription. Sci Signal 5(205):ra1. https://doi.org/10.1126/scisignal.2001906

Daly K, Hsu A, Nichol K, Horvat J, Mayall J, Hansbro P et al (2022) Histone deacetylase-6 (HDAC6) promotes protective responses during murine influenza A virus (IAV) infection. Eur Respir J 60(suppl 66):1243. https://doi.org/10.1183/13993003.congress-2022.1243

Daly RE, Myasnikov I, Gaglia MM (2024) N-terminal acetylation separately promotes nuclear localization and host shutoff activity of the influenza A virus ribonuclease PA-X bioRxiv:2023.12.01.569683. https://doi.org/10.1101/2023.12.01.569683

Dang Y, Li J, Li Y, Wang Y, Zhao Y, Zhao N et al (2024) N-acetyltransferase 10 regulates alphavirus replication via N4-acetylcytidine (ac4C) modi fi cation of the lymphocyte antigen six family member E (LY6E) mRNA. J Virol 98(1):e0135023. https://doi.org/10.1128/jvi.01350-23

Das D, Smith N, Wang X, Morgan IM (2017) The deacetylase SIRT1 regulates the replication properties of human papillomavirus 16 E1 and E2. J Virol 91(10). https://doi.org/10.1128/JVI.00102-17

De Angelis M, Amatore D, Checconi P, Zevini A, Fraternale A, Magnani M et al (2021) Influenza virus Down-modulates G6PD expression and activity to induce oxidative stress and promote its replication. Front Cell Infect Microbiol 11:804976. https://doi.org/10.3389/fcimb.2021.804976

de la Higuera I, Lazaro E (2022) Viruses in astrobiology. Front Microbiol 13:1032918. https://doi.org/10.3389/fmicb.2022.1032918

de Souza WM, Weaver SC (2024) Effects of climate change and human activities on vector-borne diseases. Nat Rev Microbiol 22(8):476–491. https://doi.org/10.1038/s41579-024-01026-0

Deeks SG (2012) Shock and kill. Nature 487(7408):439–440. https://doi.org/10.1038/487439a

Deng L, de la Fuente C, Fu P, Wang L, Donnelly R, Wade JD et al (2000) Acetylation of HIV-1 tat by CBP/P300 increases transcription of integrated HIV-1 genome and enhances binding to core histones. Virology 277(2):278–295. https://doi.org/10.1006/viro.2000.0593

Desuzinges Mandon E, Traversier A, Champagne A, Benier L, Audebert S, Balme S et al (2017) Expression and purification of native and functional influenza A virus matrix 2 proton selective ion channel. Protein Expr Purif 131:42–50. https://doi.org/10.1016/j.pep.2016.11.001

Diallo I, Seve M, Cunin V, Minassian F, Poisson JF, Michelland S et al (2019) Current trends in protein acetylation analysis. Expert Rev Proteomics 16(2):139–159. https://doi.org/10.1080/14789450.2019.1559061

Dienstag JL, Delemos AS (2015) Viral hepatitis. In: Bennett JE, Dolin R, Blaser MJ (eds) Mandell, Douglas, and Bennett's principles and practice of infectious diseases. W.B. Saunders, Philadelphia, pp 1439–1468.e7. https://doi.org/10.1016/b978-1-4557-4801-3.00119-3

Ding X, Xu J, Wang C, Feng Q, Wang Q, Yang Y et al (2019) Suppression of the SAP18/HDAC1 complex by targeting TRIM56 and Nanog is essential for oncogenic viral FLICE-inhibitory protein-induced acetylation of p65/RelA, NF-kappaB activation, and promotion of cell invasion and angiogenesis. Cell Death Differ 26(10):1970–1986. https://doi.org/10.1038/s41418-018-0268-3

Doll S, Burlingame AL (2015) Mass spectrometry-based detection and assignment of protein post-translational modifications. ACS Chem Biol 10(1):63–71. https://doi.org/10.1021/cb500904b

Dormeyer W, Dorr A, Ott M, Schnolzer M (2003) Acetylation of the HIV-1 tat protein: an in vitro study. Anal Bioanal Chem 376(7):994–1005. https://doi.org/10.1007/s00216-003-2058-z

Dorr A, Kiermer V, Pedal A, Rackwitz HR, Henklein P, Schubert U et al (2002) Transcriptional synergy between tat and PCAF is dependent on the binding of acetylated tat to the PCAF bromodomain. EMBO J 21(11):2715–2723. https://doi.org/10.1093/emboj/21.11.2715

Drazic A, Myklebust LM, Ree R, Arnesen T (2016) The world of protein acetylation. Biochim Biophys Acta 1864(10):1372–1401. https://doi.org/10.1016/j.bbapap.2016.06.007

Duckworth DH (1976) Who discovered bacteriophage? Bacteriol Rev 40(4):793–802. https://doi.org/10.1128/br.40.4.793-802.1976

Editorial (2024) Mpox: apply COVID lessons to control outbreak in Africa. Nature 633(8029):253–254. https://doi.org/10.1038/d41586-024-02912-6

Elliott G, O'Hare P (1998) Herpes simplex virus type 1 tegument protein VP22 induces the stabilization and hyperacetylation of microtubules. J Virol 72(8):6448–6455. https://doi.org/10.1128/JVI.72.8.6448-6455.1998

Fedor MJ, Daniell E (1980) Acetylation of histone-like proteins of adenovirus type 5. J Virol 35(3):637–643. https://doi.org/10.1128/JVI.35.3.637-643.1980

Fehr AR, Perlman S (2015) Coronaviruses: an overview of their replication and pathogenesis. Methods Mol Biol 1282:1–23. https://doi.org/10.1007/978-1-4939-2438-7_1

Flint SJ, Racaniello VR, Rall GF, Skalka AM, Enquist LW (2015) Principles of virology. ASM Press, Washington, DC

Gadad SS, Rajan RE, Senapati P, Chatterjee S, Shandilya J, Dash PK et al (2011) HIV-1 infection induces acetylation of NPM1 that facilitates tat localization and enhances viral transactivation. J Mol Biol 410(5):997–1007. https://doi.org/10.1016/j.jmb.2011.04.009

Galloway DA, Laimins LA (2015) Human papillomaviruses: shared and distinct pathways for pathogenesis. Curr Opin Virol 14:87–92. https://doi.org/10.1016/j.coviro.2015.09.001

Galvin HD, Husain M (2019) Influenza A virus-induced host caspase and viral PA-X antagonize the antiviral host factor, histone deacetylase 4. J Biol Chem 294(52):20207–20221. https://doi.org/10.1074/jbc.RA119.010650

Gao X, Lei J, Zhu Y, Chen X, Mao F, Miao M et al (2021) Role of the Bombyx mori nucleopolyhedrovirus LEF3 acetylation on viral replication. Microb Pathog 158:105109. https://doi.org/10.1016/j.micpath.2021.105109

Gao X, Xu S, Mo Y, Zhu Y, Chen X, Miao M et al (2022) Acetylation of fructose-bisphosphate aldolase-mediated glycolysis is essential for Bombyx mori nucleopolyhedrovirus infection. Microb Pathog 170:105695. https://doi.org/10.1016/j.micpath.2022.105695

Gerle B, Koroknai A, Fejer G, Bakos A, Banati F, Szenthe K et al (2007) Acetylated histone H3 and H4 mark the upregulated LMP2A promoter of Epstein-Barr virus in lymphoid cells. J Virol 81(23):13242–13247. https://doi.org/10.1128/JVI.01396-07

Ghosh SK, Perrine SP, Williams RM, Faller DV (2012) Histone deacetylase inhibitors are potent inducers of gene expression in latent EBV and sensitize lymphoma cells to nucleoside antiviral agents. Blood 119(4):1008–1017. https://doi.org/10.1182/blood-2011-06-362434

Giese S, Ciminski K, Bolte H, Moreira EA, Lakdawala S, Hu Z et al (2017) Role of influenza A virus NP acetylation on viral growth and replication. Nat Commun 8(1):1259. https://doi.org/10.1038/s41467-017-01112-3

Gong Y, Li Y, Liu D, Jiang L, Liang H, Wu Y et al (2023) Analysis of lysine acetylation in tomato spot wilt virus infection in Nicotiana benthamiana (original research). Front Microbiol 14:1046163. https://doi.org/10.3389/fmicb.2023.1046163

Gruffat H, Manet E, Sergeant A (2002) MEF2-mediated recruitment of class II HDAC at the EBV immediate early gene BZLF1 links latency and chromatin remodeling. EMBO Rep 3(2):141–146. https://doi.org/10.1093/embo-reports/kvf031

Guise AJ, Budayeva HG, Diner BA, Cristea IM (2013) Histone deacetylases in herpesvirus replication and virus-stimulated host defense. Viruses 5(7):1607–1632. https://doi.org/10.3390/v5071607

Guo Y, Meng X, Wang Q, Wang Y, Shang H (2013) The ING4 binding with p53 and induced p53 acetylation were attenuated by human papillomavirus 16 E6. PLoS One 8(8):e71453. https://doi.org/10.1371/journal.pone.0071453

Han X, Xu H, Weng Y, Chen R, Xu J, Cao T et al (2024) N(pro) of classical swine fever virus enhances HMGB1 acetylation and its degradation by lysosomes to evade from HMGB1-mediated antiviral immunity. Virus Res 339:199280. https://doi.org/10.1016/j.virusres.2023.199280

Han Y, Wang L, Cui J, Song Y, Luo Z, Chen J et al (2016) SIRT1 inhibits EV71 genome replication and RNA translation by interfering with the viral polymerase and 5'UTR RNA. J Cell Sci 129(24):4534–4547. https://doi.org/10.1242/jcs.193698

Hao H, Liu W, Miao Y, Ma L, Yu B, Liu L et al (2022) N4-acetylcytidine regulates the replication and pathogenicity of enterovirus 71. Nucleic Acids Res 50(16):9339–9354. https://doi.org/10.1093/nar/gkac675

Harper JW, Bennett EJ (2016) Proteome complexity and the forces that drive proteome imbalance. Nature 537(7620):328–338. https://doi.org/10.1038/nature19947

Hatakeyama D, Masuda T, Miki R, Ohtsuki S, Kuzuhara T (2021) In-vitro acetylation of SARS-CoV and SARS-CoV-2 nucleocapsid proteins by human PCAF and GCN5. Biochem Biophys Res Commun 557:273–279. https://doi.org/10.1016/j.bbrc.2021.03.173

Hatakeyama D, Ohmi N, Saitoh A, Makiyama K, Morioka M, Okazaki H et al (2018a) Acetylation of lysine residues in the recombinant nucleoprotein and VP40 matrix protein of Zaire ebolavirus by eukaryotic histone acetyltransferases. Biochem Biophys Res Commun 504(4):635–640. https://doi.org/10.1016/j.bbrc.2018.09.007

Hatakeyama D, Shoji M, Ogata S, Masuda T, Nakano M, Komatsu T et al (2022) Acetylation of the influenza A virus polymerase subunit PA in the N-terminal domain positively regulates its endonuclease activity. FEBS J 289(1):231–245. https://doi.org/10.1111/febs.16123

Hatakeyama D, Shoji M, Yamayoshi S, Yoh R, Ohmi N, Takenaka S et al (2018b) Influenza A virus nucleoprotein is acetylated by histone acetyltransferases PCAF and GCN5. J Biol Chem 293(19):7126–7138. https://doi.org/10.1074/jbc.RA117.001683

Haverkos B, Alpdogan O, Baiocchi R, Brammer JE, Feldman TA, Capra M et al (2023) Targeted therapy with nanatinostat and valganciclovir in recurrent EBV-positive lymphoid malignancies: a phase 1b/2 study. Blood Adv 7(20):6339–6350. https://doi.org/10.1182/bloodadvances.2023010330

He X, Li T, Qin K, Luo S, Li Z, Ji Q et al (2021) Demalonylation of DDX3 by Sirtuin 5 promotes antiviral innate immune responses. Theranostics 11(15):7235–7246. https://doi.org/10.7150/thno.52934

Hebner C, Beglin M, Laimins LA (2007) Human papillomavirus E6 proteins mediate resistance to interferon-induced growth arrest through inhibition of p53 acetylation. J Virol 81(23):12740–12747. https://doi.org/10.1128/JVI.00987-07

Hornikova L, Brustikova K, Ryabchenko B, Zhernov I, Fraiberk M, Marinicova Z et al (2020) The major capsid protein, VP1, of the mouse polyomavirus stimulates the activity of tubulin acetyltransferase 1 by microtubule stabilization. Viruses 12(2). https://doi.org/10.3390/v12020227

Hou S, Wang X, Ren S, Meng X, Yin X, Zhang J et al (2022) Knockout of HDAC9 gene enhances foot-and-mouth disease virus replication. Front Microbiol 13:805606. https://doi.org/10.3389/fmicb.2022.805606

Hsu CH, Chang MD, Tai KY, Yang YT, Wang PS, Chen CJ et al (2004) HCMV IE2-mediated inhibition of HAT activity downregulates p53 function. EMBO J 23(11):2269–2280. https://doi.org/10.1038/sj.emboj.7600239

Hsu E, Pennella MA, Zemke NR, Eng C, Berk AJ (2018) Adenovirus E1A activation domain regulates H3 acetylation affecting varied steps in transcription at different viral promoters. J Virol 92(18). https://doi.org/10.1128/jvi.00805-18

Hu D, Xue S, Zhao C, Wei M, Yan H, Quan Y et al (2018) Comprehensive profiling of lysine Acetylome in Baculovirus infected silkworm (Bombyx mori) cells. Proteomics 18(1):1700133. https://doi.org/10.1002/pmic.201700133

Hu M, Zhu Y, Mo Y, Gao X, Miao M, Yu W (2022) Acetylation of citrate synthase inhibits Bombyx mori nucleopolyhedrovirus propagation by affecting energy metabolism. Microb Pathog 173(Pt B):105890. https://doi.org/10.1016/j.micpath.2022.105890

Huang R, Qian D, Hu M, Zhang X, Song J, Li L et al (2015) Association between human cytomegalovirus infection and histone acetylation level in various histological types of glioma. Oncol Lett 10(5):2812–2820. https://doi.org/10.3892/ol.2015.3638

Hui KF, Cheung AK, Choi CK, Yeung PL, Middeldorp JM, Lung ML et al (2016) Inhibition of class I histone deacetylases by romidepsin potently induces Epstein-Barr virus lytic cycle and mediates enhanced cell death with ganciclovir. Int J Cancer 138(1):125–136. https://doi.org/10.1002/ijc.29698

Huo L, Li D, Sun L, Liu M, Shi X, Sun X et al (2011a) Tat acetylation regulates its actions on microtubule dynamics and apoptosis in T lymphocytes. J Pathol 223(1):28–36. https://doi.org/10.1002/path.2768

Huo L, Li D, Sun X, Shi X, Karna P, Yang W et al (2011b) Regulation of tat acetylation and transactivation activity by the microtubule-associated deacetylase HDAC6. J Biol Chem 286(11):9280–9286. https://doi.org/10.1074/jbc.M110.208884

Husain M (2024) Influenza A virus and acetylation: the picture is becoming clearer. Viruses 16(1):131. https://doi.org/10.3390/v16010131

Husain M, Cheung CY (2014) Histone deacetylase 6 inhibits influenza A virus release by downregulating the trafficking of viral components to the plasma membrane via its substrate, acetylated microtubules. J Virol 88(19):11229–11239. https://doi.org/10.1128/JVI.00727-14

Husain M, Harrod KS (2009) Influenza A virus-induced caspase-3 cleaves the histone deacetylase 6 in infected epithelial cells. FEBS Lett 583(15):2517–2520. https://doi.org/10.1016/j.febslet.2009.07.005

Husain M, Harrod KS (2011) Enhanced acetylation of alpha-tubulin in influenza A virus infected epithelial cells. FEBS Lett 585(1):128–132. https://doi.org/10.1016/j.febslet.2010.11.023

Hussain M, Ahmed F, Henzeler B, Husain M (2023) Anti-microbial host factor HDAC6 is antagonised by the influenza A virus through host caspases and viral PA. FEBS J 290(10):2744–2759. https://doi.org/10.1111/febs.16703

Jean Beltran PM, Cook KC, Cristea IM (2017) Exploring and exploiting proteome organization during viral infection. J Virol 91(18). https://doi.org/10.1128/JVI.00268-17

Jeng MY, Ali I, Ott M (2015) Manipulation of the host protein acetylation network by human immunodeficiency virus type 1. Crit Rev Biochem Mol Biol 50(4):314–325. https://doi.org/10.3109/10409238.2015.1061973

Jenkins PJ, Binne UK, Farrell PJ (2000) Histone acetylation and reactivation of Epstein-Barr virus from latency. J Virol 74(2):710–720. https://doi.org/10.1128/jvi.74.2.710-720.2000

Jiang G, Espeseth A, Hazuda DJ, Margolis DM (2007) C-Myc and Sp1 contribute to proviral latency by recruiting histone deacetylase 1 to the human immunodeficiency virus type 1 promoter. J Virol 81(20):10914–10923. https://doi.org/10.1128/JVI.01208-07

Jie J, Xu X, Xia J, Tu Z, Guo Y, Li C et al (2018) Memory impairment induced by Borna disease virus 1 infection is associated with reduced H3K9 acetylation. Cell Physiol Biochem 49(1):381–394. https://doi.org/10.1159/000492890

Jin Q, Zhuang L, Lai B, Wang C, Li W, Dolan B et al (2014) Gcn5 and PCAF negatively regulate interferon-beta production through HAT-independent inhibition of TBK1. EMBO Rep 15(11):1192–1201. https://doi.org/10.15252/embr.201438990

Jouvenet N, Monaghan P, Way M, Wileman T (2004) Transport of African swine fever virus from assembly sites to the plasma membrane is dependent on microtubules and conventional kinesin. J Virol 78(15):7990–8001. https://doi.org/10.1128/JVI.78.15.7990-8001.2004

Kaehlcke K, Dorr A, Hetzer-Egger C, Kiermer V, Henklein P, Schnoelzer M et al (2003) Acetylation of tat defines a cyclinT1-independent step in HIV transactivation. Mol Cell 12(1):167–176. https://doi.org/10.1016/s1097-2765(03)00245-4

Kalamvoki M, Roizman B (2010) Circadian CLOCK histone acetyl transferase localizes at ND10 nuclear bodies and enables herpes simplex virus gene expression. Proc Natl Acad Sci USA 107(41):17721–17726. https://doi.org/10.1073/pnas.1012991107

Kamel H, Shete V, Gadamsetty S, Graves D, Bachus S, Akkerman N et al (2024) HBO1/KAT7/MYST2 HAT complex regulates human adenovirus replicative cycle. Heliyon 10(7):e28827. https://doi.org/10.1016/j.heliyon.2024.e28827

Kannan H, Fan S, Patel D, Bossis I, Zhang YJ (2009) The hepatitis E virus open reading frame 3 product interacts with microtubules and interferes with their dynamics. J Virol 83(13):6375–6382. https://doi.org/10.1128/JVI.02571-08

Kawamura Y, Hua L, Gurtner A, Wong E, Kiyokawa J, Shah N et al (2022) Histone deacetylase inhibitors enhance oncolytic herpes simplex virus therapy for malignant meningioma. Biomed Pharmacother 155:113843. https://doi.org/10.1016/j.biopha.2022.113843

Kellogg C, Kouznetsova VL, Tsigelny IF (2021) Implications of viral infection in cancer development. Biochim Biophys Acta Rev Cancer 1876(2):188622. https://doi.org/10.1016/j.bbcan.2021.188622

Kiernan RE, Vanhulle C, Schiltz L, Adam E, Xiao H, Maudoux F et al (1999) HIV-1 tat transcriptional activity is regulated by acetylation. EMBO J 18(21):6106–6118. https://doi.org/10.1093/emboj/18.21.6106

Kim H, Lee SY, Choi YM, Kim BJ (2018) HBV polymerase-derived peptide exerts an anti-HIV-1 effect by inhibiting the acetylation of viral integrase. Biochem Biophys Res Commun 501(2):541–546. https://doi.org/10.1016/j.bbrc.2018.05.033

Kim JA, Seong RK, Shin OS (2016) Enhanced viral replication by cellular replicative senescence. Immune Netw 16(5):286–295. https://doi.org/10.4110/in.2016.16.5.286

Kimura A, Matsubara K, Horikoshi M (2005) A decade of histone acetylation: marking eukaryotic chromosomes with specific codes. J Biochem 138(6):647–662. https://doi.org/10.1093/jb/mvi184

Knipe DM, Cliffe A (2008) Chromatin control of herpes simplex virus lytic and latent infection. Nat Rev Microbiol 6(3):211–221. https://doi.org/10.1038/nrmicro1794

Koumbi L, Pollicino T, Raimondo G, Stampoulis D, Khakoo S, Karayiannis P (2016) Hepatitis B virus basal core promoter mutations show lower replication fitness associated with cccDNA acetylation status. Virus Res 220:150–160. https://doi.org/10.1016/j.virusres.2016.04.022

Koyuncu E, Budayeva HG, Miteva YV, Ricci DP, Silhavy TJ, Shenk T et al (2014) Sirtuins are evolutionarily conserved viral restriction factors. mBio 5(6). https://doi.org/10.1128/mbio.02249-14

Krump NA, You J (2018) Molecular mechanisms of viral oncogenesis in humans. Nat Rev Microbiol 16(11):684–698. https://doi.org/10.1038/s41579-018-0064-6

Kruppel U, Muller-Schiffmann A, Baldus SE, Smola-Hess S, Steger G (2008) E2 and the co-activator p300 can cooperate in activation of the human papillomavirus type 16 early promoter. Virology 377(1):151–159. https://doi.org/10.1016/j.virol.2008.04.006

Kuai Q, Lu X, Qiao Z, Wang R, Wang Y, Ye S et al (2018) Histone deacetylase inhibitor chidamide promotes reactivation of latent human immunodeficiency virus by introducing histone acetylation. J Med Virol 90(9):1478–1485. https://doi.org/10.1002/jmv.25207

Kubat NJ, Amelio AL, Giordani NV, Bloom DC (2004a) The herpes simplex virus type 1 latency-associated transcript (LAT) enhancer/rcr is hyperacetylated during latency independently of LAT transcription. J Virol 78(22):12508–12518. https://doi.org/10.1128/JVI.78.22.12508-12518.2004

Kubat NJ, Tran RK, McAnany P, Bloom DC (2004b) Specific histone tail modification and not DNA methylation is a determinant of herpes simplex virus type 1 latent gene expression. J Virol 78(3):1139–1149. https://doi.org/10.1128/jvi.78.3.1139-1149.2004

Kwon HS, Brent MM, Getachew R, Jayakumar P, Chen LF, Schnolzer M et al (2008) Human immunodeficiency virus type 1 tat protein inhibits the SIRT1 deacetylase and induces T cell hyperactivation. Cell Host Microbe 3(3):158–167. https://doi.org/10.1016/j.chom.2008.02.002

Langsfeld ES, Bodily JM, Laimins LA (2015) The deacetylase Sirtuin 1 regulates human papillomavirus replication by modulating histone acetylation and recruitment of DNA damage factors NBS1 and Rad51 to viral genomes. PLoS Pathog 11(9):e1005181. https://doi.org/10.1371/journal.ppat.1005181

Lee D, Hwang SG, Kim J, Choe J (2002) Functional interaction between p/CAF and human papillomavirus E2 protein. J Biol Chem 277(8):6483–6489. https://doi.org/10.1074/jbc.M105085200
Lee D, Lee B, Kim J, Kim DW, Choe J (2000) cAMP response element-binding protein-binding protein binds to human papillomavirus E2 protein and activates E2-dependent transcription. J Biol Chem 275(10):7045–7051. https://doi.org/10.1074/jbc.275.10.7045
Lehrman G, Hogue IB, Palmer S, Jennings C, Spina CA, Wiegand A et al (2005) Depletion of latent HIV-1 infection in vivo: a proof-of-concept study. Lancet 366(9485):549–555. https://doi.org/10.1016/S0140-6736(05)67098-5
Levrero M, Pollicino T, Petersen J, Belloni L, Raimondo G, Dandri M (2009) Control of cccDNA function in hepatitis B virus infection. J Hepatol 51(3):581–592. https://doi.org/10.1016/j.jhep.2009.05.022
Li J, Ma J, Kang W, Wang C, Bai F, Zhao K et al (2020) The histone deacetylase inhibitor chidamide induces intermittent viraemia in HIV-infected patients on suppressive antiretroviral therapy. HIV Med 21(11):747–757. https://doi.org/10.1111/hiv.13027
Li R, Chen F, Li S, Yuan L, Zhao L, Tian S et al (2023a) Comparative acetylomic analysis reveals differentially acetylated proteins regulating fungal metabolism in hypovirus-infected chestnut blight fungus. Mol Plant Pathol 24(9):1126–1138. https://doi.org/10.1111/mpp.13358
Li X, Sun R, Guo Y, Zhang H, Xie R, Fu X et al (2023b) N-acetyltransferase 9 inhibits porcine reproductive and respiratory syndrome virus proliferation by N-terminal acetylation of the structural protein GP5. Microbiol Spectr 11(1):e0244222. https://doi.org/10.1128/spectrum.02442-22
Li Y, Bie J, Song C, Li Y, Zhang T, Li H et al (2023c) SIRT2 negatively regulates the cGAS-STING pathway by deacetylating G3BP1. EMBO Rep 24(12):e57500. https://doi.org/10.15252/embr.202357500
Li Z, Duan P, Qiu R, Fang L, Fang P, Xiao S (2023d) HDAC6 degrades nsp8 of porcine deltacoronavirus through deacetylation and ubiquitination to inhibit viral replication. J Virol 97(5):e0037523. https://doi.org/10.1128/jvi.00375-23
Li Z, Fang P, Duan P, Chen J, Fang L, Xiao S (2022) Porcine Deltacoronavirus infection cleaves HDAC2 to attenuate its antiviral activity. J Virol 96(16):e0102722. https://doi.org/10.1128/jvi.01027-22
Li Z, Lai Y, Qiu R, Tang W, Ren J, Xiao S et al (2024a) Hyperacetylated microtubules assist porcine deltacoronavirus nsp8 to degrade MDA5 via SQSTM1/p62-dependent selective autophagy. J Virol 98(3):e0000324. https://doi.org/10.1128/jvi.00003-24
Li Z, Xiao W, Yang Z, Guo J, Zhou J, Xiao S et al (2024b) Cleavage of HDAC6 to dampen its antiviral activity by nsp5 is a common strategy of swine enteric coronaviruses. J Virol 98(2):e0181423. https://doi.org/10.1128/jvi.01814-23
Liao FT, Lee YJ, Ko JL, Tsai CC, Tseng CJ, Sheu GT (2009) Hepatitis delta virus epigenetically enhances clusterin expression via histone acetylation in human hepatocellular carcinoma cells. J Gen Virol 90(Pt 5):1124–1134. https://doi.org/10.1099/vir.0.007211-0
Lion T (2014) Adenovirus infections in immunocompetent and immunocompromised patients. Clin Microbiol Rev 27(3):441–462. https://doi.org/10.1128/CMR.00116-13
Liu HM, Jiang F, Loo YM, Hsu S, Hsiang TY, Marcotrigiano J et al (2016) Regulation of retinoic acid inducible gene-I (RIG-I) activation by the histone deacetylase 6. EBioMedicine 9:195–206. https://doi.org/10.1016/j.ebiom.2016.06.015
Liu P, Zhang S, Ma J, Jin D, Qin Y, Chen M (2022) Vimentin inhibits alpha-tubulin acetylation via enhancing alpha-TAT1 degradation to suppress the replication of human parainfluenza virus type 3. PLoS Pathog 18(9):e1010856. https://doi.org/10.1371/journal.ppat.1010856
Liu X, Zhao L, Yang Y, Bode L, Huang H, Liu C et al (2014) Human Borna disease virus infection impacts host proteome and histone lysine acetylation in human oligodendroglia cells. Virology 464-465:196–205. https://doi.org/10.1016/j.virol.2014.06.040
Locatelli M, Faure-Dupuy S (2023) Virus hijacking of host epigenetic machinery to impair immune response. J Virol 97(9):e0065823. https://doi.org/10.1128/jvi.00658-23

Lodewick J, Lamsoul I, Polania A, Lebrun S, Burny A, Ratner L et al (2009) Acetylation of the human T-cell leukemia virus type 1 tax oncoprotein by p300 promotes activation of the NF-kappaB pathway. Virology 386(1):68–78. https://doi.org/10.1016/j.virol.2008.12.043

Lodewick J, Sampaio C, Boxus M, Rinaldi AS, Coulonval K, Willems L et al (2013) Acetylation at lysine 346 controls the transforming activity of the HTLV-1 tax oncoprotein in the Rat-1 fibroblast model. Retrovirology 10(1):75. https://doi.org/10.1186/1742-4690-10-75

Long Y, Ren Y, Wei Q, Mobet Y, Liu Y, Zhao H et al (2023) NAT10-mediated RNA acetylation enhances HNRNPUL1 mRNA stability to contribute cervical cancer progression. Int J Med Sci 20(8):1079–1090. https://doi.org/10.7150/ijms.83828

Lou JX, Liu YY, Bai JS, Cheng Y, Zhang J, Liu CC et al (2023) Kinesin-1 regulates endocytic trafficking of classical swine fever virus along acetylated microtubules. J Virol 97(1):e0192922. https://doi.org/10.1128/jvi.01929-22

Lu F, Day L, Gao SJ, Lieberman PM (2006) Acetylation of the latency-associated nuclear antigen regulates repression of Kaposi's sarcoma-associated herpesvirus lytic transcription. J Virol 80(11):5273–5282. https://doi.org/10.1128/JVI.02541-05

Lu T, Song Z, Li Q, Li Z, Wang M, Liu L et al (2017) Overexpression of histone deacetylase 6 enhances resistance to porcine reproductive and respiratory syndrome virus in pigs. PLoS One 12(1):e0169317. https://doi.org/10.1371/journal.pone.0169317

Lu Y, Stuart JH, Talbot-Cooper C, Agrawal-Singh S, Huntly B, Smid AI et al (2019) Histone deacetylase 4 promotes type I interferon signaling, restricts DNA viruses, and is degraded via vaccinia virus protein C6. Proc Natl Acad Sci USA 116(24):11997–12006. https://doi.org/10.1073/pnas.1816399116

Lucifora J, Arzberger S, Durantel D, Belloni L, Strubin M, Levrero M et al (2011) Hepatitis B virus X protein is essential to initiate and maintain virus replication after infection. J Hepatol 55(5):996–1003. https://doi.org/10.1016/j.jhep.2011.02.015

Lum KK, Cristea IM (2016) Proteomic approaches to uncovering virus-host protein interactions during the progression of viral infection. Expert Rev Proteomics 13(3):325–340. https://doi.org/10.1586/14789450.2016.1147353

Lusic M, Marcello A, Cereseto A, Giacca M (2003) Regulation of HIV-1 gene expression by histone acetylation and factor recruitment at the LTR promoter. EMBO J 22(24):6550–6561. https://doi.org/10.1093/emboj/cdg631

Lynch KL, Gooding LR, Garnett-Benson C, Ornelles DA, Avgousti DC (2019) Epigenetics and the dynamics of chromatin during adenovirus infections. FEBS Lett 593(24):3551–3570. https://doi.org/10.1002/1873-3468.13697

Ma J, Wu R, Xu G, Cheng Y, Wang Z, Wang H et al (2020) Acetylation at K108 of the NS1 protein is important for the replication and virulence of influenza virus. Vet Res 51(1):20. https://doi.org/10.1186/s13567-020-00747-3

MacDonald VE, Howe LJ (2009) Histone acetylation: where to go and how to get there. Epigenetics 4(3):139–143. https://doi.org/10.4161/epi.4.3.8484

Mantyla E, Salokas K, Oittinen M, Aho V, Mantysaari P, Palmujoki L et al (2016) Promoter-targeted histone acetylation of chromatinized parvoviral genome is essential for the progress of infection. J Virol 90(8):4059–4066. https://doi.org/10.1128/JVI.03160-15

Marrero-Hernandez S, Marquez-Arce D, Cabrera-Rodriguez R, Estevez-Herrera J, Perez-Yanes S, Barroso-Gonzalez J et al (2019) HIV-1 Nef targets HDAC6 to assure viral production and virus infection. Front Microbiol 10:2437. https://doi.org/10.3389/fmicb.2019.02437

Martini M, Gazzaniga V, Bragazzi NL, Barberis I (2019) The Spanish influenza pandemic: a lesson from history 100 years after 1918. J Prev Med Hyg 60(1):E64–E67. https://doi.org/10.15167/2421-4248/jpmh2019.60.1.1205

Marzio G, Tyagi M, Gutierrez MI, Giacca M (1998) HIV-1 tat transactivator recruits p300 and CREB-binding protein histone acetyltransferases to the viral promoter. Proc Natl Acad Sci USA 95(23):13519–13524. https://doi.org/10.1073/pnas.95.23.13519

McGarry RC, Barron YD, Carvalho MF, Hill JE, Gold D, Cheung E et al (2003) A novel Arabidopsis acetyltransferase interacts with the geminivirus movement protein NSP. Plant Cell 15(7):1605–1618. https://doi.org/10.1105/tpc.012120

Mu JJ, Tsay YG, Juan LJ, Fu TF, Huang WH, Chen DS et al (2004) The small delta antigen of hepatitis delta virus is an acetylated protein and acetylation of lysine 72 may influence its cellular localization and viral RNA synthesis. Virology 319(1):60–70. https://doi.org/10.1016/j.virol.2003.10.024

Mujtaba S, He Y, Zeng L, Farooq A, Carlson JE, Ott M et al (2002) Structural basis of lysine-acetylated HIV-1 tat recognition by PCAF bromodomain. Mol Cell 9(3):575–586. https://doi.org/10.1016/s1097-2765(02)00483-5

Mullen PJ, Christofk HR (2022) The metabolic relationship between viral infection and cancer. Ann Rev Cancer Biol 6:1–15. https://doi.org/10.1146/annurev-cancerbio-070120-090423

Muller A, Ritzkowsky A, Steger G (2002) Cooperative activation of human papillomavirus type 8 gene expression by the E2 protein and the cellular coactivator p300. J Virol 76(21):11042–11053. https://doi.org/10.1128/jvi.76.21.11042-11053.2002

Murphy JC, Fischle W, Verdin E, Sinclair JH (2002) Control of cytomegalovirus lytic gene expression by histone acetylation. EMBO J 21(5):1112–1120. https://doi.org/10.1093/emboj/21.5.1112

Murray LA, Combs AN, Rekapalli P, Cristea IM (2019) Chapter twenty-four – methods for characterizing protein acetylation during viral infection. In: Garcia BA (ed) Methods in Enzymology. Academic, pp 587–620. https://doi.org/10.1016/bs.mie.2019.06.030

Murray LA, Sheng X, Cristea IM (2018) Orchestration of protein acetylation as a toggle for cellular defense and virus replication. Nat Commun 9(1):4967. https://doi.org/10.1038/s41467-018-07179-w

Nagesh PT, Husain M (2016) Influenza A virus dysregulates host histone deacetylase 1 that inhibits viral infection in lung epithelial cells. J Virol 90(9):4614–4625. https://doi.org/10.1128/JVI.00126-16

Nagesh PT, Hussain M, Galvin HD, Husain M (2017) Histone deacetylase 2 is a component of influenza A virus-induced host antiviral response. Front Microbiol 8:1315. https://doi.org/10.3389/fmicb.2017.01315

Nakashima H, Kaufmann JK, Wang PY, Nguyen T, Speranza MC, Kasai K et al (2015) Histone deacetylase 6 inhibition enhances oncolytic viral replication in glioma. J Clin Invest 125(11):4269–4280. https://doi.org/10.1172/JCI80713

Nakatake M, Kurosaki H, Nakamura T (2024) Histone deacetylase inhibitor boosts anticancer potential of fusogenic oncolytic vaccinia virus by enhancing cell-cell fusion. Cancer Sci 115(2):600–610. https://doi.org/10.1111/cas.16032

Naranatt PP, Krishnan HH, Smith MS, Chandran B (2005) Kaposi's sarcoma-associated herpesvirus modulates microtubule dynamics via RhoA-GTP-diaphanous 2 signaling and utilizes the dynein motors to deliver its DNA to the nucleus. J Virol 79(2):1191–1206. https://doi.org/10.1128/JVI.79.2.1191-1206.2005

Narita T, Weinert BT, Choudhary C (2019) Functions and mechanisms of non-histone protein acetylation. Nat Rev Mol Cell Biol 20(3):156–174. https://doi.org/10.1038/s41580-018-0081-3

Ngowo J, Gao X, Zhu Y, Mao F, Lei J, Obeng E et al (2022) Acetylation of late expression factor 4 is crucial for the transcription and proliferation of Bombyx mori nucleopolyhedrovirus. Acta Virol 66(1):77–84. https://doi.org/10.4149/av_2022_109

Norris KL, Lee JY, Yao TP (2009) Acetylation goes global: the emergence of acetylation biology. Sci Signal 2(97):pe76. https://doi.org/10.1126/scisignal.297pe76

Nutsford AN, Galvin HD, Ahmed F, Husain M (2019) The class IV human deacetylase, HDAC11, exhibits anti-influenza A virus properties via its involvement in host innate antiviral response. Cell Microbiol 21(4):e12989. https://doi.org/10.1111/cmi.12989

Nyborg JK, Egan D, Sharma N (2010) The HTLV-1 tax protein: revealing mechanisms of transcriptional activation through histone acetylation and nucleosome disassembly. Biochim Biophys Acta 1799(3–4):266–274. https://doi.org/10.1016/j.bbagrm.2009.09.002

Oishi K, Yamayoshi S, Kozuka-Hata H, Oyama M, Kawaoka Y (2018) N-terminal acetylation by NatB is required for the shutoff activity of influenza A virus PA-X. Cell Rep 24(4):851–860. https://doi.org/10.1016/j.celrep.2018.06.078

Ott M, Schnolzer M, Garnica J, Fischle W, Emiliani S, Rackwitz HR et al (1999) Acetylation of the HIV-1 tat protein by p300 is important for its transcriptional activity. Curr Biol 9(24):1489–1492. https://doi.org/10.1016/s0960-9822(00)80120-7

Oussaief L, Hippocrate A, Ramirez V, Rampanou A, Zhang W, Meyers D et al (2009) Phosphatidylinositol 3-kinase/Akt pathway targets acetylation of Smad3 through Smad3/CREB-binding protein interaction: contribution to transforming growth factor beta1-induced Epstein-Barr virus reactivation. J Biol Chem 284(36):23912–23924. https://doi.org/10.1074/jbc.M109.036483

Park HH, Kim HN, Kim H, Yoo Y, Shin H, Choi EY et al (2020) Acetylated K676 TGFBIp as a severity diagnostic blood biomarker for SARS-CoV-2 pneumonia. Sci Adv 6(31). https://doi.org/10.1126/sciadv.abc1564

Park SY, Kim JS (2020) A short guide to histone deacetylases including recent progress on class II enzymes. Exp Mol Med 52(2):204–212. https://doi.org/10.1038/s12276-020-0382-4

Pérez-Yanes S, Lorenzo-Sánchez I, Cabrera-Rodríguez R, García-Luis J, Trujillo-González R, Estévez-Herrera J et al (2024) The ZIKV NS5 protein aberrantly alters the tubulin cytoskeleton, induces the accumulation of Autophagic p62 and affects IFN production: HDAC6 has emerged as an anti-NS5/ZIKV factor. Cells 13(7):598. https://www.mdpi.com/2073-4409/13/7/598

Peterson JJ, Lewis CA, Burgos SD, Manickam A, Xu Y, Rowley AA et al (2023) A histone deacetylase network regulates epigenetic reprogramming and viral silencing in HIV-infected cells. Cell Chem Biol 30(12):1617–1633 e9. https://doi.org/10.1016/j.chembiol.2023.11.009

Phillips DM (1963) The presence of acetyl groups of histones. Biochem J 87(2):258–263. https://doi.org/10.1042/bj0870258

Pinto SM, Subbannayya Y, Kim H, Hagen L, Gorna MW, Nieminen AI et al (2023) Multi-OMICs landscape of SARS-CoV-2-induced host responses in human lung epithelial cells. iScience 26(1):105895. https://doi.org/10.1016/j.isci.2022.105895

Plummer M, de Martel C, Vignat J, Ferlay J, Bray F, Franceschi S (2016) Global burden of cancers attributable to infections in 2012: a synthetic analysis. Lancet Glob Health 4(9):e609–e616. https://doi.org/10.1016/S2214-109X(16)30143-7

Poon AP, Gu H, Roizman B (2006) ICP0 and the US3 protein kinase of herpes simplex virus 1 independently block histone deacetylation to enable gene expression. Proc Natl Acad Sci USA 103(26):9993–9998. https://doi.org/10.1073/pnas.0604142103

Poulin DL, Kung AL, DeCaprio JA (2004) p53 targets simian virus 40 large T antigen for acetylation by CBP. J Virol 78(15):8245–8253. https://doi.org/10.1128/JVI.78.15.8245-8253.2004

Prabhakar AT, James CD, Youssef AH, Hossain RA, Hill RD, Bristol ML et al (2024) A human papillomavirus 16 E2-TopBP1 dependent SIRT1-p300 acetylation switch regulates mitotic viral and human protein levels and activates the DNA damage response. MBio 15(6):e0067624. https://doi.org/10.1128/mbio.00676-24

Qin Y, Ren J, Yu H, He X, Cheng S, Chen W et al (2024) HOXA-AS2 epigenetically inhibits HBV transcription by recruiting the MTA1-HDAC1/2 deacetylase complex to cccDNA Minichromosome. Adv Sci (Weinh) 11(24):e2306810. https://doi.org/10.1002/advs.202306810

Qin YP, Yu HB, Yuan SY, Yang Z, Ren F, Wang Q et al (2021) KAT2A promotes hepatitis B virus transcription and replication through epigenetic regulation of cccDNA Minichromosome (original research). Front Microbiol 12:795388. https://doi.org/10.3389/fmicb.2021.795388

Quinlan EJ, Culleton SP, Wu SY, Chiang CM, Androphy EJ (2013) Acetylation of conserved lysines in bovine papillomavirus E2 by p300. J Virol 87(3):1497–1507. https://doi.org/10.1128/JVI.02771-12

Ran X, Ao Z, Olukitibi T, Yao X (2019) Characterization of the role of host cellular factor histone deacetylase 10 during HIV-1 replication. Viruses 12(1). https://doi.org/10.3390/v12010028

Rasmussen TA, Tolstrup M, Brinkmann CR, Olesen R, Erikstrup C, Solomon A et al (2014) Panobinostat, a histone deacetylase inhibitor, for latent-virus reactivation in HIV-infected patients on suppressive antiretroviral therapy: a phase 1/2, single group, clinical trial. Lancet HIV 1(1):e13–e21. https://doi.org/10.1016/S2352-3018(14)70014-1

Ravindran V, Wagoner J, Athanasiadis P, Den Hartigh AB, Sidorova JM, Ianevski A et al (2022) Discovery of host-directed modulators of virus infection by probing the SARS-CoV-2-host protein-protein interaction network. Brief Bioinform 23(6). https://doi.org/10.1093/bib/bbac456

Ree R, Varland S, Arnesen T (2018) Spotlight on protein N-terminal acetylation. Exp Mol Med 50(7):1–13. https://doi.org/10.1038/s12276-018-0116-z

Ren JH, Hu JL, Cheng ST, Yu HB, Wong VKW, Law BYK et al (2018) SIRT3 restricts hepatitis B virus transcription and replication through epigenetic regulation of covalently closed circular DNA involving suppressor of variegation 3-9 homolog 1 and SET domain containing 1A histone methyltransferases. Hepatology 68(4):1260–1276. https://doi.org/10.1002/hep.29912

Ren JH, Tao Y, Zhang ZZ, Chen WX, Cai XF, Chen K et al (2014) Sirtuin 1 regulates hepatitis B virus transcription and replication by targeting transcription factor AP-1. J Virol 88(5):2442–2451. https://doi.org/10.1128/JVI.02861-13

Riviere L, Gerossier L, Ducroux A, Dion S, Deng Q, Michel ML et al (2015) HBx relieves chromatin-mediated transcriptional repression of hepatitis B viral cccDNA involving SETDB1 histone methyltransferase. J Hepatol 63(5):1093–1102. https://doi.org/10.1016/j.jhep.2015.06.023

Romeo MM, Ko B, Kim J, Brady R, Heatley HC, He J et al (2015) Acetylation of the c-MYC oncoprotein is required for cooperation with the HTLV-1 p30(II) accessory protein and the induction of oncogenic cellular transformation by p30(II)/c-MYC. Virology 476:271–288. https://doi.org/10.1016/j.virol.2014.12.008

Sabo Y, Walsh D, Barry DS, Tinaztepe S, de Los Santos K, Goff SP et al (2013) HIV-1 induces the formation of stable microtubules to enhance early infection. Cell Host Microbe 14(5):535–546. https://doi.org/10.1016/j.chom.2013.10.012

Saenz Robles MT, Shivalila C, Wano J, Sorrells S, Roos A, Pipas JM (2013) Two independent regions of simian virus 40 T antigen increase CBP/p300 levels, alter patterns of cellular histone acetylation, and immortalize primary cells. J Virol 87(24):13499–13509. https://doi.org/10.1128/JVI.02658-13

Santiago-Rodriguez TM, Hollister EB (2022) Unraveling the viral dark matter through viral metagenomics. Front Immunol 13:1005107. https://doi.org/10.3389/fimmu.2022.1005107

Savelyeva I, Dobbelstein M (2011) Infection with E1B-mutant adenovirus stabilizes p53 but blocks p53 acetylation and activity through E1A. Oncogene 30(7):865–875. https://doi.org/10.1038/onc.2010.461

Schaffhausen BS, Benjamin TL (1976) Deficiency in histone acetylation in nontransforming host range mutants of polyoma virus. Proc Natl Acad Sci USA 73(4):1092–1096. https://doi.org/10.1073/pnas.73.4.1092

Scholthof KB (2004) Tobacco mosaic virus: a model system for plant biology. Annu Rev Phytopathol 42:13–34. https://doi.org/10.1146/annurev.phyto.42.040803.140322

Serman T, Chiang C, Liu G, Sayyad Z, Pandey S, Volcic M et al (2023) Acetylation of the NS3 helicase by KAT5gamma is essential for flavivirus replication. Cell Host Microbe 31(8):1317–1330 e10. https://doi.org/10.1016/j.chom.2023.06.013

Seto E, Yoshida M (2014) Erasers of histone acetylation: the histone deacetylase enzymes. Cold Spring Harb Perspect Biol 6(4):a018713. https://doi.org/10.1101/cshperspect.a018713

Sheng X, Cristea IM (2021) The antiviral sirtuin 3 bridges protein acetylation to mitochondrial integrity and metabolism during human cytomegalovirus infection. PLoS Pathog 17(4):e1009506. https://doi.org/10.1371/journal.ppat.1009506

Shimazu T, Komatsu Y, Nakayama KI, Fukazawa H, Horinouchi S, Yoshida M (2006) Regulation of SV40 large T-antigen stability by reversible acetylation. Oncogene 25(56):7391–7400. https://doi.org/10.1038/sj.onc.1209731

Shin HJ, DeCotiis J, Giron M, Palmeri D, Lukac DM (2014) Histone deacetylase classes I and II regulate Kaposi's sarcoma-associated herpesvirus reactivation. J Virol 88(2):1281–1292. https://doi.org/10.1128/JVI.02665-13

Snaider S, Zheng Y, Hearing P (2022) Rb-E2F-HDAC repressor complexes control interferon-induced repression of adenovirus to promote persistent infection. J Virol 96(11):e0044222. https://doi.org/10.1128/jvi.00442-22

Soday L, Lu Y, Albarnaz JD, Davies CTR, Antrobus R, Smith GL et al (2019) Quantitative temporal proteomic analysis of vaccinia virus infection reveals regulation of histone deacetylases by an interferon antagonist. Cell Rep 27(6):1920–1933 e7. https://doi.org/10.1016/j.celrep.2019.04.042

Song L, Wang D, Abbas G, Li M, Cui M, Wang J et al (2023) The main protease of SARS-CoV-2 cleaves histone deacetylases and DCP1A, attenuating the immune defense of the interferon-stimulated genes. J Biol Chem 299(3):102990. https://doi.org/10.1016/j.jbc.2023.102990

Song L, Zhan H, Wang Y, Lin Z, Li B, Shen L et al (2022) Cross-talk of protein expression and lysine acetylation in response to TMV infection in Nicotiana benthamiana. ACS Omega 7(36):32496–32511. https://doi.org/10.1021/acsomega.2c03917

Subramanian C, Hasan S, Rowe M, Hottiger M, Orre R, Robertson ES (2002) Epstein-Barr virus nuclear antigen 3C and prothymosin alpha interact with the p300 transcriptional coactivator at the CH1 and CH3/HAT domains and cooperate in regulation of transcription and histone acetylation. J Virol 76(10):4699–4708. https://doi.org/10.1128/jvi.76.10.4699-4708.2002

Szulc-Dabrowska L, Palusinski M, Struzik J, Gregorczyk-Zboroch KP, Toka FN, Schollenberger A et al (2019) Ectromelia virus induces tubulin cytoskeletal rearrangement in immune cells accompanied by a loss of the microtubule organizing center and increased alpha-tubulin acetylation. Arch Virol 164(2):559–565. https://doi.org/10.1007/s00705-018-4030-4

Tang JL, Yang Q, Xu CH, Zhao H, Liu YL, Liu CY et al (2021) Histone deacetylase 3 promotes innate antiviral immunity through deacetylation of TBK1. Protein Cell 12(4):261–278. https://doi.org/10.1007/s13238-020-00751-5

Tang Q, Maul GG (2003) Mouse cytomegalovirus immediate-early protein 1 binds with host cell repressors to relieve suppressive effects on viral transcription and replication during lytic infection. J Virol 77(2):1357–1367. https://doi.org/10.1128/jvi.77.2.1357-1367.2003

Taura A, Taura K, Choung YH, Masuda M, Pak K, Chavez E et al (2010) Histone deacetylase inhibition enhances adenoviral vector transduction in inner ear tissue. Neuroscience 166(4):1185–1193. https://doi.org/10.1016/j.neuroscience.2009.12.064

Terreni M, Valentini P, Liverani V, Gutierrez MI, Di Primio C, Di Fenza A et al (2010) GCN5-dependent acetylation of HIV-1 integrase enhances viral integration. Retrovirology 7:18. https://doi.org/10.1186/1742-4690-7-18

Thomas MC, Chiang CM (2005) E6 oncoprotein represses p53-dependent gene activation via inhibition of protein acetylation independently of inducing p53 degradation. Mol Cell 17(2):251–264. https://doi.org/10.1016/j.molcel.2004.12.016

Thomas Y, Androphy EJ (2018) Human papillomavirus replication regulation by acetylation of a conserved lysine in the E2 protein. J Virol 92(3). https://doi.org/10.1128/jvi.01912-17

Thomas Y, Androphy EJ (2019) Acetylation of E2 by P300 mediates topoisomerase entry at the papillomavirus replicon. J Virol 93(7). https://doi.org/10.1128/jvi.02224-18

Topper M, Luo Y, Zhadina M, Mohammed K, Smith L, Muesing MA (2007) Posttranslational acetylation of the human immunodeficiency virus type 1 integrase carboxyl-terminal domain is dispensable for viral replication. J Virol 81(6):3012–3017. https://doi.org/10.1128/JVI.02257-06

Trionfetti F, Alonzi T, Bontempi G, Terri M, Battistelli C, Montaldo C et al (2023) HDAC1-3 inhibition increases SARS-CoV-2 replication and productive infection in lung mesothelial and epithelial cells. Front Cell Infect Microbiol 13:1257683. https://doi.org/10.3389/fcimb.2023.1257683

Tsai K, Jaguva Vasudevan AA, Martinez Campos C, Emery A, Swanstrom R, Cullen BR (2020) Acetylation of cytidine residues boosts HIV-1 gene expression by increasing viral RNA stability. Cell Host Microbe 28(2):306–312 e6. https://doi.org/10.1016/j.chom.2020.05.011

Tsukuda S, Watashi K (2020) Hepatitis B virus biology and life cycle. Antivir Res 182:104925. https://doi.org/10.1016/j.antiviral.2020.104925

Valenzuela-Fernandez A, Alvarez S, Gordon-Alonso M, Barrero M, Ursa A, Cabrero JR et al (2005) Histone deacetylase 6 regulates human immunodeficiency virus type 1 infection. Mol Biol Cell 16(11):5445–5454. https://doi.org/10.1091/mbc.e05-04-0354

van Lint AL, Knipe DM (2014) Herpesviruses☆. In: Schmidt TM (ed) Reference module in biomedical sciences. Academic Press, Oxford, pp 565–579. https://doi.org/10.1016/b978-0-12-801238-3.02509-5

Van Lint C, Emiliani S, Ott M, Verdin E (1996) Transcriptional activation and chromatin remodeling of the HIV-1 promoter in response to histone acetylation. EMBO J 15(5):1112–1120. https://www.ncbi.nlm.nih.gov/pubmed/8605881

Vann KR, Acharya A, Jang SM, Lachance C, Zandian M, Holt TA et al (2022) Binding of the SARS-CoV-2 envelope E protein to human BRD4 is essential for infection. Structure 30(9):1224–1232 e5. https://doi.org/10.1016/j.str.2022.05.020

Varanda C, Felix MDR, Campos MD, Materatski P (2021) An overview of the application of viruses to biotechnology. Viruses 13(10). https://doi.org/10.3390/v13102073

Verdin E, Ott M (2015) 50 years of protein acetylation: from gene regulation to epigenetics, metabolism and beyond. Nat Rev Mol Cell Biol 16(4):258–264. https://doi.org/10.1038/nrm3931

Wang D, Meng Q, Huo L, Yang M, Wang L, Chen X et al (2015) Overexpression of Hdac6 enhances resistance to virus infection in embryonic stem cells and in mice. Protein Cell 6(2):152–156. https://doi.org/10.1007/s13238-014-0120-6

Wang L, Grossman SR, Kieff E (2000) Epstein-Barr virus nuclear protein 2 interacts with p300, CBP, and PCAF histone acetyltransferases in activation of the LMP1 promoter. Proc Natl Acad Sci USA 97(1):430–435. https://doi.org/10.1073/pnas.97.1.430

Wang Q, Peng W, Yang Y, Wu Y, Han R, Ding T et al (2023) Global analysis of lysine acetylation in the brain cortex of K18-hACE2 mice infected with SARS-CoV-2. Proteomics 23(16):e2300096. https://doi.org/10.1002/pmic.202300096

Wang Y, Han M, Liu S, Yuan X, Zhao J, Lu H et al (2021) S6K1 inhibits HBV replication through inhibiting AMPK-ULK1 pathway and disrupting acetylation modification of H3K27. Life Sci 265:118848. https://doi.org/10.1016/j.lfs.2020.118848

Warren JC, Rutkowski A, Cassimeris L (2006) Infection with replication-deficient adenovirus induces changes in the dynamic instability of host cell microtubules. Mol Biol Cell 17(8):3557–3568. https://doi.org/10.1091/mbc.e05-09-0850

WHO (2024) Coronavirus (COVID-19) dashboard. https://data.who.int/dashboards/covid19/. Accessed 12 July 2024

Willey J, Sandman K, Wood D (2022) Prescott's microbiology ISE. McGraw-Hill Higher Education (International)

Winans S, Goff SP (2020) Mutations altering acetylated residues in the CTD of HIV-1 integrase cause defects in proviral transcription at early times after integration of viral DNA. PLoS Pathog 16(12):e1009147. https://doi.org/10.1371/journal.ppat.1009147

Wolfe ND, Dunavan CP, Diamond J (2007) Origins of major human infectious diseases. Nature 447(7142):279–283. https://doi.org/10.1038/nature05775

Woolhouse M, Scott F, Hudson Z, Howey R, Chase-Topping M (2012) Human viruses: discovery and emergence. Philos Trans R Soc Lond Ser B Biol Sci 367(1604):2864–2871. https://doi.org/10.1098/rstb.2011.0354

Wu J, Shen S, Gao X, Miao M, Quan Y, Yu W (2023) Effects of LEF-11 acetylation modification on the regulation of baculovirus infection. Acta Virol 67(1):42–50. https://doi.org/10.4149/av_2023_104

Wu L, Zheng Z, Xun J, Liu L, Wang J, Zhang X et al (2024) Anti-PD-L1 antibody ASC22 in combination with a histone deacetylase inhibitor chidamide as a "shock and kill" strategy for ART-free virological control: a phase II single-arm study. Signal Transduct Target Ther 9(1):231. https://doi.org/10.1038/s41392-024-01943-9

Wurm T, Wright DG, Polakowski N, Mesnard JM, Lemasson I (2012) The HTLV-1-encoded protein HBZ directly inhibits the acetyl transferase activity of p300/CBP. Nucleic Acids Res 40(13):5910–5925. https://doi.org/10.1093/nar/gks244

Xia B, Lu J, Wang R, Yang Z, Zhou X, Huang P (2018) miR-21-3p regulates influenza A virus replication by targeting histone Deacetylase-8 (original research). Front Cell Infect Microbiol 8:175. https://doi.org/10.3389/fcimb.2018.00175

Xie S, Zhang L, Dong D, Ge R, He Q, Fan C et al (2020) HDAC6 regulates antibody-dependent intracellular neutralization of viruses via deacetylation of TRIM21. J Biol Chem 295(42):14343–14351. https://doi.org/10.1074/jbc.RA119.011006

Xu J, Mao J, Han X, Shi F, Gao Q, Wang T et al (2021) Porcine epidemic diarrhea virus inhibits HDAC1 expression to facilitate its replication via binding of its Nucleocapsid protein to host transcription factor Sp1. J Virol 95(18):e0085321. https://doi.org/10.1128/JVI.00853-21

Xue J, Cao Z, Cheng Y, Wang J, Liu Y, Yang R et al (2020) Acetylation of alpha-fetoprotein promotes hepatocellular carcinoma progression. Cancer Lett 471:12–26. https://doi.org/10.1016/j.canlet.2019.11.043

Xue M, Feng T, Chen Z, Yan Y, Chen Z, Dai J (2022) Protein acetylation going viral: implications in antiviral immunity and viral infection. Int J Mol Sci 23(19):11308. https://doi.org/10.3390/ijms231911308

Yan Q, Zhou J, Wang Z, Ding X, Ma X, Li W et al (2023) NAT10-dependent N(4)-acetylcytidine modification mediates PAN RNA stability, KSHV reactivation, and IFI16-related inflammasome activation. Nat Commun 14(1):6327. https://doi.org/10.1038/s41467-023-42135-3

Yang H, Fan T, Xun M, Wu B, Guo S, Li X et al (2024) N-terminal acetyltransferase 6 facilitates enterovirus 71 replication by regulating PI4KB expression and replication organelle biogenesis. J Virol 98(2):e0174923. https://doi.org/10.1128/jvi.01749-23

Yang L, Chen S, Zhao Q, Pan C, Peng L, Han Y et al (2022) Histone deacetylase 3 contributes to the antiviral innate immunity of macrophages by interacting with FOXK1 to regulate STAT1/2 transcription. Cell Rep 38(4):110302. https://doi.org/10.1016/j.celrep.2022.110302

Yang Y, Chen H, Zhang C, Shin HJ, Qian Y, Jung YS (2023) HDAC-specific inhibitors induce the release of porcine epidemic diarrhea virus via the COPII-coated vesicles. Viruses 15(9):1874. https://doi.org/10.3390/v15091874

Yang Z, Tang H, Huang H, Deng H (2009) RTA promoter demethylation and histone acetylation regulation of murine gammaherpesvirus 68 reactivation. PLoS One 4(2):e4556. https://doi.org/10.1371/journal.pone.0004556

Yuan B, Liu T, Cheng Y, Gao S, Li L, Cai L et al (2021) Comprehensive proteomic analysis of lysine acetylation in Nicotiana benthamiana after sensing CWMV infection (original research). Front Microbiol 12:672559. https://doi.org/10.3389/fmicb.2021.672559

Yuan Y, Zhao K, Yao Y, Liu C, Chen Y, Li J et al (2019) HDAC11 restricts HBV replication through epigenetic repression of cccDNA transcription. Antivir Res 172:104619. https://doi.org/10.1016/j.antiviral.2019.104619

Zanin M, DeBeauchamp J, Vangala G, Webby RJ, Husain M (2020) Histone deacetylase 6 knockout mice exhibit higher susceptibility to influenza A virus infection. Viruses 12(7):728. https://doi.org/10.3390/v12070728

Zhang D, Wang Y, Zhang HY, Jiao FZ, Zhang WB, Wang LW et al (2019) Histone deacetylases and acetylated histone H3 are involved in the process of hepatitis B virus DNA replication. Life Sci 223:1–8. https://doi.org/10.1016/j.lfs.2019.03.010

Zhang H, Chen J, Yu C, Pan Y, Ma W, Feng H et al (2024a) Innate immune evasion of PRRSV nsp11 through degradation of the HDAC2 by its endoribonuclease activity. Viruses 16(5):678. https://doi.org/10.3390/v16050678

Zhang H, Liu T, Liu X, You F, Yang J, Zhang N et al (2024b) Histone deacetylase 8 promotes innate antiviral immunity through deacetylation of RIG-I. Front Cell Infect Microbiol 14:1415695. https://doi.org/10.3389/fcimb.2024.1415695

Zhang L, Zhou S, Chen M, Yan J, Yang Y, Wu L et al (2021) P300-mediated NEDD4 acetylation drives ebolavirus VP40 egress by enhancing NEDD4 ligase activity. PLoS Pathog 17(6):e1009616. https://doi.org/10.1371/journal.ppat.1009616

Zhang S, Jiang Y, Cheng Q, Zhong Y, Qin Y, Chen M (2017) Inclusion body fusion of human parainfluenza virus type 3 regulated by acetylated alpha-tubulin enhances viral replication. J Virol 91(3). https://doi.org/10.1128/jvi.01802-16

Zhang X, Guo J, Xu H, Ding S, Liu L, Chen Z et al (2024c) NS1-mediated enhancement of MVC transcription and replication promoted by KAT5/H4K12ac. J Virol 98(3):e0169523. https://doi.org/10.1128/jvi.01695-23

Zhang X, Ma S, Gu C, Hu M, Miao M, Quan Y et al (2024d) K64 acetylation of heat shock protein 90 suppresses nucleopolyhedrovirus replication in Bombyx mori. Arch Insect Biochem Physiol 115(1):e22079. https://doi.org/10.1002/arch.22079

Zhang X, Zhang E, Ma Z, Pei R, Jiang M, Schlaak JF et al (2011) Modulation of hepatitis B virus replication and hepatocyte differentiation by MicroRNA-1. Hepatology 53(5):1476–1485. https://doi.org/10.1002/hep.24195

Zhang Z, Fang X, Wu X, Ling L, Chu F, Li J et al (2020) Acetylation-dependent Deubiquitinase OTUD3 controls MAVS activation in innate antiviral immunity. Mol Cell 79(2):304–319 e7. https://doi.org/10.1016/j.molcel.2020.06.020

Zhao D, Fukuyama S, Sakai-Tagawa Y, Takashita E, Shoemaker JE, Kawaoka Y (2015) C646, a novel p300/CREB-binding protein-specific inhibitor of histone acetyltransferase, attenuates influenza A virus infection. Antimicrob Agents Chemother 60(3):1902–1906. https://doi.org/10.1128/AAC.02055-15

Zhao LN, Yuan HF, Wang YF, Yun HL, Zheng W, Yuan Y et al (2022a) IFN-alpha inhibits HBV transcription and replication by promoting HDAC3-mediated de-2-hydroxyisobutyrylation of histone H4K8 on HBV cccDNA minichromosome in liver. Acta Pharmacol Sin 43(6):1484–1494. https://doi.org/10.1038/s41401-021-00765-7

Zhao X, Yuan H, Yang H, Liu Y, Xun M, Li X et al (2022b) N-acetyltransferase 8 promotes viral replication by increasing the stability of enterovirus 71 nonstructural proteins. J Virol 96(6):e0011922. https://doi.org/10.1128/jvi.00119-22

Zhong M, Zheng K, Chen M, Xiang Y, Jin F, Ma K et al (2014) Heat-shock protein 90 promotes nuclear transport of herpes simplex virus 1 capsid protein by interacting with acetylated tubulin. PLoS One 9(6):e99425. https://doi.org/10.1371/journal.pone.0099425

Zhong Q, Xiao X, Qiu Y, Xu Z, Chen C, Chong B et al (2023) Protein posttranslational modifications in health and diseases: functions, regulatory mechanisms, and therapeutic implications. MedComm 4(3):e261. https://doi.org/10.1002/mco2.261

Zhou L, He X, Gao B, Xiong S (2015) Inhibition of histone deacetylase activity aggravates coxsackievirus B3-induced myocarditis by promoting viral replication and myocardial apoptosis. J Virol 89(20):10512–10523. https://doi.org/10.1128/JVI.01028-15

Zhou Y, He C, Wang L, Ge B (2017) Post-translational regulation of antiviral innate signaling. Eur J Immunol 47(9):1414–1426. https://doi.org/10.1002/eji.201746959

Zhu J, Coyne CB, Sarkar SN (2011) PKC alpha regulates Sendai virus-mediated interferon induction through HDAC6 and beta-catenin. EMBO J 30(23):4838–4849. https://doi.org/10.1038/emboj.2011.351

Zhu L, Fung SY, Xie G, Wong LR, Jin DY, Cai Z (2020a) Identification of lysine acetylation sites on MERS-CoV replicase pp1ab. Mol Cell Proteomics 19(8):1303–1309. https://doi.org/10.1074/mcp.RA119.001897

Zhu N, Zhang D, Wang W, Li X, Yang B, Song J et al (2020b) A novel coronavirus from patients with pneumonia in China, 2019. N Engl J Med 382(8):727–733. https://doi.org/10.1056/NEJMoa2001017

Chapter 13
Reversible Acetylation of Non-histone Proteins in Human Cancers

Chunran Feng, Yiqun Zeng, Edward V. Prochownik, Congqin Jiang, and Youjun Li

Abstract Over the last three decades, we have witnessed great progress in uncovering the scope of reversible acetylation of non-histone proteins and understanding its mechanisms and functional consequences. In this review, we summarize the histone acetyltransferases (HATs)/deacetylases (HDACs) and their inhibitors, focusing on the role of reversible acetylation modification of non-histone proteins in tumor development while also exploring the application of HAT and HDAC inhibitors in cancer therapy.

13.1 Introduction

Post-translational modification (PTM) refers to the covalent attachment of functional chemical groups to specific amino acid residues of proteins. These changes alter protein activity, stability, subcellular localization, and interaction with other cellular molecules (Jayaprakash and Surolia 2017) thereby diversifying protein

---

Authors Chunran Feng and Yiqun Zeng have equally contributed to this chapter.

C. Feng · Y. Zeng · C. Jiang · Y. Li (✉)
Department of Colorectal and Anal Surgery, Zhongnan Hospital of Wuhan University, Hubei Key Laboratory of Cell Homeostasis, College of Life Sciences, Frontier Science Center for Immunology and Metabolism, Medical Research Institute, TaiKang Center for Life and Medical Sciences, Wuhan University, Wuhan, China
e-mail: liy7@whu.edu.cn

E. V. Prochownik
Division of Hematology/Oncology, Children's Hospital of Pittsburgh of UPMC, The Department of Microbiology and Molecular Genetics, The Pittsburgh Liver Research Center and The Hillman Cancer Center of UPMC, The University of Pittsburgh Medical Center, Pittsburgh, PA, USA

M. Halasa, A. Wawruszak (eds.), *Histone and Non-Histone Reversible Acetylation in Development, Aging and Disease*, Results and Problems in Cell Differentiation 75,
https://doi.org/10.1007/978-3-031-91459-1_13

functions and dynamically coordinating intracellular signaling networks. Due to its rapidity and low energy consumption, reversible PTM can precisely regulate protein function, which is crucial for the function of biological systems (Wang et al. 2014).

As an important PTM, acetylation and deacetylation are important aspects of signal transduction and cellular metabolism, and play key roles in regulating the functional diversity of proteins (Choudhary and Mann 2010). Acetylation is catalyzed by a variety of acetyltransferases that transfer acetyl groups (-CO-CH3) from acetylated donor molecules (such as acetyl-CoA) to specific amino acids (Polevoda and Sherman 2002). The reverse reaction of acetylation, deacetylation reduces the level of protein acetylation with similar specificity, and together with acetylation, can dynamically and transiently regulate protein function. Protein acetylation and deacetylation are catalyzed by histone acetyltransferases (HATs) and histone deacetylases (HDACs), respectively. Given the discovery of numerous non-histone substrates and the fact that acetylation predominantly occurs on lysine residues, it has been proposed that HATs and HDACs be renamed as lysine acetyltransferases (KATs) and lysine deacetylases (KDACs).

Acetylation plays important roles in a variety of biological processes, including RNA splicing, cell cycle progression, chromatin remodeling, DNA replication, transcription, and nuclear transport (Xia et al. 2020). Protein acetylation affects cellular physiological and pathological states in cancer, metabolic, inflammatory and neurological disorders, and infectious diseases (Timmermann et al. 2001). In cancer, acetylation patterns affect cell growth, the activities of oncoproteins and tumor suppressors, and both aerobic and anaerobic metabolic pathways (Diallo et al. 2019). For example, the regulation and activation of important tumor suppressors such as TP53 are strongly affected by acetylation and play a key role in cancer (Pearson et al. 2000). Lysine acetylation in particular plays a causal role in tumorigenesis and maintenance by altering homeostasis of a variety of metabolic pathways, and networks (Xu et al. 2014).

13.2 Protein Acetylation Modification and Acetyltransferases

Protein acetylation falls into three general categories: (a) N-terminal acetylation, commonly referred to as N-ter acetylation or Nα-acetylation; (b) O-acetylation; and (c) lysine acetylation, commonly referred to as K-acetylation (KAc) or Nε-acetylation (Diallo et al. 2019). N-ter acetylation is an irreversible modification of the α-amino group, which usually occurs early during translation and involves the penultimate amino acid residue or the N-terminal one following the cleavage of the initiator methionine (Driessen et al. 1985). Functionally, N-ter acetylation can affect protein stability, degradation, protein-protein interaction, and inhibition of translocation into the endoplasmic reticulum (Starheim et al. 2012). This modification is also implicated in a variety of pathologies, such as neurodegenerative diseases and tumorigenesis (Drazic et al. 2016). O-acetylation adds acetyl groups to the hydroxyl side chains of serine or threonine and plays an important role in signal transduction mechanisms, cell cycle progression, and cell pluripotency (Mukherjee et al. 2007). K-acetylation refers to the addition of acetyl groups to the amino site of lysine's R group. This modification neutralizes lysine's positive charge and regional hydrophobicity, and increases its size (Alonso and Serra 2012). Correspondingly, this may affect a variety of protein properties including DNA-protein and protein-protein interactions, transcriptional activity, protein stability, metabolism, cell cycle, migration, differentiation, and apoptosis (Narita et al. 2019). Among these three types of acetylation, K-acetylation is the most prevalent form of PTM and exhibits the broadest range of physiological functions. Unless otherwise specified, in this Review, acetylation refers only to K-acetylation.

Based on structural and sequence similarity, HATs can also be divided into three independent families: the GCN5-associated N-acetyltransferase (GNAT) family, the p300/CREB-binding protein (CBP) family, and the monocytic leukemia zinc finger protein, Ybf2/Sas3, Sas2, and Tip60 (MYST) family (Furdas et al. 2012). The GNAT family of HATs includes PCAF, Gcn5, Hat1, Hpa2, Elp3, ATAT1, NTA10, and several others whereas the p300/CBP family includes p300 and CBP. The MYST family includes Tip60, MOF, Esa1, Sas2, Sas3, Hbo1, and MORF (Roth et al. 2001). Based upon their subcellular localization, HATs can also be divided into type A (nuclear) HATs and type B (cytoplasmic) HATs. The main function of type A HATs is to acetylate nucleosomal histones and thereby participate in transcriptional regulation. However, despite their cytoplasmic location, type B HATs, particularly HatB3.1, Rtt109, HAT1, HAT2, and HAT4, can indirectly regulate transcription as well by virtue of their ability to acetylate nascent histones prior to their nuclear transport (Yang et al. 2011). In addition to histones, non-histone proteins such as the p53 tumor suppressor can also be regulated by HATs (Gu and Roeder 1997).

13.3 Acetyltransferase Inhibitors

Reported CBP/p300 inhibitors include Lys-CoA (Lau et al. 2000), CCT077791, CCT077792, C646, A-485, B026, B029–2, Curcumin, Plumbagin, Garcinol, Epigallocatechin-3-gallate, and Oridonin. These inhibitors can act in combination with bromodomain inhibitors of CBP/p300 to downregulate oncogene transcription, inhibit cancer cell growth, induce cell death, activate immune responses, overcome drug resistance, and suppress tumor progression (Chen et al. 2022) (Table 13.1).

PCAF, a member of the GNAT family of acetyltransferases, is inhibited by natural compounds such as anacardic acid, Embelin, Garcinol, and Oridonin, as well as derivatives of isothiazolones, pyridoisothiazolones, and isothiazolidinones. Recently, F2209–0381 has been identified as a novel, potent, and cell-permeable PCAF inhibitor (Suryanarayanan et al. 2018).

The human MYST family enzymes—Tip60 (KAT5), MOZ (KAT6A), MORF (KAT6B), HBO1 (KAT7), and MOF (KAT8)—are embedded in specific, large multiprotein complexes. Highly potent and selective inhibitors of KAT6A and KAT6B,

Table 13.1 Classification of HAT inhibitors

HAT		Inhibitor	References
CBP(KAT3A) p300(KAT3B)		Lys-CoA	Lau et al. (2000)
		CCT077791	Stimson et al. (2005)
		CCT077792	Stimson et al. (2005)
		C646	Bowers et al. (2010)
		A-485	Lasko et al. (2017)
		B026	Yang et al. (2020)
		B029–2	Cai et al. (2021)
		Curcumin	Marcu et al. (2006)
		Plumbagin	Ravindra et al. (2009)
		Garcinol	Balasubramanyam et al. (2004)
		Epigallocatechin- 3-gallate	Choi et al. (2009)
		Oridonin	Shi et al. (2016)
		PU139	Gajer et al. (2015)
		BF1	Secci et al. (2014)
		Georgedonin	Shi et al. (2016)
GNAT	PCAF (KAT2B)	Anacardic acid	Balasubramanyam et al. (2003)
		Curcumin	
		Lys-CoA	
		Embelin	Modak et al. (2013)
		Garcinol	Balasubramanyam et al. (2004)
		Derivatives of isothiazolones	Stimson et al. (2005)
		pyridoisothiazolones	Furdas et al. (2011)
		isothiazolidinones	Furdas et al. (2014)
		F2209-0381	
		Oridonin	Shi et al. (2016)
		HTS03381	
		H3-CoA-20	Zheng et al. (2005)
		PU139	Gajer et al. (2015)
		5-Cloroisothiazolone	Chimenti et al. (2009)
		CPTH2	Chimenti et al. (2009)
		CPTH6	Di Martile et al. (2016)
		Georgedonin	Shi et al. (2016)
	GCN5 (KAT2A)	Oridonin	Shi et al. (2016)
		PU139	Gajer et al. (2015)
		CPTH2	Chimenti et al. (2009)
		CPTH6	Di Martile et al. (2016)
		BF1	Secci et al. (2014)
		Georgedonin	Shi et al. (2016)

(continued)

Table 13.1 (continued)

HAT		Inhibitor	References
MYST	MOZ (KAT6A)	WM-8014	Baell et al. (2018)
		WM-1119	
	MORF (KAT6B)	WM-8014	Baell et al. (2018)
		WM-1119	
	MOF (KAT8)	Anacardic acid	Wapenaar et al. (2015)
		Anacardic acid derivatives	
		Fragment 4-amino-1-naphtol	Wapenaar et al. (2017)
	TIP60 (KAT5)	Anacardic acid III	Sun et al. (2006)
		Curcumin	Wu et al. (2009)
		Pentamidine	Kobayashi et al. (2010)
		NU9056	Coffey et al. (2012)
		Lys-CoA	Wu et al. (2009)
		Garcinol	Balasubramanyam et al. (2004)
		MG-149	Ghizzoni et al. (2012)
		TH1834	Gao et al. (2014)
		Bisubstrate Inhibitor A	Wu et al. (2011)
		Oridonin	Shi et al. (2016)
		Georgedonin	Shi et al. (2016)

Abbreviations (alphabetical order): *CBP (KAT3A)* CREB-Binding Protein, *GCN5 (KAT2A)* General Control Non-Derepressible 5, *MOF (KAT8)* Males Absent on the First, *MORF (KAT6B)* MOZ-Related Factor (also known as MYST4), *MOZ (KAT6A)* Monocytic Leukemia Zinc Finger Protein, *P300 (KAT3B)* E1A-Associated Protein p300, *PCAF (KAT2B)* P300/CBP-Associated Factor, *TIP60 (KAT5)* Tat-Interactive Protein 60 kDa

WM-8014 and WM-1119, inhibit MYST-catalyzed histone acetylation by reversibly competing with acetyl-CoA, thereby inducing cell cycle exit and senescence without causing DNA damage (Baell et al. 2018). Reported KAT8 inhibitors include anacardic acid (Wapenaar et al. 2015) and its derivative 4-amino-1-naphtol (Wapenaar et al. 2017). HAT inhibitors of Tip60 can be divided into two main categories: (1) designed small molecule inhibitors, such as bisubstrate inhibitor A, MG-149, and TH1834. (2) inhibitors derived from library screening, such as Lys-CoA, garcinol, curcumin, oridonin, anacardic acid, pentamidine, and NU9056.

13.4 Acetylation of Non-histone Proteins and Tumor Development

Classically, HATs are known to be crucial for the acetylation of histones, a modification that plays a key role in gene regulation. However, HATs also have the capacity to acetylate non-histone proteins, which can significantly impact various aspects of tumor cell biology (Table 13.2).

13.4.1 Lipid Metabolism

GNPAT plays a crucial role in lipid metabolism by inhibiting Tripartite Motif Containing 21 (TRIM21)-mediated fatty acid synthase (FASN) degradation, and its dysregulation is closely associated with hepatocellular carcinoma. Acetyl-CoA Acetyltransferase 1 (ACAT1), upregulated in response to excess palmitic acid, acetylates GNPAT, thereby preventing its TRIM21-mediated ubiquitination and degradation, which subsequently promotes lipid metabolism and tumor progression in liver cancer (Gu et al. 2020). Additionally, ACAT1 mediates the acetylation of malic enzyme 1 (ME1) at lysine 337 (K337), leading to ME1 dimerization and activation. This process enhances Nicotinamide Adenine Dinucleotide Phosphate (NADPH) production, lipogenesis, and contributes to the progression of colorectal cancer (Zhu et al. 2020).

13.4.2 Cell Cycle

As a tyrosine kinase, WEE1 G2 Checkpoint Kinase (WEE1) is a critical regulator of the intra-S-phase and G2/M cell-cycle checkpoints. In response to DNA damage, WEE1 undergoes sequential phosphorylation at serine 642 (S642) mediated by Checkpoint Kinase 1 (CHK1) and acetylation at lysine 177 (K177) mediated by GCN5. These modifications facilitate the dissociation of inhibitory fragments from its kinase domain, thereby activating its enzymatic activity. Activated WEE1 phosphorylates Cyclin-Dependent Kinase 1 (CDK1)/Cyclin-Dependent Kinase 2 (CDK2), leading to cell cycle arrest (Zhu et al. 2023).

13.4.3 Ferroptosis

p300-mediated acetylation of Heat Shock Protein Family A (Hsp70) Member 5 (HSPA5) enhances its ability to suppress Glutathione Peroxidase 4 (GPX4), leading to excessive lipid peroxidation and subsequently promoting ferroptosis in human pancreatic ductal adenocarcinoma (PDAC) cells (Wang et al. 2023b). Nicotinamide nucleotide transhydrogenase (NNT) catalyzes the conversion of Nicotinamide Adenine Dinucleotide Phosphate (Oxidized Form, NADP+) to NADPH, which maintains the reduced state of iron-sulfur clusters and inhibits cellular ferroptosis. Under IL-1β stimulation, PCAF translocates to the mitochondria in tumor cells, where it acetylates Nicotinamide Nucleotide Transhydrogenase (NNT), thereby promoting NADPH production and inhibiting tumor cell ferroptosis (Han et al. 2023).

Table 13.2 Acetylation of non-histone proteins in human cancers

HAT	Substrate/site	Cancer type	Function	References
CBP(KAT3A)/ p300(KAT3B)	HSPA5 K353	Pancreatic adenocarcinoma	Mediate GPX4 inhibition and induce ferroptosis	Wang et al. (2023b)
	CDC42 K153	Colorectal cancer	Promote CDC42 binding to the downstream effector PAK4	Wang et al. (2023a)
	HOXB13 K277	Breast cancer	Accumulation of HOXB13	Liu et al. (2018)
	KLF5 K358	Prostate cancer	Inhibit FGFR1 signaling	Zhang et al. (2024a)
	PHF5A K29	Colorectal cancer	Stabilize KDM3A mRNA	Wang et al. (2019b)
	KLF5 K369	Prostate cancer	Activate CXCR4 and SHH/IL-6 paracrine signals	Zhang et al. (2021)
	MOB1 K11	Non-small cell lung cancer	Stabilize MOB1, promote LATS1 activation and YAP/TAZ phosphorylation	Jin et al. (2022)
	MDM2 K182 K185	–	Switch MDM2 self-ubiquitination to p53 ubiquitination	Nihira et al. (2017)
	RB1CC1 K276	Breast cancer	Inhibit RB1CC1 ubiquitination and degradation	Yi et al. (2023)

(continued)

Table 13.2 (continued)

HAT		Substrate/site	Cancer type	Function	References
GNAT	ACAT1	GNPAT K128	Hepatocellular carcinoma	Inhibit GNPAT ubiquitination and degradation	Gu et al. (2020)
		ME1 K337	Colorectal cancer	Dimerization of ME1	Zhu et al. (2020)
	GCN5 (KAT2A)	WEE1 K177	Pancreatic cancer	Activate WEE1 and the cell cycle checkpoint	Zhu et al. (2023)
		LIFR K620	Prostate cancer	Promote LIFR phosphorylation and activate AKT signaling	Dine et al. (2022)
		SOX4 K118	Non-small cell lung cancer	Promote SOX4 to activate MMP9 transcription	Ge et al. (2023)
		KLF5	Non-small cell lung cancer	Promote KLF5 to activate GDF15 transcription	Zhao et al. (2018)
	HAT1 (KAT1)	CLIC1 K131	Cervical cancer	Inhibit CLIC1 ubiquitination and degradation	Wang et al. (2021)
		PLZF	–	Acetylated PLZF recruits HDAC3 and NF-κB p50 to inhibit NF-κB activity	Sadler et al. (2015)
	PCAF (KAT2B)	TPD52 K163	Prostate cancer	Impair interaction between TPD52 and HSPA8 to inhibit autophagy	Fan et al. (2021)
		NNT K1042	Gastric cancer	Enhance NNT activity and resistance to ferroptosis	Han et al. (2023)
		ATIC K266	Colorectal cancer	Inactivate ATIC and reduce cellular purine biosynthesis.	Zhao et al. (2023a)
		Fascin K471	Esophageal cancer	Impair its binding ability to actin to inhibit cell migration	Cheng et al. (2021)
		Lamin A K542	Breast cancer	Inhibit integration between Lamin A and SUN2 to disrupt immune synapse formation	Li et al. (2024)
		HOXB9	Non-small cell lung cancer	Promote HOXB9 ubiquitination and degradation	Zhang et al. (2024b)
		ISX	Lung cancer	Promote ISX recruitment of BRD4 to promote epithelial-mesenchymal transition	Wang et al. (2020b)
		RACK1 K130	Ovarian cancer	Inhibit RACK1 ubiquitination degradation	Pi et al. (2023)
		Stathmin1 K9	Gallbladder cancer	Upregulate stathmin1 protein stability and microtubule instability	Fan et al. (2022)
		PGK1 K323	Liver cancer	Upregulate PGK1 enzyme activity and glycolysis	Hu et al. (2017)
		EZH2 K348	Lung cancer	Inhibit EZH2 phosphorylation and enhance its stability	Wan et al. (2015)

MYST	MOZ (KAT6A)	SMAD3 K20 / K117	Breast cancer	Promote the binding of SMAD3 to TRIM24	Yu et al. (2021)
		COP1 K294	Ovarian cancer	Impair COP1 activity and activate the Wnt/β-catenin pathway	Liu et al. (2021)
	TIP60 (KAT5)	HMGCS1 K273	Glioblastoma	Stabilize HMGCS1 and upregulate cholesterol biosynthesis	Zhao et al. (2024)
		SPZ1 K369 ，K374 TWIST1 K73 ，K76	Liver cancer	SPZ1-TWIST1 complex is formed to recruit BRD4 to enhance VEGF expression	Wang et al. 2019a)
		p53 K120	Non-small cell lung cancer	Promote p53 to activate BAX and PUMA transcription	Sykes et al. (2006)
		cortactin	Breast cancer	Inhibit cell migration and invasion	Sun et al. (2015)
		KDM2B K758	Osteosarcoma	Impair KDM2B demethylase activity	Shi and Fan (2019)
		USP26 K134	Breast cancer	Promote USP26 binding to BAG3 and induce BAG3 deubiquitylation	Liu et al. (2024)
	HBO1 (KAT7)	GαS K28	Liver cancer	Promote GαS depalmitoylation and its cytoplasmic translocation	Zhou et al. (2023)
		Vimentin K104	Lung adenocarcinoma	Inhibit Vimentin ubiquitination and degradation	Yang et al. (2023)
		LDHA K118	Head and neck squamous carcinoma	Enhance LDHA enzyme activity to promote lactate production	Lu et al. (2024)
	MOF (KAT8)	IRF1 K78	Lung cancer	Activate IRF1 transcriptional activity to upregulate PD-L1 expression	Wu et al. (2023)
		SIRT6	Non-small cell lung cancer	Block the interaction between SIRT6 and FOXA2	Zhao et al. 2023b)
		FASN	Hepatocellular carcinoma	Enhance FASN ubiquitination degradation	Lin et al. (2016)
		WSTF K426	–	Enhance WSTF kinase and transcriptional regulatory activities	Liu et al. (2020)
		LSD1	–	Downregulate LSD1 demethylase activity	Luo et al. (2016)

(continued)

Table 13.2 (continued)

Abbreviations (alphabetical order): *ACAT1* Acetyl-CoA Acetyltransferase 1, *ATIC* 5-Aminoimidazole-4-Carboxamide Ribonucleotide Formyltransferase/IMP Cyclohydrolase, *CDC42* Cell Division Cycle 42, *CLIC1* Chloride Intracellular Channel 1, *COP1* E3 Ubiquitin-Protein Ligase COP1, *Cortactin* Actin-Binding Protein Cortactin, *EZH2* Enhancer of Zeste Homolog 2, *FASN* Fatty Acid Synthase, *Fascin* Fascin Actin-Bundling Protein 1, *GαS* G Protein Alpha Subunit S, *GNPAT* Glyceronephosphate O-Acyltransferase, *HMGCS1* 3-Hydroxy-3-Methylglutaryl-CoA Synthase 1, *HOXB13* Homeobox B13, *HOXB9* Homeobox B9, *HSPA5* Heat Shock Protein Family A (HSP70) Member 5, *IRF1* Interferon Regulatory Factor 1, *ISX* Intestine-Specific Homeobox, *KDM2B* Lysine Demethylase 2B, *KLF5* Kruppel-Like Factor 5, *Lamin A* Lamin A/C, *LDHA* Lactate Dehydrogenase A, *LIFR* Leukemia Inhibitory Factor Receptor, *LSD1* Lysine-Specific Histone Demethylase 1A, *MDM2* Mouse Double Minute 2 Homolog, *ME1* Malic Enzyme 1, *MOB1* MOB Kinase Activator 1, *NNT* Nicotinamide Nucleotide Transhydrogenase, *PGK1* Phosphoglycerate Kinase 1, *PHF5A* PHD Finger Protein 5A, *PLZF* Promyelocytic Leukemia Zinc Finger Protein, *RACK1* Receptor for Activated C Kinase 1, *RB1CC1* RB1 Inducible Coiled-Coil 1, *SIRT6* Sirtuin 6, *SMAD3* SMAD Family Member 3, *SOX4* SRY-Box Transcription Factor 4, *SPZ1* Spermatogenic Leucine Zipper Protein 1, *Stathmin1* Stathmin 1, *TPD52* Tumor Protein D52, *TWIST1* Twist-Related Protein 1, *USP26* Ubiquitin-Specific Peptidase 26, *Vimentin* Vimentin, *WEE1* WEE1 G2 Checkpoint Kinase, *WSTF* Williams Syndrome Transcription Factor

13.4.4 Apoptosis

The ability of p53 to induce apoptosis plays a critical role in tumor suppression. Following DNA damage, the MYST family acetyltransferases hMOF and TIP60 catalyze the acetylation of p53 at lysine 120 (K120). This modification enhances the transcription of p53 downstream pro-apoptotic target genes, such as BCL2-Associated X Protein (BAX) and p53 Upregulated Modulator of Apoptosis (PUMA), thereby promoting apoptosis (Sykes et al. 2006).

13.4.5 Tumor Metastasis

Fascin is essential for filopodia formation in cancer cells and tumor metastasis. PCAF acetylates Fascin and weakens its binding ability to actin, thereby reducing the length and density of filopodia in esophageal cancer cells and weakening the migration ability of esophageal cancer cells (Cheng et al. 2021). Another HAT, GCN5 can upregulate the expression of matrix metalloproteinase 9 (MMP9) by acetylating SRY-Box Transcription Factor 4 (SOX4), and subsequently promote the migration and invasion of non-small cell lung cancer cells (Ge et al. 2023).

13.4.6 Immune Escape

Acetylation is a critical post-translational modification that significantly influences tumor immunity. Various acetyltransferases and their non-histone substrates are implicated in the immune escape mechanisms of tumor cells. For instance, in the presence of N-acetylaspartate (NAA), the acetyltransferase PCAF acetylates lamin A, thereby inhibiting the interaction between lamin A and Sad1 and UNC84 Domain Containing 2 (SUN2). This disruption impairs the polarization of lytic granules and consequently disrupts the formation of immune synapses, facilitating immune evasion.

KAT6A, another acetyltransferase, acetylates SMAD Family Member 3 (SMAD3), enhancing its binding to the oncogenic chromatin modifier Tripartite Motif Containing 24 (TRIM24). The recruitment of TRIM24-SMAD3 complexes to chromatin increases SMAD3 activation and the expression of immune-related cytokines. This cascade of events leads to enhanced stemness of breast cancer stem-like cells, recruitment of myeloid-derived suppressor cells (MDSCs), and metastasis in triple-negative breast cancer (TNBC).

Notably, phase separation also plays a pivotal role in non-histone acetylation. In lung cancer cells exposed to Interferon-Gamma (IFN-γ), KAT8 dissociates from Interferon Regulatory Factor 1 (IRF1) and forms a biomolecular condensate. The condensation of KAT8-IRF1 promotes the acetylation of IRF1 at lysine 78 (K78), targeting the PD-L1 promoter and upregulating PD-L1 expression. Disrupting the phase separation of the KAT8-IRF1 complex inhibits PD-L1 expression, thereby enhancing antitumor immunity.

13.4.7 Others

Certain HATs can acetylate non-histone proteins and affect their stability, such as Homeobox B13 (HOXB13) and Kruppel-like factor 5 (Klf5), thereby improving drug resistance and promoting cancer cell proliferation. For example, the acetylation of PHD Finger Protein 5A (PHF5A), increases the expression and accumulation of proteins that enhance the anti-stress capabilities of cancer cells, contributing to cancer development (Wang et al. 2019b). Additionally, acetylation can competitively regulate protein ubiquitination, influencing the activation and expression of tumor-related factors, including p53. An example is the acetylation of RB1 Inducible Coiled-Coil 1 (RB1CC1), which competes with ubiquitination sites and modulates the autophagy function in cancer cells (Yi et al. 2023). Further details on additional acetylated substrate proteins and their associated functions mediated by HATs are summarized in Table 13.2.

13.5 Targeting Non-histone Protein Acetylation for Cancer Treatment

Acetyltransferase inhibitors are often used in combination with other treatments to achieve therapeutic goals. These agents can simultaneously impact cell metabolism, apoptosis, autophagy, and cancer-related signaling pathways (IL-6, PI3K/Akt, and MAPK), which further affect oncogene transcription, cancer cell proliferation, survival, tumorigenesis, metastasis, and immune escape. For example, CBP/p300 inhibitors enhance the anticancer efficacy of chemotherapy, radiotherapy, and epigenetic anticancer agents. Combination therapies that employ acetyltransferase inhibitors have shown substantial anticancer effects in mouse models of human cancers, including drug-resistant cancers (Chen et al. 2022). The KAT6A/B inhibitors WM-8014 and WM-1119 induce senescence and inhibit tumor growth (Baell et al. 2018) whereas Tip60 inhibitors in combination with ionizing radiation induce apoptosis in cervical, breast, and prostate cancer cells (Gao et al. 2014).

Georgedonin is a novel lysine acetyltransferase inhibitor that inhibits gastric cancer cell proliferation and induces p53- and caspase-3-mediated apoptosis (Shi et al. 2016). IL-1β neutralization and blockade of NNT acetylation synergistically with anti-PD-1 therapy can significantly induce ferroptosis of tumor cells (Han et al. 2023). Hepatitis B Virus X-Interacting Protein (HBXIP) enhances the acetylation of HOXB13 to prevent HOXB13 degradation and co-activates HOXB13 to promote tamoxifen resistance in breast cancer. In treatment, aspirin can be used as a potential candidate drug to reverse tamoxifen resistance by inhibiting HBXIP expression (Liu et al. 2018). ACAT1-mediated GNPAT acetylation stabilizes FASN to promote lipid metabolism and hepatocarcinogenesis, and ACAT1 inhibitors in combination with sorafenib greatly delay HCC formation in mice (Gu et al. 2020). KAT6A inhibitor (WM-1119), in combination with anti-PD-L1 monoclonal antibody therapy, is a potential epigenetic drug for the treatment of metastatic triple-negative breast cancer (Wang et al. 2019a) (Table 13.3).

13.6 Histone Deacetylases

The histone deacetylase (HDAC) superfamily encompasses 18 distinct members, sequentially numbered in accordance with their order of discovery (Thiagalingam et al. 2003). HDACs are categorized into four classes—Class I, II, III, and IV—based on their homology to yeast proteins (Table 13.4). Notably, Class I, II, and IV HDACs necessitate the presence of Zn^{2+} for their catalytic function, whereas Class III HDACs are uniquely dependent on nicotinamide adenine dinucleotide (NAD^+) (Hai et al. 2022).

Class I HDACs include HDAC1, 2, 3, and 8, and they are predominantly nuclear and homologous to the yeast Rpd3 protein (Yang and Seto 2008). Class II HDACs are further subdivided into two groups: Class IIa, comprised of HDAC4, 5, 7, and 9 and Class IIb, comprised of HDAC6 and 10. These two subclasses demonstrate a close relationship with the yeast Hda1 protein and are capable of shuttling between the cytoplasm and nucleus, thereby endowing class II HDACs with the capacity to modulate both nuclear and cytoplasmic substrates (Ma et al. 2019). Class III HDACs, also known as sirtuins (includes SIRT1 to SIRT7), share homology with the yeast Sir2 proteins (Lombard and Zwaans 2014). Class IV HDACs include only HDAC 11, which is mainly localized in the nucleus and has a structural similarity to both class I and II HDACs (Chiaradonna et al. 2015).

Table 13.3 Targeting acetylation of non-histone proteins for cancer treatment

Mechanism	Treatment	Cancer type	References
Ferroptosis	Blockade of NNT acetylation is synergistic with anti-PD-1 therapy	Pancreatic cancer	Han et al. (2023)
Apoptosis	Georgedonin	Gastric cancer	Shi et al. (2016)
	Tip60 inhibitor combined with ionizing radiation	Cervical cancer, prostate cancer, colon cancer	Gao et al. (2014)
	WM-8014, WM-1119	Liver cancer, lymphatic cancer	Baell et al. (2018)
Autophagy	Aspirin	Breast cancer	Liu et al. (2018)
Metabolism	Combine ACAT1 inhibitors with sorafenib	Liver cancer	Gu et al. (2020)

Table 13.4 Classification of HDACs

Class	Type	Homologue	Catalytic domain	Biological function	References
HDAC I	HDAC1,2,3,8	Rpd3	Zn^{2+}-dependent	Cell proliferation and apoptosis	Yang and Seto (2008)
HDAC II	HDAC IIa: HDAC4,5,7,9; HDAC IIb: HDAC6,10	Hda1	Zn^{2+}-dependent	Angiogenesis	Ma et al. (2019)
HDAC III	SIRT1–7	Sir2	NAD^+-dependent	Cell metabolism and cell growth	Lombard and Zwaans (2014)
HDAC IV	HDAC11		Zn^{2+}-dependent	Cell migration	Chiaradonna et al. (2015)

Class I HDACs exhibit a broad tissue distribution, with HDACs 1, 2, and 3 typically functioning within multiprotein complexes. HDACs 1 and 2 are co-expressed alongside key components of multiprotein corepressor complexes, such as Switch-Independent 3 Transcription Regulator Family (Sin3), Nucleosome Remodeling and Deacetylase Complex (NuRD), Corepressor of RE1-Silencing Transcription Factor (CoREST), Mitotic Deacetylase Complex (MiDAC), and Mesoderm Induction Early Response Protein (MIER) family members (Gonneaud et al. 2019). In contrast, HDAC3 serves as the catalytic subunit of the Silencing Mediator for Retinoid and Thyroid Hormone Receptors (SMRT) and Nuclear Receptor Corepressor 1 (N-CoR) complexes (Li et al. 2000). All three of these HDACs localize predominantly to the nucleus. HDAC8, however, stands out within Class I HDACs due to its ability to operate independently of multiprotein complexes and its capacity to interact with and deacetylate non-histone proteins, including the M2 pyruvate kinase (PKM2). This interaction facilitates the nuclear translocation of PKM2, thereby amplifying glycolysis and subsequently fueling the proliferative activities of cancer cells (Zhang et al. 2020).

Class II HDACs are expressed in a more tissue-specific pattern. In addition to possessing catalytic deacetylation domains, class IIa HDACs also feature myocyte enhancer factor 2 (MEF2) and 14–3-3 binding domains. Through the assistance of MEF2 and 14–3-3 proteins, class IIa HDACs undergo shuttling between the nucleus and cytoplasm for the purpose of deacetylating various substrates and exerting diverse functions. MEF2 binds to class IIa HDACs, facilitating their translocation into the nucleus; on the other hand, phosphorylation-dependent binding of 14–3-3 proteins to class IIa HDACs inhibits the binding of MEF2 to class IIa HDACs, thereby retaining them in the cytoplasm. Class IIb HDACs are predominantly found in the cytoplasm and act through various non-histone proteins as their substrates, including HDAC6 and 10. HDAC6 is known to deacetylate both α-tubulin (Ge et al. 2021) and heat shock protein 90 (HSP90) (de Ruijter et al. 2003) and has a crucial role in maintaining cell division, migration, and angiogenesis.

Distinct from the other HDACs, class III sirtuins constitute a family of pleiotropic enzymes that serve dual roles as NAD^+-dependent protein deacetylases and ADP-ribosyltransferases. The seven sirtuins are located to different subcellular compartments. SIRT1, 6, and 7, which are predominantly nuclear, modulate gene expression by targeting the deacetylation of specific substrates (Guarente 2013). SIRT2 plays a role in cell cycle regulation, exhibiting functionality across both the cytoplasmic and nuclear compartments (Serrano et al. 2013). In parallel, SIRT3, 4, and 5 are mitochondrial, orchestrating the activity of pivotal metabolic enzymes and fine-tuning the adaptive response of cell to oxidative stress (Verdin et al. 2010).

13.7 HDAC Inhibitors

Histone deacetylase inhibitors comprise a diverse class of compounds that exert their effects by impeding the activity of histone deacetylases, predominantly targeting classes I, II, and IV HDACs, thereby increasing the acetylation levels of an array of cellular proteins. A plethora of naturally occurring and synthetically derived molecules have been shown to inhibit HDACs. The former typically exhibit three principal structural domains: a surface recognition domain, a linker domain, and a metal-binding domain. The classification of small molecule HDAC inhibitors is based on their underlying chemical scaffolds and can be categorized into five distinct groups: (1) Hydroxamic acids, including suberoylanilide hydroxamic acid (SAHA), panobinostat, and belinostat; (2) Benzamides, exemplified by chidamide; (3) Cyclic peptides, such as depsipeptide; (4) Short-chain fatty acids, comprising butyric acid and valproic acid; and (5) Electrophilic ketones.

13.8 Deacetylation of Non-histone Proteins and Tumor Development

Deacetylation of non-histone proteins is involved in many biological processes such as metabolic transition, cell cycle regulation, and immune surveillance. Previous studies have detected aberrant expression of HDACs in multiple cancers. The abnormal activity of HDACs leads to the disruption of the acetylation/deacetylation balance of a variety of proteins in cells, which affects the initiation and progression of tumors (Table 13.5).

13.8.1 PD-L1

The immunosuppressive tumor microenvironment is regulated by many proteins, including programmed PD-L1 and programmed death-1 (PD-1). Overexpression of PD-L1 has been found in many solid tumors, such as melanoma, non-small-cell lung cancer, and breast cancer (Ohaegbulam et al. 2015). HDAC2 can reduce the acetylation level of PD-L1 and promote its transfer from the plasma membrane to the nucleus, thereby regulating the transcription of immune-related genes,

Table 13.5 Deacetylation of non-histone protein and tumor development

Protein	Enzymes	Cancer types	Functions	Reference
PD-L1	HDAC2	–	Increase the nuclear PD-L1 level	Gao et al. (2020)
PD-L1	HDAC6	Melanoma	Activate STAT3 to increase PD-L1 expression	Lienlaf et al. (2016)
P53	HADC1, SIRT1	–	Enhance p53 stability to promote cell autophagy and apoptosis Inhibit the expression of PD-1	Roy et al. 2005 Cao et al. (2021)
EGR1	HDAC11	Liver cancer	Reduce p53 transcription to inhibit cell apoptosis	Gong et al. (2019)
α-Tubulin	HDAC6	Breast cancer	Promote microtubule movement	Ge et al. (2021)
Cortactin	HDAC6	Breast cancer	Enhance the binding ability of cortactin to F-actin	Ge et al. (2021)
Cortactin	HDAC6, SIRT2	Bladder cancer	Promote bladder cancer cell migration	Zuo et al. (2012)
HIF1α	HDAC4	Liver cancer	Reduce HIF-1 transcriptional activity and its targets expression	Geng et al. (2011)
PKM2	HDAC8	Liver cancer	Facilitate PKM2 nuclear translocation to amplify glycolysis	Zhang et al. (2020)
LDHB	SIRT5	Colorectal cancer	Promote LDHB enzymatic activity to over-activate autophagy	Shi et al. (2019)
IDH1	SIRT2	Colorectal cancer	Enhance IDH1 enzymatic activity and α-KG production	Wang et al. (2020a)
P62	SIRT1	Liver cancer	Interfere with p62 polyubiquitin degradation	Feng et al. (2021)
LSD1	HDAC5	Hepatocellular carcinoma	Stabilize LSD1 to reduce p65 methylation and activate NF-κB signaling	Hu et al. (2020)
IRF4	HDAC11	Multiple myeloma	Enhance IRF4 transcriptional activity to suppress cell death	Mostofa et al. (2021)
CDC42	SIRT2	Colorectal cancer	Impair the binding ability of CDC42 to PAK4	Wang et al. (2023a)
WEE1	SIRT1	Pancreatic cancer	Maintain WEE1 in an inactive state	Zhu et al. (2023)

Abbreviations (alphabetical order): *α-Tubulin* Alpha-Tubulin, *CDC42* Cell Division Cycle 42, *Cortactin* Cortical Actin-Binding Protein, *EGR1* Early Growth Response 1, *HIF1α* Hypoxia-Inducible Factor 1-Alpha, *IDH1* Isocitrate Dehydrogenase 1, *IRF4* Interferon Regulatory Factor 4, *LDHB* Lactate Dehydrogenase B, *LSD1* Lysine-Specific Histone Demethylase 1A, *P53* Tumor Protein P53, *P62* Sequestosome 1, *PAK4* p21-Activated Kinase 4, *PD-L1* Programmed Death-Ligand 1, *PKM2* Pyruvate Kinase M2, *SIRT1* Sirtuin 1, *SIRT2* Sirtuin 2, *SIRT5* Sirtuin 5, *WEE1* WEE1 G2 Checkpoint Kinase

establishing a positive feedback loop of increased PD-L1 expression and promoting tumor cells to evade immune surveillance (Gao et al. 2020). In primary melanoma, deacetylation of Signal Transducer and Activator of Transcription 3 (STAT3) by HDAC6 maintains STAT3 phosphorylation and thus its transcriptional activity, downregulating the expression of PD-L1, and ultimately inhibiting the formation of immunosuppressive microenvironment (Lienlaf et al. 2016).

13.8.2 P53

The transcription factor and tumor suppressor protein p53 has been dubbed the "guardian of the genome" and plays important roles in the induction of cell cycle arrest, apoptosis, senescence, and autophagy. Acetylation and deacetylation of p53 are controlled by the CBP/p300 or MYST families of HAT and the HDAC1/mSin3 complex or SIRT1, respectively. The induced hyperacetylation of key residues of p53 by HDAC inhibitors can enhance its stability, induce cell cycle arrest, and up-regulate the expression of pro-apoptotic genes (Roy et al. 2005). P53 acetylation also affects the expression of PD-1. Acetylation of P53 helps to recruit it to the PD-1 promoter, thus enhancing its transcriptional activity and upregulating PD-1 in cancer cells (Cao et al. 2021). Moreover, HDAC11 binds to and deacetylates Early Growth Response 1 (Egr1), a transcription factor that regulates the p53-coding gene *TP53*, thereby inhibiting its transcription (Gong et al. 2019).

13.8.3 Cytoskeletal Proteins

The invasion and migration of tumor cells are at least in part controlled by cytoskeletal proteins. Previous studies have shown that HDAC6 can deacetylate a number of these and enhance tumor cell migration. HDAC6 promotes the deacetylation of α-tubulin and cortical actin-binding protein (cortactin), accelerates microtubule movement, increases cortactin binding to F-actin and thus enhancing the invasiveness of breast cancer cells (Ge et al. 2021). Additionally, HDAC6 cooperates with SIRT2 to deacetylate cortactin to promote bladder cancer cell migration and invasion (Zuo et al. 2012).

13.8.4 HIF1α

Hypoxia-inducible factor 1α (HIF1α) is an essential component of the HIF-1 transcription complex that regulates angiogenesis, cell metabolism, and cancer development. Previous studies have shown that inhibition of HDAC4 by shRNA increases HIF1α acetylation and de-stabilizes it in von Hippel-Lindau (VHL)-positive cancer cell lines. This reduces HIF-1 transcriptional activity and the expression of HIF-1 target genes, and suppresses the high levels of glycolytic activity of tumor cells mediated by hypoxia (Geng et al. 2011).

13.8.5 Metabolic Enzymes

Deacetylation affects the activities of a variety of metabolic enzymes, thereby leading to the metabolic reprogramming of tumor cells and tumor progression. Pyruvate kinase M2 (PKM2) is a key rate-limiting enzyme in glycolysis that is bound by and deacetylated by HDAC8. This not only enhances PKM2's enzymatic activity and the glycolytic flux of tumors, but also promotes PKM2 transport to the nucleus where it binds β-catenin and promotes cancer cell proliferation, thereby balancing the anabolic advantage of the Warburg effect (Zhang et al. 2020).

Lactate dehydrogenase B (LDHB) is associated with autophagy in tumor cells. It is bound by and deacetylated by SIRT5, thus promoting its enzymatic activity, activating autophagy, and accelerating the growth of colorectal cancer cells (Shi et al. 2019).

Isocitrate dehydrogenase 1 (IDH1) is hyperacetylated in primary colorectal carcinomas (CRC) and liver metastases. Overexpression of SIRT2 in CRC cells can deacetylate IDH1, enhance its enzymatic activity and α-KG production, and inhibit the proliferation, migration, and invasion of CRC cells (Wang et al. 2020a).

13.8.6 Others

p62 expression is upregulated in a variety of tumors. In hepatocellular carcinoma, high levels of SIRT1 deacetylate p62, which inhibits its binding to E3 ligase Keap1, blocks the degradation of p62 by polyubiquitin-dependent proteasomes and ultimately promotes tumor growth (Feng et al. 2021). HDAC5 mediates lysine-specific demethylase 1 (LSD1) deacetylation and stabilization. In turn, LSD1 reduces the methylation of Nuclear Factor Kappa-Light-Chain-Enhancer of Activated B Cells (NF-κB) catalytic unit p65, stabilizing p65 protein and activates the NF-κB signaling pathway to promote HCC proliferation, invasion, and resistance toward sorafenib (Hu et al. 2020). HDAC11 mediates the deacetylation of IFN response factor 4 (IRF4) and enhances its transcriptional activity. Correspondingly, targeted inhibition of HDAC11 in multiple myeloma (MM) cell lines leads to IRF4 hyperacetylation, which promotes its nuclear localization and impairs its binding to its target gene promoters leading to cell death (Mostofa et al. 2021).

Cell Division Cycle 42 (CDC42), a member of the Rho family of GTPases, plays a pivotal role in numerous critical signaling pathways in cancer cells. CDC42 undergoes acetylation by p300/CBP and deacetylation by SIRT2. Deacetylated CDC42 exhibits impaired binding to its downstream effector p21-Activated Kinase 4

(PAK4), leading to reduced phosphorylation of p38 and c-Jun N-terminal Kinase (JNK), thereby downregulates apoptosis and enhances the migratory and invasive capabilities of colon cancer cells (Wang et al. 2023a). Additionally, SIRT1 interacts with the cell cycle checkpoint protein WEE1 and deacetylates it, maintaining WEE1 in an inactive state. Conversely, knockdown of SIRT1 induces hyperacetylation and activation of WEE1, rendering cancer cells resistant to WEE1 inhibitors (Zhu et al. 2023).

13.9 Targeting Non-histone Protein Deacetylation for Cancer Treatment

HDACs extensively participate in the regulation of proliferation, migration, angiogenesis, immune escape, and therapeutic resistance of cancer cells, thus emerging as critical targets for clinical cancer therapy.

At this time, five HDAC inhibitors are approved for clinical use: vorinostat, belinostat, romidepsin, tucidinostat, and panobinostat. These HDAC inhibitors exhibit a significant therapeutic impact on hematological malignancies, including T-cell lymphoma and MM. The first HDAC inhibitor to achieve clinical success was vorinostat, a hydroxamic acid derivative also known as SAHA, which was utilized for the treatment of cutaneous T-cell lymphoma (CTCL). Other hydroxamic acid derivatives that have been approved include belinostat, indicated for peripheral T-cell lymphoma (PTCL) and panobinostat, which targets Multiple Myeloma (MM). Romidepsin, a cyclic peptide HDAC inhibitor, is also employed in the clinical setting for the management of both CTCL and PTCL. Tucidinostat, a benzenamide HDAC inhibitor also referred to as chidamide, has been approved for the treatment of PTCL. Furthermore, it is used in combination with exemestane for the treatment of postmenopausal advanced breast cancer thus highlighting its potential for the treatment of solid tumors (Bondarev et al. 2021).

Two major ongoing issues that surround the use of and research into HDAC inhibitors are treatment resistance and lack of progress in targeting solid tumors. The combined application of HDAC inhibitors and other chemotherapeutics and immunotherapeutic strategies have recently emerged as significant areas of focus. Immune checkpoints can block the hyperactive immune response by stimulating immunosuppressive pathways. However, tumor cells can also escape from the immune system by upregulating the expression of immunosuppressive molecules. Blocking immune checkpoints in tumors, especially PD-1/PD-L1 and CTLA-4, has become a key focus of tumor immunotherapy. Combining HDAC inhibitors with immune checkpoint inhibitors (ICI) may be an effective way to treat solid tumors and avoid the development of drug resistance. Many preclinical studies have reported the efficacy of HDAC inhibitors in combination with ICI for cancer therapy. For example, nexturastat (an HDAC6 inhibitor) in combination with anti-PD-1 agents has been shown to be effective in treating MM by reducing levels of protumorigenic M2 macrophages and upregulating effector T cells (Knox et al. 2019). Etinostat (a class I HDAC inhibitor) inhibits the immunosuppressive function of both polymorphonuclear (PMN)- and monocytic-myeloid-derived suppressor cell (M-MDSC) populations. This in turn enhances the antitumor effect of PD-1 inhibition in mouse models of lung and renal-cell carcinoma (Orillion et al. 2017). Finally, the combination of HDAC inhibitors and anti-CTLA-4 antibody can synergistically promote $CD4^+$ T cell infiltration and enhance the immune response in several tumor types (Cao et al. 2015).

13.10 Perspective

Over the past three decades, significant progress has been made in elucidating the mechanisms by which reversible acetylation of non-histone proteins contributes to oncogenesis and in applying this knowledge to cancer therapy. HATs and HDACs play crucial roles in various biological processes in tumor cells, including signal transduction, metabolic reprogramming, cell proliferation, immune evasion, migration, and invasion, by dynamically regulating the acetylation levels of numerous non-histone substrates.

Notably, most previous studies have focused on the role of reversible acetylation in regulating a single protein, however, a specific biological process, such as apoptosis, may involve multiple proteins subjected to reversible acetylation. A more comprehensive analysis of the reversible acetylation of different proteins in a given biological process can greatly aid in understanding the sequence and diverse impacts of targeting multiple proteins through reversible acetylation. Furthermore, previous researches have reported that a single HAT or HDAC may exert opposing effects in the progression of different types of cancer. The fact that a single HAT or HDAC typically has multiple protein substrates may explain these discrepancies. More research is needed to systematically dissect the roles of individual HATs and HDACs in different cancer types and even at different stages of tumorigenesis.

HAT and HDAC inhibitors have been successfully applied in the treatment of some hematological malignancies, but challenges remain in the treatment of solid tumors, namely, the toxicity of non-selective inhibitors and their applicability to different types of cancer. Taking HDAC inhibitors (HDACis) as an example, the most common HDACis currently under clinical evaluation are broad-spectrum non-selective HDACis. The efficacy of non-selective HDACis in cancer treatment relies on their broad-spectrum inhibition of HDACs, which is also the primary cause of their toxicity. Additionally, non-selective HDACis are ineffective in treating tumors where HDACs do not play a pro-oncogenic role. Therefore, the current focus is on developing HAT and HDAC inhibitors with higher target specificity, which may be more effective and less toxic. Moreover, combining HAT and HDAC inhibitors with immune checkpoint inhibitors, tumor vaccines, anti-tumor antibodies, or other chemotherapeutic agents may represent another important direction for enhancing the therapeutic effects of cancer treatment.

References

Alonso VL, Serra EC (2012) Lysine acetylation: elucidating the components of an emerging global signaling pathway in trypanosomes. J Biomed Biotechnol 2012(452934):1

Baell JB, Leaver DJ, Hermans SJ, Kelly GL, Brennan MS, Downer NL, Nguyen N, Wichmann J, Mcrae HM, Yang YQ, Cleary B, Lagiakos HR, Mieruszynski S, Pacini G, Vanyai HK, Bergamasco MI, May RE, Davey BK, Morgan KJ, Sealey AJ, Wang BN, Zamudio N, Wilcox S, Garnham AL, Sheikh BN, Aubrey BJ, Doggett K, Chung MC, de Silva M, Bentley J, Pilling P, Hattarki M, Dolezal O, Dennis ML, Falk H, Ren B, Charman SA, White KL, Rautela J, Newbold A, Hawkins ED, Johnstone RW, Huntington ND, Peat TS, Heath JK, Strasser A, Parker MW, Smyth GK, Street IP, Monahan BJ, Voss AK, Thomas T (2018) Inhibitors of histone acetyltransferases KAT6A/B induce senescence and arrest tumour growth. Nature 560:253–257

Balasubramanyam K, Swaminathan V, Ranganathan A, Kundu TK (2003) Small molecule modulators of histone acetyltransferase p300. J Biol Chem 278:19134–19140

Balasubramanyam K, Altaf M, Varier RA, Swaminathan V, Ravindran A, Sadhale PP, Kundu TK (2004) Polyisoprenylated benzophenone, garcinol, a natural histone acetyltransferase inhibitor, represses chromatin transcription and alters global gene expression. J Biol Chem 279:33716–33726

Bondarev AD, Attwood MM, Jonsson J, Chubarev VN, Tarasov VV, Schiöth HB (2021) Recent developments of HDAC inhibitors: emerging indications and novel molecules. Br J Clin Pharmacol 87:4577–4597

Bowers EM, Yan G, Mukherjee C, Orry A, Wang L, Holbert MA, Crump NT, Hazzalin CA, Liszczak G, Yuan H, Larocca C, Saldanha SA, Abagyan R, Sun Y, Meyers DJ, Marmorstein R, Mahadevan LC, Alani RM, Cole PA (2010) Virtual ligand screening of the p300/CBP histone acetyltransferase: identification of a selective small molecule inhibitor. Chem Biol 17:471–482

Cai LY, Chen SJ, Xiao SH, Sun QJ, Ding CH, Zheng BN, Zhu XY, Liu SQ, Yang F, Yang YX, Zhou B, Luo C, Zhang X, Xie WF (2021) Targeting p300/CBP attenuates hepatocellular carcinoma progression through epigenetic regulation of metabolism. Cancer Res 81:860–872

Cao K, Wang G, Li W, Zhang L, Wang R, Huang Y, Du L, Jiang J, Wu C, He X, Roberts AI, Li F, Rabson AB, Wang Y, Shi Y (2015) Histone deacetylase inhibitors prevent activation-induced cell death and promote anti-tumor immunity. Oncogene 34:5960–5970

Cao ZJ, Kon N, Liu YJ, Xu WB, Wen J, Yao H, Zhang M, Wu Z, Yan XJ, Zhu WG, Gu W, Wang DL (2021) An unexpected role for p53 in regulating cancer cell-intrinsic PD-1 by acetylation. Sci Adv 7:eabf4148

Chen QJ, Yang BH, Liu XC, Zhang XD, Zhang LR, Liu T (2022) Histone acetyltransferases CBP/p300 in tumorigenesis and CBP/p300 inhibitors as promising novel anticancer agents. Theranostics 12:4935–4948

Cheng YW, Zeng FM, Li DJ, Wang SH, He JZ, Guo ZC, Nie PJ, Wu ZY, Shi WQ, Wen B, Xu XE, Liao LD, Li ZM, Wu JY, Zhan J, Zhang HQ, Chang ZJ, Zhang K, Xu LY, Li EM (2021) P300/CBP-associated factor (PCAF)-mediated acetylation of Fascin at lysine 471 inhibits its actin-bundling activity and tumor metastasis in esophageal cancer. Cancer Commun 41:1398–1416

Chiaradonna F, Cirulli C, Palorini R, Votta G, Alberghina L (2015) New insights into the connection between histone deacetylases, cell metabolism, and cancer. Antioxid Redox Signal 23:30–50

Chimenti F, Bizzarri B, Maccioni E, Secci D, Bolasco A, Chimenti P, Fioravanti R, Granese A, Carradori S, Tosi F, Ballario P, Vernarecci S, Filetici P (2009) A novel histone acetyltransferase inhibitor modulating Gcn5 network: cyclopentylidene-(4-(4′-chlorophenyl)thiazol-2-yl)hydrazone. J Med Chem 52:530–536

Choi KC, Jung MG, Lee YH, Yoon JC, Kwon SH, Kang HB, Kim MJ, Cha JH, Kim YJ, Jun WJ, Lee JM, Yoon HG (2009) Epigallocatechin-3-gallate, a histone acetyltransferase inhibitor, inhibits EBV-induced B lymphocyte transformation via suppression of RelA acetylation. Cancer Res 69:583–592

Choudhary C, Mann M (2010) Decoding signalling networks by mass spectrometry-based proteomics. Nat Rev Mol Cell Biol 11:427–439

Coffey K, Blackburn TJ, Cook S, Golding BT, Griffin RJ, Hardcastle IR, Hewitt L, Huberman K, Mcneill HV, Newell DR, Roche C, Ryan-Munden CA, Watson A, Robson CN (2012) Characterisation of a Tip60 specific inhibitor, NU9056, in prostate cancer. PLoS One 7:e45539

de Ruijter AJM, van Gennip AH, Caron HN, Kemp S, van Kuilenburg ABP (2003) Histone deacetylases (HDACs): characterization of the classical HDAC family. Biochem J 370:737–749

Di Martile M, Desideri M, DE Luca T, Gabellini C, Buglioni S, Eramo A, Sette G, Milella M, Rotili D, Mai A, Carradori S, Secci D, de Maria R, del Bufalo D, Trisciuoglio D (2016) Histone acetyltransferase inhibitor CPTH6 preferentially targets lung cancer stem-like cells. Oncotarget 7:11332–11348

Diallo I, Seve M, Cunin V, Minassian F, Poisson JF, Michelland S, Bourgoin-Voillard S (2019) Current trends in protein acetylation analysis. Expert Rev Proteomics 16:139–159

Dine YF, Chi HG, Shao JL, Sh TZ, Yu H, Wan X, Wang XJ (2022) Leukemia inhibitory factor receptor homodimerization mediated by acetylation of extracellular lysine promotes prostate cancer progression through the PDPK1/AKT/GCN5 axis. Clin Transl Med 12:e676

Drazic A, Myklebust LM, Ree R, Arnesen T (2016) The world of protein acetylation. Biochimica Et Biophys Acta-Prot Proteom 1864:1372–1401

Driessen HPC, Dejong WW, Tesser GI, Bloemendal H (1985) The mechanism of N-terminal acetylation of proteins. CRC Crit Rev Biochem 18:281–325

Fan YZ, Hou T, Gao Y, Dan WC, Liu TJ, Liu B, Chen YL, Xie HJ, Yang Z, Chen JQ, Zeng J, Li L (2021) Acetylation-dependent regulation of TPD52 isoform 1 modulates chaperone-mediated autophagy in prostate cancer. Autophagy 17:4386–4400

Fan K, Ni XJ, Shen S, Gong ZJ, Wang JW, Xin YL, Zheng BH, Sun WT, Liu H, Suo T, Ni XL, Liu HB (2022) Acetylation stabilizes stathmin1 and promotes its activity contributing to gallbladder cancer metastasis. Cell Death Discov 8:265

Feng LF, Chen MQ, Li YL, Li MC, Hu SM, Zhou BL, Zhu LY, Yu L, Zhou QY, Tan LH, An HM, Wang X, Jin HC (2021) Sirt1 deacetylates and stabilizes p62 to promote hepato-carcinogenesis. Cell Death Dis 12:405

Furdas SD, Shekfeh S, Bissinger EM, Wagner JM, Schlimme S, Valkov V, Hendzel M, Jung M, Sippl W (2011) Synthesis and biological testing of novel pyridoisothiazolones as histone acetyltransferase inhibitors. Bioorg Med Chem 19:3678–3689

Furdas SD, Kannan S, Sippl W, Jung M (2012) Small molecule inhibitors of histone acetyltransferases as epigenetic tools and drug candidates. Arch Pharm 345:7–21

Furdas SD, Hoffmann I, Robaa D, Herquel B, Malinka W, Swiatek P, Akhtar A, Sippl W, Jung M (2014) Pyrido- and benzisothiazolones as inhibitors of histone acetyltransferases (HATs). Medchemcomm 5:1856–1862

Gajer JM, Furdas SD, Gründer A, Gothwal M, Heinicke U, Keller K, Colland F, Fulda S, Pahl H, Fichtner I, Sippl W, Jung M (2015) Histone acetyltransferase inhibitors block neuroblastoma cell growth in vivo. Oncogenesis 4:e137

Gao CX, Bourke E, Scobie M, Famme MA, Koolmeister T, Helleday T, Eriksson LA, Lowndes NF, Brown JAL (2014) Rational design and validation of a Tip60 histone acetyltransferase inhibitor. Sci Rep 4:5372

Gao Y, Nihira NT, Bu X, Chu C, Zhang JF, Kolodziejczyk A, Fan YZ, Chan NT, Ma LA, Liu J, Wang D, Dai XM, Liu HD, Ono M, Nakanishi A, Inuzuka H, North BJ, Huang YH, Sharma S, Geng Y, Xu W, Liu XS, Li L, Miki Y, Sicinski P, Freeman GJ, Wei WY (2020) Acetylation-dependent regulation of PD-L1 nuclear translocation dictates the efficacy of anti-PD-1 immunotherapy. Nat Cell Biol 22:1064–1075

Ge LP, Jin X, Yang YS, Liu XY, Shao ZM, Di GH, Jiang YZ (2021) Tektin4 loss promotes triple-negative breast cancer metastasis through HDAC6-mediated tubulin deacetylation and increases sensitivity to HDAC6 inhibitor. Oncogene 40:2323–2334

Ge W, Gong YJ, Li Y, Wu NX, Ruan YT, Xu TP, Shu YQ, Qiu W, Wang YW, Zhao CH (2023) IL-17 induces non-small cell lung cancer metastasis via GCN5-dependent SOX4 acetylation enhancing MMP9 gene transcription and expression. Mol Carcinog 62:1399–1416

Geng H, Harvey CT, Pittsenbarger J, Liu Q, Beer TM, Xue CH, Qian DZ (2011) HDAC4 protein regulates HIF1α protein lysine acetylation and cancer cell response to hypoxia. J Biol Chem 286:38095–38102

Ghizzoni M, Wu J, Gao TL, Haisma HJ, Dekker FJ, Zheng YG (2012) 6-alkylsalicylates are selective Tip60 inhibitors and target the acetyl-CoA binding site. Eur J Med Chem 47:337–344

Gong D, Zeng ZW, Yi FM, Wu JB (2019) Inhibition of histone deacetylase 11 promotes human liver cancer cell apoptosis. Am J Transl Res 11:983–990

Gonneaud A, Turgeon N, Jones C, Couture C, Lévesque D, Boisvert FM, Boudreau F, Asselin C (2019) HDAC1 and HDAC2 independently regulate common and specific intrinsic responses in murine enteroids. Sci Rep 9:5363

Gu W, Roeder RG (1997) Activation of p53 sequence-specific DNA binding by acetylation of the p53 C-terminal domain. Cell 90:595–606

Gu L, Zhu YH, Lin X, Tan XY, Lu BJ, Li YJ (2020) Stabilization of FASN by ACAT1-mediated GNPAT acetylation promotes lipid metabolism and hepatocarcinogenesis. Oncogene 39:2437–2449

Guarente L (2013) Calorie restriction and sirtuins revisited. Genes Dev 27:2072–2085

Hai RH, Yang DY, Zheng FF, Wang WQ, Han X, Bode AM, Luo XJ (2022) The emerging roles of HDACs and their therapeutic implications in cancer. Eur J Pharmacol 931:175216

Han Y, Zhang YY, Pan YQ, Zheng XJ, Liao K, Mo HY, Sheng H, Wu QN, Liu ZX, Zeng ZL, Yang W, Yuan SQ, Huang P, Ju HQ, Xu RH (2023) IL-1β-associated NNT acetylation orchestrates iron-sulfur cluster maintenance and cancer immunotherapy resistance. Mol Cell 83:1887–1902

Hu HG, Zhu WW, Qin J, Chen M, Gong LY, Li L, Liu XY, Tao YZ, Yin HY, Zhou H, Zhou L, Ye D, Ye QH, Gao D (2017) Acetylation of PGK1 promotes liver cancer cell proliferation and tumorigenesis. Hepatology 65:515–528

Hu B, Xu Y, Li YC, Huang JF, Cheng JW, Guo W, Yin Y, Gao Y, Wang PX, Wu SY, Zhou J, Fan J, Yang XR (2020) CD13 promotes hepatocellular carcinogenesis and sorafenib resistance by activating HDAC5-LSD1-NF-κB oncogenic signaling. Clin Transl Med 10:e233

Jayaprakash NG, Surolia A (2017) Role of glycosylation in nucleating protein folding and stability. Biochem J 474:2333–2347

Jin JQ, Zhang L, Li XY, Xu WZ, Yang SY, Song JG, Zhang WH, Zhan J, Luo JY, Zhang HQ (2022) Oxidative stress-CBP axis modulates MOB1 acetylation and activates the Hippo signaling pathway. Nucleic Acids Res 50:3817–3834

Knox T, Sahakian E, Banik D, Hadley M, Palmer E, Noonepalle S, Kim J, Powers J, Gracia-Hernandez M, Oliveira V, Cheng FD, Chen J, Barinka C, Pinilla-Ibarz J, Lee NH, Kozikowski A, Villagra A (2019) Selective HDAC6 inhibitors improve anti-PD-1 immune checkpoint blockade therapy by decreasing the anti-inflammatory phenotype of macrophages and down-regulation of immunosuppressive proteins in tumor cells. Sci Rep 9:6136

Kobayashi J, Kato A, Ota Y, Ohba R, Komatsu K (2010) Bisbenzamidine derivative, pentamidine represses DNA damage response through inhibition of histone H2A acetylation. Mol Cancer 9:34

Lasko LM, Jakob CG, Edalji RP, Qiu W, Montgomery D, Digiammarino EL, Hansen TM, Risi RM, Frey R, Manaves V, Shaw B, Algire M, Hessler P, Lam LT, Uziel T, Faivre E, Ferguson D, Buchanan FG, Martin RL, Torrent M, Chiang GG, Karukurichi K, Langston JW, Weinert BT, Choudhary C, de Vries P, van Drie JH, Mcelligott D, Kesicki E, Marmorstein R, Sun CH, Cole PA, Rosenberg SH, Michaelides MR, Lai A, Bromberg KD (2017) Discovery of a selective catalytic p300/CBP inhibitor that targets lineage-specific tumours. Nature 550:128–132

Lau OD, Kundu TK, Soccio RE, Ait-Si-Ali S, Khalil EM, Vassilev A, Wolffe AP, Nakatani Y, Roeder RG, Cole PA (2000) HATs off: selective synthetic inhibitors of the histone acetyltransferases p300 and PCAF. Mol Cell 5:589–595

Li JW, Wang J, Wang JX, Nawaz Z, Liu JM, Qin J, Wong JM (2000) Both corepressor proteins SMRT and N-CoR exist in large protein complexes containing HDAC3. EMBO J 19:4342–4350
Li YH, Huang M, Wang ME, Wang Y, Deng P, Li CN, Huang JY, Chen H, Wei ZH, Ouyang Q, Zhao JH, Lu YW, Su SC (2024) Tumor cells impair immunological synapse formation via central nervous system-enriched metabolite. Cancer Cell 42:985–1002
Lienlaf M, Perez-Villarroel P, Knox T, Pabon M, Sahakian E, Powers J, Woan KV, Lee C, Cheng F, Deng S, Smalley KSM, Montecino M, Kozikowski A, Pinilla-Ibarz J, Sarnaik A, Seto E, Weber J, Sotomayor EM, Villagra A (2016) Essential role of HDAC6 in the regulation of PD-L1 in melanoma. Mol Oncol 10:735–750
Lin HP, Cheng ZL, He RY, Song L, Tian MX, Zhou S, Groh BS, Liu WR, Ji MB, Ding C, Shi YH, Guan KL, Ye D, Xiong Y (2016) Destabilization of fatty acid synthase by acetylation inhibits lipogenesis and tumor cell growth. Cancer Res 76:6924–6936
Liu BW, Wang TJ, Wang HW, Zhang L, Xu FF, Fang RP, Li LL, Cai XL, Wu Y, Zhang WY, Ye LH (2018) Oncoprotein HBXIP enhances HOXB13 acetylation and co-activates HOXB13 to confer tamoxifen resistance in breast cancer. J Hematol Oncol 11:26
Liu Y, Zhang YY, Wang SQ, Li M, Long YH, Li YF, Liu YK, Li YH, Wang YQ, Mi JS, Yu CH, Li DY, Zhang JH, Zhang XJ (2020) WSTF acetylation by MOF promotes WSTF activities and oncogenic functions. Oncogene 39:5056–5067
Liu WX, Zhan ZY, Zhang MY, Sun BW, Shi QQ, Luo F, Zhang MD, Zhang WW, Hou YL, Xiao XY, Li YX, Feng HZ (2021) KAT6a, a novel regulator of β-catenin, promotes tumorigenicity and chemoresistance in ovarian cancer by acetylating COP1. Theranostics 11:6278–6292
Liu JZ, Zhai M, Chen YY, Wei YX, Li F, Chen YR, Duan BX, Xing L, Du HM, Jiang M, Li HZ, Ren GS (2024) Acetylation-dependent deubiquitinase USP26 stabilizes BAG3 to promote breast cancer progression. Cancer Lett 597:217005
Lombard DB, Zwaans BMM (2014) SIRT3: as simple as it seems? Gerontology 60:56–64
Lu Y, Wang Y, Zhang LL, Ma ZF, Yu KT, Shu Y, Zou X, Yang JJ, Liu X, Wang CL, Du YM, Li QH (2024) KAT7 enhances the proliferation and metastasis of head and neck squamous carcinoma by promoting the acetylation level of LDHA. Cancer Lett 590:216869
Luo HC, Shenoy AK, Li XH, Jin Y, Jin LH, Cai QS, Tang M, Liu Y, Chen H, Reisman D, Wu LZ, Seto E, Qiu Y, Dou YL, Casero RA, Lu JR (2016) MOF acetylates the histone demethylase LSD1 to suppress epithelial-to-mesenchymal transition. Cell Rep 15:2665–2678
Ma ZQ, Liu D, Di SY, Zhang ZP, Li WM, Zhang J, Xu LQ, Guo K, Zhu YF, Li XF, Han J, Yan XL (2019) Histone deacetylase 9 downregulation decreases tumor growth and promotes apoptosis in non-small cell lung cancer after melatonin treatment. J Pineal Res 67:e12587
Marcu MG, Jung YJ, Lee S, Chung EJ, Lee MJ, Trepel J, Neckers L (2006) Curcumin is an inhibitor of p300 histone acetylatransferase. Med Chem 2:169–174
Modak R, Basha J, Bharathy N, Maity K, Mizar P, Bhat AV, Vasudevan M, Rao VK, Kok WK, Natesh N, Taneja R, Kundu TK (2013) Probing p300/CBP Associated Factor (PCAF)-dependent pathways with a small molecule inhibitor. ACS Chem Biol 8:1311–1323
Mostofa AGM, Distler A, Meads MB, Sahakian E, Powers JJ, Achille A, Noyes D, Wright G, Fang B, Izumi V, Koomen J, Rampakrishnan R, Nguyen TP, de Avila G, Silva AS, Sudalagunta P, Canevarolo RR, Silva M, Alugubelli RR, Dai HA, Kulkarni A, Dalton WS, Hampton OA, Welsh EA, Teer JK, Tungesvik A, Wright KL, Pinilla-Ibarz J, Sotomayor EM, Shain KH, Brayer J (2021) Plasma cell dependence on histone/protein deacetylase11 reveals a therapeutic target in multiple myeloma. JCI Insight 6:e151713
Mukherjee S, Hao YH, Orth K (2007) A newly discovered post-translational modification - the acetylation of serine and threonine residues. Trends Biochem Sci 32:210–216
Narita T, Weinert BT, Choudhary C (2019) Functions and mechanisms of non-histone protein acetylation. Nat Rev Mol Cell Biol 20:156–174
Nihira NT, Ogura K, Shimizu K, North BJ, Zhang JF, Gao DM, Inuzuka H, Wei WY (2017) Acetylation-dependent regulation of MDM2 E3 ligase activity dictates its oncogenic function. Sci Signal 10:e151713

Ohaegbulam KC, Assal A, Lazar-Molnar E, Yao Y, Zang XX (2015) Human cancer immunotherapy with antibodies to the PD-1 and PD-L1 pathway. Trends Mol Med 21:24–33

Orillion A, Hashimoto A, Damayanti N, Shen L, Adelaiye-Ogala R, Arisa S, Chintala S, Ordentlich P, Kao CA, Elzey B, Gabrilovich D, Pili R (2017) Entinostat neutralizes myeloid-derived suppressor cells and enhances the antitumor effect of PD-1 inhibition in murine models of lung and renal cell carcinoma. Clin Cancer Res 23:5187–5201

Pearson M, Carbone R, Sebastiani C, Cioce M, Fagioli M, Saito S, Higashimoto Y, Appella E, Minucci S, Pandolfi PP, Pelicci PG (2000) PML regulates p53 acetylation and premature senescence induced by oncogenic Ras. Nature 406:207–210

Pi YN, Feng QS, Sun FS, Wang ZQ, Zhao Y, Chen DJ, Liu YM, Lou G (2023) Loss of SMURF2 expression enhances RACK1 stability and promotes ovarian cancer progression. Cell Death Differ 30:2382–2392

Polevoda B, Sherman F (2002) The diversity of acetylated proteins. Genome Biol 3:reviews0006

Ravindra KC, Selvi BR, Arif M, Reddy BAA, Thanuja GR, Agrawal S, Pradhan SK, Nagashayana N, Dasgupta D, Kundu TK (2009) Inhibition of lysine acetyltransferase KAT3B/p300 activity by a naturally occurring hydroxynaphthoquinone, plumbagin. J Biol Chem 284:24453–24464

Roth SY, Denu JM, Allis CD (2001) Histone acetyltransferases. Annu Rev Biochem 70:81–120

Roy S, Packman K, Jeffrey R, Tenniswood M (2005) Histone deacetylase inhibitors differentially stabilize acetylated p53 and induce cell cycle arrest or apoptosis in prostate cancer cells. Cell Death Differ 12:482–491

Sadler AJ, Suliman BA, Yu L, Yuan XL, Wang D, Irving AT, Sarvestani ST, Banerjee A, Mansell AS, Liu JP, Gerondakis S, Williams BRG, Xu DK (2015) The acetyltransferase HAT1 moderates the NF-κB response by regulating the transcription factor PLZF. Nat Commun 6:6795

Secci D, Carradori S, Bizzarri B, Bolasco A, Ballario P, Patramani Z, Fragapane P, Vernarecci S, Canzonetta C, Filetici P (2014) Synthesis of a novel series of thiazole-based histone acetyltransferase inhibitors. Bioorg Med Chem 22:1680–1689

Serrano L, Martínez-Redondo P, Marazuela-Duque A, Vazquez BN, Dooley SJ, Voigt P, Beck DB, Kane-Goldsmith N, Tong Q, Rabanal RM, Fondevila D, Muñoz P, Krüger M, Tischfield JA, Vaquero A (2013) The tumor suppressor SirT2 regulates cell cycle progression and genome stability by modulating the mitotic deposition of H4K20 methylation. Genes Dev 27:639–653

Shi X, Fan MF (2019) Tip60-dependent acetylation of KDM2B promotes osteosarcoma carcinogenesis. J Cell Mol Med 23:6154–6163

Shi M, Lu XJ, Zhang J, Diao H, Li GM, Xu L, Wang T, Wei J, Meng WY, Ma JL, Yu HG, Wang YG (2016) Oridonin, a novel lysine acetyltransferases inhibitor, inhibits proliferation and induces apoptosis in gastric cancer cells through p53-and caspase-3-mediated mechanisms. Oncotarget 7:22623–22631

Shi L, Yan H, An SX, Shen MQ, Jia WZ, Zhang RX, Zhao L, Huang G, Liu JJ (2019) SIRT5-mediated deacetylation of LDHB promotes autophagy and tumorigenesis in colorectal cancer. Mol Oncol 13:358–375

Starheim KK, Gevaert K, Arnesen T (2012) Protein N-terminal acetyltransferases: when the start matters. Trends Biochem Sci 37:152–161

Stimson L, Rowlands MG, Newbatt YM, Smith NF, Raynaud FI, Rogers P, Bavetsias V, Gorsuch S, Jarman M, Bannister A, Kouzarides T, Mcdonald E, Workman P, Aherne GW (2005) Isothiazolones as inhibitors of PCAF and p300 histone acetyltransferase activity. Mol Cancer Ther 4:1521–1532

Sun YL, Jiang XF, Chen SJ, Price BD (2006) Inhibition of histone acetyltransferase activity by anacardic acid sensitizes tumor cells to ionizing radiation. FEBS Lett 580:4353–4356

Sun YF, Sun JW, Lungchukiet P, Quarni W, Yang SY, Zhang XH, Bai WL (2015) Fe65 suppresses breast cancer cell migration and invasion through Tip60 mediated cortactin acetylation. Sci Rep 5:11529

Suryanarayanan V, Rajavel T, Devi KP, Singh SK (2018) Structure based identification and biological evaluation of novel and potent inhibitors of PCAF catalytic domain. Int J Biol Macromol 120:823–834

Sykes SM, Mellert HS, Holbert MA, Li K, Marmorstein R, Lane WS, Mcmahon SB (2006) Acetylation of the p53 DNA-binding domain regulates apoptosis induction. Mol Cell 24:841–851

Thiagalingam S, Cheng KH, Lee HJ, Mineva N, Thiagalingam A, Ponte JF (2003) Histone deacetylases: Unique players in shaping the epigenetic histone code. Ann N Y Acad Sci 983:84–100

Timmermann S, Lehrmann H, Polesskaya A, HAREL-Bellan, A. (2001) Histone acetylation and disease. Cell Mol Life Sci 58:728–736

Verdin E, Hirschey MD, Finley LWS, Haigis MC (2010) Sirtuin regulation of mitochondria: energy production, apoptosis, and signaling. Trends Biochem Sci 35:669–675

Wan JH, Zhan J, Li S, Ma J, Xu WZ, Liu C, Xue XW, Xie YP, Fang WG, Chin YE, Zhang HQ (2015) PCAF-primed EZH2 acetylation regulates its stability and promotes lung adenocarcinoma progression. Nucleic Acids Res 43:3591–3604

Wang YC, Peterson SE, Loring JF (2014) Protein post-translational modifications and regulation of pluripotency in human stem cells. Cell Res 24:143–160

Wang LT, Wang SN, Chiou SS, Liu KY, Chai CY, Chiang CM, Huang SK, Yokoyama KK, Hsu SH (2019a) TIP60-dependent acetylation of the SPZ1-TWIST complex promotes epithelial-mesenchymal transition and metastasis in liver cancer. Oncogene 38:518–532

Wang Z, Yang X, Liu C, Li X, Zhang BY, Wang B, Zhang Y, Song C, Zhang TZ, Liu MH, Liu BY, Ren MM, Jiang HP, Zou JH, Liu XY, Zhang HQ, Zhu WG, Yin YX, Zhang Z, Gu W, Luo JY (2019b) Acetylation of PHF5A modulates stress responses and colorectal carcinogenesis through alternative splicing-mediated upregulation of KDM3A. Mol Cell 74:1250–1263

Wang B, Ye YJ, Yang X, Liu BY, Wang Z, Chen SY, Jiang KW, Zhang W, Jiang HP, Mustonen H, Puolakkainen P, Wang S, Luo JY, Shen ZL (2020a) SIRT2-dependent IDH1 deacetylation inhibits colorectal cancer and liver metastases. EMBO Rep 21:e48183

Wang LT, Liu KY, Jeng WY, Chiang CM, Chai CY, Chiou SS, Huang MS, Yokoyama KK, Wang SN, Huang SK, Hsu SH (2020b) PCAF-mediated acetylation of ISX recruits BRD4 to promote epithelial-mesenchymal transition. EMBO Rep 21:e48795

Wang WY, Li X, Xu Y, Guo WK, Yu H, Zhang L, Wang YX, Chen XW (2021) Acetylation-stabilized chloride intracellular channel 1 exerts a tumor-promoting effect on cervical cancer cells by activating NF-κB. Cell Oncol 44:557–568

Wang DN, Ni JJ, Li JH, Gao YQ, Ni FJ, Zhang ZZ, Fang JY, Lu J, Yao YF (2023a) Bacterial infection promotes tumorigenesis of colorectal cancer via regulating CDC42 acetylation. PLoS Pathog 19:e1011189

Wang Y, Liu Y, Wang C, Kang R, Tang DL, Liu J (2023b) EP300 promotes ferroptosis via HSPA5 acetylation in pancreatic cancer. Sci Rep 13:15004

Wapenaar H, van der Wouden PE, Groves MR, Rotili D, Mai A, Dekker FJ (2015) Enzyme kinetics and inhibition of histone acetyltransferase KAT8. Eur J Med Chem 105:289–296

Wapenaar H, van den Bosch T, Leus NGJ, van der Wouden PE, Eleftheriadis N, Hermans J, Hailu GS, Rotili D, Mai A, Dömling A, Bischoff R, Haisma HJ, Dekker FJ (2017) The relevance of K_i calculation for bi-substrate enzymes illustrated by kinetic evaluation of a novel lysine (K) acetyltransferase 8 inhibitor. Eur J Med Chem 136:480–486

Wu J, Xie N, Wu ZK, Zhang Y, Zheng YG (2009) BisubstrateInhibitors of the MYST HATs Esa1 and Tip60. Bioorg Med Chem 17:1381–1386

Wu J, Wang JX, Li MY, Yang YT, Wang BH, Zheng YG (2011) Small molecule inhibitors of histone acetyltransferase Tip60. Bioorg Chem 39:53–58

Wu YZ, Zhou LW, Zou YZ, Zhang YJ, Zhang MF, Xu LP, Zheng LS, He WT, Yu K, Li T, Zhang X, Chen ZX, Zhang RH, Zhou PH, Zhang N, Zheng LM, Kang TB (2023) Disrupting the phase separation of KAT8-IRF1 diminishes PD-L1 expression and promotes antitumor immunity. Nature Cancer 4:382–400

Xia C, Tao Y, Li MS, Che TJ, Qu J (2020) Protein acetylation and deacetylation: an important regulatory modification in gene transcription (review). Exp Ther Med 20:2923–2940

Xu W, Li Y, Liu C, Zhao S (2014) Protein lysine acetylation guards metabolic homeostasis to fight against cancer. Oncogene 33:2279–2285

Yang XJ, Seto E (2008) The Rpd3/Hda1 family of lysine deacetylases: from bacteria and yeast to mice and men. Nat Rev Mol Cell Biol 9:206–218

Yang XH, Yu WH, Shi L, Sun LY, Liang J, Yi X, Li Q, Zhang Y, Yang F, Han X, Zhang D, Yang J, Yao Z, Shang YF (2011) HAT4, a golgi apparatus-anchored B-type histone acetyltransferase, acetylates free histone H4 and facilitates chromatin assembly. Mol Cell 44:39–50

Yang YX, Zhang RK, Li ZJ, Mei LH, Wan SL, Ding H, Chen ZF, Xing J, Feng HJ, Han J, Jiang HL, Zheng MY, Luo C, Zhou B (2020) Discovery of highly potent, selective, and orally efficacious p300/CBP histone acetyltransferases inhibitors. J Med Chem 63:1337–1360

Yang R, Peng WL, Shi S, Peng X, Cai QD, Zhao ZY, He BX, Tu GX, Yin W, Chen YC, Zhang YQ, Liu F, Wang X, Xiao DS, Tao YG (2023) The NLRP11 protein bridges the histone lysine acetyltransferase KAT7 to acetylate vimentin in the early stage of lung adenocarcinoma. Adv Sci 10:e2300971

Yi F, Cai CM, Ruan BZ, Hao MA, Yeo SK, Haas M, Yang FC, Zhang XT, Guan JL (2023) Regulation of RB1CC1/FIP200 stability and autophagy function by CREBBP-mediated acetylation in an intrinsically disordered region. Autophagy 19:1662–1677

Yu B, Luo F, Sun BW, Liu WX, Shi QQ, Cheng SY, Chen CS, Chen GQ, Li YX, Feng HZ (2021) KAT6A acetylation of SMAD3 regulates myeloid-derived suppressor cell recruitment, metastasis, and immunotherapy in triple-negative breast cancer. Adv Sci 8:e2100014

Zhang RX, Shen MQ, Wu CH, Chen YM, Lu JN, Li JJ, Zhao L, Meng HN, Zhou X, Huang G, Zhao XP, Liu JJ (2020) HDAC8-dependent deacetylation of PKM2 directs nuclear localization and glycolysis to promote proliferation in hepatocellular carcinoma. Cell Death Dis 11:1036

Zhang BT, Li YX, Wu Q, Xie L, Barwick B, Fu CY, Li X, Wu DQ, Xia SY, Chen J, Qian WP, Yang L, Osunkoya AO, Boise L, Vertino PM, Zhao YC, Li ML, Chen HR, Kowalski J, Kucuk O, Zhou W, Dong JT (2021) Acetylation of KLF5 maintains EMT and tumorigenicity to cause chemoresistant bone metastasis in prostate cancer. Nat Commun 12:1714

Zhang BT, Liu MC, Mai FY, Li XW, Wang WZ, Huang QQ, Du XC, Ding WJ, Li YX, Barwick BG, Ni JJ, Osunkoya AO, Chen YL, Zhou W, Xia SY, Dong JT (2024a) Interruption of KLF5 acetylation promotes PTEN deficient prostate cancer progression by reprogramming cancer-associated fibroblasts. J Clin Invest 134:e175949

Zhang CX, Liu HY, Li XH, Xiao N, Chen HX, Feng HR, Li Y, Yang Y, Zhang RK, Zhao XZ, Du YM, Bai L, Ma RN, Wan JH (2024b) Cold atmospheric plasma enhances SLC7A11-mediated ferroptosis in non-small cell lung cancer by regulating PCAF mediated HOXB9 acetylation. Redox Biol 75:103299

Zhao CH, Li YT, Qiu W, He FX, Zhang WM, Zhao D, Zhang ZW, Zhang EB, Ma P, Liu YQ, Ma L, Yang FM, Wang YW, Shu YQ (2018) C5a induces A549 cell proliferation of non-small cell lung cancer via GDF15 gene activation mediated by GCN5-dependent KLF5 acetylation. Oncogene 37:4821–4837

Zhao JH, Zhou XJ, Chen BX, Lu MZ, Wang GX, Elumalai N, Tian CH, Zhang JM, Liu YL, Chen ZQ, Zhou XY, Wu MZ, Li MJ, Prochownik EV, Tavassoli A, Jiang CQ, Li YJ (2023a) p53 promotes peroxisomal fatty acid β-oxidation to repress purine biosynthesis and mediate tumor suppression. Cell Death Dis 14:87

Zhao KQ, Zheng MY, Su ZZ, Ghosh S, Zhang C, Zhong WZ, Ho JWK, Jin GX, Zhou ZJ (2023b) MOF-mediated acetylation of SIRT6 disrupts SIRT6-FOXA2 interaction and represses SIRT6 tumor- suppressive function by upregulating ZEB2 in NSCLC. Cell Rep 42:112939

Zhao LJ, Qiu ZX, Yang ZN, Xu L, Pearce TM, Wu QL, Yang KL, Li FL, Saulnier O, Fei F, Yu HX, Gimple RC, Varadharajan V, Liu JX, Hendrikse LD, Fong V, Wang W, Zhang J, Lv DG, Lee DR, Lehrich BM, Jin CY, Ouyang L, Dixit D, Wu HX, Wang X, Sloan AE, Wang XX, Huan T, Brown JM, Goldman SA, Taylor MD, Zhou ST, Rich JN (2024) Lymphatic endothelial-like cells promote glioblastoma stem cell growth through cytokine-driven cholesterol metabolism. Nature Cancer 5:147–166

Zheng YJ, Balasubramanyam K, Cebrat M, Buck D, Guidez F, Zelent A, Alani RM, Cole PA (2005) Synthesis and evaluation of a potent and selective cell-permeable p300 histone acetyltransferase inhibitor. J Am Chem Soc 127:17182–17183

Zhou Y, Jia KW, Wang SY, Li ZY, Li YH, Lu S, Yang YY, Zhang LY, Wang M, Dong Y, Zhang LX, Zhang WN, Li N, Yu YZ, Cao XT, Hou J (2023) Malignant progression of liver cancer progenitors requires lysine acetyltransferase 7-acetylated and cytoplasm-translocated G protein GαS. Hepatology 77:1106–1121

Zhu YH, Gu L, Lin X, Liu C, Lu BJ, Cui KS, Zhou F, Zhao Q, Prochownik EV, Fan CP, Li YJ (2020) Dynamic regulation of ME1 phosphorylation and acetylation affects lipid metabolism and colorectal tumorigenesis. Mol Cell 77:138–149

Zhu XM, Su QS, Xie HY, Song LZ, Yang F, Zhang DD, Wang BH, Lin SX, Huang J, Wu MJ, Liu T (2023) SIRT1 deacetylates WEE1 and sensitizes cancer cells to WEE1 inhibition. Nat Chem Biol 19:585–595

Zuo QQ, Wu WJ, Li X, Zhao L, Chen W (2012) HDAC6 and SIRT2 promote bladder cancer cell migration and invasion by targeting cortactin. Oncol Rep 27:819–824

Chapter 14
Lysine Acetyltransferase 6 Complexes in Neurodevelopmental Disorders and Different Types of Cancer

Negar Mousavi and Xiang-Jiao Yang

Abstract Lysine acetylation is a critical post-translational modification that regulates gene expression and cellular functions. The MYST family lysine acetyltransferases KAT6A (also known as MOZ and MYST3) and KAT6B (a.k.a. MORF and MYST4), in complex with the multivalent epigenetic regulator BRPF1, play key roles in hematopoietic and neural development. Dysregulation of these complexes is implicated in neurodevelopmental disorders, such as Genitopatellar and Say-Barber-Biesecker-Young-Simpson syndromes, as well as in various cancers, including leukemia and medulloblastoma. The evolutionary conservation of these complexes in *Drosophila melanogaster* and *Caenorhabditis elegans* underscores their fundamental biological significance. Understanding the structural and functional mechanisms of KAT6-BRPF1 complexes provides insight into their pathological roles and therapeutic potential.

N. Mousavi (✉)
Rosalind and Morris Goodman Cancer Institute, McGill University, Montreal, Quebec, Canada

Department of Medicine, McGill University, Montreal, Quebec, Canada
e-mail: seyedeh.mousavifirouzabadi@mail.mcgill.ca

X.-J. Yang (✉)
Rosalind and Morris Goodman Cancer Institute, McGill University, Montreal, Quebec, Canada

Department of Medicine, McGill University, Montreal, Quebec, Canada

Department of Biochemistry, McGill University, Montreal, Quebec, Canada

McGill University Health Center, Montreal, Quebec, Canada
e-mail: xiang-jiao.yang@mcgill.ca

M. Halasa, A. Wawruszak (eds.), *Histone and Non-Histone Reversible Acetylation in Development, Aging and Disease*, Results and Problems in Cell Differentiation 75,
https://doi.org/10.1007/978-3-031-91459-1_14

14.1 Introduction

Histone acetylation, a pivotal post-translational modification, critically regulates gene expression by modulating chromatin structure and accessibility (Cruickshank et al. 2010; Kleff et al. 1995). This acetylation process, facilitated by HATs (histone acetyltransferases), involves the transfer of acetyl groups to lysine residues on histone tails, resulting in an open chromatin configuration conducive to transcriptional activation (Blakey and Litt 2015). The regulation of histone acetylation is essential for various cellular processes, including DNA repair, replication, and transcription (Struhl 1998).

The discovery of lysine acetylation as a post-translational modification in the 1960s marked a transformative era in molecular biology, highlighting the significance of acetylation in chromatin regulation (Kleff et al. 1995; Brownell et al. 1996; Yang et al. 1996). Later, Gcn5 (general control nonderepressible 5) was identified as the first histone acetyltransferase, cementing the link between acetylation and transcriptional control (Dent 2024). The subsequent discovery of the MYST family, named after its founding members MOZ (monocytic leukemia zinc finger protein), Ybf2/Sas3 (something about silencing 3), Sas2, and Tip60 (HIV Tat-interacting 60 kDa protein), further underscored the diverse regulatory roles of histone acetylation (Sapountzi and Côté 2011). The MYST family is the largest family of HATs and includes key members such as KAT5 (lysine Acetyltransferase 5; also known as Tip60), KAT6A (lysine acetyltransferase 6A; a.k.a. MOZ and MYST3), KAT6B (a.k.a. lysine acetyltransferase 6B, MORF and MYST4), KAT7 (a.k.a. HBO1 and MYST2), and KAT8 (a.k.a. MOF and MYST1). These proteins are conserved across all eukaryotes, from yeast to humans, highlighting their fundamental biological importance (Liang et al. 2023). The MYST family is implicated in crucial cellular functions, such as hematopoiesis, neural development, and the maintenance of stem cell pluripotency (Wiesel-Motiuk and Assaraf 2020). The MYST family exemplifies the evolutionary conservation of chromatin regulators (Yang 2015). While MOZ and MORF are vertebrate-specific, their functional and structural homologs, such as *Enok* in *Drosophila melanogaster* and *LIN-49* in *Caenorhabditis elegans*, underscore their fundamental roles in chromatin dynamics across species in the animal kingdom (Yang and Ullah 2007; O'Meara et al. 2010; Chamberlin and Thomas 2000; Chang et al. 2003; Scott et al. 2001). *Enok*, the *Drosophila melanogaster* ortholog, plays a critical role in neuroblast proliferation and oocyte development, highlighting its significance in developmental processes (Yang and Ullah 2007; O'Meara et al. 2010; Chamberlin and Thomas 2000; Chang et al. 2003; Scott et al. 2001). Similarly, *LIN-49* in *Caenorhabditis elegans* governs neuron asymmetry, hindgut development, and reproductive processes, further emphasizing the conserved regulatory functions of this protein family (Yang and Ullah 2007; O'Meara et al. 2010; Chamberlin and Thomas 2000; Chang et al. 2003; Scott et al. 2001).

Bromodomain and PHD finger-containing protein 1 (BRPF1) acts as a critical activator for KAT6A and KAT6B, serving as a scaffold protein that promotes the assembly of the KAT6-BRPF1 complex (Zu et al. 2022). BRPF1 itself is a multivalent chromatin regulator, equipped with multiple histone-binding domains, including double PHD fingers, a bromodomain, and a PWWP domain. These domains enable BRPF1 to interact with various histone marks, orchestrating the localization and function of the KAT6 acetyltransferases (Fig. 14.2) (Zu et al. 2022). The domain organization of BRPF1 is conserved in its paralogs BRPF2 and BRPF3, which share similar domains but exhibit distinct interactions with different MYST family acetyltransferases (Yang 2015). The formation of the KAT6-BRPF1 complex is crucial for its stability and enzymatic activity, highlighting the significance of BRPF1 in chromatin regulation (Zu et al. 2022).

BRPF1 belongs to a larger family of proteins, including BRPF2 (also known as BRD1), BRPF3, and JADE proteins (Yan et al. 2016). These epigenetic regulators share structural similarities with BRPF1 but exhibit distinct functional roles and interactions (Yan et al. 2016). BRPF2/BRD1 primarily interacts with HBO1 (or KAT7), another MYST family acetyltransferase, and is involved in erythropoiesis and thymocyte development (Mishima et al. 2011, 2014; Yokoyama et al. 2024; Su et al. 2024). BRPF3, although less studied, is also thought to play a role in chromatin regulation through interaction with KAT7 (Yan et al. 2016).

The functional roles of the KAT6-BRPF1 complexes extend to several biological processes, including transcriptional regulation through histone acetylation (see "Histone Acylation by KAT6-BRPF1 Complexes" for details on specific histone targets) (Viita and Côté 2023). These modifications are associated with active transcriptional regions, influencing cell cycle progression, differentiation, and development (Viita and Côté 2023). For instance, in hematopoiesis, KAT6A is known to regulate the expression of genes essential for the self-renewal and differentiation of

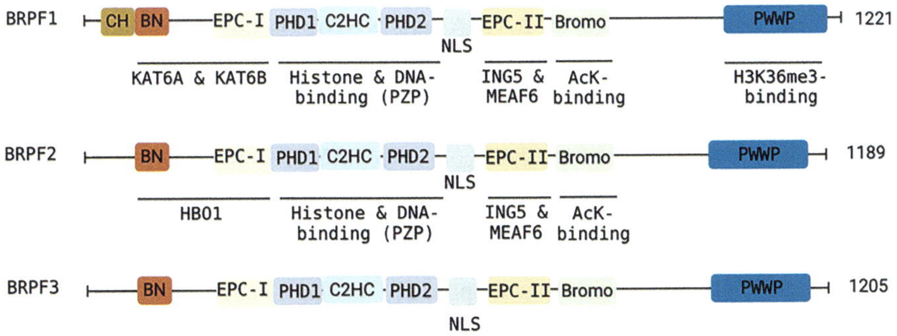

Fig. 14.2 Domain organization of BRPF1 and its paralogs. This figure provides a detailed schematic of the domain organization of the BRPF (Bromodomain and PHD finger-containing) proteins: BRPF1, BRPF2, and BRPF3. Key domains such as the bromodomain, PWWP, and PHD fingers are highlighted, showcasing their roles in DNA/histone binding and regulation of chromatin structure. The bromodomain, in particular, suggests an epigenetic regulatory function through interaction with acetylated lysine residues. Each protein's amino acid length is indicated at right, allowing a comparison of their sizes. The unique domain architectures underscore their diverse roles in epigenetic regulation, particularly in recognizing and binding histone modifications, DNA, and other proteins involved in chromatin remodeling

hematopoietic stem cells (HSCs) (Sheikh et al. 2016). Similarly, KAT6B has been implicated in neural development and function (Cosentino et al. 2019).

Dysregulation of the KAT6-BRPF1 complex has been linked to various pathological conditions. Mutations and aberrant expression of KAT6A and KAT6B are associated with developmental disorders, such as Genitopatellar syndrome and Say-Barber-Biesecker-Young-Simpson syndrome, respectively (Simpson et al. 2012; Davarnia et al. 2024). Furthermore, alterations in the components of this complex have been implicated in oncogenesis, with KAT6A and KAT6B identified as fusion partners in chromosomal translocations observed in acute myeloid leukemia (AML) and other cancers (Troke et al. 2006; Sheikh et al. 2015). The complex's role in these diseases underscores its potential as a therapeutic target, highlighting the importance of understanding its regulatory mechanisms and interactions (Su et al. 2021; Mukohara et al. 2024a; Cristofanilli et al. 2024).

This mini-review aims to provide a comprehensive overview of the KAT6-BRPF1 complex, emphasizing its structural characteristics, functional roles, and implications in health and disease. By synthesizing current knowledge and recent research findings, we aim to outline the critical functions of this complex and identify future research directions that could lead to novel therapeutic strategies.

14.2 Structure of KAT6A and KAT6B

KAT6A and KAT6B are characterized by several distinct structural domains: the NEMM domain, tandem PHD fingers, the MYST domain, an acidic region, and a serine/methionine (SM)-rich domain. Each of these domains plays a critical role in the function of these histone acetyltransferases (Fig. 14.1) (Zu et al. 2022). The MYST domain architecture of KAT6A and KAT6B is conserved among other MYST family members, such as KAT5, KAT7, and KAT8, highlighting their structural and functional significance in chromatin regulation.

KAT6A and KAT6B are integral members of the MYST family of histone acetyltransferases, distinguished by their conserved MYST domain (Zu et al. 2022). This domain includes a zinc finger motif and an acetyl-CoA-binding site, both critical for their catalytic activity (Zu et al. 2022). The zinc finger interacts with DNA, facilitating the enzyme's positioning on chromatin, while the acetyl-CoA-binding site binds acetyl-CoA, positioning it optimally for the transfer of acetyl groups to lysine residues on histone tails (Zu et al. 2022) (Fig. 14.2).

The NEMM (N-terminal part of Enok, MOZ, and MORF) domain is involved in protein-protein interactions essential for their function (Klein et al. 2019). This domain shares significant homology with *Enok* in *Drosophila melanogaster*, particularly in the MYST domain responsible for acetyl-CoA binding (Scott et al. 2001; Huang et al. 2014). *Enok*'s acetylation of lysine 23 on histone H3 (H3K23ac) mirrors MOZ/MORF's function in vertebrates, underscoring conserved mechanisms in chromatin dynamics (Scott et al. 2001; Huang et al. 2014). This domain plays a role in forming multiprotein complexes necessary for the acetyltransferase

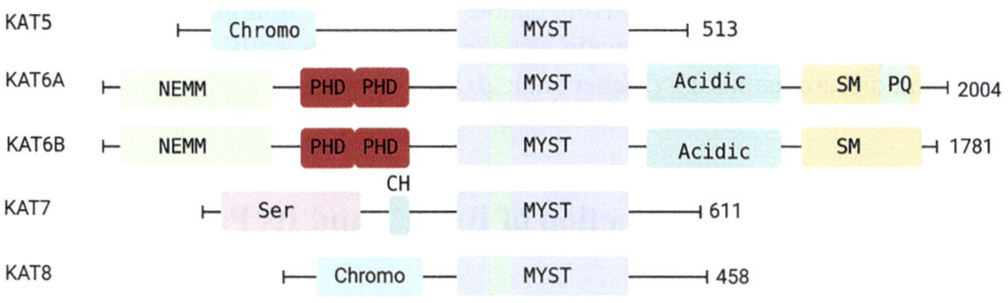

Fig. 14.1 Domain architecture of the MYST family histone acetyltransferases. This figure illustrates the domain organization of MYST family histone acetyltransferases (HATs): KAT5, KAT6A, KAT6B, KAT7, and KAT8. Each enzyme is shown with distinct functional domains, including chromodomains for binding methylated histones, NEMM domains, plant homeodomain (PHD) fingers for recognizing histone tail modifications, serine-rich (Ser) regions, MYST domains with acetyltransferase activity, and acidic regions. The total amino acid count for each protein is indicated. The schematic highlights shared features among MYST HATs and their unique elements, underscoring their specificity in histone acetylation and gene regulation. The structural variations, such as additional domains and length differences, contribute to their functional diversity and regulatory roles

activity of KAT6A and KAT6B (Zu et al. 2022; Klein et al. 2019). The NEMM domain also shows some sequence similarity to histones H1 and H5, suggesting a regulatory function for this region (Zu et al. 2022). These zinc finger motifs are involved in recognizing and binding to histone H3 tails, particularly when they are methylated (Huang et al. 2016). This recognition is critical for recruiting the acetyltransferase activity of KAT6A and KAT6B to specific genomic regions, facilitating precise chromatin modifications (Huang et al. 2016).

The MYST domain is a defining feature of KAT6A and KAT6B, encompassing a zinc finger motif and an acetyl-CoA-binding site (Huang et al. 2016; Shukla et al. 2018). The zinc finger motif interacts with DNA, positioning the enzyme on chromatin, while the acetyl-CoA-binding site binds acetyl-CoA, facilitating the transfer of acetyl groups to lysine residues on histone tails (Marmorstein 2001). This acetylation is essential for chromatin remodeling and gene expression regulation (Marmorstein 2001). The MYST domain catalyzes histone acetylation and interacts with BRPF1, ensuring that acetylation is targeted to specific sites within the chromatin (Zu et al. 2022).

The acidic region is associated with leukemia and developmental disorders (Wiesel-Motiuk and Assaraf 2020; Zu et al. 2022). It interacts with other chromatin-remodeling proteins, further facilitating the formation of functional multiprotein complexes (Zu et al. 2022). This region aids in coordinating the activities of various proteins involved in chromatin remodeling, ensuring that the acetyltransferase activity is properly regulated (Zu et al. 2022). The SM domain has transcriptional activation potential, contributing to the regulation of gene expression (Wiesel-Motiuk and Assaraf 2020). This domain's role in transcriptional activation underscores the multifaceted functions of KAT6A and KAT6B in regulating gene expression through chromatin modification (Wiesel-Motiuk and Assaraf 2020).

These structural features collectively enable KAT6A and KAT6B to regulate gene expression by modifying chromatin structure, playing essential roles in development and disease pathology (Weber et al. 2023).

14.3 Structure and Function of BRPF1 and Its Paralogs

Bromodomain and PHD finger-containing protein 1 (BRPF1) is a multifaceted chromatin regulator that plays a pivotal role in the assembly and function of the KAT6 complex (You et al. 2015a). The primary structural elements of BRPF1 include double PHD fingers, a bromodomain, and a PWWP domain, each contributing to its function as a scaffold protein in chromatin regulation (You et al. 2015a).

The double PHD fingers of BRPF1 are essential for recognizing and binding to specific histone marks, particularly methylated lysine residues (Poplawski et al. 2014). Structural studies have revealed that these PHD fingers interact with trimethylated lysine 4 on histone H3 (H3K4me3), a mark associated with active transcription (Zhang et al. 2023). This interaction enhances the specificity of BRPF1 for transcriptionally active regions of the genome, facilitating the targeted recruitment of the KAT6 complex to these sites (Zhang et al. 2023).

Adjacent to the PHD fingers is the bromodomain, a motif known for its affinity for acetylated lysine residues (Zhang et al. 2023). The bromodomain of BRPF1 binds to acetylated histones, further stabilizing its association with chromatin and promoting the acetylation activity of the KAT6 complex (Viita and Côté 2023). This dual recognition of both methylated and acetylated histones underscores the versatile binding capabilities of BRPF1 and its critical role in orchestrating histone modifications (You et al. 2016).

Additionally, BRPF1 contains a PWWP domain, which is implicated in recognizing methylated histones and DNA (Rona et al. 2016). The PWWP domain binds to di- and trimethylated lysine residues on histones, contributing to the localization of BRPF1 to chromatin regions marked by these modifications (Rona et al. 2016). This domain's ability to interact with both histone and DNA modifications enhances the chromatin-binding affinity and specificity of BRPF1, ensuring precise regulation of gene expression (Rona et al. 2016).

The structural integrity and functionality of BRPF1 are further supported by its C-terminal regions, which mediate interactions with other proteins in the KAT6 complex (Viita and Côté 2023). These interactions are crucial for the formation and stability of the complex, enabling effective acetylation of histones and regulation of chromatin structure (Viita and Côté 2023).

BRPF2 (also known as BRD1) and BRPF3 share significant structural similarities with BRPF1, as they also possess multiple histone-binding domains that contribute to their roles in chromatin regulation (Barman et al. 2022). BRPF2 contains tandem PHD fingers, a bromodomain, and a PWWP domain, similar to BRPF1 (Barman et al. 2022). The tandem PHD fingers of BRPF2 bind to methylated histones, specifically recognizing H3K4me3 marks, which is essential for its recruitment to active chromatin regions (Gatchalian and Kutateladze 2015; Sun et al. 2015). The bromodomain of BRPF2, like that of BRPF1, binds to acetylated lysine residues on histones, facilitating chromatin association and enhancing acetyltransferase activity (Gatchalian and Kutateladze 2015; Sun et al. 2015).

BRPF3 also features PHD fingers, a bromodomain, and a PWWP domain (Yan et al. 2016). Although BRPF3 is less studied, its structural domains suggest a role in recognizing and binding to specific histone modifications (Tao et al. 2017). The PHD fingers of BRPF3 likely interact with methylated histones, while the bromodomain binds to acetylated lysines, stabilizing its interaction with chromatin (Tao et al. 2017). The PWWP domain of BRPF3, similar to those in BRPF1 and BRPF2, contributes to its localization to chromatin regions with methylated histone marks (Yan et al. 2016).

Understanding the structures of BRPF1, BRPF2, and BRPF3 is fundamental to comprehending their roles in gene regulation and their implications in various diseases.

14.4 KAT6-BRPF1 Tetrameric Complexes

The formation of the KAT6-BRPF1 complex is a meticulously regulated process essential for chromatin modification and gene expression regulation. KAT6A and KAT6B form a functional acetyltransferase complex through direct interactions with BRPF1 (Sharma et al. 2023). This interaction is crucial for stabilizing the KAT6 enzymes and enhancing their catalytic efficiency, thus enabling precise histone acetylation (Yan et al. 2017).

BRPF1 acts as a scaffold protein, facilitating the assembly of the complex by simultaneously binding to multiple regions of KAT6 proteins (Yan et al. 2017). The double PHD fingers of BRPF1 are critical for recognizing and binding to H3K4me3 and trimethylated lysine 36 on histone H3 (H3K36), a histone mark associated with active transcription (Viita and Côté 2023). This interaction stabilizes BRPF1's association with chromatin and positions the KAT6 enzymes in proximity to their histone substrates (Viita and Côté 2023). Additionally, the bromodomain of BRPF1 binds to acetylated lysine residues on histones, anchoring the complex to regions of active transcription and enhancing its acetylation activity (Yan et al. 2017). Moreover, BRPF1 interacts with acetylated lysine 9 on histone H3 (H3K9ac), further anchoring the complex to specific genomic loci and facilitating precise chromatin modifications (Yan et al. 2017).

The PWWP domain of BRPF1 further contributes to the complex's specificity and stability by interacting with methylated histones and DNA (van Nuland et al. 2013). The synergistic actions of these domains ensure that the KAT6-BRPF1 complex is accurately targeted to specific genomic loci, where it can effectively perform its acetyltransferase functions (van Nuland et al. 2013).

The BRPF1-KAT6 complex also includes inhibitor of growth protein 5 (ING5) and MYST/Esa1-associated factor 6 (MEAF6) as integral components (Mattioli et al. 2017). ING5, through its C-terminal PHD domain, binds to histone, while MEAF6 enhances the stability and activity of the complex (Huang et al. 2016; Mattioli et al. 2017; Champagne et al. 2008). These interactions further refine the targeting and functionality of the KAT6-BRPF1 complex, ensuring precise histone acetylation and chromatin remodeling activities (Mattioli et al. 2017).

The formation of the KAT6-BRPF1 complex not only increases the enzymatic activity of KAT6A and KAT6B but also regulates their substrate specificity (White et al. 2024). Through interaction with BRPF1, the KAT6 enzymes can acetylate specific lysine residues on histone H3, such as H3K9ac, acetylated lysine 14 on histone H3 (H3K14ac), and H3K23ac (Yan et al. 2020). These acetylation events result in the relaxation of chromatin structure, thereby facilitating the transcriptional activation of target genes (Yan et al. 2020).

14.5 Histone Acylation by KAT6-BRPF1 Complexes

The KAT6-BRPF1 complex is a critical regulator of gene expression through histone acetylation, a key epigenetic modification affecting chromatin structure and transcriptional activity (Yan et al. 2020). The complex involves KAT6A and KAT6B, which transfer acetyl groups from acetyl-CoA to specific lysine residues on histone tails, notably H3K9ac, H3K14, and H3K23ac (Yan et al. 2020).

H3K9ac is associated with euchromatin formation and transcriptional initiation, H3K14ac is frequently found at promoters and enhancers of actively transcribed genes, where it facilitates chromatin remodeling and nucleosome displacement, and H3K23ac contributes to higher-order chromatin structure and transcriptional activation (Igolkina et al. 2019; Liu et al. 2023). These modifications create a permissive chromatin environment, enhancing transcription factor recruitment and gene expression regulation (Igolkina et al. 2019; Liu et al. 2023). Dysregulation of these histone marks has been implicated in various pathological conditions, including neurodevelopmental disorders, where BRPF1 mutations result in reduced H3K14ac and H3K23ac, leading to impaired neural gene expression (Viita and Côté 2023; Zhang et al. 2023; You et al. 2016; Yan et al. 2020). Similarly, in leukemia, loss-of-function mutations in BRPF1 lead to decreased H3K14ac and H3K23ac, underscoring their critical role in maintaining chromatin accessibility during hematopoiesis (Shima et al. 2014).

The process begins with the binding of the KAT6-BRPF1 complex to chromatin. KAT6A and KAT6B interact with chromatin through their zinc finger motifs, ensuring site-specific histone acetylation (Wiesel-Motiuk and Assaraf 2020; Weber et al. 2023). BRPF1 enhances binding specificity through its histone-binding domains, such as double PHD fingers and bromodomain, which recognize H3K4me3 and acetylated lysines (Wiesel-Motiuk and Assaraf 2020; Weber et al. 2023). Once anchored to chromatin, the acetyl-CoA-binding site in the MYST domain binds acetyl-CoA, allowing KAT6 to transfer the acetyl group to lysine residues on histone tails (Wiesel-Motiuk and Assaraf 2020). This acetylation neutralizes the positive charge of lysines, reducing histone-DNA interactions and relaxing chromatin structure to facilitate transcription (Wiesel-Motiuk and Assaraf 2020; Weber et al. 2023).

Histone acetylation by the KAT6-BRPF1 complex increases chromatin accessibility, allowing transcription factors and other regulatory proteins to access DNA (Wiesel-Motiuk and Assaraf 2020; Weber et al. 2023). Acetylated histones serve as binding sites for bromodomain-containing proteins, which recruit transcriptional coactivators and components of the transcription machinery, further enhancing gene expression (Wiesel-Motiuk and Assaraf 2020; Weber et al. 2023). Beyond direct histone modification, the KAT6-BRPF1 complex recruits and stabilizes other chromatin-modifying enzymes, coordinating the deposition of multiple histone marks and establishing specific chromatin states that regulate gene expression (Wiesel-Motiuk and Assaraf 2020; Weber et al. 2023).

The KAT6-BRPF1 complex plays a crucial role in chromatin remodeling. The acetylation of histone H3 at lysine residues 9, 14, and 23 by this complex is associated with active transcription (You et al. 2016). H3K9ac is linked to transcriptional activation and euchromatin formation (Igolkina et al. 2019). H3K14ac is often found at promoters and enhancers of actively transcribed genes, facilitating chromatin remodeling and nucleosome displacement (Liu et al. 2023). H3K23ac promotes a transcriptionally permissive chromatin state, influencing higher-order chromatin structure (Klein et al. 2019). BRPF1's interaction with specific chromatin regions ensures precise targeting of these histone marks (You et al. 2016; Barman et al. 2021). The tandem PHD fingers and bromodomain of BRPF1 recognize and bind to existing histone modifications, such as H3K4me3 and acetylated lysine residues, anchoring the complex at sites of active transcription (You et al. 2016). This targeted binding ensures that acetylation is confined to specific genomic loci, allowing precise control over gene expression (You et al. 2016).

In addition to histone acetylation, the KAT6-BRPF1 complex acetylates non-histone proteins, affecting various cellular processes, including DNA repair, cell cycle regulation, and apoptosis (Yang et al. 2022). By modifying transcription factors and other regulatory proteins, the complex influences their activity, stability, and interactions, thereby modulating multiple signaling pathways. Precise regulation of histone acetylation by the KAT6-BRPF1 complex is vital for cellular homeostasis and normal development (Viita and Côté 2023; Zhong et al. 2023). Dysregulation of this complex, whether due to mutations, altered expression, or disrupted chromatin interactions, can lead to aberrant gene expression patterns associated with diseases such as cancer, neurodevelopmental disorders, and inflammatory conditions (Yang et al. 2022; Zhao et al. 2024; Colson et al. 2024; Morison et al. 2024). Understanding the mechanisms of action in histone acetylation by the KAT6-BRPF1 complex provides essential insights into its broader role in gene regulation and its potential as a therapeutic target (Weber et al. 2023). Therapeutic strategies targeting this complex could offer significant promise for treating diseases linked to epigenetic dysregulation. Research by Salvati et al. identified BRPF1 as a critical epigenetic regulator in endocrine therapy-resistant breast cancers. Their study demonstrated that pharmacological inhibition of BRPF1 disrupts estrogen receptor alpha (ERα)-mediated transcriptional programs, reduces cell proliferation, induces apoptosis, and overcomes resistance in both hormone-sensitive and resistant breast cancer models, providing new therapeutic avenues for targeting BRPF1 in luminal-like breast cancers (Salvati et al. 2024).

14.6 Biological Functions of KAT6-BRPF1 Complexes

14.6.1 Role in Cellular Processes

The KAT6-BRPF1 complex plays a crucial role in regulating chromatin structure and gene expression across various cellular processes (Yokoyama et al. 2024). In cell cycle regulation, the complex acetylates histones at promoters of key regulatory genes, facilitating their transcription and ensuring proper cell cycle progression (Yang et al. 2022). Disruption of this acetylation activity can lead to cell cycle arrest or uncontrolled proliferation, which are hallmarks of oncogenesis (Yang et al. 2022).

In differentiation, the KAT6-BRPF1 complex selectively acetylates histones at specific gene loci, activating transcriptional programs essential for cell fate decisions (Yan et al. 2020). For instance, in hematopoiesis, KAT6A and KAT6B regulate the expression of genes critical for the differentiation and maturation of blood cells (Bergamasco et al. 2024; Kalle and Naik 2024). Similarly, during neurogenesis, the complex modulates the transcription of genes involved in neural differentiation and development (Bergamasco et al. 2025; Liu et al. 2024a).

The complex is also integral to DNA repair processes (Olbromski et al. 2024). By acetylating histones at sites of DNA damage, it facilitates the recruitment of DNA repair proteins, thereby maintaining genomic integrity and preventing genomic instability (Burke and Grant 2010).

14.6.2 Genetic Studies and Animal Models

Genetic studies and animal models have provided significant insights into the functions of the KAT6-BRPF1 complex. Mutations in KAT6A, KAT6B, or BRPF1 are linked to various developmental disorders and cancers (Yan et al. 2020; Liu et al. 2021, 2024b; Gu et al. 2025; Cevatemre et al. 2024). Mouse models with targeted deletions or mutations in these genes exhibit phenotypes that mimic human diseases, offering valuable models for studying the underlying mechanisms and exploring therapeutic strategies (You et al. 2016; Yan et al. 2020; Su et al. 2019).

In hematopoietic research, KAT6A and KAT6B knockout mice show defects in blood cell differentiation and an increased propensity for leukemogenesis (Bergamasco et al. 2024; Yan et al. 2022; Khokhar et al. 2020). Neurodevelopmental studies using KAT6 and BRPF1-deficient mice reveal impairments in neural development and cognitive functions, underscoring its role in brain function (Liu et al. 2024a; You et al. 2015b). These models are crucial for elucidating the contributions of the KAT6-BRPF1 complex to development, disease, and potential therapeutic interventions. Mattioli et al. identified multiple BRPF1 mutations leading to intellectual disability and disrupted histone acetylation, underscoring the protein's role in neurodevelopment and disease (Mattioli et al. 2017). Additionally, Yan et al. and Colson et al. demonstrated that BRPF1 mutations result in impaired histone propionylation, contributing to syndromic intellectual disabilities (Yan et al. 2020; Colson et al. 2024). Mutations in KAT6A, KAT6B, and BRPF1 result in significant pathological conditions. KAT6A and KAT6B mutations are linked to Genitopatellar syndrome and Say-Barber-Biesecker-Young-Simpson syndrome, which involve developmental delays and congenital abnormalities (Lonardo et al. 2019). Studies also highlighted that specific KAT6A mutations disrupt critical domains necessary for chromatin binding and acetylation, leading to defective gene regulation and tumorigenesis (Arboleda et al. 2015).

14.7 KAT6-BRPF1 Complexes in Diseases

14.7.1 KAT6-BRPF1 Complexes in Neurodevelopmental Disorders

KAT6A, KAT6B, and BRPF1 play crucial roles in neurodevelopment, primarily through their regulation of histone acetylation, which modulates chromatin accessibility and gene expression (Yan et al. 2020). Dysregulation of this complex has been implicated in multiple neurodevelopmental disorders, including syndromic intellectual disabilities (ID), developmental delays, and structural brain anomalies (Zu et al. 2022).

Mutations in *KAT6A* disrupt histone acetylation and gene regulation in neural progenitors, leading to severe neurodevelopmental consequences, and patients with *KAT6A* mutations present with syndromic forms of intellectual disability, craniofacial abnormalities, growth delays, and cardiac defects (Tham et al. 2015; Urreizti et al. 2020). These mutations impair the catalytic activity of KAT6A, disrupting its ability to acetylate histones and regulate genes critical for neural progenitor maintenance and differentiation (Liu et al. 2024a; Tham et al. 2015; Nava et al. 2023). Structural studies suggest that loss-of-function mutations in *KAT6A* diminish its chromatin-binding affinity and destabilize transcriptional programs involved in neuronal development (Weber et al. 2023; Arboleda et al. 2015; Tham et al. 2015). In mouse models, *kat6a* knockout results in abnormal cortical development, reduced neural progenitor populations, and impaired neuronal migration, underscoring its essential role in brain formation (Liu et al. 2024a; Tham et al. 2015; Kennedy et al. 2019).

Similarly, mutations in *KAT6B* lead to distinct but overlapping syndromic phenotypes, including Genitopatellar Syndrome and Say-Barber-Biesecker-Young-Simpson Syndrome, characterized by intellectual disability, skeletal abnormalities, and structural brain defects (Lonardo et al. 2019; Szakszon et al. 2013; Li et al. 2017). Functional analyses reveal that *KAT6B* mutations disrupt acetylation of histone residues such as H3K9ac and H3K14ac, leading to impaired transcriptional regulation of neural differentiation genes (Wiesel-Motiuk and Assaraf 2020; Bergamasco et al. 2025). Emerging evidence suggests that *KAT6B* mutations also disrupt promoter-level epigenetic landscapes, further contributing to deficits in neurogenesis and synaptic connectivity (Park et al. 2022).

BRPF1, a critical scaffold protein in KAT6A and KAT6B complexes, modulates histone acetylation and transcription during brain development (Mattioli et al. 2017). Loss-of-function mutations in *BRPF1* are associated with syndromic intellectual disability and epileptic encephalopathy, characterized by cortical thinning, neuronal apoptosis, and learning deficits (You et al. 2015b; Xian et al. 2021; Li et al. 2019; Cao et al. 2021). *BRPF1* mutations disrupt histone acetylation, impairing chromatin remodeling and transcriptional activation of genes essential for neural circuit formation (see "Histone Acylation by KAT6-BRPF1 Complexes") (Viita and Côté 2023; Zhang et al. 2023; You et al. 2016; Yan et al. 2020). Mouse models of *brpf1* deficiency exhibit severe neurodevelopmental defects, reinforcing its critical role in brain formation (Zhao et al. 2024; Su et al. 2019).

Collectively, dysregulation of the KAT6A, KAT6B, and BRPF1 impairs neurodevelopment by altering chromatin accessibility, disrupting transcriptional programs essential for neural differentiation, and leading to developmental disorders (Yan et al. 2020; Yang et al. 2022). These findings highlight shared epigenetic mechanisms underlying syndromic intellectual disabilities associated with this complex.

14.7.2 KAT6-BRPF1 Complexes in Cancer

Dysregulation of the KAT6A, KAT6B, and BRPF1 complexes is also heavily implicated in oncogenesis, where its aberrant activity alters chromatin states and drives tumorigenesis (Yang et al. 2022). Both hematologic malignancies and solid tumors have been linked to genetic mutations or aberrant expression of these factors, leading to uncontrolled proliferation and tumor progression (Cristofanilli et al. 2024; Partynska et al. 2022; Baell et al. 2018; Mukohara et al. 2024b).

In hematologic cancers, chromosomal translocations involving KAT6A and KAT6B are recurrently observed in acute myeloid leukemia (AML) (Bergamasco et al. 2024; Yan et al. 2022; Smolko et al. 2024). These translocations generate oncogenic fusion proteins, such as KAT6A-CREBBP, which aberrantly recruit transcriptional coactivators to inappropriate gene loci (Xie et al. 2019; Sramkova et al. 2020). This misdirected acetylation disrupts normal hematopoietic differentiation and drives leukemogenesis (Xie et al. 2019; Sramkova et al. 2020). Loss-of-function mutations in BRPF1 in leukemia disrupt histone acetylation, altering chromatin accessibility and transcription (see "Histone Acylation by KAT6-BRPF1 Complexes") (Shima et al. 2014). This disrupts gene expression patterns critical for normal hematopoiesis, contributing to leukemogenesis (Shima et al. 2014). While BRPF1 facilitates *MOZ-dependent HOX* gene activation in MOZ–TIF2 leukemia, its loss highlights a tumor-suppressive role in maintaining normal histone acetylation and differentiation programs (Shima et al. 2014). These findings underscore the critical balance of BRPF1 activity in hematopoietic regulation (Shima et al. 2014).

Solid tumors also exhibit significant links to KAT6-BRPF1 complexes (Aiello et al. 2019; Drozdowicz 2017). KAT6A is significantly upregulated in ovarian cancer, where it promotes tumor proliferation, invasion, and metastasis by stabilizing β-catenin through acetylation of COP1 (Liu et al. 2021). In addition, elevated KAT6A expression in non-small cell lung cancer (SCLC) suggests its potential oncogenic role and highlights the need for further exploration of KAT6A-targeted therapies (Partynska et al. 2022).

In SCLC, KAT6B functions as a tumor suppressor by acetylating H3K23ac (Simó-Riudalbas et al. 2015). Genomic loss of *KAT6B* in SCLC leads to decreased H3K23ac levels, resulting in chromatin condensation and transcriptional repression of tumor suppressor genes, thereby promoting tumorigenesis (Simó-Riudalbas et al. 2015). Additionally, genetic evaluations have identified *KAT6B* as a novel susceptibility gene in neuroblastoma, suggesting its involvement in the development of this pediatric cancer (Gu et al. 2025). These findings underscore the complex role of KAT6B in cancer, acting as a tumor suppressor in certain contexts while contributing to oncogenesis in others (Gu et al. 2025).

BRPF1 mutations have been implicated in adult medulloblastoma, the most common malignant brain tumor in adults (Aiello et al. 2019; Drozdowicz 2017). Elevated BRPF1 expression has been observed in hepatocellular carcinoma (HCC), suggesting its involvement in liver cancer, while in breast cancer, BRPF1 plays a critical role in estrogen receptor-positive cells, where its inhibition leads to growth arrest and apoptosis (Salvati et al. 2024; Cheng et al. 2021).

14.8 Therapeutic Potential

Targeting the KAT6-BRPF1 complexes offer significant therapeutic potential, particularly in cancer treatment (White et al. 2024; Zheng et al. 2025). Inhibitors of KAT6A and KAT6B have shown promise in preclinical models by reducing tumor growth through the restoration of normal acetylation patterns and gene expression profiles (Mukohara et al. 2024a; Sharma et al. 2023; Zheng et al. 2025). Sharma et al. identified a selective KAT6A/B inhibitor, effective in ER+ breast cancer by suppressing H3K23ac acetylation (Sharma et al. 2023). Baell et al. further demonstrated the potential of KAT6A/B inhibitors through the development of WM-8014 and WM-1119, compounds that induce senescence and arrest tumor progression (Baell et al. 2018). WM-8014 selectively inhibits KAT6A/B activity by competing with acetyl-CoA, resulting in irreversible cell cycle exit via the INK4A/ARF pathway (Baell et al. 2018). WM-1119, with improved bioavailability, effectively halted lymphoma progression in vivo, underscoring its therapeutic promise (Baell et al. 2018). Xia et al. identified BRPF1 as a potential drug target for lower-grade gliomas. Using transcriptomic and inflammatory signature analyses, they demonstrated that BRPF1 inhibition significantly reduced glioma cell proliferation and colony formation, highlighting its role in tumorigenesis and potential as a therapeutic target (Xia et al. 2021). These findings exemplify how targeting acetyltransferase activity can reprogram epigenetic states and disrupt oncogenic transcriptional programs.

14.9 Conclusions

The KAT6-BRPF1 complexes play a dual yet interwoven role in human health and disease. As a master regulator of chromatin architecture, this complex is pivotal in the transcriptional control of genes essential for neurodevelopment and hematopoiesis (Zu et al. 2022; Yan et al. 2020; Liu et al. 2024b). Dysregulation and mutation of its components results in profound consequences, including syndromic intellectual disabilities, neurodevelopmental disorders with brain malformations, and malignancies such as pediatric leukemia and adult medulloblastoma (Su et al. 2021; Yan et al. 2017; Colson et al. 2024; Liu et al. 2024b; Aiello et al. 2019; Farria et al. 2015).

In neurodevelopment, mutations in KAT6A and KAT6B disrupt histone acetylation patterns, impairing neural differentiation and circuit formation (Bergamasco et al. 2025; Matsushita 2023; Fallah et al. 2021). Meanwhile, BRPF1 mutations have emerged as a critical factor in both intellectual disabilities and cancers, illustrating its centrality in developmental and oncogenic pathways (Zu et al. 2022; You et al. 2016; Liu et al. 2024b; Aiello et al. 2019). This complex's evolutionary conservation across species further underscores its fundamental role in regulating biological processes (Yang and Ullah 2007; O'Meara et al. 2010; Chamberlin and Thomas 2000; Chang et al. 2003; Scott et al. 2001).

The oncogenic implications of the KAT6-BRPF1 complexes are equally significant. Targeting this complex offers a promising therapeutic avenue, as inhibitors of BRPF1 and the acetyltransferase activities of KAT6A and KAT6B demonstrate efficacy in preclinical models of cancer (Salvati et al. 2024; Cevatemre et al. 2024; Cheng et al. 2021; Zheng et al. 2025).

Future research shall focus on unraveling the intricate interplay between the KAT6-BRPF1 complexs' neurodevelopmental and oncogenic roles. By bridging these domains, we can develop tailored therapies that address not only the molecular underpinnings of neurodevelopmental disorders but also the epigenetic vulnerabilities of cancers, establishing these complexes as a cornerstone of translational epigenetic medicine.

References

Aiello G, Ballabio C, Ruggeri R, Fagnocchi L, Anderle M, Morassut I, Caron D, Garilli F, Gianno F, Giangaspero F (2019) Truncated BRPF1 cooperates with smoothened to promote adult Shh medulloblastoma. Cell Rep 29(4036–4052):e4010

Arboleda VA, Lee H, Dorrani N, Zadeh N, Willis M, Macmurdo CF, Manning MA, Kwan A, Hudgins L, Barthelemy F (2015) De novo nonsense mutations in KAT6A, a lysine acetyltransferase gene, cause a syndrome including microcephaly and global developmental delay. Am J Hum Genet 96:498–506

Baell JB, Leaver DJ, Hermans SJ, Kelly GL, Brennan MS, Downer NL, Nguyen N, Wichmann J, McRae HM, Yang Y (2018) Inhibitors of histone acetyltransferases KAT6A/B induce senescence and arrest tumour growth. Nature 560:253–257

Barman S, Roy A, Bardhan I, Kandasamy T, Shivani S, Sudhamalla B (2021) Insights into the molecular mechanisms of histone code recognition by the BRPF3 bromodomain. Chemist Asian J 16:3404–3412

Barman S, Roy A, Padhan J, Sudhamalla B (2022) Molecular insights into the recognition of acetylated histone modifications by the BRPF2 bromodomain. Biochemistry 61:1774–1789

Bergamasco MI, Ranathunga N, Abeysekera W, Li-Wai-Suen CS, Garnham AL, Willis SN, McRae HM, Yang Y, D'Amico A, Di Rago L (2024) The histone acetyltransferase KAT6B is required for hematopoietic stem cell development and function. Stem Cell Reports 19:469–485

Bergamasco MI, Abeysekera W, Garnham AL, Hu Y, Li-Wai-Suen CS, Sheikh BN, Smyth GK, Thomas T, Voss AK (2025) KAT6B is required for histone 3 lysine 9 acetylation and SOX gene expression in the developing brain. Life Sci Alliance 8

Blakey CA, Litt MD (2015) Histone modifications—models and mechanisms. In: Epigenetic gene expression and regulation. Elsevier, pp 21–42

Brownell JE, Zhou J, Ranalli T, Kobayashi R, Edmondson DG, Roth SY, Allis CD (1996) Tetrahymena histone acetyltransferase A: a homolog to yeast Gcn5p linking histone acetylation to gene activation. Cell 84:843–851

Burke TL, Grant PA (2010) Histone acetylation complexes. In: Handbook of cell signaling. Elsevier, pp 2369–2378

Cao J, Xian W, Palihati M, Zhu Y, Wang G, Xie Y, Zhou G, You L (2021) Deficiency of intellectual disability-related gene Brpf1 reduced inhibitory neurotransmission in MGE-derived GABAergic interneurons. G3 11:jkab090

Cevatemre B, Bulut I, Dedeoglu B, Isiklar A, Syed H, Bayram OY, Bagci-Onder T, Acilan C (2024) Exploiting epigenetic targets to overcome taxane resistance in prostate cancer. Cell Death Dis 15:132

Chamberlin HM, Thomas JH (2000) The bromodomain protein LIN-49 and trithorax-related protein LIN-59 affect development and gene expression in Caenorhabditis elegans. Development 127:713–723

Champagne KS, Saksouk N, Peña PV, Johnson K, Ullah M, Yang X-J, Côté J, Kutateladze TG (2008) The crystal structure of the ING5 PHD finger in complex with an H3K4me3 histone peptide. Proteins 72:1371

Chang S, Johnston RJ, Hobert O (2003) A transcriptional regulatory cascade that controls left/right asymmetry in chemosensory neurons of *C. elegans*. Genes Dev 17:2123–2137

Cheng CL-H, Tsang FH-C, Wei L, Chen M, Chin DW-C, Shen J, Law C-T, Lee D, Wong CC-L, Ng IO-L (2021) Bromodomain-containing protein BRPF1 is a therapeutic target for liver cancer. Commun Biol 4:888

Colson C, Tessarech M, Boucher-Brischoux E, Boute-Benejean O, Vincent-Delorme C, Vanlerberghe C, Boussion S, Cunff JL, Duban-Bedu B, Faivre L (2024) The phenotypic and genotypic Spectrum of BRPF1-related disorder: 29 new patients and literature review. Clin Genet

Cosentino MS, Oses C, Echegaray CV, Solari C, Waisman A, Álvarez Y, Petrone MV, Francia M, Schultz M, Sevlever G (2019) Kat6b modulates Oct4 and Nanog binding to chromatin in embryonic stem cells and is required for efficient neural differentiation. J Mol Biol 431:1148–1159
Cristofanilli M, Liu L, Deng S, Huang X, André F, Loibl S, DeMichele A, Gauthier E, Liu Y, Turner NC (2024) Association of KAT6A expression with clinical outcomes in previously treated HR-positive and HER2-negative metastatic breast cancer. Cancer Res 84:7613–7613
Cruickshank MN, Besant P, Ulgiati D (2010) The impact of histone post-translational modifications on developmental gene regulation. Amino Acids 39:1087–1105
Davarnia B, Panahi M, Rahimi B, Anari H, Farajollahi R, Rodbaneh EA, Jeddi F (2024) De novo KAT6B mutation causes say–barber–Biesecker–Young–Simpson variant of Ohdo syndrome in an Iranian boy: a case report. J Med Case Rep 18:4
Dent SY (2024) KAT tales: functions of Gcn5 and PCAF lysine acetyltransferases in SAGA and ATAC. J Biol Chem 107744
Drozdowicz K (2017) Investigate the role bromodomain-and plant homeodomain-linked zinc finger-containing protein 1 (BRPF1) plays in medulloblastoma. Boston University
Fallah MS, Szarics D, Robson CM, Eubanks JH (2021) Impaired regulation of histone methylation and acetylation underlies specific neurodevelopmental disorders. Front Genet 11:613098
Farria A, Li W, Dent SY (2015) KATs in cancer: functions and therapies. Oncogene 34:4901–4913
Gatchalian J, Kutateladze TG (2015) PHD fingers as histone readers. Histone Recognition, pp 27–47
Gu H, Yu Y, Sisoudiya SD, Mishra P, Li H, Schraw JM, Scheurer ME, Muzny DM, Mitchell D, Taylor O (2025) An evaluation of genetic predisposition to congenital anomalies and pediatric cancer supports KAT6B as a novel neuroblastoma susceptibility gene. Genet Medicine Open 3:101901
Huang F, Paulson A, Dutta A, Venkatesh S, Smolle M, Abmayr SM, Workman JL (2014) Histone acetyltransferase Enok regulates oocyte polarization by promoting expression of the actin nucleation factor spire. Genes Dev 28:2750–2763
Huang F, Abmayr SM, Workman JL (2016) Regulation of KAT6 acetyltransferases and their roles in cell cycle progression, stem cell maintenance, and human disease. Mol Cell Biol 36:1900–1907
Igolkina AA, Zinkevich A, Karandasheva KO, Popov AA, Selifanova MV, Nikolaeva D, Tkachev V, Penzar D, Nikitin DM, Buzdin A (2019) H3K4me3, H3K9ac, H3K27ac, H3K27me3 and H3K9me3 histone tags suggest distinct regulatory evolution of open and condensed chromatin landmarks. Cells 8:1034
Kalle AM, Naik D (2024) "epigenetics of hematopoiesis: role of HATs and HDACs" in hematopoiesis. Apple Academic Press, pp 109–123
Kennedy J, Goudie D, Blair E, Chandler K, Joss S, McKay V, Green A, Armstrong R, Lees M, Kamien B (2019) KAT6A syndrome: genotype–phenotype correlation in 76 patients with pathogenic KAT6A variants. Genet Med 21:850–860
Khokhar ES, Borikar S, Eudy E, Stearns T, Young K, Trowbridge JJ (2020) Aging-associated decrease in the histone acetyltransferase KAT6B is linked to altered hematopoietic stem cell differentiation. Exp Hematol 82(43–52):e44
Kleff S, Andrulis ED, Anderson CW, Sternglanz R (1995) Identification of a gene encoding a yeast histone H4 acetyltransferase (∗). J Biol Chem 270:24674–24677
Klein BJ, Jang SM, Lachance C, Mi W, Lyu J, Sakuraba S, Krajewski K, Wang WW, Sidoli S, Liu J (2019) Histone H3K23-specific acetylation by MORF is coupled to H3K14 acylation. Nat Commun 10:4724
Li G, Li N, Li J, Ding Y, Yu T, Wang X, Wang J (2017) De novo mutation of KAT6B gene causing atypical Say–Barber–Biesecker–Young–Simpson syndrome or genitopatellar syndrome. Fetal Pediatr Pathol 36:130–138
Li L, Jin J, Yang X-J (2019) Histone deacetylase 3 governs perinatal cerebral development via neural stem and progenitor cells. Iscience 20:148–167

Liang F, Li X, Shen X, Yang R, Chen C (2023) Expression profiles and functional prediction of histone acetyltransferases of the MYST family in kidney renal clear cell carcinoma. BMC Cancer 23:586

Liu W, Zhan Z, Zhang M, Sun B, Shi Q, Luo F, Zhang M, Zhang W, Hou Y, Xiao X (2021) KAT6A, a novel regulator of β-catenin, promotes tumorigenicity and chemoresistance in ovarian cancer by acetylating COP1. Theranostics 11:6278

Liu X, Guo C, Leng T, Fan Z, Mai J, Chen J, Xu J, Li Q, Jiang B, Sai K (2023) Differential regulation of H3K9/H3K14 acetylation by small molecules drives neuron-fate-induction of glioma cell. Cell Death Dis 14:142

Liu Y, Fan M, Yang J, Mihaljević L, Chen KH, Ye Y, Sun S, Qiu Z (2024a) KAT6A deficiency impairs cognitive functions through suppressing RSPO2/Wnt signaling in hippocampal CA3. *Science*. Advances 10:eadm9326

Liu Q, Li F, Wang N, Fan Z (2024b) Genetic analysis of BRPF1 exon deletion variant causing intellectual developmental disorder with dysmorphic facies and ptosis in a Chinese family. Egypt J Med Human Genet 25:67

Lonardo F, Lonardo M, Acquaviva F, Della Monica M, Scarano F, Scarano G (2019) Say-barber-Biesecker-young-Simpson syndrome and Genitopatellar syndrome: lumping or splitting? Clin Genet 95:253–261

Marmorstein R (2001) Structure of histone acetyltransferases. J Mol Biol 311:433–444

Matsushita N (2023) Dysregulated histone acetylation causes congenital diseases. Gene Reports 31:101778

Mattioli F, Schaefer E, Magee A, Mark P, Mancini GM, Dieterich K, Von Allmen G, Alders M, Coutton C, van Slegtenhorst M (2017) Mutations in histone acetylase modifier BRPF1 cause an autosomal-dominant form of intellectual disability with associated ptosis. Am J Hum Genet 100:105–116

Mishima Y, Miyagi S, Saraya A, Negishi M, Endoh M, Endo TA, Toyoda T, Shinga J, Katsumoto T, Chiba T (2011) The Hbo1-Brd1/Brpf2 complex is responsible for global acetylation of H3K14 and required for fetal liver erythropoiesis. Blood J Am Soc Hematol 118:2443–2453

Mishima Y, Wang C, Miyagi S, Saraya A, Hosokawa H, Mochizuki-Kashio M, Nakajima-Takagi Y, Koide S, Negishi M, Sashida G (2014) Histone acetylation mediated by Brd1 is crucial for Cd8 gene activation during early thymocyte development. Nat Commun 5:5872

Morison LD, Van Reyk O, Baker E, Ruaud L, Couque N, Verloes A, Amor DJ, Morgan AT (2024) Beyond'speech delay': expanding the phenotype of BRPF1-related disorder. Eur J Med Genet 68:104923

Mukohara T, Park YH, Sommerhalder D, Yonemori K, Hamilton E, Kim S-B, Kim JH, Iwata H, Yamashita T, Layman RM (2024a) Inhibition of lysine acetyltransferase KAT6 in ER+ HER2− metastatic breast cancer: a phase 1 trial. Nat Med:1–9

Mukohara T, Park YH, Sommerhalder D, Yonemori K, Kim S-B, Kim JH, Iwata H, Yamashita T, Layman RM, Kim GM (2024b) A phase 1 dose expansion study of a first-in-class KAT6 inhibitor (PF-07248144) in patients with advanced or metastatic ER+ HER2− breast cancer. American Society of Clinical Oncology

Nava AA, Jops CT, Vuong CK, Niles-Jensen SL, Bondhus L, Ong CJ, de la Torre-Ubieta L, Gandal MJ, Arboleda VA (2023) KAT6A mutations drive transcriptional dysregulation of cell cycle and autism risk genes in an arboleda-tham syndrome cerebral organoid model. bioRxiv 2006:545322

Olbromski M, Mrozowska M, Smolarz B, Romanowicz H, Rusak A, Piotrowska A (2024) ERα status of invasive ductal breast carcinoma as a result of regulatory interactions between lysine deacetylases KAT6A and KAT6B. Sci Rep 14:26935

O'Meara MM, Zhang F, Hobert O (2010) Maintenance of neuronal laterality in Caenorhabditis elegans through MYST histone acetyltransferase complex components LSY-12, LSY-13 and LIN-49. Genetics 186:1497–1502

Park J, Lee K, Kim K, Yi S-J (2022) The role of histone modifications: from neurodevelopment to neurodiseases. Signal Transduct Target Ther 7:217

Partynska A, Piotrowska A, Pawelczyk K, Rzechonek A, Ratajczak-Wielgomas K, Podhorska-Okolow M, Dziegiel P (2022) The expression of histone acetyltransferase KAT6A in non-small cell lung cancer. Anticancer Res 42:5731–5741

Poplawski A, Hu K, Lee W, Natesan S, Peng D, Carlson S, Shi X, Balaz S, Markley JL, Glass KC (2014) Molecular insights into the recognition of N-terminal histone modifications by the BRPF1 bromodomain. J Mol Biol 426:1661–1676

Rona GB, Eleutherio EC, Pinheiro AS (2016) PWWP domains and their modes of sensing DNA and histone methylated lysines. Biophys Rev 8:63–74

Salvati A, Giurato G, Lamberti J, Terenzi I, Crescenzo L, Melone V, Palo L, Giordano A, Sabbatino F, Roscigno G (2024) Essential gene screening identifies the bromodomain-containing protein BRPF1 as a new actionable target for endocrine therapy-resistant breast cancers. Mol Cancer 23:160

Sapountzi V, Côté J (2011) MYST-family histone acetyltransferases: beyond chromatin. Cell Mol Life Sci 68:1147–1156

Scott EK, Lee T, Luo L (2001) Enok encodes a drosophila putative histone acetyltransferase required for mushroom body neuroblast proliferation. Curr Biol 11:99–104

Sharma S, Chung C-Y, Uryu S, Petrovic J, Cao J, Rickard A, Nady N, Greasley S, Johnson E, Brodsky O (2023) Discovery of a highly potent, selective, orally bioavailable inhibitor of KAT6A/B histone acetyltransferases with efficacy against KAT6A-high ER+ breast cancer. Cell Chem Biol 30(1191–1210):e1120

Sheikh BN, Lee SC, El-Saafin F, Vanyai HK, Hu Y, Pang SHM, Grabow S, Strasser A, Nutt SL, Alexander WS (2015) MOZ regulates B-cell progenitors and, consequently, Moz haploinsufficiency dramatically retards MYC-induced lymphoma development. Blood J Am Soc Hematol 125:1910–1921

Sheikh BN, Yang Y, Schreuder J, Nilsson SK, Bilardi R, Carotta S, McRae HM, Metcalf D, Voss AK, Thomas T (2016) MOZ (KAT6A) is essential for the maintenance of classically defined adult hematopoietic stem cells. Blood J Am Soc Hematol 128:2307–2318

Shima H, Yamagata K, Aikawa Y, Shino M, Koseki H, Shimada H, Kitabayashi I (2014) Bromodomain-PHD finger protein 1 is critical for leukemogenesis associated with MOZ–TIF2 fusion. Int J Hematol 99:21–31

Shukla S, Levine C, Sripathi RP, Elson G, Lutz CS, Leibovich SJ (2018) The kat in the HAT: the histone acetyl transferase Kat6b (MYST4) is downregulated in murine macrophages in response to LPS. Mediat Inflamm 2018:7852742

Simó-Riudalbas L, Pérez-Salvia M, Setien F, Villanueva A, Moutinho C, Martínez-Cardús A, Moran S, Berdasco M, Gomez A, Vidal E (2015) KAT6B is a tumor suppressor histone H3 lysine 23 acetyltransferase undergoing genomic loss in small cell lung cancer. Cancer Res 75:3936–3945

Simpson MA, Deshpande C, Dafou D, Vissers LE, Woollard WJ, Holder SE, Gillessen-Kaesbach G, Derks R, White SM, Cohen-Snuijf R (2012) De novo mutations of the gene encoding the histone acetyltransferase KAT6B cause Genitopatellar syndrome. Am J Hum Genet 90:290–294

Smolko AE, Sullivan DW, Olsen SN, Kang H, Whedon SD, Baell JB, Cole PA, Armstrong SA, Kuroda MI (2024) A MOZ-TIF2 leukemia mouse model displays KAT6-dependent H3K23 propionylation and overexpression of a set of active developmental genes. Proc Natl Acad Sci 121:e2405905121

Sramkova L et al (2020) Rapidly progressing acute myeloid leukemia with KAT6A-LEUTX fusion in a newborn. Pediatr Blood Cancer 28663

Struhl K (1998) Histone acetylation and transcriptional regulatory mechanisms. Genes Dev 12:599–606

Su Y, Liu J, Yu B, Ba R, Zhao C (2019) Brpf1 haploinsufficiency impairs dendritic arborization and spine formation, leading to cognitive deficits. Front Cell Neurosci 13:249

Su J, Wang X, Bai Y, Sun M, Yao Y, Duan Y (2021) The role of MOZ/KAT6A in hematological malignancies and advances in MOZ/KAT6A inhibitors. Pharmacol Res 174:105930

Su Z, Zhang Y, Tang J, Zhou Y, Long C (2024) Multifunctional acyltransferase HBO1: a key regulatory factor for cellular functions. Cell Mol Biol Lett 29:141

Sun X-J, Man N, Tan Y, Nimer SD, Wang L (2015) The role of histone acetyltransferases in normal and malignant hematopoiesis. Front Oncol 5:108

Szakszon K, Salpietro C, Kakar N, Knegt AC, Oláh É, Dallapiccola B, Borck G (2013) De novo mutations of the gene encoding the histone acetyltransferase KAT6B in two patients with say-barber/Biesecker/Young-Simpson syndrome. Am J Med Genet A 161:884–888

Tao Y, Zhong C, Zhu J, Xu S, Ding J (2017) Structural and mechanistic insights into regulation of HBO1 histone acetyltransferase activity by BRPF2. Nucleic Acids Res 45:5707–5719

Tham E, Lindstrand A, Santani A, Malmgren H, Nesbitt A, Dubbs HA, Zackai EH, Parker MJ, Millan F, Rosenbaum K (2015) Dominant mutations in KAT6A cause intellectual disability with recognizable syndromic features. Am J Hum Genet 96:507–513

Troke PJ, Kindle KB, Collins HM, Heery DM (2006) MOZ fusion proteins in acute myeloid leukaemia. In: Biochemical society symposia. Portland Press Ltd, pp 23–39

Urreizti R, Lopez-Martin E, Martinez-Monseny A, Pujadas M, Castilla-Vallmanya L, Pérez-Jurado LA, Serrano M, Natera-de Benito D, Martínez-Delgado B, Posada-De-La-Paz M (2020) Five new cases of syndromic intellectual disability due to KAT6A mutations: widening the molecular and clinical spectrum. Orphanet J Rare Dis 15:1–14

van Nuland R, van Schaik FM, Simonis M, van Heesch S, Cuppen E, Boelens R, Timmers HM, van Ingen H (2013) Nucleosomal DNA binding drives the recognition of H3K36-methylated nucleosomes by the PSIP1-PWWP domain. Epigenetics Chromatin 6:1–12

Viita T, Côté J (2023) The MOZ-BRPF1 acetyltransferase complex in epigenetic crosstalk linked to gene regulation, development, and human diseases. Front Cell Develop Biol 10:1115903

Weber LM, Jia Y, Stielow B, Gisselbrecht SS, Cao Y, Ren Y, Rohner I, King J, Rothman E, Fischer S (2023) The histone acetyltransferase KAT6A is recruited to unmethylated CpG islands via a DNA binding winged helix domain. Nucleic Acids Res 51:574–594

White J, Derheimer FA, Jensen-Pergakes K, O'Connell S, Sharma S, Spiegel N, Paul TA (2024) Histone lysine acetyltransferase inhibitors: an emerging class of drugs for cancer therapy. Trends Pharmacol Sci

Wiesel-Motiuk N, Assaraf YG (2020) The key roles of the lysine acetyltransferases KAT6A and KAT6B in physiology and pathology. Drug Resist Updat 53:100729

Xia M, Chen H, Chen T, Xue P, Dong X, Lin Y, Ma D, Zhou W, Shi W, Li H (2021) Transcriptional networks identify BRPF1 as a potential drug target based on inflammatory signature in primary lower-grade gliomas. Front Oncol 11:766656

Xian W, Cao J, Yuan X, Wang G, Jin Q, Zhang H, Zhou G, You L (2021) Deficiency of intellectual disability-related gene brpf1 attenuated hippocampal excitatory synaptic transmission and impaired spatial learning and memory ability. Front Cell Develop Biol 9:711792

Xie W, Hu S, Xu J, Chen Z, Medeiros LJ, Tang G (2019) Acute myeloid leukemia with t (8; 16) (p11. 2; p13. 3)/KAT6A-CREBBP in adults. Ann Hematol 98:1149–1157

Yan K, You L, Degerny C, Ghorbani M, Liu X, Chen L, Li L, Miao D, Yang X-J (2016) The chromatin regulator BRPF3 preferentially activates the HBO1 acetyltransferase but is dispensable for mouse development and survival. J Biol Chem 291:2647–2663

Yan K, Rousseau J, Littlejohn RO, Kiss C, Lehman A, Rosenfeld JA, Stumpel CT, Stegmann AP, Robak L, Scaglia F (2017) Mutations in the chromatin regulator gene BRPF1 cause syndromic intellectual disability and deficient histone acetylation. Am J Hum Genet 100:91–104

Yan K, Rousseau J, Machol K, Cross LA, Agre KE, Gibson CF, Goverde A, Engleman KL, Verdin H, De Baere E (2020) Deficient histone H3 propionylation by BRPF1-KAT6 complexes in neurodevelopmental disorders and cancer. Sci Adv 6:eaax0021

Yan F, Li J, Milosevic J, Petroni R, Liu S, Shi Z, Yuan S, Reynaga JM, Qi Y, Rico J (2022) KAT6A and ENL form an epigenetic transcriptional control module to drive critical leukemogenic gene-expression programs. Cancer Discov 12:792–811

Yang X-J (2015) MOZ and MORF acetyltransferases: molecular interaction, animal development and human disease. Biochimica et Biophys Acta (BBA)-Mol Cell Res 1853:1818–1826
Yang X, Ullah M (2007) MOZ and MORF, two large MYSTic HATs in normal and cancer stem cells. Oncogene 26:5408–5419
Yang X-J, Ogryzko VV, Nishikawa J-I, Howard BH, Nakatani Y (1996) A p300/CBP-associated factor that competes with the adenoviral oncoprotein E1A. Nature 382:319–324
Yang Y, Zhang M, Wang Y (2022) The roles of histone modifications in tumorigenesis and associated inhibitors in cancer therapy. J Nat Cancer Center 2:277–290
Yokoyama A, Niida H, Kutateladze TG, Côté J (2024) HBO1, a MYSTerious KAT and its links to cancer. Biochim Biophys Acta (BBA) Gene Regulat Mech 195045
You L, Yan K, Zou J, Zhao H, Bertos NR, Park M, Wang E, Yang X-J (2015a) The chromatin regulator Brpf1 regulates embryo development and cell proliferation. J Biol Chem 290:11349–11364
You L, Zou J, Zhao H, Bertos NR, Park M, Wang E, Yang X-J (2015b) Deficiency of the chromatin regulator BRPF1 causes abnormal brain development. J Biol Chem 290:7114–7129
You L, Li L, Zou J, Yan K, Belle J, Nijnik A, Wang E, Yang X-J (2016) BRPF1 is essential for development of fetal hematopoietic stem cells. J Clin Invest 126:3247–3262
Zhang C, Lin H, Zhang Y, Xing Q, Zhang J, Zhang D, Liu Y, Chen Q, Zhou T, Wang J (2023) BRPF1 bridges H3K4me3 and H3K23ac in human embryonic stem cells and is essential to pluripotency. Iscience 26
Zhao B, Zhang H, Liu Y, Zu G, Zhang Y, Hu J, Liu S, You L (2024) Forebrain excitatory neuron-specific loss of Brpf1 attenuates excitatory synaptic transmission and impairs spatial and fear memory. Neural Regen Res 19:1133–1141
Zheng T, Wang S, Liu W, Lu Y (2025) Targeting KAT6A/B as a new therapeutic strategy for cancer therapy. J Med Chem
Zhong Q, Xiao X, Qiu Y, Xu Z, Chen C, Chong B, Zhao X, Hai S, Li S, An Z (2023) Protein posttranslational modifications in health and diseases: functions, regulatory mechanisms, and therapeutic implications. Med Comm 4:e261
Zu G, Liu Y, Cao J, Zhao B, Zhang H, You L (2022) BRPF1-KAT6A/KAT6B complex: molecular structure, biological function and human disease. Cancers 14:4068

Chapter 15
Bromodomain and PHD Finger-Containing Protein 1: From Functions to a Developmental Disorder, Cancer, and Therapeutics

Arezousadat Razavi, Paulina Varela-Castillo, and Xiang-Jiao Yang

Abstract Bromodomain and PHD finger-containing protein 1 (BRPF1) is an essential epigenetic regulator and plays a key role in post-translational modification of histones. It is a chromatin reader that recognizes acetylated histones and interacts with the paralogous lysine acetyltransferases KAT6A and KAT6B to promote histone acetylation and related acylations, such as propionylation, at lysine 23 of histone H3, thereby influencing gene expression and regulating developmental programs. BRPF1 contributes to a variety of cellular processes such as cell cycle progression, cell proliferation, cell differentiation, and responses to cellular stresses, including DNA damage. Moreover, BRPF1 is implicated in hematopoiesis, embryonic development, skeletal development, neurodevelopment, neurogenesis, learning, and memory. BRPF1 gene knockout in mice leads to severe bone marrow failure, anemia, and eventual death in a few weeks after birth. This review provides a brief overview of BRPF1 and its contribution to the molecular structure and biological functions of KAT6A and KAT6B complexes. We will explore the emerging evidence linking BRPF1 dysfunction to human diseases, particularly cancer and

A. Razavi (✉) · P. Varela-Castillo
Rosalind and Morris Goodman Cancer Institute, McGill University, Montreal, Quebec, Canada

Department of Medicine, McGill University, Montreal, Quebec, Canada
e-mail: arezousadat.razavi@mail.mcgill.ca

X.-J. Yang (✉)
Rosalind and Morris Goodman Cancer Institute, McGill University, Montreal, Quebec, Canada

Department of Medicine, McGill University, Montreal, Quebec, Canada

Department of Biochemistry, McGill University, Montreal, Quebec, Canada

McGill University Health Center, Montreal, Quebec, Canada
e-mail: xiang-jiao.yang@mcgill.ca

M. Halasa, A. Wawruszak (eds.), *Histone and Non-Histone Reversible Acetylation in Development, Aging and Disease*, Results and Problems in Cell Differentiation 75,
https://doi.org/10.1007/978-3-031-91459-1_15

abnormal neurodevelopment, to highlight promising therapeutic opportunities for treating associated pathology.

15.1 Introduction

Post-translational modification of histones is a key regulator in chromatin structure and gene expression (Allfrey et al. 1964; Wolffe 2000). Bromodomains are reader protein domains that specifically recognize and bind to modified histones and play a critical role in epigenetic regulation (Grunstein 1997). The human genome encodes 57 bromodomain-containing proteins categorized into 8 subfamilies based on sequence homology and domain structure (Filippakopoulos et al. 2010; Schultz et al. 2000). Bromodomain and PHD finger-containing protein 1 (BRPF1) is a member of family IV bromodomain-containing proteins (Qin et al. 2011; Lloyd and Glass 2018). Other members of this family include BRPF2, BRPF3, BRD7, BRD9, ATAD2, and ATAD2b, all of which contain tandem PHD fingers that recognize acetylated lysine residues on the N-terminal region of histone tails, a bromodomain, and a PWWP domain (Perry 2006; Zeng and Zhou 2002). BRPF1, BRPF2, and BRPF3 are homologous scaffold proteins that share a high degree of sequence similarity and domain organization (Yang and Ullah 2007). However, BRPF1 has distinct biological functions, expression patterns, and disease associations that differ from BRPF2 and BRPF3 (Yang and Ullah 2007).

BRPF1 is highly conserved from worms to humans, suggesting its evolutionarily preserved interaction with histone acetyltransferases (Yang and Ullah 2007). BRPF1 interacts with different histone acetyltransferases, including KAT6A, KAT6B (in vivo and in vitro), and KAT7 (only reported in vitro), which are crucial for histone acetylation (Lalonde et al. 2013). BRPF1 is a subunit of a multiunit protein complex including BRPF1, KAT6A/KAT6B, ING4/5, and EAF6 and contributes to diverse biological processes such as cell cycle regulation, embryonic development, hematopoiesis, and neurodevelopment (Lloyd and Glass 2018; Ullah et al. 2008; Carlson and Glass 2014; Yang 2015). Mutations or dysregulation of BRPF1 are associated with developmental disorders, including intellectual disability and cancer (Yang 2015). Aberrant expression of BRPF1 leads to leukemia and solid tumors, indicating its importance in oncogenesis (Yang 2015). This review aims to explore the molecular and biological function of BRPF1 and its involvement in disease pathogenesis and emerging therapeutic approaches targeting BRPF1.

15.2 Molecular Structure and Biological Function of BRPF1 in KAT6 Complexes

15.2.1 BRPF1 Domain Composition

BRPF1 is a multifaceted epigenetic regulator characterized by functionally distinct domains. Each domain has a specialized role in mediating interaction with histones, DNA, and associated proteins, enabling BRPF1 to participate in chromatin remodeling and gene regulation (Yang and Ullah 2007). From the uncharacterized N-terminal zinc finger to the C-terminal PWWP domain, these structural motifs facilitate the assembly, targeting, and activity of the BRPF1-KAT6A/KAT6B histone acetyltransferase complex (Yang and Ullah 2007).

The SZ (Sfp1-like zinc finger) domain within BRPF1 is a specific type of C2H2 zinc finger that resembles the zinc finger of the yeast transcription factor Sfp1. Zinc finger domains are small protein motifs capable of binding to DNA, RNA, proteins, or even small molecules, and are mainly involved in transcriptional regulation, DNA recognition, and repair, among other functions (Lubula et al. 2014). The SZ domain located at the N-terminus is unique to BRPF1 as this domain is absent in its two paralogs (BRPF2 and BRPF3), suggesting it may confer specific functions (Yan et al. 2020; Cho et al. 2020; Barman et al. 2021; Zhang et al. 2021). Although the function of the SZ domain is not fully characterized, it might contribute to DNA binding and protein interaction (Lubula et al. 2014). Zinc fingers are often capable of binding to specific DNA sequences, suggesting a role in the regulation of gene expression (Zu et al. 2022). According to The Cancer Genome Atlas (TCGA), BRPF1 is frequently mutated in various cancer types (Yan et al. 2020). Some cancer-associated somatic mutations have been identified in the SZ domain, specifically Pro20Leu, Arg29Cys, and Ser36Ile. Notably, R29 substitutions (R29Q or R29*) have been reported three times in patients with endometrial cancer and esophagogastric cancer (Fig. 15.1c). These mutations may impact complex formation and histone H3K23 acetylation (Zu et al. 2022).

NLS1 (nuclear localization signal 1) is a short amino acid sequence in the N-terminal region of BRPF1 that is recognized by the cellular nuclear import machinery. This recognition allows proteins containing these signals to be transported from the cytoplasm into the nucleus, an active process requiring GTP hydrolysis. After releasing the protein into the nucleus, importins are recycled back into the cytoplasm for further transport (Christodoulou and Yokoyama 2015). NLS1 is crucial for the proper localization and function of nuclear proteins like BRPF1, which is involved in chromatin modification and gene regulation. Nuclear localization signals often consist of short sequences of positively charged amino acids, such as lysine and arginine, which are essential for their recognition by nuclear import receptors (Zu et al. 2022).

BRPF1 contains another NLS sequence, NLS2. The presence of the two NLS sequences may provide redundancy to ensure efficient nuclear targeting. On the other hand, the two NLS domains might be regulated differently or recognized by

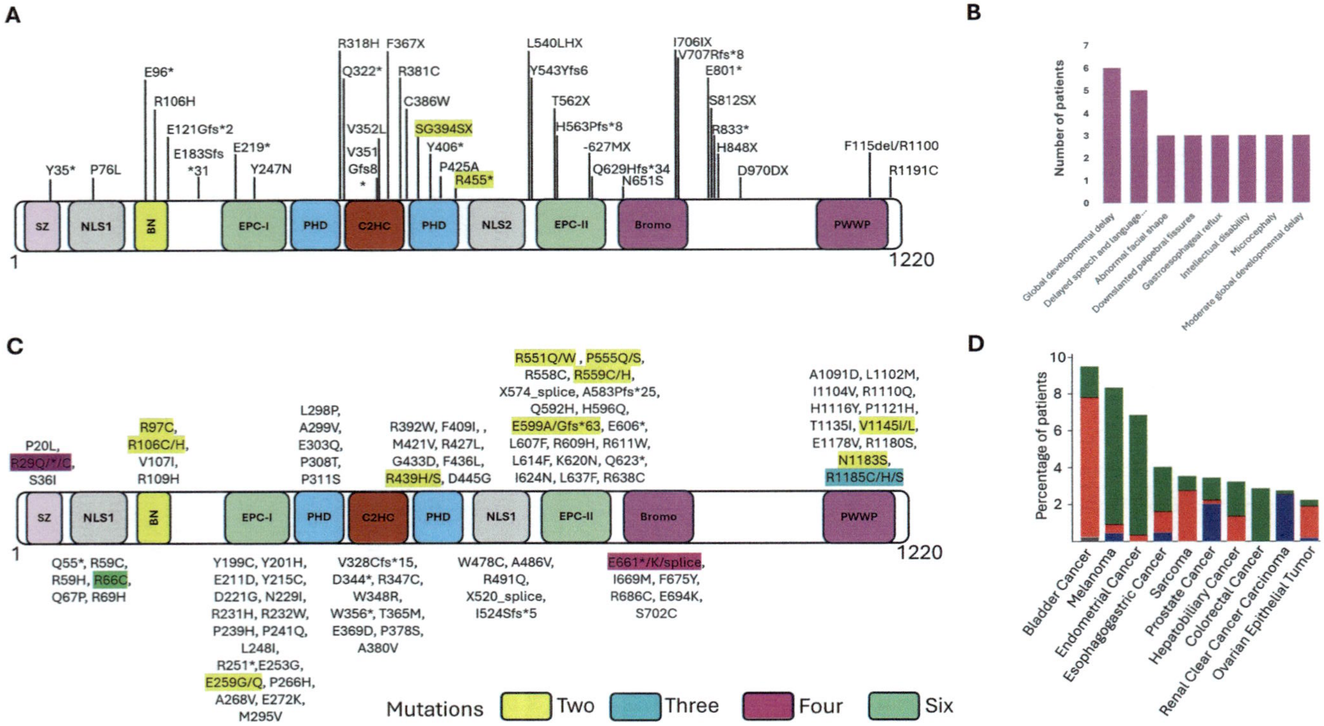

Fig. 15.1 Mutation profile of *BRPF1* in a developmental disorder and different types of cancer. (**a**) *BRPF1* mutations in associated with different phenotypes of a developmental disorder. Mutations are visualized across the BRPF1 protein and its domains. Yellow indicates two mutations in specific amino acids of the protein. (**b**) Phenotype presence in patients with *BRPF1* mutations. Global developmental delay is the most common phenotype across the individuals with *BRPF1* mutations. All analyses and visualizations in all panels were done based on data downloaded from DECIPHER v11.27. (**c**) Representation of mutation presence in different protein domains of BRPF1 across 33 cancers. Mutations are profiled and visualized based on the domain and their location in the BRPF1 proteins. Yellow, blue, purple, and green indicate two, three, four, and six mutations in specific amino acids of the proteins, respectively. (**d**) Presence of different genetic abnormalities in the *BRPF1* gene in different cancers. Data from all genetic abnormalities for different types of cancer were downloaded from cBioPortal. The top 10 cancer types based on the percentage of patients (Y-axis) carrying *BRPF1* genetic abnormalities are visualized. The color key highlights the type of genetic abnormalities in each cancer

distinct nuclear import pathways, allowing precise control of BRPF1 localization in the nucleus in response to cellular signals or conditions (Zu et al. 2022). One of the most frequently observed mutations in the NLS domain is R66C, which has been reported six times in patients with endometrial, colorectal, and esophagogastric cancers (Fig. 15.1c).

The BN (BRPF N-terminal) domain is a conserved region located near the N-terminus of BRPF1, between the NLS1 and EPC-I domains. It is a defining feature of the BRPF family proteins, which include BRPF1, BRD1, and BRPF3. The BN domain is highly conserved across the BRPF family, suggesting its functional importance. The conservation of the BN domain also indicates that the interaction between BRPF proteins and their associated histone acetyltransferases is likely to be evolutionarily conserved (Zu et al. 2022). The BN domain, along with the EPC-I domain, is essential for the formation of the BRPF1-KAT6A/KAT6B complex, which is involved in histone H3 acetylation and transcriptional regulation. The BN domain may also direct the BRPF1-KAT6A/KAT6B complex to specific chromatin regions. Other domains within BRPF1, such as the bromodomain and the PZP domain, directly recognize histone modifications (Zu et al. 2022).

The EPC-I (enhancer of polycomb-like motif 1) domain is located in the middle portion of the BRPF1 protein, adjacent to the PZP (PHD-Zinc knuckle-PHD) domain. The EPC-I domain in human BRPF1 contains approximately 80 amino acid residues, including several conserved residues such as a zinc-chelating motif (Cys-X-X-Cys, where X is any amino acid). This suggests that zinc binding may be important for the structural integrity or function of the domain (Zu et al. 2022). The EPCI domain, together with the BN domain, is required for BRPF1's association with the MYST domain of KAT6A or KAT6B, facilitating the formation of the tetrameric complex. By mediating the interaction between BRPF1 and KAT6A/KAT6B, the EPCI domain enables BRPF1 to serve as a scaffold protein that bridges the histone acetyltransferases with the accessory proteins ING5/ING4 and MEAF6. The EPC-I domain is highly conserved across species, suggesting its important functional role. This conservation also indicates that the interaction between BRPF proteins and their associated histone acetyltransferases is likely to be evolutionarily conserved (Zu et al. 2022). Additionally, the EPC-I domain might influence the catalytic activity of the histone acetyltransferases. Given its essential role in the assembly and function of the BRPF1-KAT6A/KAT6B complex, this domain could potentially be targeted by small molecule inhibitors to disrupt the complex for therapeutic purposes in cancers or other diseases involving aberrant activity of the complex (Fig. 15.1a–c).

The PZP (PHD–zinc knuckle–PHD) domain in BRPF1 is a specialized domain that includes two plant homeodomain fingers separated by a "zinc knuckle". This domain is implicated in histone recognition and plays a key role in chromatin regulation. The BRPF1 PZP domain (residues 271–454) binds to the unmodified tail of histone H3 (residues 1–12 from the N-terminal tail) through PHD1 (Klein et al. 2016, 2021). This interaction is part of a bipartite binding mechanism where BRPF1 recognizes specific epigenetic marks or their absence on the histone tail. Additionally, the PZP domain can bind DNA through PHD2 (Klein et al. 2016). This dual ability to bind both the histone tail and DNA is crucial for its function, enabling BRPF1 to tightly bind to the nucleosome core particle and position itself at specific chromatin locations. The interaction of BRPF1 PZP with H3 and DNA is necessary for the acetyltransferase activity of the complex consisting of BRPF1, MOZ/MORF, ING5, and MEAF6 and further gene expression regulation (Klein et al. 2016). While both histone and DNA interactions are important, binding to extranucleosomal DNA appears to be dominant. This suggests that BRPF1 PZP domain might be particularly important in chromatin regions with accessible DNA, influencing the localization and function of the MOZ/MORF complexes (Klein et al. 2020).

NLS2 (Nuclear Localization Signal 2) is a domain in BRPF1 that, as the name suggests, plays a role in the nuclear targeting of the protein. The NLS2 domain is situated between the PZP and EPC2 domains in the central region of the BRPF1 protein. It consists of a short stretch of amino acids rich in basic residues (lysine and arginine), a typical feature of nuclear localization signals. The positive charge of these residues facilitates interactions with the negatively charged surface of nuclear pore complexes during nuclear import (Christodoulou and Yokoyama 2015). The primary function of the NLS2 domain is to mediate the nuclear import of BRPF1. As a chromatin regulator, BRPF1 needs to be localized to the nucleus to exert its functions in histone modification and transcriptional regulation. Together with NLS1 in the N-terminal region of BRPF1, NLS2 may ensure efficient nuclear targeting of the protein. The basic residues in the NLS2 domain are likely recognized by importin-α, a key adaptor protein in the classical nuclear import pathway. Importin-α binds to the NLS and links BRPF1 to importin-β, which then facilitates the transfer of the complex through the nuclear pore (Christodoulou and Yokoyama 2015). The activity of the NLS2 domain may be regulated by post-translational modifications or interactions with other proteins. For instance, phosphorylation of residues near the NLS2 domain could modulate its interaction with importins and affect the nuclear localization of BRPF1. To date, no disease-causing mutations have been specifically reported in the NLS2 domain of BRPF1. However, given its importance for nuclear localization, mutations that disrupt the NLS2 function could potentially impair the chromatin regulatory activities of BRPF1 and contribute to developmental disorders or other pathologies (Fig. 15.1a, c). This is supported by the conserved presence of a nuclear localization signal in BRPF1 across species, highlighting the importance of proper nuclear targeting for the function of this protein (Lu et al. 2021; Bernhofer et al. 2018).

The EPC-II (Enhancer of Polycomb II) domain is another conserved protein motif found in BRPF1. It consists of approximately 50 amino acid residues and is located at the middle part of the protein. EPC-II domain is primarily involved in interaction with the ING5 (or ING4) and MEAF6 subunits of the tetrameric complex. This interaction is crucial for the assembly and stability of the complete BRPF1-KAT6A/KAT6B-ING/ING5 complexes. It has been shown that deletion of the EPCII domain in the mouse homolog of BRPF1 disrupts complex formation, leading to reduced histone H3K23 acetylation and defects in fetal liver erythropoiesis (Mishima et al. 2011). Sequence analysis and structural prediction suggest that EPC-II may adopt a helical conformation with potential protein-protein interaction surfaces. The EPC-II domain appears to specifically bind to ING5/ING4 and MEAF6, as it does not interact with other ING family members or unrelated proteins, enabling a selective assembly of the BRPF1-containing complex. By recruiting ING5/ING4 and MEAF6 to the complex, the EPC-II domain allows these accessory subunits to contribute to chromatin targeting and regulatory functions. ING5/ING4 can bind to trimethylated histone H3 lysine 4 (H3K4me3), a mark of active promoters, while MEAF6 may stabilize the complex or mediate additional protein-protein interactions.

Compared to the PZP domain, fewer mutations have been reported in the EPC-II domain of BRPF1 in patients with intellectual disability or developmental disorders (Fig. 15.1a). However, some truncating mutations that remove the EPC-II domain have been identified, which may disrupt the interaction with ING5/ING4 and MEAF6 and impair complex function. The EPC-II domain is conserved across BRPF1 orthologs in various species, suggesting its important role in the assembly and function of the BRPF1-containing complex. However, the level of conservation is somewhat lower than that of the EPC-I domain, indicating potential differences in the evolutionary constraints on these two domains (Zu et al. 2022).

While the EPC-II domain of BRPF1 is the primary mediator of the interaction with ING5 and MEAF6, there is evidence for functional redundancy with other domains or regions of BRPF1. For example, deletion of the EPC-II domain in some studies did not completely abolish the interaction with ING5/MEAF6, suggesting that other parts of BRPF1 may also contribute to the interaction. In yeast, the EPC-II domain of Epl1 (the homolog of BRPF1) has been shown to interact with the NuA4 histone acetyltransferase complex, which is related to the human KAT6A/KAT6B complexes. Specifically, the EPC-II domain of Epl1 interacts with the Yng2 subunit (homologous to human ING5/ING4) of the NuA4 complex, and this interaction is important for the recruitment of NuA4 to chromatin (Searle et al. 2017). In addition to interacting with ING5/ING4 and MEAF6, the EPC-II domain of BRPF1 also interacts with the MYST domain of KAT6A/KAT6B. Studies have demonstrated that the EPC-II domain and the C-terminal region of BRPF1 can bind to the MYST domain of acetyltransferases in vitro, suggesting multiple points of contact between BRPF1 and the catalytic subunit (Ullah et al. 2008).

The bromodomain is a highly conserved protein module of about 110 amino acids. As a reader of acetylated lysine on histones H4 and H3, the bromodomain helps target the BRPF1-containing complex to chromatin regions with these histone acetylation marks, particularly on histone tails, and is a key player in epigenetic regulation (Zu et al. 2022). Structurally, bromodomains form a distinctive left-handed four-helix bundle, with a hydrophobic acetyl-lysine binding pocket located at one end, formed by the ZA and BC loops (Filippakopoulos et al. 2012). Bromodomains specifically recognize acetylated lysine residues on histone tails, targeting their host proteins to specific chromatin regions (Viita and Côté 2023). This allows bromodomain-containing proteins to "read" the histone acetylation and recruit other chromatin modifiers or transcriptional regulators to specific genomic regions. By recruiting these proteins to acetylated chromatin, bromodomains facilitate gene activation and transcription. Dysregulation of bromodomain-containing proteins is implicated in various diseases, particularly cancer (Zu et al. 2022). For example, the BET (Bromodomain and Extra-Terminal) family proteins (BRD2, BRD3, BRD4, BRDT) are overexpressed in several cancers and promote oncogene expression. Mutations in other bromodomain proteins, such as BRPF1, are linked to developmental disorders and leukemia (Fig. 15.1a, b). Given their role in disease, bromodomains have emerged as attractive drug targets. Several small-molecule inhibitors targeting bromodomains (particularly BET bromodomains) have been developed and are in clinical trials for cancer and other diseases. These inhibitors disrupt the interaction between bromodomains and acetylated histones, leading to the suppression of oncogene expression and tumor growth (Zu et al. 2022).

The chromo/Tudor-related Pro-Tyr-Tyr-Pro (PWWP) domain is located at the C-terminal end of BRPF1 (Qin et al. 2011; Laue et al. 2008; Wu et al. 2011). The PWWP domain is named after the conserved Pro-Trp-Trp-Pro motif and is approximately 100–150 amino acids in size. This domain adopts a conserved fold consisting of a five-stranded β-barrel followed by one or two α-helices (Vezzoli et al. 2010). This domain is necessary for the association of BRPF1 with condensed chromatin, as it recognizes trimethylated lysine 36 of histone H3 (H3K36me3) (Vezzoli et al. 2010). Structural studies showed that the PWWP domain of BRPF1 binds to H3K36me3, providing a molecular basis for this recognition. Along with other histone-binding domains in BRPF1 (the PZP domain and bromodomain), the PWWP domain facilitates the recruitment of the KAT6A/KAT6B histone acetyltransferases to specific sites of chromatin marked by H3K36 trimethylation. Mutations in the PWWP domain, such as the Phe1154del variant, likely inactivate this domain and contribute to the neurodevelopmental disorders caused by BRPF1 mutations (You et al. 2015c) (Fig. 15.1a). In zebrafish, mutation of the PWWP domain causes developmental defects, highlighting its in vivo importance (Laue et al. 2008).

The PWWP domains from other proteins such as HDGF, MSH6, DNMT3A, and DNMT3B, also bind to H3K36me3 (Wu et al. 2011). However, the PWWP domains of BRPF2, BRPF3, and WHSC1 do not bind H3K36me3, suggesting functional diversity among PWWP domains (Wu et al. 2011). BRPF1 colocalizes with H3K36me3, and this colocalization is lost when the PWWP domain is deleted, supporting its role in chromatin targeting. Mouse embryos lacking BRPF1 exhibit reduced global H3K36me3 levels, suggesting that BRPF1 plays a role in regulating this modification, likely through the recruitment of histone methyltransferases (Yan et al. 2017).

15.2.2 ING4 and ING5

ING4 (Inhibitor of Growth 4) and ING5 are members of the ING family of tumor suppressor proteins. In the context of the BRPF1-KAT6A/KAT6B complex, ING4 and ING5 serve as essential subunits that contribute to the complex's chromatin targeting and regulatory functions (Champagne et al. 2001). ING4 and ING5 interact directly with the EPC-II domain of BRPF1. This interaction is crucial for the incorporation of ING4 or ING5 into the BRPF1-KAT6A/KAT6B complex. Both ING4 and ING5 contain a conserved C-terminal PHD (Plant Homeodomain) finger domain. The PHD finger is a well-established histone recognition module that binds specifically to trimethylated lysine 4 on histone H3 (H3K4me3) (Champagne et al. 2001; Qiu et al. 2012). The binding of the ING4 or ING5 PHD finger to H3K4me3 helps to target the BRPF1-KAT6A/KAT6B complex to open chromatin regions with active transcription. This targeting is complementary to the chromatin recognition functions of the BRPF1 PZP and bromodomains, which bind to unmodified histone H3 tails and acetylated histones, respectively. As members of the ING family of tumor suppressors, ING5 and ING4 have been implicated in various cellular processes related to tumor suppression, such as cell cycle regulation, apoptosis, and senescence (Doyon et al. 2006). Alterations in ING5 and ING4 expression and function have been reported in various types of cancer (Du et al. 2019; Li et al. 2014; Zhang et al. 2018). Reduced expression or loss of function of these proteins may contribute to tumor progression by disrupting the tumor-suppressive functions of the BRPF1-KAT6 complexes and other ING-containing complexes (Zu et al. 2022; Du et al. 2019). In addition to their tumor suppressor functions, ING4 and ING5 have been implicated in developmental processes, such as stem cell differentiation and embryonic development (Doyon et al. 2006).

15.2.3 Functions of BRPF1 in the KAT6A or KAT6B Complex

BRPF1 is a multiunit chromatin regulator that plays crucial roles in diverse biological processes, from embryonic development to adult tissue homeostasis. As a member of the bromodomain-containing protein subfamily IV, BRPF1 functions as both an activator and a scaffold protein within multiunit histone acetyltransferase complexes (Ullah et al. 2008). These complexes, which include BRPF1, KAT6A/KAT6B, ING4/5, and MEAF6, are integral to the regulation of gene expression through post-translational modifications of histones, particularly acetylation of H3K23 (Lalonde et al. 2013; You et al. 2015c). It has been shown that BRPF1 can interact with KAT7 (HBO1) in vitro. However, its interaction with KAT6A/B is more significant (Lalonde et al. 2013).

BRPF1 is particularly important for regulating acetylation of histone H3 lysine 23 (H3K23). This specific modification is frequent across various organisms, suggesting its conserved importance in cellular processes. Mutations in BRPF1 disrupt its ability to interact with KAT6A and KAT6B, leading to deficient H3K23 acetylation and changes in gene expression patterns. Some BRPF1 variants can still bind to KAT6A/B but fail to promote the formation of the complete HAT complex, including ING4/5 and MEAF6, resulting in a reduced HAT activity. BRPF1 mutations significantly reduce global H3K23 acetylation levels, unlike mutations in KAT6A and KAT6B, which often preserve the HAT (MYST) domain and thus may have a less drastic impact on overall H3K23 acetylation (Yan et al. 2017).

As stated above, BRPF1 is a member of family IV of bromodomain-containing proteins that contributes to the recognition of modified histone tails at specific regions of the chromatin for chromatin remodeling (Zu et al. 2022). BRPF1 functions as a reader bromodomain and is a key member of the BRPF1-KAT6A/B--ING4/5-MEAF6 complexes (Ullah et al. 2008; Zu et al. 2022; Klein et al. 2014, 2020; Feng et al. 2016). This bromodomain is encoded by the BRPF1 gene on the short arm of chromosome 3 (3p25.3) and contains 1220 amino acids (Zu et al. 2022; Champagne et al. 1999). BRPF1 includes several domains such as the PZP (PHD-zinc knuckle-PHD) domain, a bromodomain, and PWWP domain (Zu et al. 2022).

Additionally, there is an interaction between the N-terminal region (BN-EPC-I domain) and the MYST domain of MOZ/MORF (Lalonde et al. 2013; Ullah et al. 2008). The EPC-II domain is responsible for binding to ING4/5 and MEAF6 proteins (Lalonde et al. 2013; Ullah et al. 2008). These two proteins increase the acetyltransferase activity of MOZ/MORF (Ullah et al. 2008). The PHD domain of ING4/5 recognizes H3K4me3, which is an epigenetic mark typically found near gene transcription sites (Lalonde et al. 2013; Champagne et al. 2008; Shi et al. 2006). In comparison to unmodified histones, the HAT activity of the MOZ/MORF complex notably increases when encountering H3K4me3 (Lalonde et al. 2013). Some of the BRFP1 domains do not contribute to the acetyltransferase activity directly. However, they are responsible for PTMs recognition and interaction with chromatin. For instance, the PWWP domain mainly recognizes H3K36me3 (Lalonde et al. 2013; Klein et al. 2016; Vezzoli et al. 2010), and contributes to the stabilization of the complex at this specific chromatin region (Vezzoli et al. 2010). Another chromatin reader domain in BRPF1 is the Bromodomain, which can recognize H2AK5ac, H4K12ac, H4K8ac, H4K5ac, and H3K14ac marks with a preference for H2AK5ac (Lloyd and Glass 2018; Poplawski et al. 2014). Another crucial and highly conserved chromatin reader domain within BRPF1 is the PZP domain, which recognizes the unmodified N-terminal tail of histone H3 (Zu et al. 2022). Moreover, both PHD1 and PHD2 domains within PZP are important for H3 recognition. PHD2 is responsible for binding to a nucleosome and PHD1 stabilizes the N-terminal tail of histone H3 (Lalonde et al. 2013; Klein et al. 2016, 2020). Interestingly, the PZP domain is conserved in BRPF2/3 and JADE1/2/3, which are part of different HBO1 complexes, suggesting their critical roles (Mishima et al. 2011; Feng et al. 2016; Saksouk et al. 2009; Avvakumov and Côté 2007). The intricate interplay between these domains and their binding partners underlines BRPF1's diverse biological functions, which span embryogenesis, neurogenesis, hematopoiesis, and cellular differentiation (You et al. 2015a).

Moreover, BRPF1-KAT6A/B complexes play a crucial role in DNA repair by recognizing and responding to DNA double-strand breaks (DSBs). When a DSB occurs, the surrounding chromatin is modified, notably through the phosphorylation of histone H2AX to form γ-H2AX, which signals the recruitment of repair proteins, including KAT6A. KAT6A interacts with γ-H2AX and other DNA damage response proteins, facilitating its localization to the DSBs. Once at the break site, MOZ promotes chromatin remodeling, making the DNA more accessible to other repair proteins. This process is essential for homologous recombination (HR), an error-free repair pathway that uses a homologous sequence as a template to accurately repair the break, thereby maintaining genomic stability and preventing mutations that could lead to cancer (Kowalczykowski 2015; Ito and Shinohara 2023; Modesti and Kanaar 2001; Gursoy-Yuzugullu et al. 2015). Additionally, the involvement of BRPF1 in this process has been demonstrated. BRPF1 depletion leads to a decrease in H3 acetylation levels in response to DNA damage, suggesting that BRPF1 is important for MOZ-mediated histone acetylation during the DNA repair process (You et al. 2016).

The importance of the *BRPF1* gene in human health and disease has increasingly been appreciated in recent years. Germline mutations in *BRPF1* have been identified as the cause of a syndromic form of intellectual disability, known as intellectual developmental disorder with dysmorphic facies and ptosis (or BRPF1 Syndrome), characterized by developmental delay, language impairment, and distinct facial features (Yan et al. 2017). Research done in mice has shown that BRPF1 is crucial for the development of the forebrain, with its loss leading to neocortical disorganization, corpus callosum hypoplasia, and hippocampal dentate gyrus agenesis (You et al. 2015a). Moreover, somatic alterations in *BRPF1* have been implicated in various cancers, including leukemia and medulloblastoma, highlighting its potential role in tumorigenesis (Huether et al. 2014). In leukemia, BRPF1 has been shown to be essential for fetal hematopoietic stem cells by regulating the expression of key multipotency genes, such as Slamf1, Mecom, and Hoxa9 (You et al. 2016).

15.3 BRPF1 in Mouse Development

15.3.1 BRPF1 in Hematopoiesis

BRPF1 is an essential regulator in the hematopoietic stem cells (HSC) and hematopoietic progenitor compartments (You et al. 2016). Loss of BRPF1 results in the transcriptional downregulation of genes important for HSCs and hematopoietic progenitors, including *Slamf1, Evi1, Mds1, Sox17, Hoxa9, Hoxa7, Meis1, Hlf, Gfi1, Rora*, and *Gata3* (You et al. 2016). In the bone marrow, BRPF1 loss leads to a dramatic reduction of histone acetylation at lysines 9, 14, and 23 (You et al. 2016). Hematopoietic-specific disruption of the *Brpf1* gene in mice is lethal. Mutant mice exhibited pale extremities, along with a small spleen and thymus. The bone marrow of these mutants becomes aplastic and is filled with large adipocytes (You et al. 2016).

15.3.2 BRPF1 in Embryonic Development

BRPF1 also plays an important role in embryonic development, particularly in vascular and neural tube formation. Studies on BRPF1-deficient mouse embryos revealed severe vascular defects in the placenta, yolk sac, and the embryo proper

(You et al. 2015b). BEPF1 is essential for embryonic survival, as its inactivation in mice results in embryonic lethality around embryonic day 9.5 (E9.5) (You et al. 2014). This lethality occurs earlier than that observed in *Moz* knockout mice, suggesting that BRPF1 may have functions independent of Moz (You et al. 2014, 2015b). BRPF1 is highly expressed in extraembryonic tissues, such as the placenta and yolk sac, which are critical for nutrient supply and waste removal during embryonic development (You et al. 2014).

BRPF1 is also essential for proper neural tube closure, a critical event in the formation of the central nervous system. Mutant embryos exhibit forebrain hypoplasia and an open neural tube in the cephalic region (You et al. 2015b). Neuron migration is also affected in the mutant neural tube, with fewer Tuj1-positive (a marker for post-mitotic neurons) neurons observed in the cephalic and trunk regions. These neurons fail to migrate to the outer epithelium of the neural tube. At the cellular level, BRPF1 is crucial for cell proliferation. Loss of BRPF1 inhibits the proliferation of embryonic fibroblasts and hematopoietic progenitors (You et al. 2016).

BRPF1's role in embryonic development is closely tied to its ability to regulate gene expression. BRPF1 deficiency leads to reduced histone H3 acetylation at lysine 23 (H3K23ac), a mark important for gene expression and cell proliferation (You et al. 2015b, 2016). It influences the transcription of genes involved in cell cycle control, ribosomal protein synthesis, and chromosome segregation. For example, in embryos at E8, BRPF1 inactivation reduces the expression of *Rpl10l*, a gene encoding a ribosomal protein homologous to *Rpl10*, which is essential for ribosome function. Reduced *Rpl10l* expression could impair protein synthesis and contribute to the developmental defects observed in BRPF1-deficient embryos. Conversely, BRPF1 loss dramatically increases the expression of *Scp3l*, a gene encoding a protein involved in chromosome segregation during meiosis. Elevated *Scp3l* levels might disrupt proper chromosome segregation, further harming developing embryos.

BRPF1 also regulates the expression of cell cycle inhibitors, such as p16 and p27, in mouse embryonic fibroblasts (MEFs). The altered expression of these inhibitors may contribute to the cell proliferation defects observed in BRPF1-deficient embryos and MEFs (You et al. 2015b).

BRPF1 plays a crucial role in regulating Hox gene expression, which is crucial for proper skeletal patterning, particularly maintaining the expression of anterior Hox genes required for proper craniofacial development (Hibiya et al. 2009). Hox genes are a family of transcription factors that play a fundamental role in establishing the body plan of developing embryos. It achieves this by recruiting MOZ to Hox genes, promoting histone acetylation, and potentially protecting these genes from deacetylation (Laue et al. 2008). BRPF1 loss in zebrafish and medaka fish leads to skeletal abnormalities, particularly in the craniofacial and caudal regions (Hibiya et al. 2009; Laue et al. 2008). Anterior transformations of pharyngeal arches are observed in zebrafish lacking zygotic BRPF1, caused by a progressive reduction of anterior Hox gene expression (Laue et al. 2008). The expression of posterior Hox genes like *hoxc11a*, *hoxc12a*, and *hoxc13a* gradually decreases in zebrafish BRPF1 mutants (Hibiya et al. 2009). In medaka fish, BRPF1 is essential for maintaining the expression of both anterior and posterior Hox genes, ensuring the proper development of craniofacial and caudal skeletons (Hibiya et al. 2009). Surprisingly, BRPF1 inactivation also causes elevated expression of Hox genes and other transcription factors, such as *Lhx4*, *Foxa1*, *Tbx5*, and *Twist1*, in the forebrain, suggesting a context-dependent role in both activating and silencing gene expression (You et al. 2015c).

15.3.3 BRPF1 in Neurodevelopment

BRPF1 exhibits a dynamic expression pattern throughout brain development. At prenatal stages, BRPF1 expression is detected in the developing brain, specifically in the neuroepithelium of the neural tube and after birth. BRPF1 expression significantly increases in the brain, particularly in specific regions such as the neocortex, hippocampus, cerebellum, and olfactory bulb. This suggests that BRPF1 is involved in both early neurogenesis and later stages of brain maturation and function (You et al. 2015c). BRPF1 is highly expressed in the brain during development and plays a critical role in forebrain development, which is vital for cognitive functions (You et al. 2014, 2015c). Forebrain-specific deletion of *Brpf1* in mice results in neocortical abnormalities, partial agenesis of the corpus callosum, and hypoplasia of the dentate gyrus in the hippocampus (Cho et al. 2020; You et al. 2015c). Loss of BRPF1 affects neurogenesis, the process of generating neurons, and leads to an aberrant distribution of neural stem cells and progenitors (You et al. 2015c; You et al. 2015a). BRPF1 is involved in regulating the expression of genes important for neocortical development, such as *Robo3* and *Otx1*, and its loss leads to decreased transcription of these genes (You et al. 2015c).

BRPF1 is a multidomain protein that acts as a transcriptional coactivator, recognizing different epigenetic marks and activating three histone acetyltransferases, including MOZ, MORF, and HBO1, consequently influencing histone acetylation, a key epigenetic mark associated with gene activation (You et al. 2015a, 2017). BRPF1, through its interaction with MOZ, MORF, and HBO1, regulates the transcription of genes crucial for neural development. BRPF1 loss reduces the expression of genes like *NeuroD1*, *Tbr2*, *Tlx*, *FoxG1*, and *Emx2*, all of which are implicated in dentate gyrus development (You et al. 2015a, c). While BRPF1 is traditionally considered a transcriptional activator, studies in mice suggest that BRPF1 might also act as a repressor of gene expression in specific contexts. Forebrain-specific BRPF1 deletion leads to elevated expression of various genes, including Hox genes and other transcription factors, normally silenced in the forebrain. This finding suggests a complex role for BRPF1 in fine-tuning transcriptional programs during neural development (You et al. 2015c). BRPF1 loss impairs cell cycle progression in neural progenitors. This is evidenced by the altered expression of cell cycle inhibitors such as p15 and p19 (You et al. 2015a). BRPF1 is dynamically expressed during mouse forebrain development, with strong expression in the dentate gyrus after birth. Forebrain-specific inactivation of *BRPF1* results in dentate gyrus hypoplasia. This indicates a crucial role for BRPF1 in the development of this brain structure, which is vital for learning, memory, and adult neurogenesis (You et al. 2015a, b). Moreover, BRPF1 loss disrupts neural stem cells and progenitors, affecting their proper development and migration. Specifically, Sox2+ neural stem cells and Tbr2+ intermediate neuronal progenitors fail to settle at the subgranular zone of the dentate gyrus, a region impacted by pathogenic BRPF1 variants (You et al. 2015a). BRPF1 is highly expressed in the cerebellum, particularly in the granular layer and Purkinje neurons. Its expression pattern suggests an important role in regulating cerebellar development and function (You et al. 2014). The sources note that Brpf1-deficient mice display defective myelination in the corpus callosum. Myelin is essential for efficient nerve impulse transmission. Disruptions in myelination can lead to various neurological problems, including impaired motor function and cognitive deficits (You et al. 2015c).

15.4 BRPF1 in a Genetic Disease and Different Types of Cancer

15.4.1 BRPF1 Syndrome

Clinical studies have documented over 60 cases of individuals with BRPF1 mutations exhibiting intellectual disability as a prominent symptom, alongside other developmental delays and neurological issues. BRPF1 encodes a crucial protein involved in regulating gene expression through its interaction with histone acetyltransferases, particularly KAT6A/B, and its mutations lead to deficient histone acetylation at histone H3 lysine 23 (H3K23) (Morison et al. 2024). Pathogenic variants in BRPF1, including frameshift, nonsense, missense, and intronic variants, disrupt the gene's function, leading to haploinsufficiency, which means that a single functional copy of the gene is insufficient for normal development (Morison et al. 2024; Cao et al. 2021).

BRPF1 syndrome is characterized by a broad spectrum of phenotypic features, highlighting its variable expressivity. Intellectual disability is a common feature, affecting approximately 79% of reported cases, and the severity can range from mild to severe. Moreover, individuals may exhibit borderline to average cognitive disabilities (Morison et al. 2024). Speech and language disorders are highly prevalent in BRPF1-related disorders, affecting nearly all individuals (Yan et al. 2020; Morison et al. 2024). Speech disorders have a wide range of manifestations, including phonological delay, articulation disorders, childhood apraxia of speech, expressive and receptive language disorders, and word-finding difficulties (Fig. 15.1b) (Yan et al. 2020; Morison et al. 2024).

Vision impairment is another prominent feature, affecting around 78% of individuals. Dysmorphic facial features are common, including a broad nasal bridge, high palate, ptosis (drooping eyelids), and hypertelorism (widely spaced eyes). Infant feeding difficulties are frequently reported, often related to hypotonia (low muscle tone) and oral motor skills impairments. Gross and fine motor delays are also prevalent, with some individuals experiencing gait impairments and developmental coordination disorder. While epilepsy has been reported in previous studies, it was not observed in the cohort examined in one of the studies (Fig. 15.1b) (Morison et al. 2024).

Another clinical manifestation in patients with BRPF1 syndrome is anemia and thrombocytopenia. Through whole exome sequencing, two novel pathogenic BRPF1 variants were discovered in a Turkish family (Kose et al. 2023). These two variants were located on exon 3, where variations occur most frequently, including c.1433G > A and (p.W478*) (Kose et al. 2023). Mouse studies have already established that BRPF1 is essential for the development of fetal hematopoietic stem cells, and its deficiency results in bone marrow failure (You et al. 2016). These specific hematopoietic disorders have not been previously documented in individuals with IDDDFP (Intellectual Developmental Disorder with Dysmorphic Facies and Ptosis). This study suggests the potential role of histone deacetylase (HDAC) inhibitors as a treatment strategy for the hematopoietic anomalies associated with BRPF1 variants (Kose et al. 2023).

Another study identifies nine heterozygous BRPF1 mutations in ten individuals with intellectual disability and developmental anomalies. These mutations include one missense mutation (c.1108C > T, p.Pro370Ser), three nonsense mutations (c.1363C > T, p.Arg455*; c.2497C > T, p.Arg833*; c.3298C > T, p.Arg1100*), and five frameshift mutations (c.362_363delAG, p.Glu121Glyfs*2; c.942_955del, p.Trp315Leufs*26; c.1688_1689del, p.His563Profs*8; c.1883_1886dup, p.Gln629Hisfs*34; c.2915dupC, p.Met973Asnfs*24). Most mutations were de novo, with c.1108C > T inherited from a mosaic mother and c.942_955del presumed inherited from an affected biological mother. All mutations led to a deficiency in histone H3K23 acetylation. The mutations impair BRPF1's ability to form complexes with and activate KAT6A and KAT6B acetyltransferases, affecting gene expression and developmental programs. Clinically, all individuals presented with global developmental delay, expressive language impairment, and intellectual disability, along with other features, such as hypotonia, motor delays, facial dysmorphisms, as well as epilepsy and MRI anomalies in some cases (Fig. 15.1b).

In 2022, a case report was made on BRPF1 syndrome. The study examines copy number variations (CNVs) at the terminal region of chromosome 3p, which can lead to similar clinical manifestations of BRPF1 syndrome. The report discusses an 18-year-old male patient, with non-consanguineous parents and no family history of neurodevelopmental disorders from Peru, presenting with facial dysmorphism, intellectual disability, ptosis, and congenital heart disease. A chromosomal microarray analysis revealed a previously unreported 90 kb deletion encompassing seven genes, including the *BRPF1* gene. The patient displayed clinical features such as palpebral ptosis, thick lips, and developmental delays but did not show some of the common features associated with the condition, such as motor delay or congenital spinal abnormalities. The 90 kb de novo deletion on chromosome 3p25.3 involved seven genes: CPNE9, BRPF1, OGG1 (involved in DNA repair), CAMK1 (a kinase involved in neuronal signaling), TADA3, ARPC4 (important in actin filament dynamics), and ARPC4-TTLL3. The *BRPF1* gene, a known contributor to the patient's phenotype, is involved in chromatin modification, which affects gene expression and is essential for brain development. The study highlights the concept of haploinsufficiency, where a single functional copy of a gene is insufficient for normal function, causing a pathogenic phenotype. In this study, *BRPF1* has a haploinsufficiency score of 15.77%, indicating that having only one functional copy of this gene is insufficient for normal brain function. BRPF1 is particularly implicated due to its role in histone acetylation, which is vital for brain development. Moreover, the *BRPF1* gene is known to interact with other proteins involved in chromatin remodeling, such as the MOZ/MORF and HBO1 complexes, which are also important for the development of the nervous system. Other genes in the deletion were not strongly linked to the observed clinical features. The study suggests that IDDDFP in this patient is primarily due to the deletion involving BRPF1, given its established role in neurodevelopment (Fig. 15.1b) (Abarca-Barriga et al. 2022).

Another case report describes a newly recognized genetic disorder caused by mutations in the *BRPF1* gene. The study presents a non-consanguineous family of mixed Jewish descent with three male siblings affected by varying degrees of intellectual disability (ID) and distinctive facial features, such as bilateral ptosis, blepharophimosis, down-slanted palpebral fissures, and mild retrognathia. Through whole exome sequencing (WES) and Sanger sequencing, a novel heterozygous truncating mutation (c.556C > T, p.Q186*) in the *BRPF1* gene was identified in the affected siblings and their mother, who also exhibited mild symptoms, suggesting variable expressivity of the disease. The study highlights that *BRPF1* haploinsufficiency is a significant cause of intellectual disability and specific facial dysmorphism, aligning with previous findings on the BRPF1-related phenotype (Yan et al. 2017; Mattioli et al. 2017).

These two studies emphasize the importance of considering *BRPF1* mutations in cases of intellectual disability with ptosis and/or blepharophimosis (Abarca-Barriga et al. 2022; Pode-Shakked et al. 2019).

To provide additional evidence for the correlation between ID and specific facial anomalies, a recently published case report also underscores this connection. The study, conducted in 2024, provided a detailed genetic and clinical analysis of Intellectual Developmental Disorder with Dysmorphic Facies and Ptosis (BRPF1 syndrome) in a Chinese family. The research focuses on a 10-month-old female proband exhibiting focal seizures, developmental delays, and distinctive facial features, including ptosis and blepharophimosis. Genetic testing revealed a heterozygous deletion of *BRPF1* (exons 2–14) in the proband. This deletion affects the bromodomain, plant homeodomain finger, and PWWP motif of the BRPF1 protein, likely impairing its function as a chromatin reader and scaffolding subunit of KAT6A/KAT6B histone acetyltransferase complexes. Her mother and sister also showed varying degrees of intellectual disability and facial dysmorphism. Notably, while the proband experienced seizures and developmental delays, her mother and sister only displayed mild intellectual disability without these additional symptoms. The study highlights BRPF1's crucial role in regulating H3K23 acetylation and propionylation, processes implicated in neurodevelopmental disorders (Yan et al.

2020). The study also discusses potential therapeutic approaches based on enhancing H3K23 acylation, including the use of valproate, vorinostat, propionate, and butyrate, known histone deacetylase inhibitors (HDAC inhibitors). Furthermore, the research emphasizes the importance of differential diagnosis, particularly in distinguishing IDDDFP from related syndromes caused by KAT6A or KAT6B genetic defects (Liu et al. 2024).

15.4.2 BRPF1 in Leukemia and Other Types of Cancer

BRPF1 plays a crucial role in controlling cell proliferation and cell cycle progression. For instance, induced BRPF1 inactivation in mouse embryonic fibroblasts (MEFs) led to defective cell growth and cell cycle control. This suggests that BRPF1 dysfunction could disrupt normal cell growth control mechanisms, potentially contributing to uncontrolled proliferation, a hallmark of cancer (You et al. 2015c; Cao et al. 2021). Somatic mutations in different domains' of BRPF1 have been seen frequently in various types of cancer, such as leukemia and various solid tumors. For instance, E253G, L298P, W348R, and E369D mutations within EPCI and PZP domains have been reported in medulloblastoma (Fig. 15.1a, b).

BRPF1 contributes to regulating the activity of MOZ and MORF, two histone acetyltransferases that are frequently rearranged in leukemia. These rearrangements can lead to the formation of fusion proteins that contribute to the development and progression of leukemia (You et al. 2016). For instance, BRPF1 is essential for the self-renewal of leukemic stem cells in acute myeloid leukemia associated with the MOZ-TIF2 fusion protein. This finding suggests that targeting BRPF1 could be a potential therapeutic strategy for leukemia (You et al. 2016).

Over 100 missense and truncating somatic mutations in the *BRPF1* gene have been identified in various cancer types (Fig. 15.1c). These mutations are acquired during an individual's lifetime and can accumulate in cells, potentially contributing to the development of cancer. Approximately one-third of these mutations are predicted to disrupt the normal function of BRPF1 (Yan et al. 2017). Interestingly, despite the prevalence of *BRPF1* mutations in cancer, there is no current evidence to suggest that individuals with *BRPF1* mutations are predisposed to cancer (Yan et al. 2017). Ten individuals with *BRPF1* mutations associated with intellectual disability and developmental abnormalities did not show any indications of cancer predisposition. However, it is important to note that the oldest individual in this group was only 17 years old, and cancer development may occur later in life. These individuals carry germline BRPF1 mutations, while cancer is often associated with somatic mutations acquired throughout life (Yan et al. 2017).

15.5 Therapeutic Strategies Targeting BRPF1

15.5.1 Therapeutic Role of BRPF1 Inhibitors in ER$^+$ Breast Cancer

One of the most recent studies conducted in 2024 investigated the function of BRPF1 as a significant factor in estrogen receptor-positive (ER$^+$) breast cancer and a prospective target for endocrine therapy-resistant malignancies (Salvati et al. 2024). *BRPF1* was recognized as one of the five important genes in ER+ breast cancers due to its significant interaction with ERα on the chromatin of breast cancer cells, sustaining estrogen signaling pathways. The research indicated that the inhibition of *BRPF1*, by siRNA-mediated gene silencing or small molecule inhibitors (GSK6853 and GSK5959), leads to cell cycle arrest, reduced proliferation, and triggered apoptosis in both antiestrogen-sensitive and resistant breast cancer cells. Moreover, *BRPF1* inhibition influenced chromatin accessibility at common BRPF1 + ERα binding sites, resulting in the downregulation of the *ESR1* gene (which encodes ERα) and other critical components of the estrogen signaling pathway, such as FOXA1. Analysis of The Cancer Genome Atlas (TCGA) data indicated elevated *BRPF1* expression in antiestrogen-insensitive tumors, underscoring its potential as a target in endocrine therapy-resistant breast malignancies. The research indicated that the combination of BRPF1 inhibitors with antiestrogens or CDK4/6 inhibitors yielded synergistic effects, indicating a potentially effective approach to surmount resistance to existing endocrine medications. These findings uncover a unique therapeutic target in ER+ breast cancer, especially in resistant variants, by focusing on the functional interaction between ERα and BRPF1, hence identifying BRPF1 as a new actionable target for enhancing breast cancer therapy outcomes (Salvati et al. 2024).

15.5.2 Therapeutic Role of BRPF1 Inhibitors in Liver Cancer

BRPF1 has been known as a critical epigenetic regulator in hepatocellular carcinoma (HCC) progression. Sequencing of the transcriptomes of 16 paired HCC samples revealed that *BRPF1* was the bromodomain-containing gene that was upregulated the most. *BRPF1* overexpression correlated with poor patient survival and was associated with copy number gain. As a functional component of the MOZ/MORF HAT complex, BRPF1 contributes to acetylation of H3K14. CRISPR-Cas9 mediated knockout and pharmacological inhibition of BRPF1 using GSK5959 significantly decreased HCC cell growth in vitro and in vivo, inducing G1 cell cycle arrest and cellular senescence. Mechanistically, BRPF1 was shown to regulate the expression of key oncogenes, notably *E2F2* and *EZH2*, through modulation of H3K14 acetylation at their promoter regions. RNA-seq analysis revealed that *BRPF1* inhibition affected pathways related to cell cycle, DNA replication, and p53 signaling. Moreover, BRPF1 was implicated in maintaining cancer stem cell properties, with its inhibition reducing the expression of stemness-related genes and sphere formation ability (Cheng et al. 2021).

15.5.3 Therapeutic Role of BRPF1 Inhibitors for Glioma

Inhibiting BRPF1 function using the specific inhibitor GSK6853 was found to reduce glioma cell proliferation and colony formation in vitro. This finding suggests that BRPF1 could be a promising target for developing new therapies for glioma (Xia et al. 2021).

15.5.4 Therapeutic Role of BRPF1 Inhibitors for Transplantation Therapies

OF-1 is a pan-BRPF bromodomain inhibitor that has shown promising results in enhancing the ex vivo expansion of hematopoietic stem and progenitor cells (HSPCs) for transplantation therapies (He et al. 2020). OF-1 increases the number and proportion of functional HSPCs without affecting their differentiation potential (He et al. 2020). This could address the limitations posed by the lack of HLA-matched donors and the low number of HSPCs in umbilical cord blood, which currently hinder the wider application of this life-saving therapy. BRPF1 normally functions to enhance HSPC quiescence and decrease HSPC expansion. By inhibiting BRPF1, OF-1 releases MOZ, allowing it to stimulate the expression of self-renewal-related genes, such as *Mn1* and *Runx1*, and facilitate HSPC self-renewal division (He et al. 2020). Importantly, the effects of OF-1 are reversible. When the drug is withdrawn, BRPF1 is reactivated, and HSPCs regain their normal functional properties (He et al. 2020).

15.6 Conclusion

BRPF1 is a critical epigenetic regulator that possesses a reader bromodomain for recognizing acetylated histones and another domain for recruiting histone acetyltransferases, particularly KAT6A/KAT6B, to regulate specific histone modifications, such as H3K23 acetylation and propionylation. BRPF1 contains several other specialized domains (such as the PWWP domain) and is highly conserved from worms to humans, highlighting its evolutionary importance in the animal kingdom. BRPF1 plays crucial roles in various developmental processes, including hematopoiesis (where its loss leads to bone marrow failure and severe anemia), embryonic development (particularly in vascular and neural tube formation), and neurodevelopment (essential for forebrain development and neurogenesis). Clinically, mutations in *BRPF1* cause BRPF1 syndrome, characterized by intellectual disability, dysmorphic features, and ptosis, while somatic mutations have been implicated in various types of cancer, including leukemia and solid tumors. The protein's diverse functions and involvement in multiple pathological conditions have made it an attractive therapeutic target, with promising applications in treating ER^+ breast cancer, liver cancer, gliomas, and improving transplantation therapies through BRPF1 inhibitors. Understanding BRPF1's structure, function, and clinical relevance has opened new avenues for potential therapeutic interventions in both developmental disorders and cancer treatment.

Acknowledgments This work was supported by funds from Canadian Institutes of Health Research (CIHR), Natural Sciences and Engineering Research Council of Canada (NSERC), and Compute Canada (to X.J.Y.).

References

Abarca-Barriga HH, Chavesta Velásquez F, Punil Luciano R (2022) Intellectual developmental disorder with dysmorphic facies and ptosis caused by copy number variation including the BRPF1 gene in Peruvian patient. Egypt J Med Hum Genet 23. https://doi.org/10.1186/s43042-022-00356-z

Allfrey V, Faulkner RM, Mirsky AE (1964) Acetylation and methylation of histones and their possible role in the regulation of RNA synthesis. Proc Natl Acad Sci USA 51:786–794

Avvakumov N, Côté J (2007) The MYST family of histone acetyltransferases and their intimate links to cancer. Oncogene 26:5395–5407

Barman S, Roy A, Bardhan I et al (2021) Insights into the molecular mechanisms of histone code recognition by the BRPF3 bromodomain. Chem Asian J 16:3404–3412. https://doi.org/10.1002/asia.202100793

Bernhofer M, Goldberg T, Wolf S et al (2018) NLSdb-major update for database of nuclear localization signals and nuclear export signals. Nucleic Acids Res 46:D503–D508. https://doi.org/10.1093/nar/gkx1021

Cao J, Xian W, Palihati M et al (2021) Deficiency of intellectual disability-related gene Brpf1 reduced inhibitory neurotransmission in MGE-derived GABAergic interneurons. G3: genes, genomes. Genetics 11. https://doi.org/10.1093/g3journal/jkab090

Carlson S, Glass KC (2014) The MOZ histone acetyltransferase in epigenetic signaling and disease. J Cell Physiol 229:1571–1574. https://doi.org/10.1002/jcp.24617

Champagne N, Bertos NR, Pelletier N et al (1999) Identification of a human histone acetyltransferase related to monocytic leukemia zinc finger protein. J Biol Chem 274:28528–28536. https://doi.org/10.1074/jbc.274.40.28528

Champagne N, Pelletier N, Yang X-J (2001) The monocytic leukemia zinc finger protein MOZ is a histone acetyltransferase. Oncogene 20:404–409. https://doi.org/10.1038/sj.onc.1204114

Champagne KS, Saksouk N, Peña PV et al (2008) The crystal structure of the ING5 PHD finger in complex with an H3K4me3 histone peptide. Prot Struct Funct Genet 72:1371–1376. https://doi.org/10.1002/prot.22140

Cheng CLH, Tsang FHC, Wei L et al (2021) Bromodomain-containing protein BRPF1 is a therapeutic target for liver cancer. Commun Biol 4:888. https://doi.org/10.1038/s42003-021-02405-6

Cho HI, Kim MS, Lee J et al (2020) BRPF3-HUWE1-mediated regulation of MYST2 is required for differentiation and cell-cycle progression in embryonic stem cells. Cell Death Differ 27:3273–3288. https://doi.org/10.1038/s41418-020-0577-1

Christodoulou A, Yokoyama H (2015) Purification of nuclear localization signal-containing proteins and its application to investigation of the mechanisms of the cell division cycle. Small GTPases 6:20–27. https://doi.org/10.4161/21541248.2014.978162

Doyon Y, Cayrou C, Ullah M et al (2006) ING tumor suppressor proteins are critical regulators of chromatin acetylation required for genome expression and perpetuation. Mol Cell 21:51–64. https://doi.org/10.1016/j.molcel.2005.12.007

Du Y, Cheng Y, Su G (2019) The essential role of tumor suppressor gene ING4 in various human cancers and non-neoplastic disorders. Biosci Rep 39

Feng Y, Vlassis A, Roques C et al (2016) BRPF 3- HBO 1 regulates replication origin activation and histone H3K14 acetylation. EMBO J 35:176–192. https://doi.org/10.15252/embj.201591293

Filippakopoulos P, Qi J, Picaud S et al (2010) Selective inhibition of BET bromodomains. Nature 468:1067–1073. https://doi.org/10.1038/nature09504

Filippakopoulos P, Picaud S, Mangos M et al (2012) Histone recognition and large-scale structural analysis of the human bromodomain family. Cell 149:214–231. https://doi.org/10.1016/j.cell.2012.02.013

Grunstein M (1997) Histone acetylation in chromatin structure and transcription. Nature 389:349–352. https://doi.org/10.1038/38664

Gursoy-Yuzugullu O, Ayrapetov MK, Price BD (2015) Histone chaperone Anp32e removes H2A.Z from DNA double-strand breaks and promotes nucleosome reorganization and DNA repair. Proc Natl Acad Sci USA 112:7507–7512. https://doi.org/10.1073/pnas.1504868112

He Q, Hong M, He J et al (2020) Isoform-specific involvement of Brpf1 in expansion of adult hematopoietic stem and progenitor cells. J Mol Cell Biol 12:359–371. https://doi.org/10.1093/jmcb/mjz092

Hibiya K, Katsumoto T, Kondo T et al (2009) Brpf1, a subunit of the MOZ histone acetyl transferase complex, maintains expression of anterior and posterior Hox genes for proper patterning of craniofacial and caudal skeletons. Dev Biol 329:176–190. https://doi.org/10.1016/j.ydbio.2009.02.021

Huether R, Dong L, Chen X et al (2014) The landscape of somatic mutations in epigenetic regulators across 1,000 paediatric cancer genomes. Nat Commun 5:3630. https://doi.org/10.1038/ncomms4630

Ito M, Shinohara A (2023) Chromosome architecture and homologous recombination in meiosis. Front Cell Dev Biol 10

Klein BJ, Lalonde ME, Côté J et al (2014) Crosstalk between epigenetic readers regulates the MOZ/MORF HAT complexes. Epigenetics 9:186. https://doi.org/10.4161/epi.26792

Klein BJ, Muthurajan UM, Lalonde ME et al (2016) Bivalent interaction of the PZP domain of BRPF1 with the nucleosome impacts chromatin dynamics and acetylation. Nucleic Acids Res 44:472–484. https://doi.org/10.1093/nar/gkv1321

Klein BJ, Cox KL, Jang SM et al (2020) Molecular basis for the PZP domain of BRPF1 association with chromatin. Structure 28:105–110.e3. https://doi.org/10.1016/j.str.2019.10.014

Klein BJ, Cox KL, Jang SM et al (2021) Structural and biophysical characterization of the nucleosome-binding PZP domain. STAR Protoc 2:100479. https://doi.org/10.1016/j.xpro.2021.100479

Kose CC, Kaya D, Akcan MB, Silan F (2023) Anemia and thrombocytopenia due to a novel BRPF1 variant in a family from Çanakkale with intellectual disability and dysmorphic facies: case report and review of the literature. Am J Med Genet A 191:2209–2214. https://doi.org/10.1002/ajmg.a.63244

Kowalczykowski SC (2015) An overview of the molecular mechanisms of Recombinational DNA repair. Cold Spring Harb Perspect Biol 7:a016410. https://doi.org/10.1101/cshperspect.a016410

Lalonde ME, Avvakumov N, Glass KC et al (2013) Exchange of associated factors directs a switch in HBO1 acetyltransferase histone tail specificity. Genes Dev 27:2009–2024. https://doi.org/10.1101/gad.223396.113

Laue K, Daujat S, Crump JG et al (2008) The multidomain protein Brpf1 binds histones and is required for hox gene expression and segmental identity. Development 135:1935–1946. https://doi.org/10.1242/dev.017160

Li M, Zhu Y, Zhang H et al (2014) Delivery of inhibitor of growth 4 (ING4) gene significantly inhibits proliferation and invasion and promotes apoptosis of human osteosarcoma cells. Sci Rep 4. https://doi.org/10.1038/srep07380

Liu Q, Li F, Wang N, Fan Z (2024) Genetic analysis of BRPF1 exon deletion variant causing intellectual developmental disorder with dysmorphic facies and ptosis in a Chinese family. Egypt J Med Hum Genet 25. https://doi.org/10.1186/s43042-024-00539-w

Lloyd JT, Glass KC (2018) Biological function and histone recognition of family IV bromodomain-containing proteins. J Cell Physiol 233:1877–1886

Lu J, Wu T, Zhang B et al (2021) Types of nuclear localization signals and mechanisms of protein import into the nucleus. Cell Commun Signal 19:60. https://doi.org/10.1186/s12964-021-00741-y

Lubula MY, Eckenroth BE, Carlson S et al (2014) Structural insights into recognition of acetylated histone ligands by the BRPF1 bromodomain. FEBS Lett 588:3844–3854. https://doi.org/10.1016/j.febslet.2014.09.028

Mattioli F, Schaefer E, Magee A et al (2017) Mutations in histone Acetylase modifier BRPF1 cause an autosomal-dominant form of intellectual disability with associated ptosis. Am J Hum Genet 100:105–116. https://doi.org/10.1016/j.ajhg.2016.11.010

Mishima Y, Miyagi S, Saraya A et al (2011) The Hbo1-Brd1/Brpf2 complex is responsible for global acetylation of H3K14 and required for fetal liver erythropoiesis. Blood 118:2443–2453. https://doi.org/10.1182/blood-2011-01-331892

Modesti M, Kanaar R (2001) Homologous recombination: from model organisms to human disease. Genome Biol 2:reviews1014.1

Morison LD, Van Reyk O, Baker E et al (2024) Beyond "speech delay": expanding the phenotype of BRPF1-related disorder. Eur J Med Genet 68:104923. https://doi.org/10.1016/j.ejmg.2024.104923

Perry J (2006) The Epc-N domain: a predicted protein-protein interaction domain found in select chromatin associated proteins. BMC Genomics 7. https://doi.org/10.1186/1471-2164-7-6

Pode-Shakked N, Barel O, Pode-Shakked B et al (2019) BRPF1-associated intellectual disability, ptosis, and facial dysmorphism in a multiplex family. Mol Genet Genomic Med 7:e665. https://doi.org/10.1002/mgg3.665

Poplawski A, Hu K, Lee W et al (2014) Molecular insights into the recognition of N-terminal histone modifications by the BRPF1 bromodomain. J Mol Biol 426:1661–1676. https://doi.org/10.1016/j.jmb.2013.12.007

Qin S, Jin L, Zhang J et al (2011) Recognition of unmodified histone H3 by the first PHD finger of bromodomain-PHD finger protein 2 provides insights into the regulation of histone acetyltransferases monocytic leukemic zinc-finger protein (MOZ) and MOZ-related factor (MORF). J Biol Chem 286:36944–36955. https://doi.org/10.1074/jbc.M111.244400

Qiu Y, Liu L, Zhao C et al (2012) Combinatorial readout of unmodified H3R2 and acetylated H3K14 by the tandem PHD finger of MOZ reveals a regulatory mechanism for HOXA9 transcription. Genes Dev 26:1376–1391. https://doi.org/10.1101/gad.188359.112

Saksouk N, Avvakumov N, Champagne KS et al (2009) HBO1 HAT complexes target chromatin throughout gene coding regions via multiple PHD finger interactions with histone H3 tail. Mol Cell 33:257–265. https://doi.org/10.1016/j.molcel.2009.01.007

Salvati A, Giurato G, Lamberti J et al (2024) Essential gene screening identifies the bromodomain-containing protein BRPF1 as a new actionable target for endocrine therapy-resistant breast cancers. Mol Cancer 23:160

Schultz J, Copley RR, Doerks T et al (2000) SMART: a web-based tool for the study of genetically mobile domains. Nucleic Acids Res 28:231–234. https://doi.org/10.1093/nar/28.1.231

Searle NE, Torres-Machorro AL, Pillus L (2017) Chromatin regulation by the NuA4 acetyltransferase complex is mediated by essential interactions between enhancer of polycomb (Epl1) and Esa1. Genetics 205:1125–1137. https://doi.org/10.1534/genetics.116.197830

Shi X, Hong T, Walter KL et al (2006) ING2 PHD domain links histone H3 lysine 4 methylation to active gene repression. Nature 442:96–99. https://doi.org/10.1038/nature04835

Ullah M, Pelletier N, Xiao L et al (2008) Molecular architecture of quartet MOZ/MORF histone acetyltransferase complexes. Mol Cell Biol 28:6828–6843. https://doi.org/10.1128/mcb.01297-08

Vezzoli A, Bonadies N, Allen MD et al (2010) Molecular basis of histone H3K36me3 recognition by the PWWP domain of Brpf1. Nat Struct Mol Biol 17:617–619. https://doi.org/10.1038/nsmb.1797

Viita T, Côté J (2023) The MOZ-BRPF1 acetyltransferase complex in epigenetic crosstalk linked to gene regulation, development, and human diseases. Front Cell Dev Biol 10

Wolffe A (2000) Chapter two - chromatin structure. In: Wolffe A (ed) Chromatin, 3rd edn. Academic, London, pp 7–172

Wu H, Zeng H, Lam R et al (2011) Structural and histone binding ability characterizations of human PWWP domains. PloS One 6. https://doi.org/10.1371/journal.pone.0018919

Xia M, Chen H, Chen T et al (2021) Transcriptional networks identify BRPF1 as a potential drug target based on inflammatory signature in primary lower-grade gliomas. Front Oncol 11. https://doi.org/10.3389/fonc.2021.766656

Yan K, Rousseau J, Littlejohn RO et al (2017) Mutations in the chromatin regulator gene BRPF1 cause syndromic intellectual disability and deficient histone acetylation. Am J Hum Genet 100:91–104. https://doi.org/10.1016/j.ajhg.2016.11.011

Yan K, Rousseau J, Machol K et al (2020) Deficient histone H3 propionylation by BRPF1-KAT6 complexes in neurodevelopmental disorders and cancer. Sci Adv 6:eaax0021. https://doi.org/10.1126/sciadv.aax0021

Yang XJ (2015) MOZ and MORF acetyltransferases: molecular interaction, animal development and human disease. Biochim Biophys Acta, Mol Cell Res 1853:1818–1826

Yang XJ, Ullah M (2007) MOZ and MORF, two large MYSTic HATs in normal and cancer stem cells. Oncogene 26:5408–5419

You L, Chen L, Penney J et al (2014) Expression atlas of the multivalent epigenetic regulator Brpf1 and its requirement for survival of mouse embryos. Epigenetics 9:860–872. https://doi.org/10.4161/epi.28530

You L, Yan K, Zhou J et al (2015a) The lysine acetyltransferase activator Brpf1 governs dentate gyrus development through neural stem cells and progenitors. 11:e1005034. https://doi.org/10.1371/journal.pgen.1005034

You L, Yan K, Zou J et al (2015b) The chromatin regulator Brpf1 regulates embryo development and cell proliferation. J Biol Chem 290:11349–11364. https://doi.org/10.1074/jbc.M115.643189

You L, Zou J, Zhao H et al (2015c) Deficiency of the chromatin regulator Brpf1 causes abnormal brain development. J Biol Chem 290:7114–7129. https://doi.org/10.1074/jbc.M114.635250

You L, Li L, Zou J et al (2016) BRPF1 is essential for development of fetal hematopoietic stem cells. J Clin Invest 126:3247–3262. https://doi.org/10.1172/JCI80711

Zeng L, Zhou MM (2002) Bromodomain: an acetyl-lysine binding domain. FEBS Lett 513:124–128. https://doi.org/10.1016/S0014-5793(01)03309-9

Zhang GJ, Zhao J, Jiang ML, Zhang LC (2018) ING5 inhibits cell proliferation and invasion in esophageal squamous cell carcinoma through regulation of the Akt/NF-κB/MMP-9 signaling pathway. Biochem Biophys Res Commun 496:387–393. https://doi.org/10.1016/j.bbrc.2018.01.045

Zhang M, Lei M, Qin S et al (2021) Crystal structure of the BRPF2 PWWP domain in complex with DNA reveals a different binding mode than the HDGF family of PWWP domains. Biochim Biophys Acta (BBA) Gene Regulat Mechan 1864:194688. https://doi.org/10.1016/j.bbagrm.2021.194688
Zu G, Liu Y, Cao J et al (2022) BRPF1-KAT6A/KAT6B complex: molecular structure, vol 14. Biol Funct Hum Dis, Cancers (Basel), p 14

GPSR Compliance

The European Union's (EU) General Product Safety Regulation (GPSR) is a set of rules that requires consumer products to be safe and our obligations to ensure this.

If you have any concerns about our products, you can contact us on ProductSafety@springernature.com

In case Publisher is established outside the EU, the EU authorized representative is:

Springer Nature Customer Service Center GmbH
Europaplatz 3
69115 Heidelberg, Germany

Batch number: 10755222

Printed by Printforce, the Netherlands